Noise and Noise Control

CRC

Noise and Noise Control

Volume I

Authors:

Malcolm J. Crocker

Ray W. Herrick Laboratories
Purdue University
Lafayette, Indiana

A. John Price

University of British Columbia
Vancouver, British Columbia

Published by

CRC PRESS

CRC PRESS, Inc.
18901 Cranwood Parkway · Cleveland, Ohio 44128

Library of Congress Cataloging

Crocker, Malcolm J.
Noise and noise control.

Vol. 2 by M. J. Crocker, A. J. Price, and F. M. Kessler.
At head of title: CRC.
Bibliography: p.
Includes index.
1. Noise. 2. Noise control. I. Price, A. John, joint author. II. Kessler, F. M., joint author. III. Chemical Rubber Company, Cleveland. IV. Title.
TD892.C76 620.2'3 75-2352
ISBN 0-87819-064-3

International Standard Book Number 0-87819-064-3

Library of Congress Card Number 75-2352

Direct all inquiries and correspondence to
CRC PRESS, INC., 18901 Cranwood Parkway, Cleveland, Ohio 44128

This book is dedicated by Malcolm J. Crocker to the memory of William E. Crocker and Robert P. Hutchinson

This book is dedicated by A. J. Price to Ernest E. Williams who was a source of inspiration to him

THE AUTHORS

Malcolm J. Crocker, Ph.D., is a Professor in the School of Mechanical Engineering at Purdue University, and conducts research at the Ray W. Herrick Laboratories.

Dr. Crocker received his higher education in England. He was graduated from Southampton University with a B. S. degree in aeronautical engineering in 1961 and received an M.S.M.E. degree in noise and vibration studies in 1963 from the same university. In 1969, Dr. Crocker obtained his Ph.D. degree in acoustics at Liverpool University.

In 1969, Dr. Crocker came to Purdue University to lead the noise control research effort at the Ray W. Herrick Laboratories and to teach courses concerned with engineering acoustics and noise control in the School of Mechanical Engineering. His research pertains to the area of machinery noise control which includes acoustic design of automobile mufflers, standards for measurement of fan sound power, microphone windscreen development, air conditioner noise, diesel engine noise source investigation, acoustic design of diesel engine enclosures, and reduction of electric clock noise.

Dr. Crocker has been active in organizing a number of conferences on acoustics and noise control. He was General Chairman of the first international conference on noise control engineering, INTER-NOISE 72, held in Washington, D.C., in 1972. Dr. Crocker is a member of the Institute of Noise Control Engineering (INCE), and served as Vice President of the executive committee in 1973 and 1975. He was also one of the six founding directors of the International Institute of Noise Control Engineering based in Geneva, Switzerland and founded in 1974.

Dr. Crocker has edited four books of proceedings of noise conferences and has written numerous articles for scientific journals. In 1973, he was named a Fellow of the Acoustical Society of America for his contributions to noise control engineering. Along with his present duties, he is Editor-in-Chief of the journal *Noise Control Engineering*.

A. John Price, Ph.D., is an Assistant Professor in the School of Architecture of the University of British Columbia at Vancouver, Canada.

Dr. Price obtained his higher education in England. He received a B.S. degree in physics from Liverpool University in 1966, and in 1969 was awarded a Ph.D. degree in acoustics at the same university.

In 1969, Dr. Price was employed as a Senior Consultant with a firm of consulting engineers in Vancouver, where he worked on a wide variety of architectural and engineering noise control problems. Dr. Price became an Assistant Professor in the School of Architecture of the University of British Columbia in 1971, where he has set up a research acoustics laboratory and presently lectures at the undergraduate and graduate levels.

Dr. Price has authored articles for numerous scientific journals.

PREFACE

The governments and public in many countries have come to realize that noise pollution threatens the quality of life itself. It is of course a by-product of the industrial revolution. All machines make noise to a greater or lesser extent. Although in an industrial society some noise must be tolerated, in many cases the application of basic principles at the design stage can reduce noise levels considerably.

This book is written more for the practitioner than the casual reader. Although a high mathematical level is not needed, for much of the material some engineering knowledge is desirable. Noise control is not easy and there are no magic answers to problems. Careful study and patience are required to produce proficiency in the field of noise control. No two noise control problems require exactly the same solution. For that reason in Volume I of this book, principles of noise control are discussed. Some mathematical and theoretical background are included where they are useful for a better understanding. Also in this volume the human ear and psychological acoustics are discussed. An understanding of the ear is necessary in order that noise reduction can be of sufficient magnitude and in suitable frequency ranges to be subjectively effective. Volume I also describes the use of noise and vibration measuring equipment and the propagation of noise in rooms. Buildings are used of course in all our working environments. Some special rooms, i.e., anechoic and reverberant, are discussed which are used in special measurements of noise from machines. Some space is devoted to the design of rooms for use as auditoria or theaters.

Volume II of this series will devote space to industrial noise, community noise, noise regulations and standards, and noise from surface transportation vehicles and aircraft.

It is hoped that the two-volume set will be helpful to the existing practitioner in noise control, to students in mechanical, environmental, civil, electrical, and aeronautical engineering, and to students in architecture, or architectural science or engineering. With so many engineers now concerned with noise control this book should bring to a wide audience an up-to-date, very badly needed, reference work in the field of noise control.

Although I was originally responsible for writing Chapters 1, 2, and 3 and A. J. Price for writing Chapters 4 and 5, Volume I of this book was a cooperative enterprise between the two authors and there has been a considerable amount of collaboration.

The authors have tried to eliminate all errors from the text, however, some will undoubtedly remain and we shall be grateful to readers who bring them to our attention.

The authors would like to thank Mr. Lewis S. Goodfriend for reading the whole manuscript and making many helpful suggestions. Thanks are also due to Professors William A. Cooper and Anshel J. Schiff of Purdue University for reading parts of Chapters 2 and 3, respectively.

Malcolm J. Crocker
Ray W. Herrick Laboratories
Purdue University
May 25, 1975

TABLE OF CONTENTS

Chapter 1

SOME FUNDAMENTALS OF SOUND AND VIBRATION

TABLE OF CONTENTS

1.1 INTRODUCTION

Sound waves behave like ripples on a pond. We have all, at one time, thrown a stone into a pond and watched the ripples spreading out from the point where the stone entered. Sound waves behave in a very similar way to these ripples. There are several points to observe here. Like the ripples, sound waves move away from the source at a constant speed. Like the ripples on the pond, sound waves simply pass through the atmosphere by exchange of momentum. There is no net transfer of matter away from the source, just as on the pond there was no net fluid flow. In the pond we can see a two-dimensional pattern of circular waves spreading out from the source. In the atmosphere, of course, the pattern is three-dimensional and the sound waves are spherical.

The speed of sound in air varies slightly with temperature and humidity, but is independent of frequency. For most purposes, it may be regarded as having a constant value of about 1,120 ft/sec. The speed of sound in water is about 4,850 ft/sec.

Most sources of sound which we encounter are due to some mechanical vibration such as the vibration of a panel covering a machine. In order to reduce the sound produced, the vibration amplitude must be decreased. This is because the sound pressure is directly proportional to the vibration amplitude.

However, we should also note that in some sources of sound there are no mechanical vibrations involved. Such sources are found in air flows such as jet flow and pipe or duct flow. Here the sound is caused by forces in the fluid itself due to the process of high velocity fluid mixing with fluid moving with lower velocity. Some reduction of the sound generated can be achieved with a proper understanding of the process involved. However, in many instances the best approach in these cases is to try to absorb the sound once it has been produced.

In Chapter 1 we will discuss some of the simpler theory which is useful in the control of noise. In Section 1.2 we start off by discussing *simple harmonic motion.* This is because very often motion is like this idealized case. Next, we introduce the ideas of *period, frequency, phase, displacement, velocity,* and *acceleration.* Then we study *free* and *forced vibration* and the influence of damping forces.

In Sections 1.3 and 1.4 we discuss how sound propagates in waves and then we study acoustic *intensity, energy density,* and finally the use of decibels to express *sound pressure levels* and *sound power levels.* In Section 1.5 we discuss the propagation of sound from some idealized spherical sources and the important ideas of *near, far, free,* and *reverberant* sound fields.

In Sections 1.6 and 1.7 we study the propagation of sound indoors and outdoors. This has applications to industrial noise control problems in buildings and to community noise problems, respectively. Finally, in Section 1.8 we study frequency analysis of sound.

1.2 SIMPLE HARMONIC MOTION

The motion of vibrating systems such as parts of machines, and the value of sound pressure with time is often said to be simple harmonic. Let us examine what we mean by simple harmonic motion.

Suppose a point P is revolving around an origin O with a constant angular velocity, ω, as shown in Figure 1.1. If the vector OP is aligned in the direction OX when time $t = 0$, then the angle between OP and OX is ωt after t sec. Suppose OP

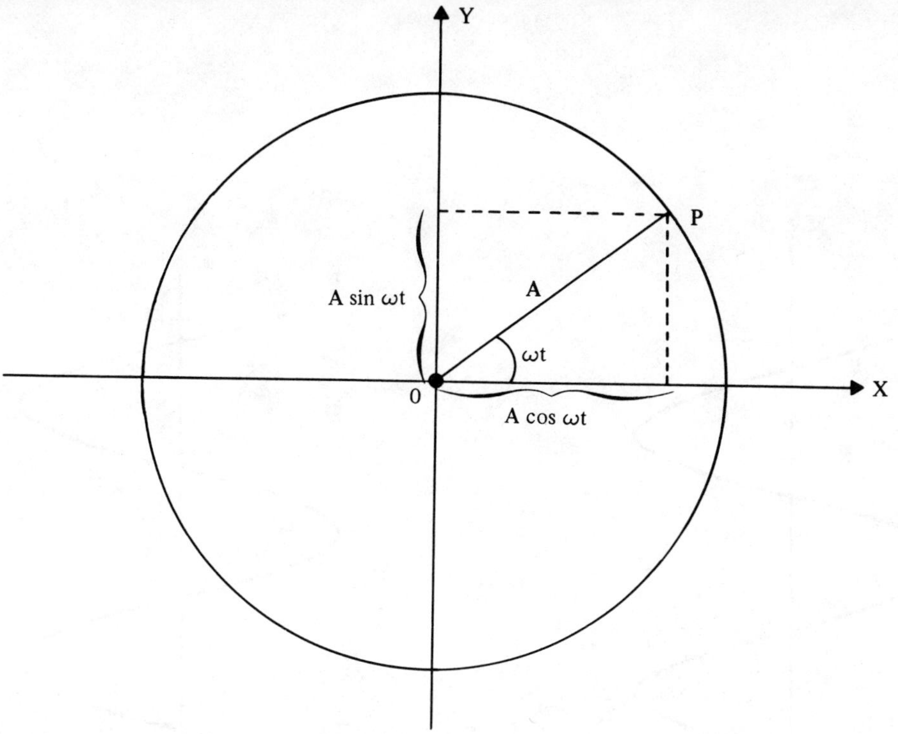

FIGURE 1.1. Representation of simple harmonic motion by projection on X- or Y-axis.

has a length A, then the projection on the X-axis is A cos ωt and on the Y-axis, A sin ωt. The variation of the projected length on either the X or the Y axes with time is said to represent simple harmonic motion.

It is easy to generate a displacement vs. time plot with this model as is shown in Figure 1.2. The projection on the Y-axis is $y = A \sin \omega t$, as before. If we move the circle to the right at a constant speed, then the point P traces out a curve $y = A \sin \omega t$ as shown.

1.2.1 Period, Frequency, Phase

The motion is seen to repeat itself every time the vector OP rotates once (in Figure 1.2A) or after time T sec (in Figure 1.2B). When the motion has repeated itself, the displacement is said to have gone through one *cycle*. The number of cycles which occur per second is called the *frequency*, f. Frequency may be expressed in cycle/second or equivalently, in hertz, Hz. The use of Hz is preferable because this has become internationally agreed upon as the unit of frequency (Note cycle/second ≡ Hz). Thus

$$f = 1/T \text{ Hz}. \tag{1.1}$$

The time taken for the motion to repeat itself, T second, is known as the *period*. From Figures 1.2A and B we see that the motion repeats itself every time ωt increases by 2π, since $\sin 0 = \sin 2\pi = \sin 4\pi$, etc. Thus $\omega T = 2\pi$ and from Equation 1.1,

$$\omega = 2\pi f. \tag{1.2}$$

The angular frequency, ω, is expressed in radian per second (rad/sec).

The motion described by the displacement, y, in Figure 1.2B or the projection OP on the X- or Y-axes in Figure 1.2A is said to be *simple harmonic*. We must now discuss something called the *initial phase angle* or sometimes just called *phase*. For the case we have chosen in Figure 1.2, the phase angle is zero. If instead we start counting time when the vector points in the direction OP shown in Figure 1.2A and we let the angle XOP = ϕ, this is equivalent to moving the time origin t sec to the right in Figure 1.2B. The result is shown in Figure 1.3. Time is started when P is at P_1 and thus the initial displacement is A sin ϕ. The *initial phase angle* is ϕ. After time t, P_1 has moved to P_2 and the displacement

$$y = A \sin(\omega t + \phi). \tag{1.3}$$

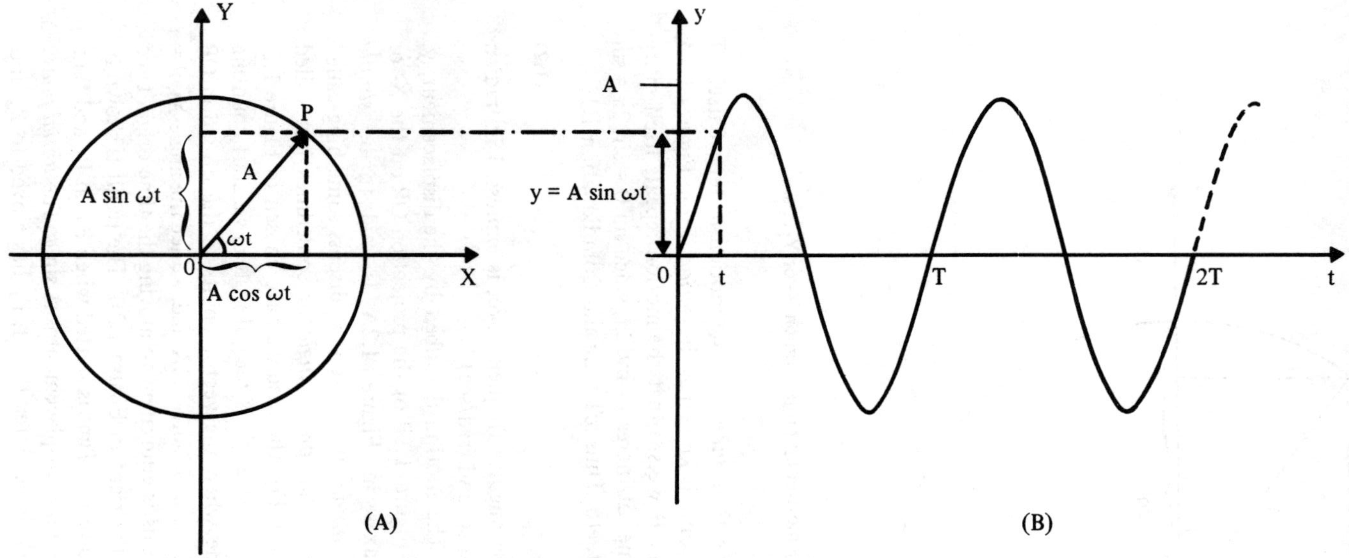

FIGURE 1.2. Simple harmonic motion.

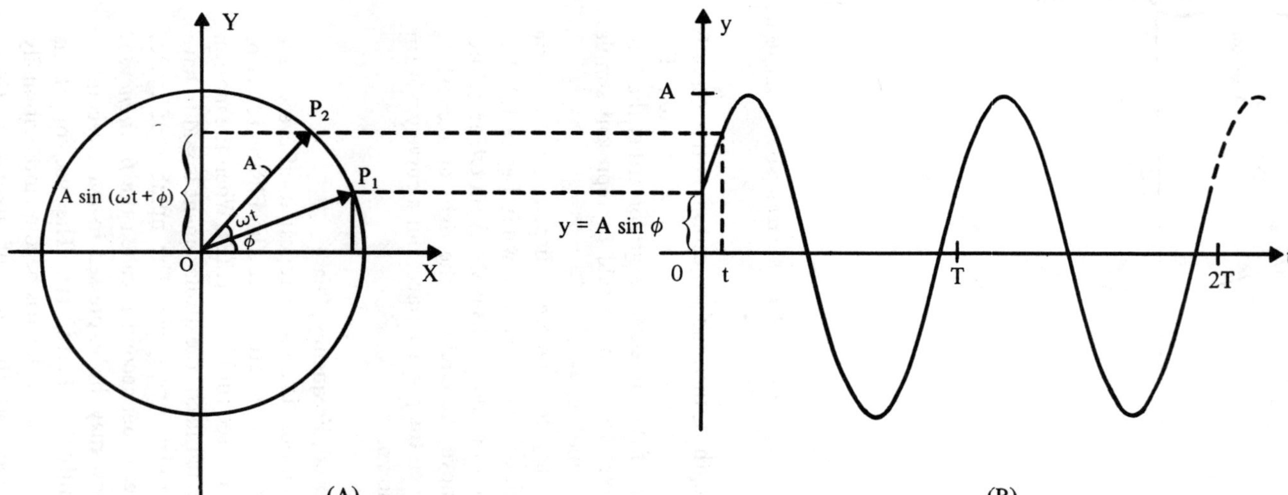

FIGURE 1.3 Simple harmonic motion with initial phase angle ϕ.

1.2.2 Velocity and Acceleration

So far we have examined the displacement, y, of a point. Note that when the displacement is in the OY direction we say it is positive, when it is in the opposite direction to OY we say it is negative. Displacement, velocity and acceleration are really vector quantities in mathematics; that is, they have magnitude and direction. The velocity, v, of a point is the rate of change of position with time of the point in ft/sec or m/sec. The acceleration, a, is the rate of change of velocity with time. Thus, using simple calculus:

$$v = \frac{dy}{dt} = \frac{d}{dt}[A \sin(\omega t + \phi)] = A\omega \cos(\omega t + \phi), \qquad (1.4)$$

and

$$a = \frac{dv}{dt} = \frac{d}{dt}[A\omega \cos(\omega t + \phi)] = -A\omega^2 \sin(\omega t + \phi). \qquad (1.5)$$

Equations 1.3, 1.4, and 1.5 are plotted in Figure 1.4.

Note, by trigonometric manipulation we can rewrite Equations 1.4 and 1.5 as 1.6 and 1.7

$$v = A\omega \cos(\omega t + \phi) = A\omega \sin(\omega t + \frac{\pi}{2} + \phi) \qquad (1.6)$$

and

$$a = -A\omega^2 \sin(\omega t + \phi) = +A\omega^2 \sin(\omega t + \pi + \phi). \qquad (1.7)$$

Equations 1.3, 1.6, and 1.7 tell us that for simple harmonic motion the *amplitude* of the velocity is ω or $2\pi f$ greater than the *amplitude* of the displacement, while the *amplitude* of the acceleration is ω^2 or $(2\pi f)^2$ greater. The *phase* of the velocity is $\pi/2$ or 90° ahead of the displacement, while the acceleration is π or 180° out of phase.

1.2.3 Mass-Spring System

A. Free Vibration – Undamped

Suppose a mass of M lb is placed on a spring of stiffness K lb_f/ft (see Figure 1.5A), and the mass is allowed to sink down a distance d ft to its equilibrium position under its own weight Mg lb_f where g is the acceleration of gravity. Taking forces and deflections to be positive upwards gives:

$$-Mg = -K(d),$$

thus

$$d = Mg/K. \qquad (1.8)$$

The distance d is normally called the static deflection.

Suppose now we displace the mass from its equilibrium position and release it, then it will oscillate about this position. We will measure the deflection from the equilibrium position of the mass (see Figure 1.5B). Newton's law gives force equal to mass × acceleration. Forces and deflections are again assumed positive upwards, and thus

$$-Ky = M\frac{d^2y}{dt^2}. \qquad (1.9)$$

Let us assume a solution to Equation 1.9 of y = A sin $(\omega t + \phi)$. Then upon substitution into Equation 1.9 we obtain

$$-KA \sin(\omega t + \phi) = M(-\omega^2 \sin(\omega t + \phi)).$$

We see our solution satisfies Equation 1.9 if

$$\omega^2 = K/M.$$

We normally signify this frequency with the subscript n,

$$\omega_n = \sqrt{K/M}$$

and

$$f_n = \frac{1}{2\pi}\sqrt{K/M}, \text{ Hz}. \qquad (1.10)$$

The frequency, f_n, is known as the *resonance* or *natural* frequency of the mass on the spring. This result, Equation 1.10, looks physically correct since if K increases (with M constant), f_n increases. If M increases with K constant, f_n decreases. These are the results we also find in practice. (Check this yourself by hanging a weight on a spring. You can even calculate K from Equation 1.8 by measuring d and weighing the mass M and you can check your calculated result for the *resonance* frequency f_n from Equation 1.10 against the value measured with a watch.)

We have seen in Section 1.2.3 that the solution to Equation 1.9 is y = A sin $(\omega t + \phi)$ or the same as Equation 1.3. Hence we know that any system which has a restoring force which is proportional to the displacement will have a displacement which is *simple harmonic.* This is an alternative definition to that given in Section 1.2.1 for *simple harmonic motion.*

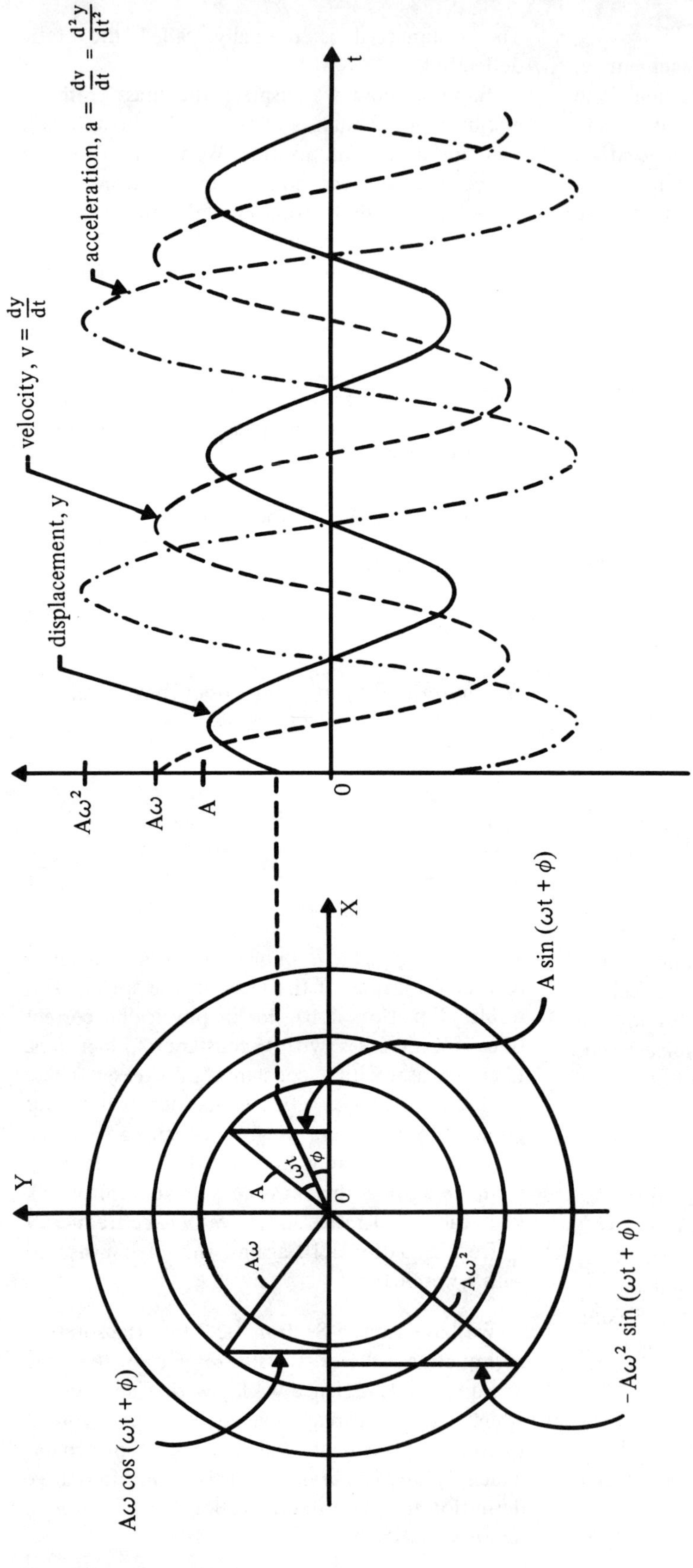

FIGURE 1.4. Displacement, velocity, and acceleration.

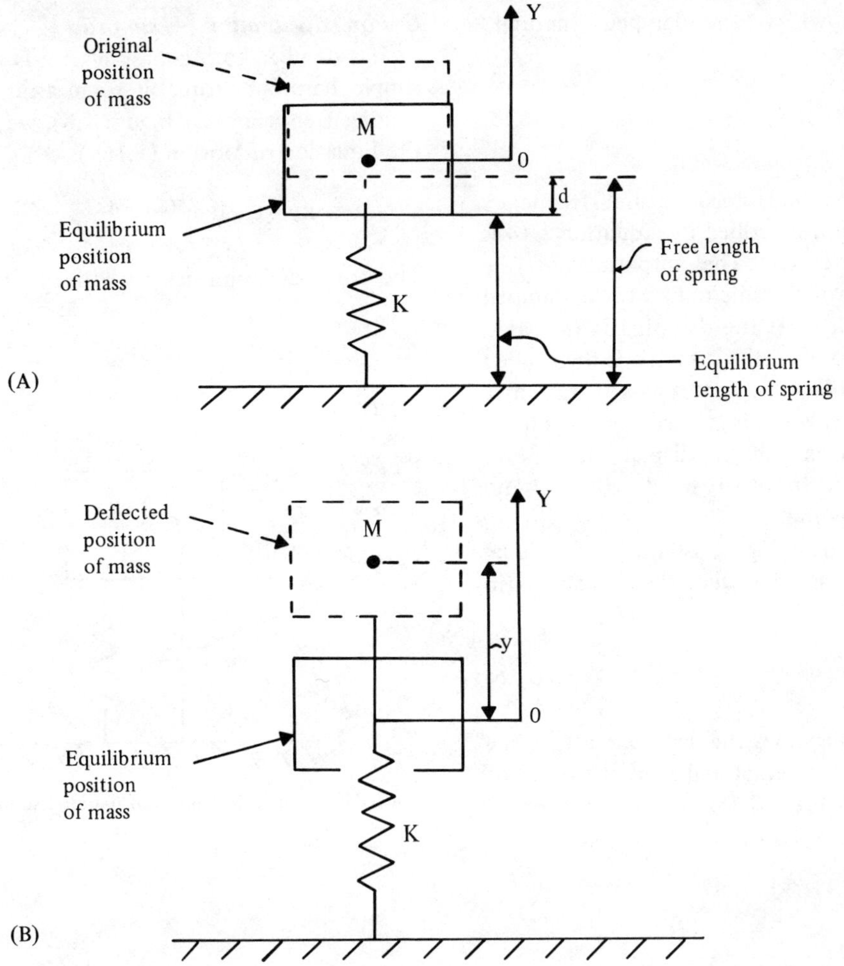

FIGURE 1.5 Movement of mass on a spring; A. Static deflection due to gravity; B. Oscillation due to a displacement x.

B. Free Vibration – Damped

Many mechanical systems can be adequately described by the simple mass-spring system just discussed above. However, for some purposes it is necessary to include the effects of losses (sometimes called damping). This is normally done by including a viscous damper into the system (see Figure 1.6). With viscous or "coulomb" damping the friction or damping force F_d is assumed to be proportional to the velocity, $\frac{dy}{dt}$. If the constant of proportionality is R, then the damping force on the mass

$$F_d = -R\frac{dy}{dt} \tag{1.11}$$

and Equation 1.9 becomes

$$-R\frac{dy}{dt} - Ky = M\frac{d^2y}{dt^2}, \tag{1.12}$$

or equivalently

$$M\ddot{y} + R\dot{y} + Ky = 0, \tag{1.13}$$

where the dots represent single and double differentiation with respect to time.

The solution of Equation 1.13 is most conveniently found by assuming a solution of the form: y is the real part or coefficient of the imaginary part of $\mathbf{A}^{i\lambda t}$ where $\mathbf{A}$ is a complex number and λ is an arbitrary constant to be determined. By substitution into Equation 1.13 and assuming that the damping constant R is small, $R < (4MK)^{1/2}$ (which is true in most engineering applications), the solution is found that:

$$y = Ae^{-(R/2M)t}\sin(\omega_d t + \phi). \tag{1.14}$$

Here ω_d is known as the damped "natural" angular frequency

$$\omega_d = \omega_n(1 - (R/2M)^2)^{1/2}, \qquad (1.15)$$

where ω_n is the undamped natural frequency $\sqrt{K/M}$. The motion described by Equation 1.14 is plotted in Figure 1.7. The amplitude of the motion decreases with time unlike for undamped motion (Figure 1.3). If the damping is increased until R equals $(4MK)^{1/2}$, the damping is then called critical, $R_{crit} = (4MK)^{1/2}$. In this case, if the mass in Figure 1.6 is displaced, it gradually returns to its equilibrium position and the displacement never becomes negative. In other words, there is no oscillation or vibration.

The ratio of damping constant R to critical damping constant R_{crit} is called the damping ratio δ:

$$\delta = R/R_{crit} = R/(2M\omega). \qquad (1.16a)$$

In most engineering cases the damping ratio δ in a structure is hard to predict and is of the order of 0.01 to 0.1. However, there are several ways to measure damping experimentally.

C. Forced Vibration – Damped

If a damped spring-mass system is excited by a simple harmonic force at some arbitrary angular forcing frequency ω (Figure 1.8), we now obtain the Equation of Motion (1.16b):

$$M\ddot{y} + R\dot{y} + Ky = \mathbf{F} = |\mathbf{F}|\, e^{i(\omega t + \phi)}. \qquad (1.16b)$$

The force **F** is normally written in the complex

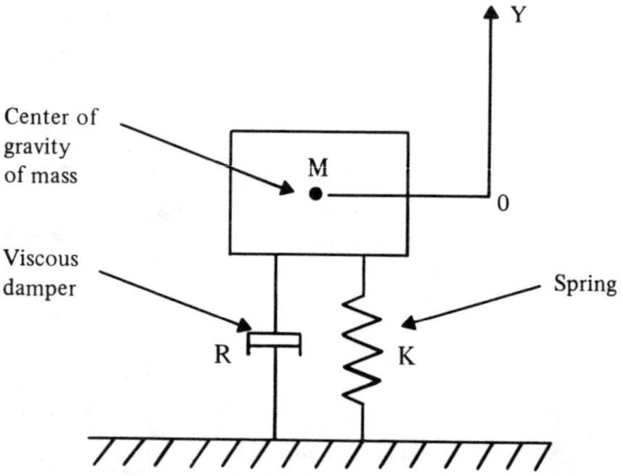

FIGURE 1.6 Movement of damped simple system.

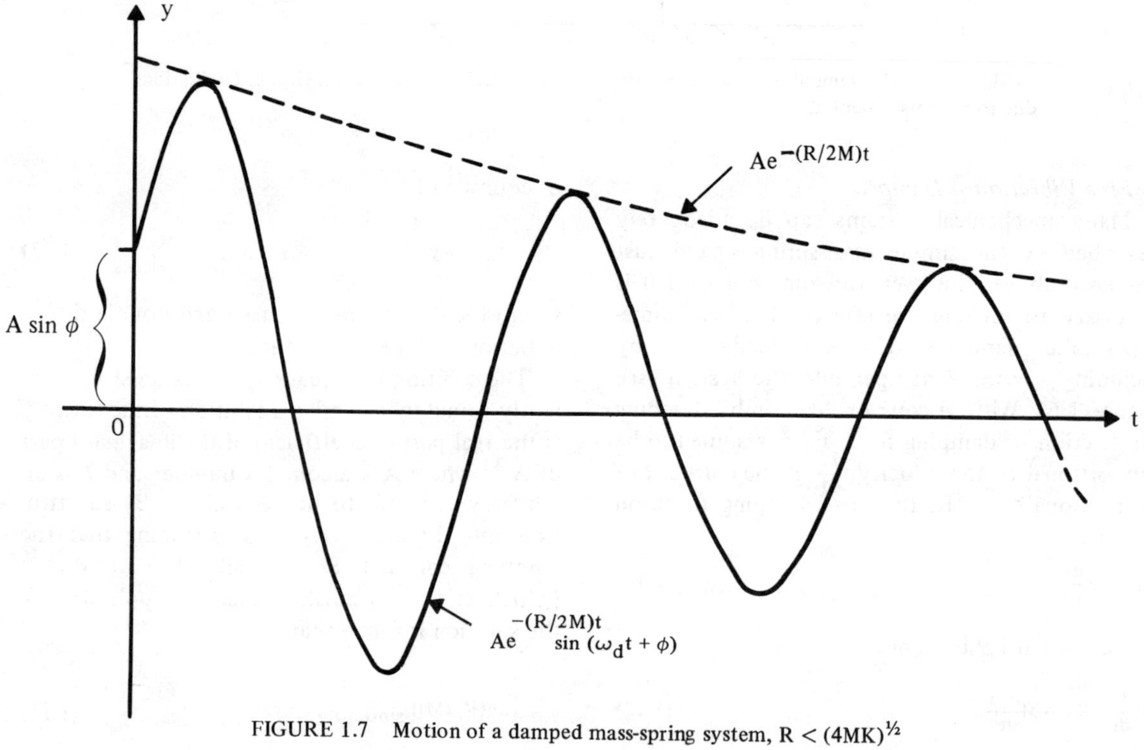

FIGURE 1.7 Motion of a damped mass-spring system, $R < (4MK)^{1/2}$

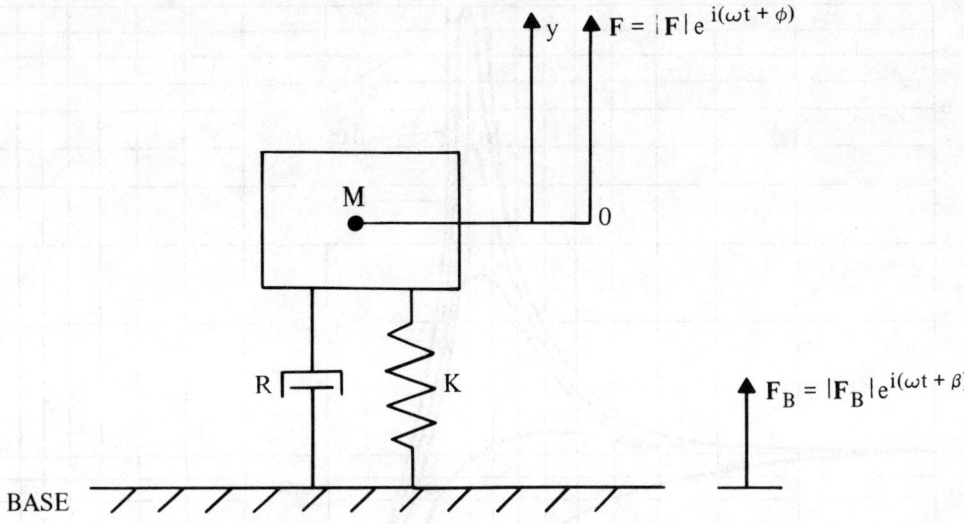

FIGURE 1.8 Forced vibration of damped simple system.

form* for mathematical convenience. The real force acting is, of course, the real part of **F** or |**F**| cos (ωt + ϕ), where |**F**| is the force amplitude.

If we assume a solution of the form: y = **y** = $\mathbf{A}e^{i\omega t}$ then we obtain from Equation 1.16b:

$$\mathbf{A} = \frac{|\mathbf{F}|}{i\omega R + K - M\omega^2}. \qquad (1.17)$$

We can write **A** = |**A**| e $^{i\alpha}$ where α is the phase angle between force and displacement. The phase, α, is not normally of much interest, but the amplitude of motion of the mass, |**A**| is. The amplitude

$$|\mathbf{A}| = \frac{|\mathbf{F}|}{[\omega^2 R^2 + (K - M\omega^2)^2]^{1/2}}. \qquad (1.18)$$

This can be expressed in alternative form:

$$\frac{|\mathbf{A}|}{|\mathbf{F}|/K} = \frac{1}{[4\delta^2(\omega/\omega_n)^2 + (1 - (\omega/\omega_n)^2)^2]^{1/2}}. \qquad (1.19)$$

Equation 1.19 is plotted in Figure 1.9. It is observed that if the forcing frequency ω is equal to the resonance frequency of the structure, ω_n, or equivalently f = f_n, then the amplitude of the motion is proportional to 1/(2δ). The ratio |**A**|/(|**F**|/K) is sometimes called the *dynamic magnification factor*, DMF. The number |**F**|/K is the *static deflection* the mass would assume if exposed to a *constant* force |**F**|. If the damping ratio, δ, is small, the displacement of a structure excited at its resonance or natural frequency is very high. For example, if a simple system has a damping ratio, δ, of 0.01, then its dynamic displacement amplitude is 50 times (when exposed to an oscillating force of 5 lb_f) its static deflection (when exposed to a static force of 5 lb_f), i.e., DMF = 50. Situations such as this should be avoided in practice wherever possible. For instance, if an oscillating force is present in some machine or structure the frequency of the force should be made different from the resonance frequencies of the machine or structure if possible. If the forcing frequency is close to or coincides with a resonance frequency, large amplitude vibrations can occur with consequent vibration and noise problems.

The force on the idealized damped simple system will create a force on the base $F_B = R\dot{y} + Ky$. Substituting this into Equation 1.16 and rearranging and finally comparing the amplitudes of the imposed force |**F**| with the force on the base |$\mathbf{F}_B$| gives

$$\frac{|\mathbf{F}_B|}{|\mathbf{F}|} = \left[\frac{1 + 4\delta^2(\omega/\omega_n)^2}{4\delta^2(\omega/\omega_n)^2 + (1 - (\omega/\omega_n)^2)^2}\right]^{1/2}. \qquad (1.20)$$

Equation 1.20 is plotted in Figure 1.10. The ratio |$\mathbf{F}_B$|/|**F**| is sometimes called the force transmissibility T_F. The force amplitude transmitted is seen to be much greater than one if the exciting frequency is at the system resonance frequency. The results in Equation 1.20 and Figure 1.10 have important applications to machinery noise problems which will be discussed again in detail in volume II of this book. Briefly, we can observe that these results can be utilized in designing vibration isolators for a machine. The resonance frequency ω_n of a machine of mass M resting on its isolators of stiffness

*Complex numbers are shown in this book by boldface or heavy type.

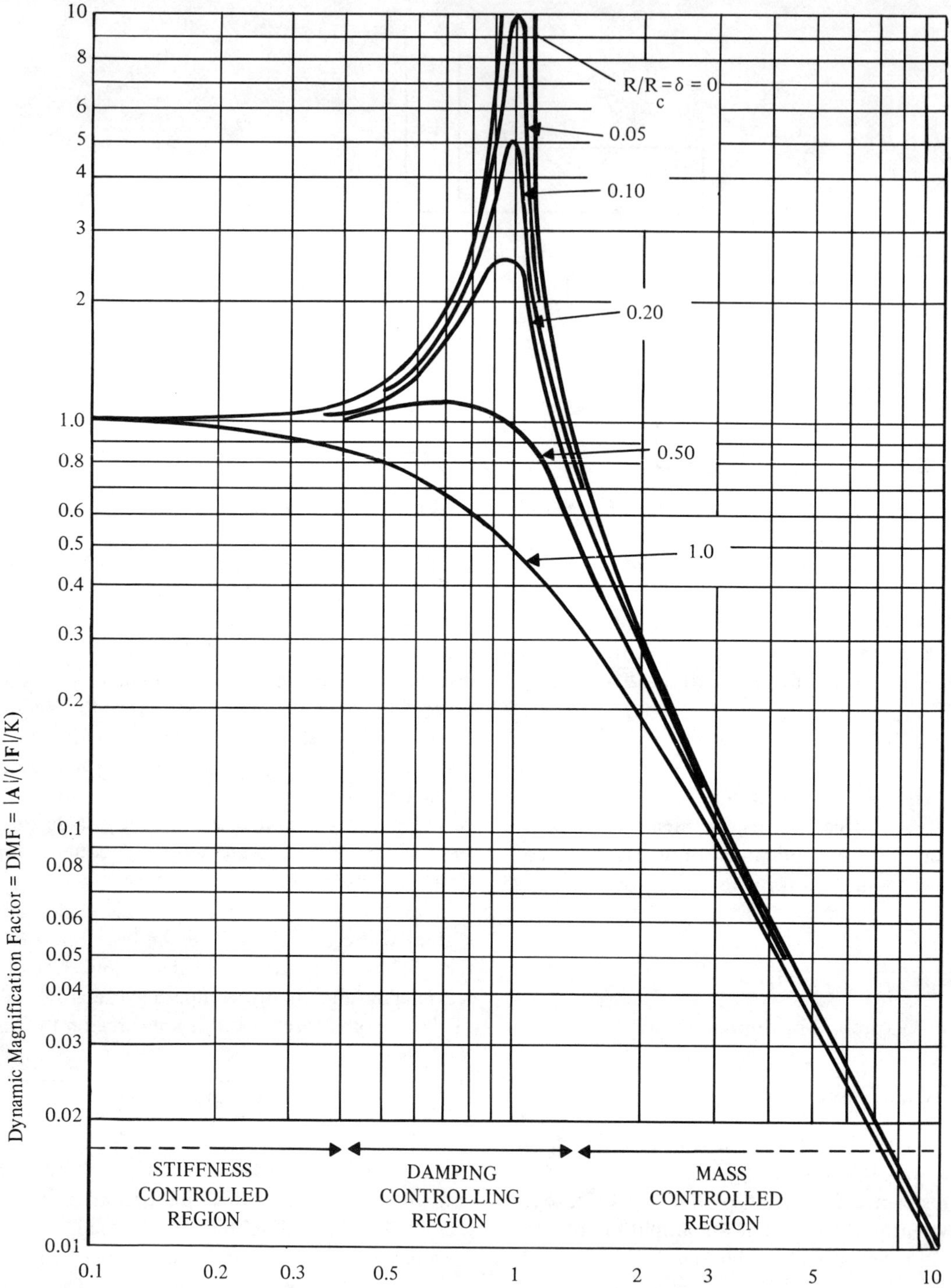

Ratio $\frac{\text{forcing frequency}}{\text{undamped natural frequency}} = \left(\frac{f}{f_n}\right) = \left(\frac{\omega}{\omega_n}\right)$

FIGURE 1.9 Dynamic magnification factor, DMF, for a damped simple system.

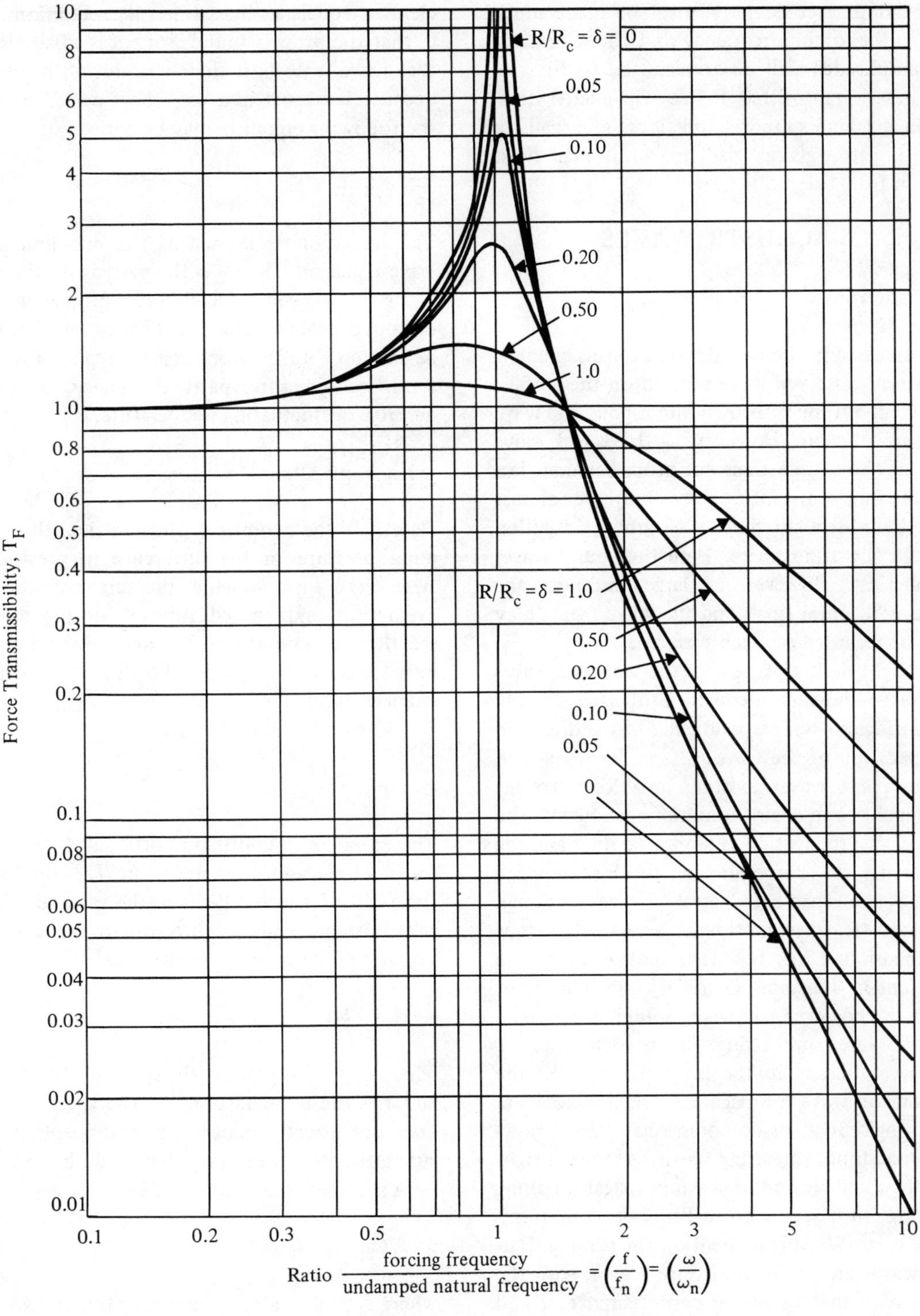

FIGURE 1.10 Force transmissibility, T_F, for a damped simple system.

K and damping constant R must be made much less than the forcing frequency ω. Otherwise, large force amplitudes will be transmitted to the machine base. Transmitted forces will excite vibrations in machine supports and floors and walls of buildings, etc. giving rise to additional noise radiation from these other areas.

1.3 ACOUSTIC WAVES

1.3.1 Plane Waves

A. Wave Motion

As discussed in the introduction to this chapter, we have all observed wave motion on the surface of water. If a stone is thrown into a pond the wave fronts are circular. However, as the waves move out from the source their curvature becomes less until at large distance, they may be almost regarded as straight lines. A similar situation occurs in the atmosphere. Here the sound waves are spherical. However, at large distances the curvature becomes small and the wave fronts may almost be regarded as plane surfaces.

We might define plane acoustic waves as waves which have the same acoustic properties on any plane surface drawn perpendicular to the direction of propagation of the wave. It is perhaps easier to visualize plane waves existing in a long straight duct (such as an air conditioning duct). The direction of propagation is along the duct and the wave fronts are perpendicular to this direction (i.e., they are represented by duct cross-sections). Such acoustic wave motion in a straight duct is one-dimensional, like the wave motion on a long string under tension. There is one important difference, however, between sound waves in a fluid and waves on a string. The particle displacement in a fluid is in the same direction as the direction of wave propagation; such waves are called longitudinal or compressional waves. However, on a string, the string particles move at right angles to the direction of wave propagation along the string; such waves are called lateral or transverse waves. (We might mention, in passing, that water waves are really transverse waves, with the complication that the water particles move up and down (and to a lesser extent forward and backward) describing somewhat elliptical paths. The wave direction is horizontal.)

By considering a thin element of fluid in a duct and considering: a. the fluid mass in the element is conserved, b. the net longitudinal force on the element equals its mass times its acceleration, and c. that the process in the element is adiabatic (i.e., that there is no heat flow in or out of the element because it occurs in a very short period of time), the following equation may be derived[1,2]

$$\frac{\partial^2 \xi}{\partial x^2} - \frac{1}{c^2}\frac{\partial^2 \xi}{\partial t^2} = 0. \tag{1.21}$$

This equation is known as the one-dimensional wave equation. Here ξ is the particle displacement and c is the speed at which waves propagate (speed of sound). Wave equations of identical form may be written with the displacement replaced with the acoustical pressure, particle velocity, fluctuating density, or fluctuating temperature, e.g.,

$$\frac{\partial^2 p}{\partial x^2} - \frac{1}{c^2}\frac{\partial^2 p}{\partial t^2} = 0. \tag{1.22}$$

Here p is the acoustical pressure, i.e., the fluctuating pressure or the difference in pressure between the instantaneous pressure and the time average or undisturbed atmospheric pressure. We see that the rate at which sound pressure changes in space is intimately related to the rate at which it changes in time.

The speed of sound is given by:

$$c = (\gamma RT)^{1/2}. \tag{1.23}$$

The speed of sound is proportional to the square root of the absolute temperature T. Note that γ is the ratio of specific heats in the gas and R is the gas constant. Thus, both γ and R are constants. Equation 1.23 may be rewritten as[3]

$$c = c_0 + 0.6t, \tag{1.24}$$

where $c_0 = 331.6$ m/sec, the speed of sound at 0°C and t is the temperature, °C. The speed of sound does not directly depend upon atmospheric pressure although it does vary slightly with humidity.[4]

A solution to Equation 1.22 is

$$p = f_1(ct - x) + f_2(ct + x), \tag{1.25}$$

where f_1 and f_2 are any arbitrary functions such as sine, cosine, exponential, log, etc., etc. To prove Equation 1.25 is a solution can easily be demonstrated by differentiation and substitution back into Equation 1.22. By varying x and t in Equation 1.25, it soon becomes apparent that $f_1(ct - x)$ represents a wave traveling in the positive

X-direction with speed c and f_2 (ct + x) represents a wave traveling in the negative X-direction with speed c (see Figure 1.11).

The waves represented in Figure 1.11 might have been created by some disturbance such as an impact. We do, however, very frequently encounter waves created by some source which is varying sinusoidally with time (e.g., a structure vibrating at some discrete angular frequency ω). In this case, simple harmonic traveling waves are created:

$$p = A_1 \sin(\omega t - kx + \phi_1) + A_2 \sin(\omega t + kx + \phi_2). \quad (1.26)$$

The similarity of the expressions in Equation 1.26 with Equation 1.3 should be noted. At any one point in space, x, the pressure is simple harmonic. The first expression in Equation 1.26 represents a wave of amplitude A_1 traveling in the positive X-direction; the second expression represents a wave of amplitude A_2 traveling in the negative X-direction. The symbols ϕ_1 and ϕ_2 are phase angles and k is the acoustic wave number. By examining the ratio of x and t in Equation 1.25, we see immediately that $c = \omega/k$ or $k = \omega/c$. The first expression in Equation 1.26 is plotted in Figure 1.12. At some time t, the pressure pattern repeats itself each time kx increases or decreases by 2π radians or 360°. Such a repetition is called a *wavelength,* λ. Hence, $k\lambda = 2\pi$ or $k = 2\pi/\lambda$.

This gives $\omega/c = 2\pi/\lambda$ or $2\pi f/c = 2\pi/\lambda$, or

$$\lambda = c/f. \quad (1.27)$$

Since the speed of sound is a constant (independent of frequency although dependent on atmospheric temperature), again, if the position x is kept constant, the pressure goes through one complete cycle when ωt increases by 2π. The time for a cycle is called the *period*,T. Hence, $\omega T = 2\pi$ or $2\pi f T = 2\pi$

$$T = 1/f. \quad (1.28)$$

B. Intensity

We now see that the wave disturbances travel at a constant wavespeed, c, although there is no net, time-averaged, movement of the air particles. The air particles simply oscillate back and forth in the direction of travel of the wave exchanging momentum. Energy also flows along with the wave. The particle velocity, u, is simply related to the particle displacement ξ by $u = \partial\xi/\partial t$. In the derivation of the wave Equation 1.21 it may be shown that considering forces on an elemental area of fluid that[5,6]

$$-\frac{\partial p}{\partial x} = \rho \frac{\partial u}{\partial t}. \quad (1.29)$$

This simply is a statement of Newton's law of motion since for an element with unit cross-sectional area, $\frac{\partial p}{\partial x} dx$ equals $p_x - p_{x+dx}$, the force acting in the x direction due to the variation in acoustic pressure over the element length dx. The term $\rho(1.dx)\frac{\partial u}{\partial x}$ is the mass times the acceleration of the element, since ρ is the fluid density.

Now $-\frac{\partial p}{\partial x}$ for the right traveling wave is $+kA_1 \cos(\omega t - kx + \phi_1)$. Thus $\partial u/\partial t = +(k/\rho)A_1 \cos(\omega t - kx + \phi_1)$ and by integration $u = +(k/\rho\omega)A_1 \sin(\omega t - kx + \phi_1)$. Thus, for the right traveling wave $p/u = \rho\omega/k$ or since $k = \omega/c$,

$$p/u = \rho c. \quad (1.30)$$

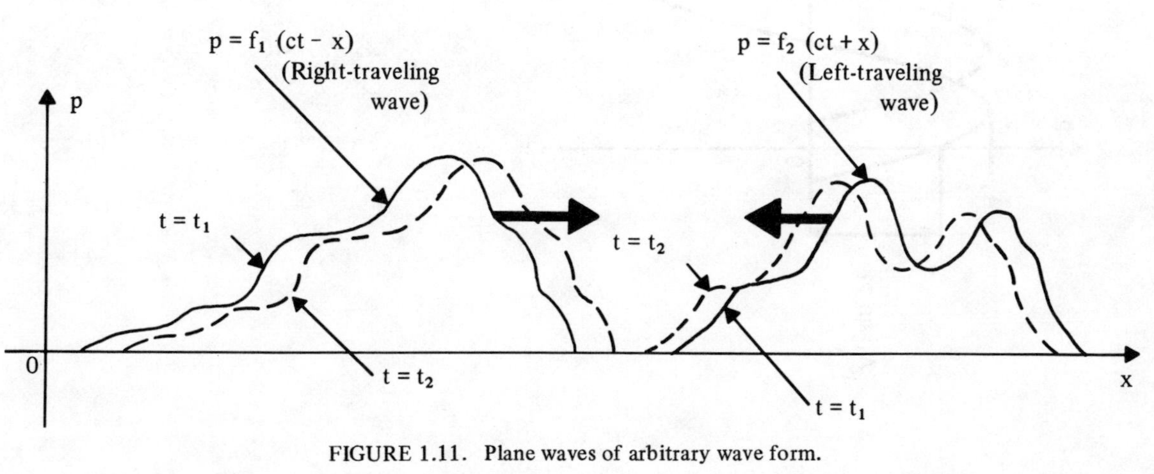

FIGURE 1.11. Plane waves of arbitrary wave form.

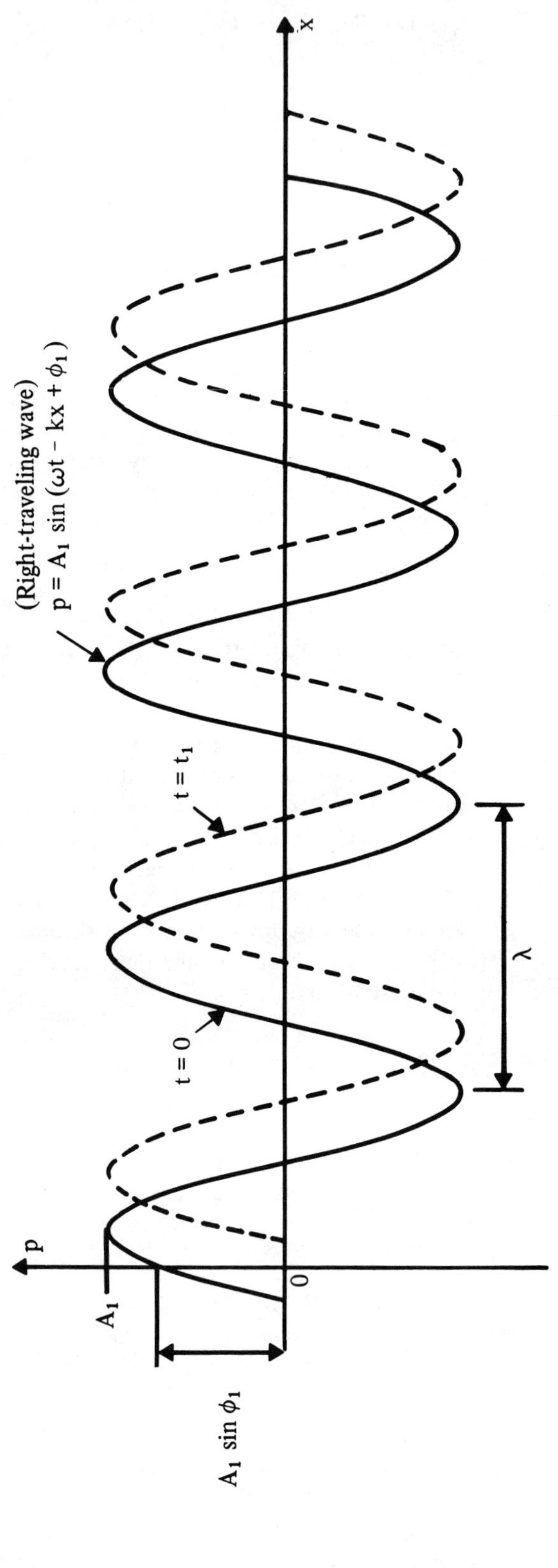

FIGURE 1.12. Simple harmonic plane wave.

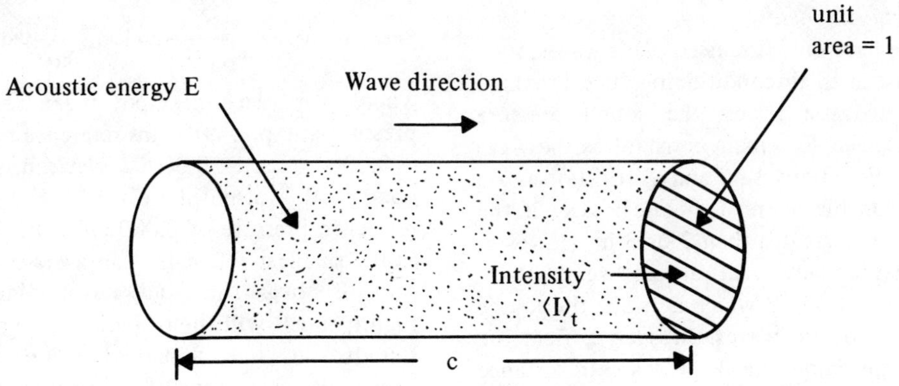

FIGURE 1.13. Energy density of acoustic waves.

Similarly by considering the left traveling wave (the second expression in Equation 1.26) we find that $p/u = -\rho c$ for this wave. The sign difference comes about from our sign convention for positive and negative displacement and velocity.

The acoustic intensity, I, is the rate at which the fluid does work on an imaginary surface of unit area placed at right angles to the direction of wave propagation. For the right traveling wave,

$$I = (\text{Force}) \times \left(\frac{\text{Distance}}{\text{Time}}\right) = (1 \cdot p) \cdot (u) =$$

$$1 \cdot p \frac{\partial \xi}{\partial t} \equiv 1 \cdot p \frac{\delta \xi}{\delta t}. \tag{1.31}$$

For unit cross-section area, $1 \cdot p$ represents the acoustic force which moves through a small distance, $\delta\xi$, in a small time, δt. Hence, $1 \cdot p \cdot \delta$ is work and $1 \cdot p \cdot \delta\xi/\delta t$, rate of doing work. Thus, acoustic intensity is the acoustic power per unit area.

By substituting for u and p from Equations 1.26 and 1.30 into 1.31 for the right traveling wave, the instantaneous intensity is

$$I = (A_1{}^2/\rho c) \sin^2 (\omega t - kx + \phi_1). \tag{1.32}$$

The time average of the intensity $\langle I \rangle_t$ is normally the quantity of most interest:

$$\langle I \rangle_t = (A_1{}^2/\rho c) \frac{1}{T} \int_{-T/2}^{T/2} \sin^2 (\omega t - kx + \phi_1) dt$$

$$\langle I \rangle_t = A_1{}^2/(2\rho c), \tag{1.33}$$

or the peak acoustic pressure A_1 squared divided by $2\rho c$. The quantity ρc is known as the *characteristic acoustic impedance* of the fluid, and for air ρc = 428 kg/(m^2 sec)* at 0°C and 415 kg/(m^2 sec) at 20°C.

If p_{rms} is the root mean square pressure $p_{rms}^2 = A_1{}^2/2$,

$$\langle I \rangle_t = p_{rms}^2/(\rho c). \tag{1.34}$$

C. Energy Density

The acoustic energy flowing through unit area in unit time is the intensity $\langle I \rangle_t$. This quantity is really a vector quantity having the same direction as the particle velocity. In unit time, energy from a fluid column of length c and unit cross-sectional area will flow through the unit area (shaded in Figure 1.13). Thus, the total acoustic energy E contained in the column of fluid is $\langle I \rangle_t$, because this energy must all flow through the area in unit time. If the energy/unit volume, ϵ, is called energy density then

$$(1 \cdot c)\, \epsilon = E = \langle I \rangle_t \cdot 1$$

$$\epsilon = (1/c) \langle I \rangle_t = A_1{}^2/(2\rho c^2) = p_{rms}^2/(\rho c^2). \tag{1.35}$$

The energy density may be derived by alternative means and is found always to be given by Equation 1.35 even for different types of acoustic fields. However, the acoustic intensity is different for different types of acoustic fields. For instance, for a diffuse acoustic field, such as may exist in a hard-walled room, (where waves have the same probability of coming from any direction at once), $\langle I \rangle_t = p_{rms}^2/(4\rho c)$.

*kg/(m^2 sec) =rayl (MKS).

D. Other Acoustic Fields

So far we have only discussed plane waves, such as might exist in an air conditioning duct. Provided we ignore acoustic losses, the sound pressure amplitude is seen to remain constant as the wave propagates, (Equations 1.25 and 1.26). Losses are normally negligible in most practical cases if the duct diameter is greater than 2 or 3 in. However, we know that in many practical cases we may have a small source of sound which acts like the source of spherical acoustic waves. In such a field the sound pressure amplitude decreases with distance from the source. This will be discussed further in Section 1.5.

1.4 SOUND PRESSURE AND POWER LEVELS

1.4.1 The Decibel

The thing to remember about the decibel is that it represents a relative measurement or ratio. Each quantity measured in decibels is expressed as a ratio relative to a reference power, pressure, intensity, or whatever quantity we are considering. For this reason whenever we use the decibel we also use the word *level* . . . sound power level, sound pressure level, etc. to remind us that we are referring to a reference quantity and that our measurement is relative, not absolute.

The Bel was originated in the 1920's to describe attenuation of signals in telephone cables. One Bel (named after Alexander Graham Bell) is the ratio R given by:

$$\log_{10} R = 1,$$

$$\therefore R = 10^1 = 10.$$

Because this is a large ratio, a smaller ratio, the decibel, is used more. One decibel is the ratio r given by:

$$\log_{10} r = 0.1,$$

or

$$10 \log_{10} r = 1, \text{dB},$$

$$\therefore r = 10^{0.1} = 1.26.$$

Thus, the decibel represents the ratio of 2 numbers equal to 1.26.

1.4.2 Sound Pressure Level

This is often abbreviated to SPL or L_p.

$$SPL = 10 \log_{10} \frac{\langle p^2 \rangle_t}{\langle p_{ref}^2 \rangle_t} = 20 \log_{10} \frac{p}{p_{ref}}, \text{ dB} \qquad (1.36)$$

where p is the rms (root mean square) sound pressure and p_{ref} is an rms reference pressure (p_{ref} = 0.0002 μbar = 0.00002 Newton/m^2). Note 1 μbar = 10^{-6} atmosphere = 14.7 × 10^{-6} psi. The reference pressure of 0.0002 μbar has been chosen as the smallest sound that an average young adult can just hear (in quiet surroundings). Perhaps Figure 1.14 will help indicate some common sounds on the Sound Pressure Level scale pictorially. We note that, above, the ratio of pressure squared terms was formed to obtain the decibel. The reason for this is that we prefer to compare power-like quantities such as intensity (which is power/unit area) and we know from Equation 1.34 that $\langle I \rangle_t \propto p^2_{rms}$.

We see that the ear responds to very small sound pressures and the range of sound pressure amplitudes we can experience varies by a factor of about 10^6. More will be said about the ear's response to sound in Chapter 2.

1.4.3 Sound Power Level

Again we note the word level suggests a relative measurement. Sound Power Level is often abbreviated as PWL or L_W.

$$PWL = 10 \log_{10} \frac{W}{W_{ref}}, \text{ dB} \qquad (1.37)$$

where W is the acoustic power of the source and W_{ref} is the reference acoustic power (W_{ref} = 10^{-12} watt). Some typical acoustic power levels are given in Figure 1.15. It is very striking that such small sound powers can have such a large effect on the ear. It is estimated that the roar of a football crowd produces only enough energy to heat a cup of coffee! A full symphony orchestra only produces about one acoustic watt!

1.4.4 Addition of Decibels

Very often we have two independent sound sources acting at the same time. Then if the two instantaneous sound pressures created by the two sources at one point in space, when each source acts on its own, are $p_1(t)$ and $p_2(t)$, the total sound pressure at this point when both sources are acting is:

$$p_{tot} = p_1(t) + p_2(t)$$

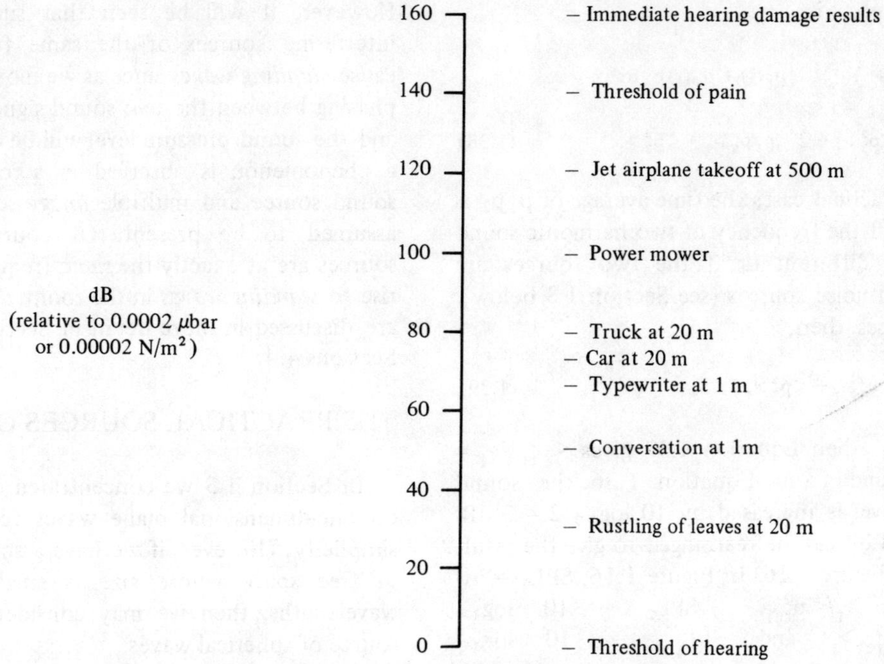

FIGURE 1.14. Some typical sound pressure levels.

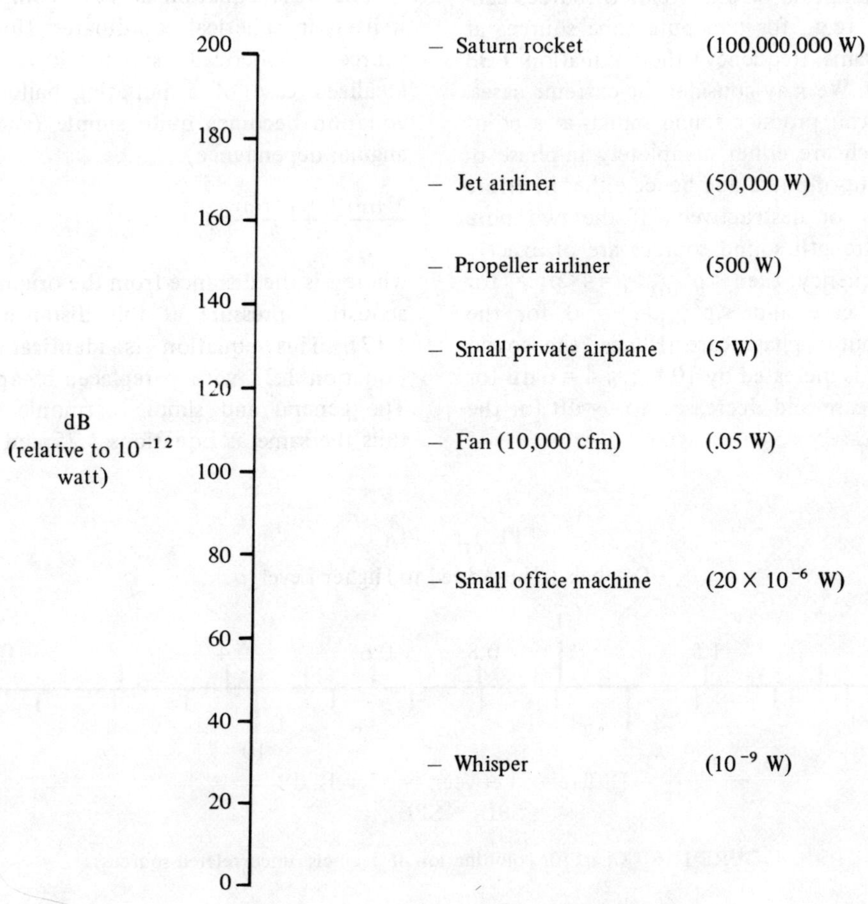

FIGURE 1.15. Some typical sound power levels.

and the mean square sound pressure (or time average of p_{tot}^2) is:

$$\langle p_{tot}^2 \rangle_t = \frac{1}{T} \int_{-T/2}^{T/2} [p_1(t) + p_2(t)]^2 \, dt$$

$$\langle p_{tot}^2 \rangle_t = \langle p_1^2 \rangle_t + 2 \langle p_1 p_2 \rangle_t + \langle p_2^2 \rangle_t. \quad (1.38)$$

In most practical cases the time average of $p_1 p_2$ is zero, e.g., if the frequency of two harmonic sound sources is different or if the two sources are broad-band noise sources (see Section 1.8 below). In most cases, then,

$$\langle p_{tot}^2 \rangle_t \equiv \langle p_1^2 \rangle_t + \langle p_2^2 \rangle_t. \quad (1.39)$$

If $p_1 = p_2$, then Equation 1.39 gives $\langle p_{tot}^2 \rangle_t = 2\langle p_1^2 \rangle_t$ and from Equation 1.36 the sound pressure level is increased by $10 \log_{10} 2 = 3$ dB. Equation 1.39 can be rearranged to give the result shown in Figure 1.16. In Figure 1.16, $SPL_1 = 10 \log_{10} \langle p_1^2 \rangle_t / \langle p_{ref}^2 \rangle_t$, $SPL_2 = 10 \log_{10} \langle p_2^2 \rangle_t / \langle p_{ref}^2 \rangle_t$ and $SPL_{tot} = 10 \log_{10} \langle p_{tot}^2 \rangle_t / \langle p_{ref}^2 \rangle_t$. We have assumed here that $SPL_1 > SPL_2$.

If the assumption of uncorrelated sources cannot be made (e.g., for two pure tone sources at exactly the same frequency) then Equation 1.38 must be used. We may consider the extreme cases. The sources can produce sound signals at a point in space which are either completely in-phase or completely out-of-phase and hence either interfere constructively or destructively. If the two pure tone equal strength sound sources are of exactly the same frequency, then $\langle p^2{}_{tot} \rangle_t = 4\langle p_1^2 \rangle_t$ for the in-phase case and $\langle p^2{}_{tot} \rangle_t = 0$ for the completely out-of-phase case. Hence, the sound pressure level is increased by $10 \log_{10} 4 = 6$ dB for the in-phase case and decreased to $-\infty$dB for the out-of-phase case.

Such cases do not often occur in practice. However, it will be seen that such pure tone interfering sources of the same frequency will cause *standing waves* since as we move in space the phasing between the two sound signals will change and the sound pressure level will be changed. Such a phenomenon is observed in a room. Here the sound source and multiple *image* sources may be assumed to be present. Of course the image sources are at exactly the *same* frequency and give rise to *standing waves* in the room. *Standing waves* are discussed in more detail in Section 1.6 and in Sections 4.4.

1.5 PRACTICAL SOURCES OF SOUND

In Section 1.3 we concentrated our discussion on one-dimensional plane waves for the sake of simplicity. However, if we have a source of sound in free space whose size is small in acoustic wavelengths, then we may consider it to be the source of spherical waves.

1.5.1 Monopole Source

The wave equation is quite complicated when written in spherical coordinates. However, if the source is spherically symmetric (such as in the idealized case of a pulsating balloon) then the equation becomes quite simple (and there is no angular dependence):

$$\frac{\partial^2 (rp)}{\partial r^2} - \frac{1}{c^2} \frac{\partial^2 (rp)}{\partial t^2} = 0, \quad (1.40)$$

where r is the distance from the origin, and p is the acoustical pressure at this distance (see Figure 1.17). This equation is identical in form to Equation 1.22 with p replaced by rp and x by r. The general and simple harmonic solutions are thus the same as Equations 1.25 and 1.26, with p

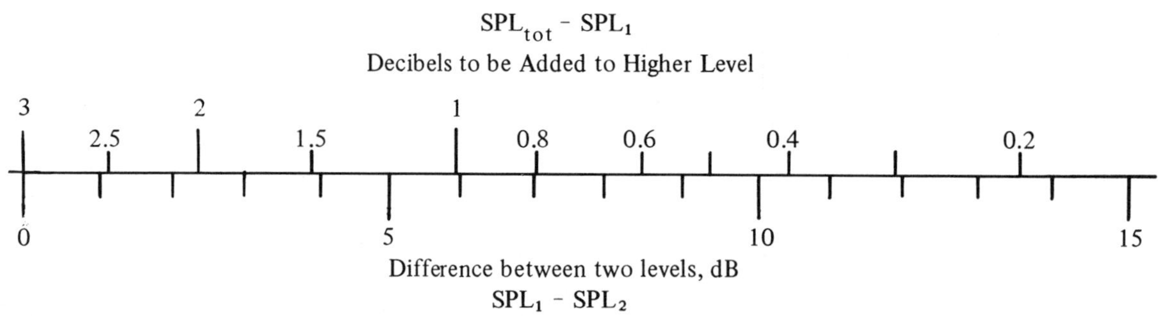

FIGURE 1.16. Chart for combination of decibels (uncorrelated sources).

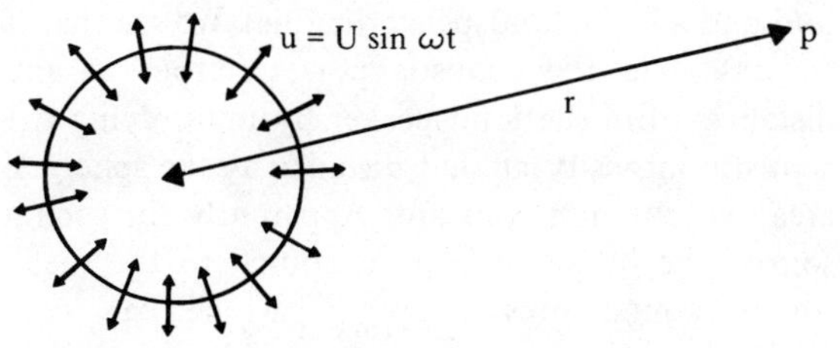

FIGURE 1.17. Idealized monopole source of sound.

replaced with rp. The general solution is

$$rp = f_1(ct - r) + f_2(ct + r),$$

or

$$p = \frac{1}{r} f_1(ct - r) + \frac{1}{r} f_2(ct + r). \tag{1.41}$$

The first term represents a wave traveling outward from the origin and the sound pressure amplitude is seen to be inversely proportional to the distance r. If the distance from the source is doubled the sound pressure is halved; from Equation 1.36 the sound pressure level, SPL, decreases by $20 \log_{10} 2 = 20\,(.301) = 6$ dB. This is known as the *inverse square law* which results from the spherical spreading. The second term in Equation 1.41 represents waves traveling toward the origin and in most practical situations can be ignored.

The simple harmonic solution of Equation 1.40 is:

$$rp = A_1 \sin(\omega t - kr + \phi_1) + A_2 \sin(\omega t + kr + \phi_2)$$

or

$$p = \frac{A_1}{r} \sin(\omega t - kr + \phi_1) + \frac{A_2}{r} \sin(\omega t + kr + \phi_2). \tag{1.42}$$

The second term in Equation 1.42 again represents waves traveling toward the origin and is of no practical interest. However, the first term represents harmonic waves of angular frequency ω traveling outward from the origin. For the outward traveling waves we may rewrite the first term of Equation 1.42 as:

$$p = \frac{\rho c k Q}{4\pi r} \sin(\omega t - kr + \phi_1), \tag{1.43}$$

where Q is known as the source strength ($Q = \frac{4\pi A_1}{\rho c k}$). The mean square pressure p^2_{rms} is

$$p^2_{rms} = \frac{(\rho c k)^2 Q^2}{16\pi^2 r^2} \frac{1}{T} \int_{-T/2}^{T/2} \sin^2(\omega t - kr + \phi_1)\, dt$$

$$p^2_{rms} = \frac{(\rho c k)^2 Q^2}{32\pi^2 r^2}. \tag{1.44}$$

The mean square pressure decreases with the inverse square of the radius (*the inverse square law*). If a sphere of radius a, pulsating with a simple harmonic velocity amplitude U, is assumed, then the same strength $Q = 4\pi a^2 U$ can be shown[7] (provided that the source radius is small in wavelengths; $a \ll \lambda$, $\frac{a}{\lambda} \ll 1$, or $ka \ll 1$). We see that the source strength Q has units of volume flow rate. If this volume flow were a constant then no sound would be generated since $Q \equiv 0$. (Note that U is a velocity amplitude and if the volume flow is constant then $U \equiv 0$.) The source strength, then, only has a nonzero value if the volume flow is changing with time.

The particle velocity may be obtained by using an equation similar to Equation 1.29:

$$-\frac{\partial p}{\partial r} = \rho \frac{\partial u}{\partial t} \tag{1.45}$$

from Equations 1.43 and 1.45:

$$\frac{\partial u}{\partial t} = \frac{\rho c k^2 Q}{\rho 4\pi r} \cos(\omega t - kr + \phi_1) + \frac{\rho c k Q}{\rho 4\pi r^2} \sin(\omega t - kr + \phi_1)$$

and by integration

$$u = \frac{ck^2 Q}{\omega 4\pi r} \sin(\omega t - kr + \phi_1) - \frac{ckQ}{\omega 4\pi r^2} \cos(\omega t - kr + \phi_1). \tag{1.46}$$

The instantaneous intensity for the outgoing wave is the product of p and u from Equations 1.43 and 1.46 (c.f. Equations 1.30 to 1.34 for the plane wave). The instantaneous intensity contains two terms, a $\sin^2$ term and a sin cos term. The second term integrates to be zero over a period, T, and the time average of the intensity becomes:

$$\langle I \rangle_t = \frac{\rho c k^2 Q^2}{16\pi^2 r^2} \frac{1}{T} \int_{-T/2}^{T/2} \sin^2(\omega t - kr + \phi_1)\, dt$$

$$\langle I \rangle_t = \frac{\rho c k^2 Q^2}{32\pi^2 r^2}, \tag{1.47}$$

and from Equation 1.44:

$$\langle I \rangle_t = p^2_{rms}/(\rho c), \tag{1.48}$$

the same result as Equation 1.34. The intensity then decreases with the inverse square of the radius (the *inverse square law*, again). The total sound power emitted by the source, however, is found by integrating over a sphere at some radius r. The sound power W_m is

$$W_m = \langle I \rangle_t 4\pi r^2 \quad (1.49)$$

$$W_m = \frac{\rho c k^2 Q^2}{8\pi}, \quad (1.50)$$

which is independent of radius as conservation of energy requires.

There are several important results which we have derived which we should interpret physically. First consider an actual acoustic source which behaves almost like the one we have considered here. Probably the simplest such source would be a loudspeaker situated in free space but backed by an absorbent enclosure (Figure 1.18a). If the loudspeaker is supplied with a voltage at angular frequency ω (where $\omega = 2\pi f$) then it will behave just like the monopole described above, provided ka is not too large (where k is the wave number ω/c and a is the loudspeaker radius). We see that if we calculate the acoustic power output at any distance from the loudspeaker by multiplying the acoustic intensity at that distance by the spherical area, we obtain a constant. Apparently for such a source the intensity is proportional to the mean square sound pressure p^2_{rms} which can conveniently be measured with a microphone (see Equations 1.47 and 1.48). This gives us a method of obtaining the sound power of such a source, then, by measurement of p_{rms} with a microphone and use of Equations 1.48 and 1.49 to calculate $\langle I \rangle_t$ and then W, provided the loudspeaker is situated in free space or in simulated free space (i.e., an anechoic room) where reflections can be ignored. Provided ka is small, ka $\ll$ 1, then the source strength Q may be shown to be Q =

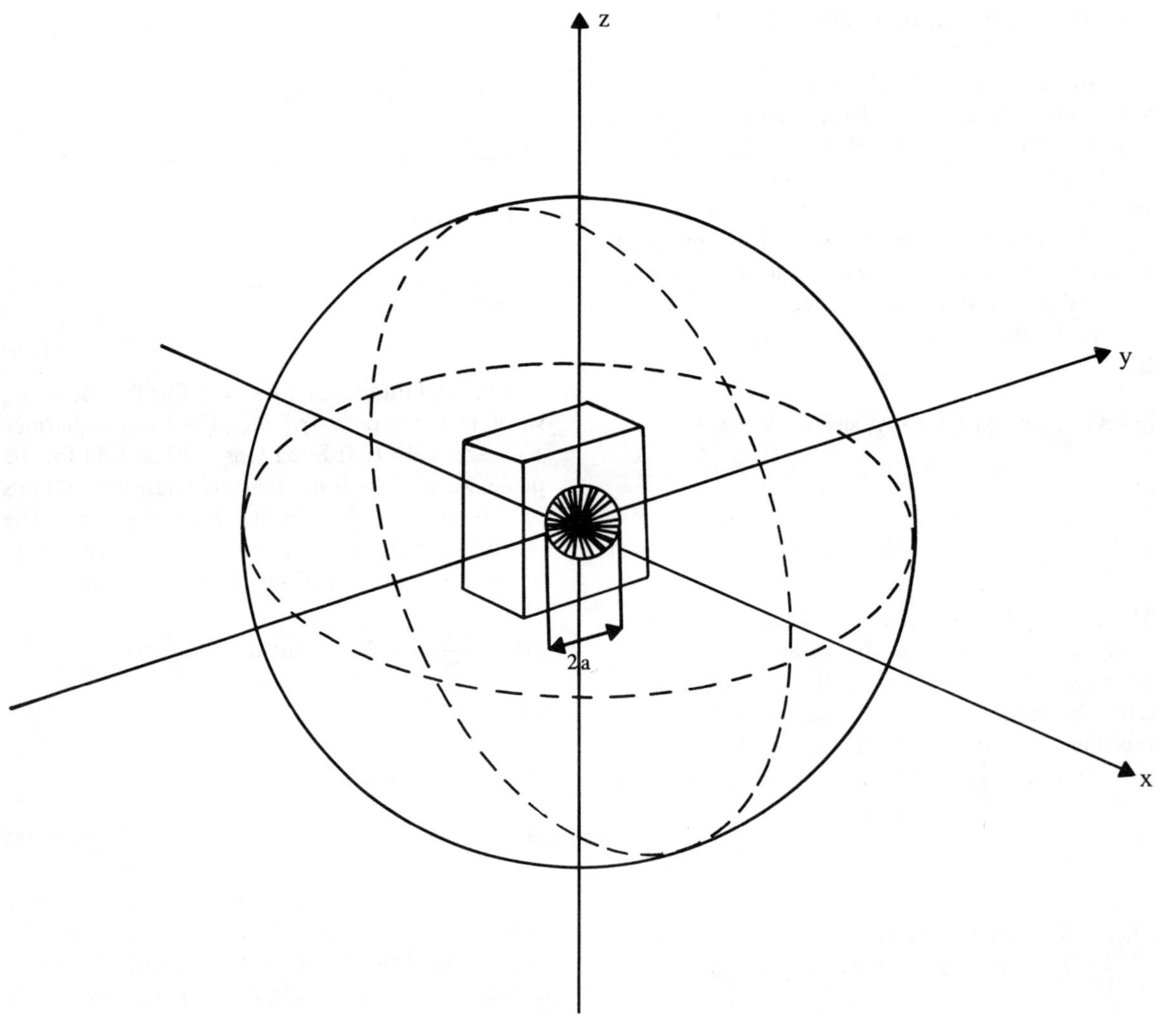

FIGURE 1.18a. Monopole source (loudspeaker enclosure with ka << 1).

$4\pi a^2 U$ where U is the velocity amplitude of the speaker cone.

We see that the sound particle velocity (Equation 1.46) is in phase with the sound pressure (Equation 1.43) when the distance, r, is large, since the sine term is much greater than the cosine term in Equation 1.46. Also at large distances from the loudspeaker, the particle velocity is inversely proportional to the distance, $u \propto 1/r$. At small distances from the loudspeaker the cosine term dominates in Equation 1.46 and the particle velocity increases more rapidly as the loudspeaker is approached, $u \propto 1/r^2$. We can easily show that in the *far acoustic field* (or more precisely $kr >> 1$) the acoustic intensity is proportional to $< u^2 >_t$ but in the *near acoustic field* ($kr << 1$) (close to the loudspeaker), the intensity is not proportional to $< u^2 >_t$. It would only be possible then to determine the acoustic intensity easily with some device such as a hot wire anemometer far from the loudspeaker where $kr >> 1$.

We can think of several other practical examples of actual acoustic sources which behave like monopoles. Such sources are fluctuating supplies of air. For example, if the exhaust velocity of a jet were completely steady (ignoring the noise caused by the mixing of the jet with the stationary atmosphere), no sound would be created by the flow emerging into the atmosphere at the nozzle exit. However, if the burning is not steady (called rough burning), sound can actually be created by the fluctuating air flow right at the jet exit. The sound radiated from the exhaust of an automobile or from an air conditioning duct exit is essentially monopole in character (nondirectional) at low frequency or more precisely if $ka << 1$, where a is the exhaust or duct radius.

1.5.2 Dipole Source

Many practical sources of sound do not behave like the simple nondirectional monopole source we have just discussed. For example, if we removed the backing enclosure of the loudspeaker discussed above, at low frequencies there would be interference between the sound produced by the back and the front of the loudspeaker. These two sources would be 180° out-of-phase with each other. We can model this system as two out-of-phase monopoles of source strength Q separated by a distance ℓ_1. The sound pressure produced by such a system is:

$$p = \frac{\rho c k Q \ell_1 \cos\theta}{4\pi r}\left[\frac{1}{r}\sin(\omega t - kr + \phi_3) + k\cos(\omega t - kr + \phi_3)\right], \tag{1.51}$$

where θ is the angle measured from the axis joining the two sources (the loudspeaker axis in the practical case). Of course the sound pressure field is symmetrical about this axis (see Figure 1.18b). Figure 1.18b represents the sound pressure level variation in space for variation in angle θ at a constant distance r from the origin O. The length of the line OP represents the sound pressure level $L_p(\theta)$ for the angle θ.

We see that with the dipole source the acoustic pressure (Equation 1.51) has a *near field* and a *far field* behavior similar to the particle velocity of a monopole (Equation 1.46). Close to the source (the *near field*), for some fixed angle θ, the pressure falls off rapidly, $p \propto 1/r^2$, while far from the source (the *far field*, $kr >> 1$), the pressure falls off more slowly, $p \propto 1/r$. The phase of the pressure also changes with distance, r, from the source, since close to the source the sine term dominates and far from the source the cosine term dominates.

The particle velocity may be obtained from Equations 1.45 and 1.51 and is found to have an even more complicated behavior with distance r. The instantaneous intensity of the sound from a dipole may be obtained from the product of the pressure and particle velocity and then the time average found as we did for a monopole (Equation 1.47). The result is:

$$\langle I \rangle_t = \frac{\rho c k^4 (Q\ell_1)^2 \cos^2\theta}{32\pi^2 r^2} \tag{1.52}$$

and

$$\langle I \rangle_t = p_{rms}^2/\rho c \tag{1.53}$$

from Equation 1.51, provided the pressure p_{rms} is measured in the far field (i.e., $kr >> 1$) and we see then that the cosine term in Equation 1.51 dominates.

It may also be shown that an oscillating force applied at a point in space gives rise to a result identical to that of Equations 1.52 and 1.53. Hence there are many sources of sound which behave like the idealized dipole source we have described above, e.g., pure tone fan noise, a tuning fork, a vibrating beam, and an unbaffled loudspeaker. We immediately see that a dipole source is

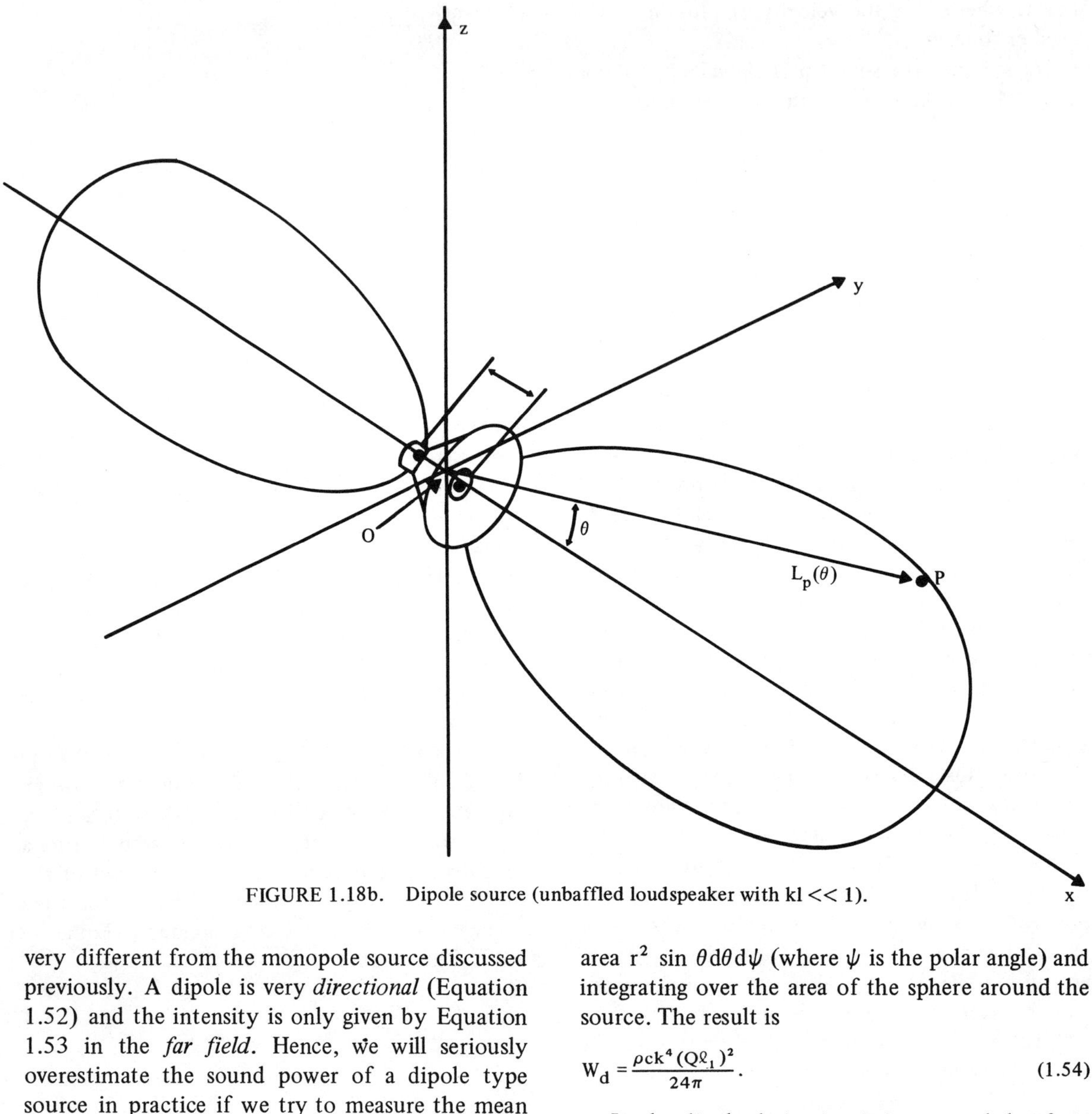

FIGURE 1.18b. Dipole source (unbaffled loudspeaker with kl << 1).

very different from the monopole source discussed previously. A dipole is very *directional* (Equation 1.52) and the intensity is only given by Equation 1.53 in the *far field*. Hence, we will seriously overestimate the sound power of a dipole type source in practice if we try to measure the mean square pressure with a microphone in its near field and then integrate over a sphere around the source at the origin. In order to obtain a correct answer by this method we must measure the mean square pressure in the *far field*. We may easily discover whether we are measuring in the *far field* by checking to see that $p_{rms}^2 \propto 1/r^2$ (see Equation 1.51 and Figure 1.19). In the "true" acoustic field, the *far field*, the sound pressure level decreases by 10 log 2^2 or 6 dB each time the distance is doubled from the dipole source. This is something which is easy to do in practice.

The acoustic power of the dipole source is found by multiplying by the elemental spherical area $r^2 \sin\theta d\theta d\psi$ (where ψ is the polar angle) and integrating over the area of the sphere around the source. The result is

$$W_d = \frac{\rho c k^4 (Q\ell_1)^2}{24\pi}. \qquad (1.54)$$

In the dipole derivation it is assumed that ℓ_1 is very small but that the dipole source strength $Q\ell_1$ remains finite as $\ell_1 \rightarrow 0$. If we compare the sound power of a dipole with that of a monopole (Equations 1.50 and 1.54) we find that

$$\frac{W_d}{W_m} = k^2 \ell_1^2/3 \qquad (1.55)$$

and since $k = \omega/c = 2\pi f/c$, we see that at low frequencies a dipole is much less efficient at radiating energy than a monopole. This explains why an unbaffled loudspeaker cannot produce low notes but a loudspeaker with a backing enclosure can (for the same velocity amplitude of the speaker cone).

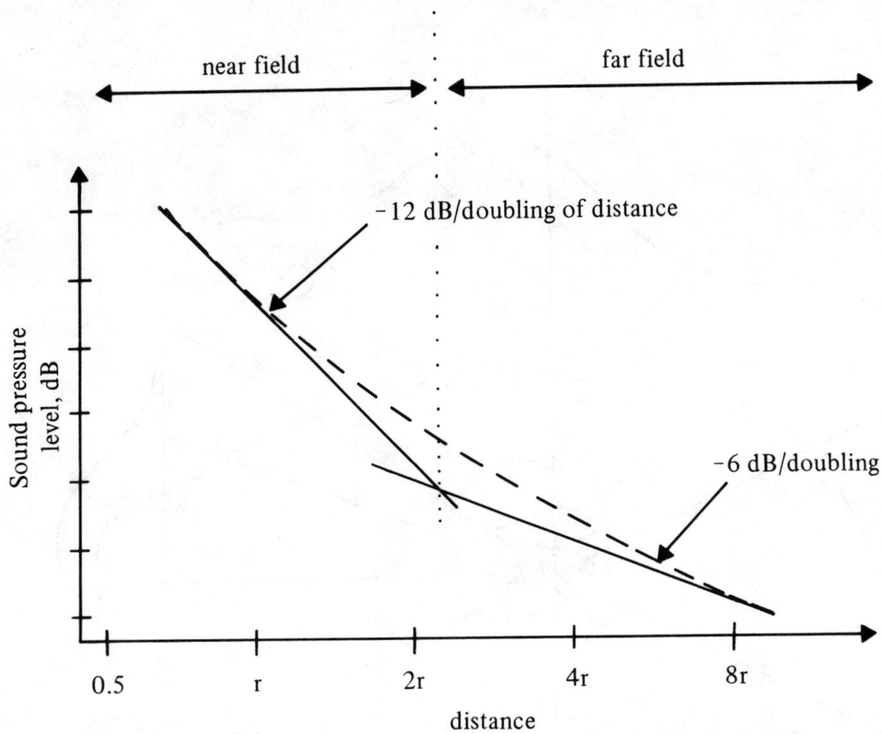

FIGURE 1.19. Sound pressure level for a dipole against distance.

1.5.3 Quadrupole Source

It is useful to study even higher order sources than *monopoles* or *dipoles.* It is thought that the sound produced by the mixing process in an air jet gives rise to stresses which are *quadrupole* in nature. Quadrupoles may be considered to consist of two opposing point forces (two opposing dipoles) or equivalently four monopoles. The expression for the sound pressure for a quadrupole is even more complicated than that for a dipole (Equation 1.51). Close to the source, in the *near field,* the pressure falls off even more rapidly than with a dipole. The pressure $p \propto 1/r^3$. Further from the source $p \propto 1/r^2$; while in the *far field,* $p \propto 1/r$.

The acoustic power of a quadrupole is given by*

$$W_{Lat} = \frac{\rho c k^6 Q_{quad}^2}{480\pi}, \qquad (1.56a)$$

or

$$W_{Long} = \frac{\rho c k^6 Q_{quad}^2}{40\pi}. \qquad (1.56b)$$

The sound power for a quadrupole is seen to be much less than for a dipole since the quadrupole source strength $Q_{quad} = Q\ell_1\ell_2$. By comparing Equations 1.56 and 1.50 we find that:

$$\frac{W_{Lat}}{W_m} = \frac{k^4 \ell_1^2 \ell_2^2}{60}, \qquad (1.57a)$$

and

$$\frac{W_{Long}}{W_m} = \frac{k^4 \ell_1^2 \ell_2^2}{5}. \qquad (1.57b)$$

Quadrupoles are very inefficient at low frequency (since $k = \omega/c = 2\pi f/c$).

1.5.4 Practical Sources

Sound sources experienced in practice may be even more complicated than dipoles or quadrupoles. Figure 1.20 shows the directivity plot of the sound pressure from a typical machine. The behavior of the sound pressure level with distance for such a machine is shown in Figure 1.21. In the shaded areas of the figure the sound pressure level

*The subscripts *Lat* and *Long* refer to Lateral and Longitudinal quadrupoles, respectively. If the two point forces are separated in the same direction as the forces, we have a Longitudinal quadrupole. If the separation is perpendicular to the force axis we have a Lateral quadrupole. In general, of course, we have an oblique quadrupole when the separation is in some other direction.

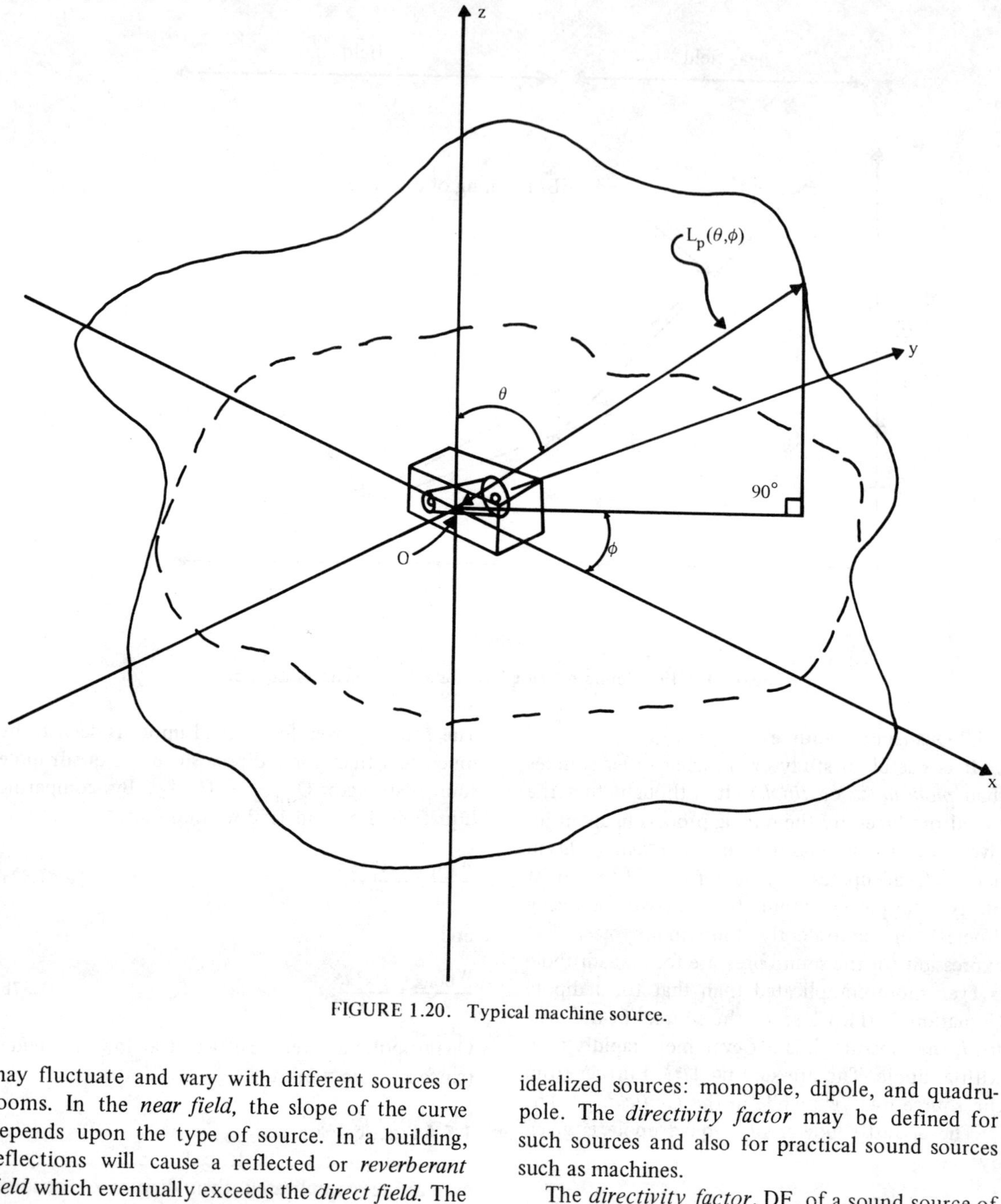

FIGURE 1.20. Typical machine source.

may fluctuate and vary with different sources or rooms. In the *near field,* the slope of the curve depends upon the type of source. In a building, reflections will cause a reflected or *reverberant field* which eventually exceeds the *direct field.* The slope of the curve in this region depends on the amount of acoustic absorption present in the room (see Section 1.6). It is normally desirable for most purposes to measure sound in the *far field* and *free field* where the *inverse square law* holds (6 dB decrease for doubling of distance).

1.5.5 Directivity Factor and Directivity Index

We have already examined the directivity of the idealized sources: monopole, dipole, and quadrupole. The *directivity factor* may be defined for such sources and also for practical sound sources such as machines.

The *directivity factor*, DF, of a sound source of power W is defined as the ratio of the mean-square sound pressure at angles θ and ϕ (see Figure 1.22) and distance r to the mean-square pressure measured at the same distance r for a nondirectional source of the same sound power.

$$DF = \langle p^2_{\theta,\phi} \rangle_t / \langle p^2 \rangle_t = I_{\theta,\phi}/I$$

where

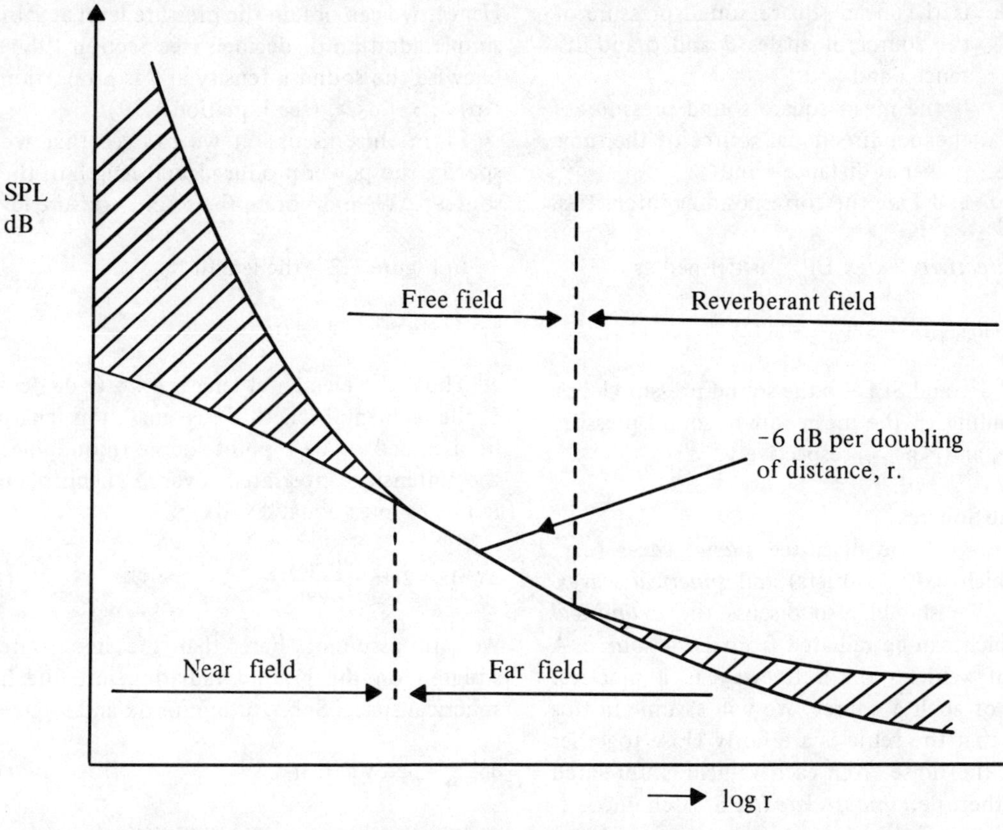

FIGURE 1.21 The dependence of sound pressure level on distance for a typical source.

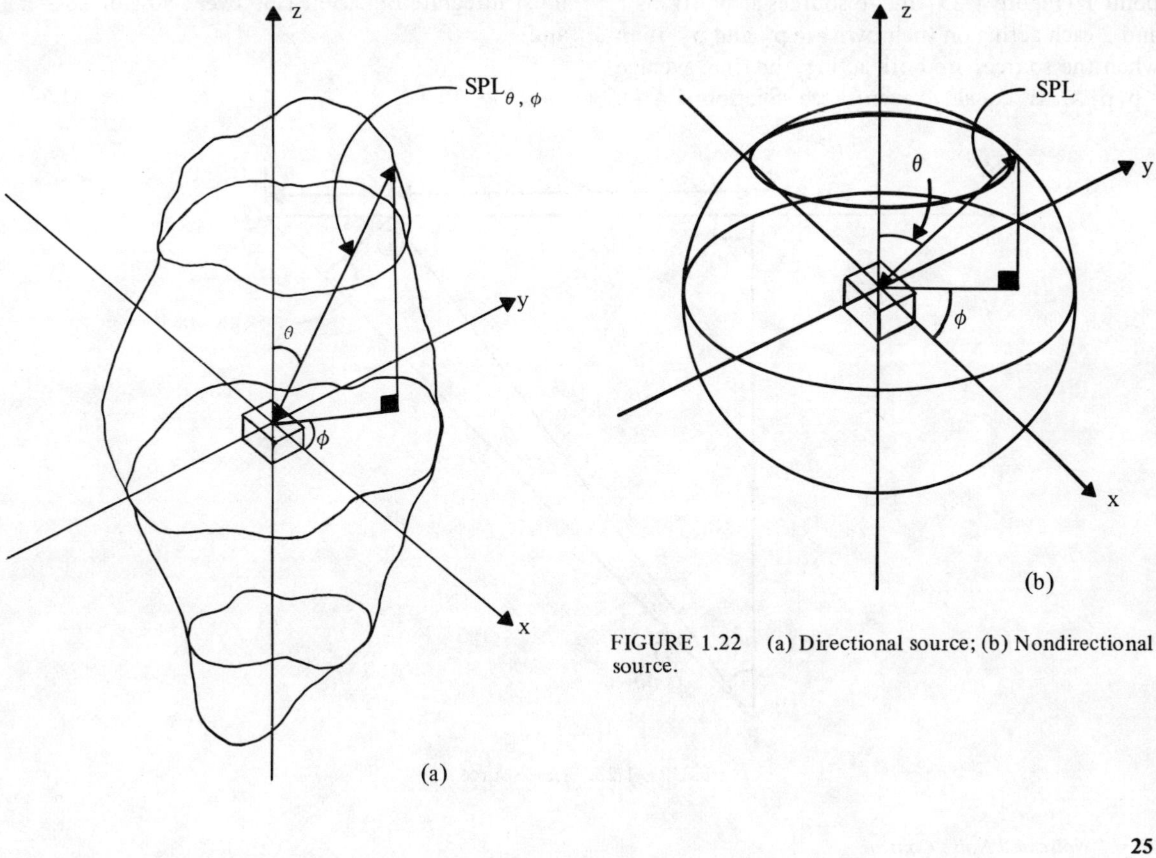

FIGURE 1.22 (a) Directional source; (b) Nondirectional source.

$\langle p^2{}_{\theta,\phi}\rangle_t$	is the mean square sound pressure of the source at angles θ and ϕ and distance r and
$\langle p^2\rangle_t$	is the mean square sound pressure of the nondirectional source of the same power at distance r and
$I_{\theta,\phi}$	and I are the corresponding intensities.

The *directivity index* $DI_{\theta,\phi}$ is defined as

$$DI_{\theta,\phi} = 10 \log_{10} DF = SPL_{\theta,\phi} - SPL.$$

Here $SPL_{\theta,\phi}$ and SPL are the sound pressure levels corresponding to the mean square sound pressure $\langle p^2{}_{\theta,\phi}\rangle_t$ and $\langle p^2\rangle_t$, respectively.

1.5.6 Line Sources

So far, we have discussed *plane waves* (e.g., waves which exist in ducts) and *spherical waves.* However, we should also discuss the *cylindrical waves* which can be radiated from a *line source.* A stream of vehicles on a freeway is a practical example of such a source. We will assume in this example that the vehicles are fairly close together and that the noise from each vehicle is unrelated to any other, i.e., *uncorrelated.* For such uncorrelated sources, if the sound pressures created at a point P (Figure 1.23) due to sources at positions 1 and 2 each acting on their own are p_1 and p_2, then when the sources are both acting, the time average $\langle p_1 p_2\rangle_t$ is equal to zero (see Section 1.4.4). Hence, we can obtain the pressure level at P by the simple addition of decibels (see Section 1.4.4) and likewise the sound intensity at P is proportional to $\langle p_1^2\rangle_t + \langle p_2^2\rangle_t$ (see Equation 1.39).

From this discussion we can see that we can specify the power produced/unit length of the line source, ΔW, and obtain the sound pressure level at P.

In Figure 1.23, the length

$$a = dx \cos\theta = r_1\, d\theta = (r/\cos\theta)\, d\theta.$$

Thus, the elemental length $dx = (r\, d\theta)/\cos^2\theta$. If the elemental length dx is small, this length can be assumed to be a point source (monopole) and the intensity integrated over a hemisphere of radius r_1 must equal $\Delta W\, dx$.

$$\Delta W\, dx = 2\pi r_1^2 \cdot \frac{dp^2_{rms}}{\rho c}. \qquad (1.58)$$

We are assuming here that the line source is situated on the ground radiating into the hemispherical space. Substituting for dx and r_1 gives

$$dp^2_{rms} = \rho c \Delta W\, d\theta / 2\pi r \qquad (1.59)$$

and if the line source is infinitely long then we must integrate Equation 1.59 over 180°, or $d\theta = \pi$ and

$$p^2_{rms} = \rho c\, \Delta W / 2r. \qquad (1.60)$$

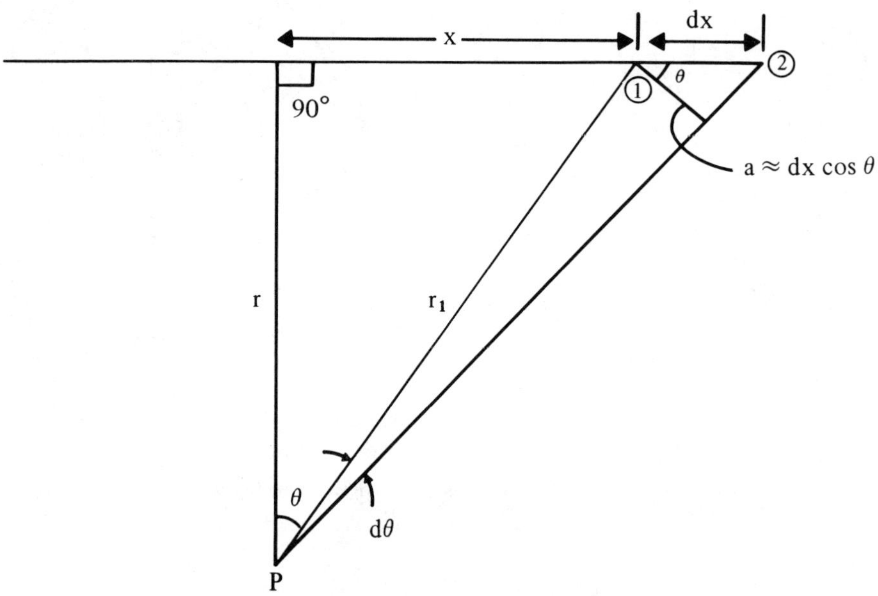

FIGURE 1.23. Line source.

We see from Equation 1.60 that with an uncorrelated line source the sound pressure is halved each time the distance increases four times. From Equations 1.36 and 1.60 the sound pressure level decreases by only 3 dB for each doubling of distance. Contrast this with the 6 dB decrease in sound pressure level for doubling of distance in the far field of simple sources (see Sections 1.5.1, 1.5.2, and 1.5.3).

1.6 SOUND IN ENCLOSURES AND ROOMS

We have already briefly discussed the effect of putting a sound source in a room (see Figure 1.21) and this will be discussed again in Chapter 4. In the early 1900s, Sabine[8] was perhaps the first to study room acoustics in a scientific manner although qualitative knowledge had existed for the use of designers of theatres, churches, cathedrals, and auditoria for centuries.

Sabine established that the *reverberation time* in a room was perhaps the most important acoustic design variable. Although modern acoustic instrumentation was unavailable to him, he showed how the reverberation time was related to the room volume and *acoustic absorption* in the room. The unit of acoustic absorption, the *sabin*, is named after him.

1.6.1 Absorption Coefficient

The absorption coefficient, α (f), which is a function of the frequency, f, is defined as the fraction of the incident acoustic energy which is absorbed by a surface.

$$\alpha = \frac{\text{acoustic power absorbed}}{\text{acoustic power incident}}. \qquad (1.61)$$

The absorption coefficient α can have a value between 0 and 1. Materials which have a high value of α are porous or fibrous. Fiberglass and open celled rubber foams are good examples. It is thought that the sound waves propagate into the narrow channels of the material and lose energy by viscous dissipation. Hence, heat is produced.

We can see that there are several requirements in order for materials to have high values of α. The air channels should all be open to the surface so that the sound waves can propagate into the material. Nonporous materials have low absorption coefficients. We can place some protective wall over the material (if it is very thin, such as mylar sheet, this should be made porous by drilling a large number of holes) so that the material channels are connected with the atmosphere. A common fault is the painting of the surface of some absorbing materials. This can have disastrous results since the air channels become sealed off and the absorbent properties are lost.

It is found that the absorption coefficient, α, of an absorbing material at a certain frequency depends upon the thickness used. This is particularly apparent at low frequency. It is found that provided the thickness of the material is less than 1/16 of the acoustic wavelength, then the absorption coefficient approximately doubles if we double the frequency. This is shown in Figure 1.24. The noise reduction coefficient, NRC, is found by averaging the absorption coefficients* at the frequencies 250, 500, 1,000, and 2,000 Hz.

Some increase in low frequency absorption can be obtained by mounting such absorption materials at a suitable distance, for example, 6 in., from the walls of a room. If the material is placed a quarter wavelength ($\lambda/4$) from the wall the absorption is greatly increased at a frequency given by:

$$\frac{1}{2} = \lambda/4 = c/4f = 1120/4f \text{ ft},$$

$$f = \frac{1120}{2} = 560 \text{ Hz}.$$

The reason for the improved acoustic absorption at this frequency is that the acoustic particle velocity must be zero at the wall while at a quarter wavelength from the wall it will be a maximum and thus the material placed there will absorb energy well. For this reason drapes should be hung away from a wall, not touching a wall, if it is desired to increase the low frequency absorption.

1.6.2 Average Absorption Coefficient and Reverberation Time

Provided that the total absorption material in a room is not too large, we may compute the average surface absorption coefficient $\bar{\alpha}$ as follows:

$$\bar{\alpha} = \frac{S_1\alpha_1 + S_2\alpha_2 + \ldots . S_n\alpha_n}{S_1 + S_2 + \ldots . S_n}, \qquad (1.62)$$

where

$S_1, S_2, \ldots . S_n$ are the areas of material with absorption coefficients $\alpha_1, \alpha_2, \ldots . \alpha_n$.

If the total area $S = S_1 + S_2 + \ldots + S_n$, then

$$S\bar{\alpha} = S_1\alpha_1 + S_2\alpha_2 + \ldots . + S_n\alpha_n. \qquad (1.63)$$

*In averaging, NRC is rounded off to the nearest multiple of 0.05.

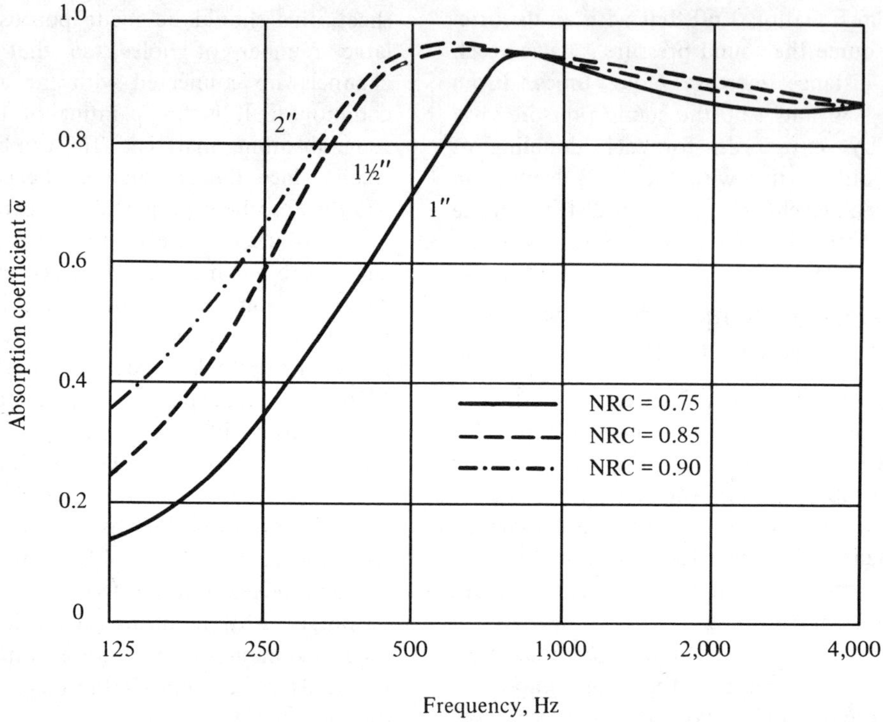

FIGURE 1.24. Absorption and noise reduction coefficients (NRC) of fiberglass formboard.

The units of $S\overline{\alpha}$ are sabins. If the area S is in ft^2, then $S\overline{\alpha}$ is in sabins (ft^2); if S is in m^2, then $S\overline{\alpha}$ is in sabins (m^2). If $\overline{\alpha} = 1.0$ then the material is perfectly absorbing and we will have S sabins of perfectly absorbing material. We normally write $A = S\overline{\alpha}$ and hence A also represents an area of perfectly absorbing material.

If we decrease the amount of absorption in a room we would expect the reverberation time to increase and we may show that:

$$T_R = \frac{0.161\ V}{S\overline{\alpha}}, \tag{1.64}$$

where:

V = room volume in m^3 and $S\overline{\alpha}$ is the total absorption in sabins (m^2). This equation is known as Sabine's Equation and is fairly accurate provided $\overline{\alpha} < 0.2$ (see Section 4.2.2).

Reverberation time is discussed in more detail in Sections 4.2.2, 4.2.3, and 4.2.4.

1.6.3 Energy Density and Intensity in a Diffuse Field

If we assume for simplicity that the energy density, ϵ, is the same at all points in the room (see Section 1.3.1.C), then we may calculate the rate at which energy is absorbed by the room. First of all let us calculate what the intensity on a wall of the room is. This requires a little integration. The energy density in a diffuse field is given by

$$\epsilon = \hat{p}^2/(2\rho c^2) = p_{rms}^2/(\rho c^2). \tag{1.65}$$

Note that this is the same expression we had for a plane wave in Equation 1.35. We may consider the diffuse field to be made up of a very large number (infinite) of plane waves. There is equal probability that these waves may come from any direction.

Let us take a little elemental volume dV in the room and let dS be a small surface area on a wall (see Figure 1.25). The fraction of the energy in the elemental volume dV traveling towards area dS is $dS \cos\theta/(4\pi r^2)$, since $dS \cos\theta$ is the projected area "seen" by the wave traveling towards dS. The energy in the volume dV is ϵ dV; hence the energy reaching dS is $(\epsilon\ dV)\ dS \cos\theta/(4\pi r^2)$. But dV is $dr\ (r d\theta)\ (r \sin\theta\ d\phi)$ (see Figure 1.25c).

The intensity is the total energy reaching dS in unit time. Hence, we must integrate over θ and ϕ (i.e., over the hemisphere) to include the energy for similar elemental volumes situated all over the

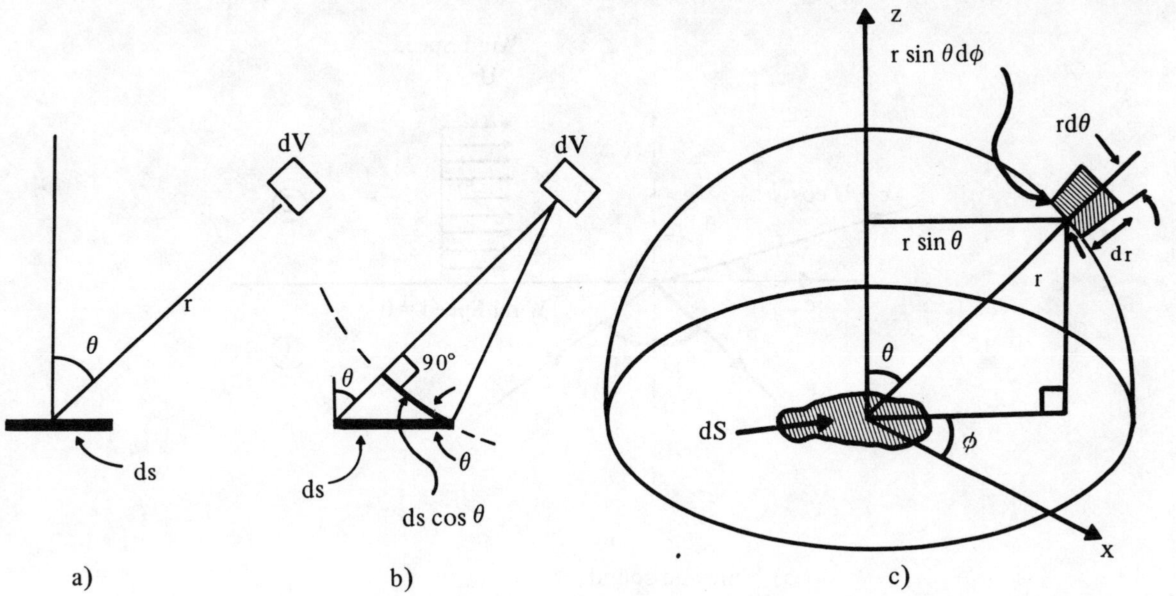

FIGURE 1.25 Acoustic intensity in a diffuse field.

hemisphere. Also in unit time waves can reach dS from elemental volumes situated on radii from a distance zero to c away. Hence, the energy reaching dS is

$$\text{energy} = \epsilon\, dr\, (r\, d\theta)(r \sin\theta\, d\phi))\, dS \cos\theta/(4\pi r^2). \quad (1.66)$$

Integrating Equation 1.66 as discussed above gives:

$$\langle I\rangle_t = \epsilon \int_{r=0}^{c} (r^2/4\pi r^2)dr \int_{\theta=0}^{2\pi} d\phi \int_{\theta=0}^{\pi/2} \sin\theta \cos\theta\, d\theta\, \frac{dS}{dS} \quad (1.67)$$

We have divided by dS because intensity is power/unit area. Integrating Equation 1.67 gives:

$$\langle I\rangle_t = \epsilon(c/4\pi)\,(2\pi)\int_{\theta=0}^{\pi/2} \frac{1}{2}\sin 2\theta\, d\theta = \epsilon\,\frac{c}{2}\left[\frac{-\cos 2\theta}{4}\right]_0^{\pi/2},$$

$$= \epsilon\, c/4 \quad (1.68)$$

and substituting for ϵ from Equation 1.65 gives:

$$\langle I\rangle_t = \epsilon\, c/4 = \hat{p}^2/(8\rho c) = p_{rms}^2/(4\rho c). \quad (1.69)$$

Notice the intensity in a diffuse field is thus four times smaller than the intensity in a plane wave (Equation 1.34).

Now we may calculate the rate at which acoustic energy is absorbed by the walls of a room. If the absorption area of a room is $S\bar{\alpha}$ (see Section 1.6.2), then the rate at which energy is absorbed is

$$W_a = S\bar{\alpha}\langle I\rangle_t = S\bar{\alpha}p_{rms}^2/(4\rho c). \quad (1.70)$$

Here S is the total wall area of the room, $\bar{\alpha}$ is the mean absorption coefficient, p_{rms} is the space-average of the sound pressure in the room, and ρc is the characteristic impedance of air (415 MKS rayls). If $\bar{\alpha} > 0.2$, there are complications and these are further discussed in Section 4.2.

1.7 SOUND IN FREE SPACE

Whenever we consider outdoor noise, for example, in traffic, aircraft, or community noise problems, we find that the additional complications of refraction caused by wind or by temperature gradients, sound absorption by the atmosphere, and other effects must be considered. These effects can be of considerable importance outdoors. Even indoors at high frequencies (if the humidity is low) air absorption should be accounted for. The attenuation caused by wind, temperature gradients, atmospheric absorption, rain, snow, or fog are in addition to the inverse square law reductions caused by spherical spreading discussed in Section 1.5.

1.7.1 Effects of Wind

We may easily see the effect of refraction caused by wind by considering the simple case of plane sound waves traveling through a stationary atmosphere and impinging on air moving with a uniform speed U (see Figures 1.26a and b). For simplicity we will assume that the direction of

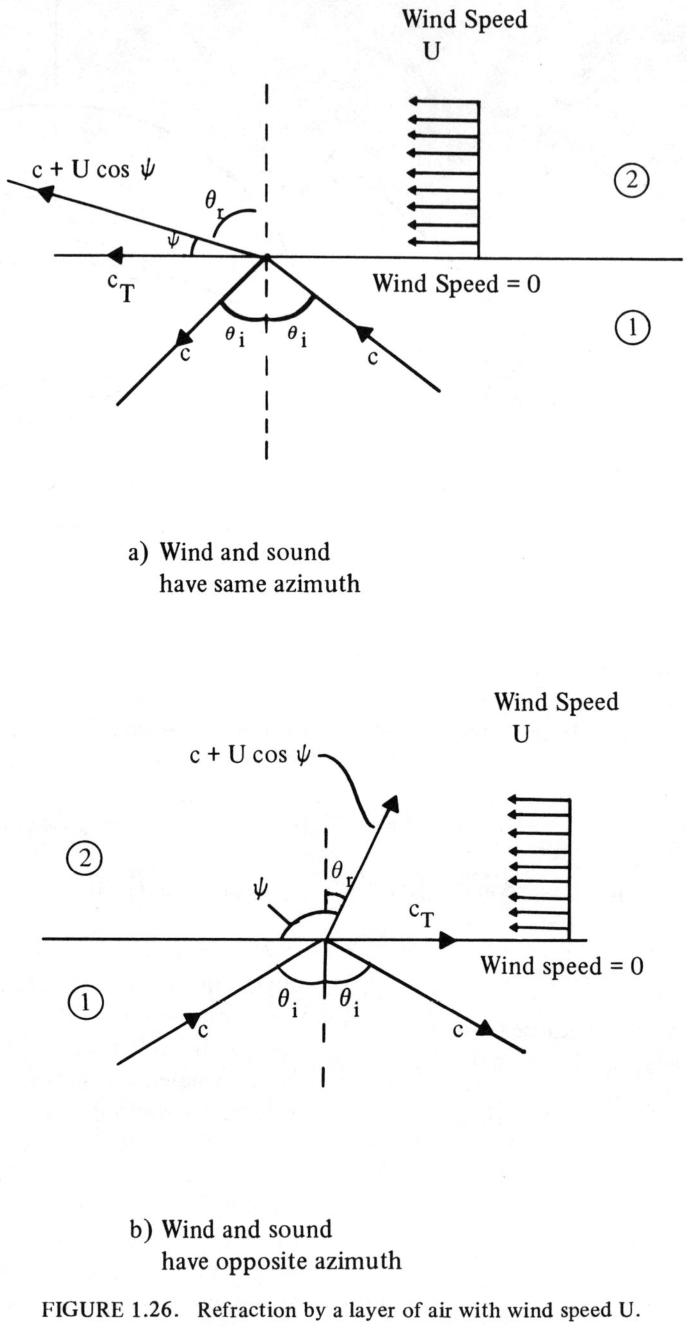

FIGURE 1.26. Refraction by a layer of air with wind speed U.

sound propagation and the wind direction have the same or the opposite azimuth (i.e., the same or opposite compass heading). We will also assume for simplicity that the temperature is the same in the stationary atmosphere and in the wind, regions 1 and 2, respectively.

There will be incident, reflected, and transmitted waves at the interface of the two regions. The condition which governs the relation between the angles of incidence and transmission is that the trace speed of the incident wave on the interface must equal the trace speed of the transmitted wave on the interface.[9] Provided that the wind speed is much less than the sound speed, $U \ll c$ (the usual case), then the wavespeed in medium 2, $c_2 \approx c + U \cos \psi$. Now $c = c_T \sin \theta_i$ and $c + U \cos \psi = c_T \sin \theta_r$. When the wind and sound have the same azimuth (Figure 1.26a), $\sin \theta_r = \cos \psi$, and

$$\frac{c}{\sin \theta_r} + U = \frac{c}{\sin \theta_i} = c_T, \tag{1.71}$$

thus

$$\sin \theta_r = \sin \theta_i / (1 - \frac{U}{c} \sin \theta_i). \tag{1.72}$$

When the wind and sound have the opposite azimuth (Figure 1.26b), $\sin \theta_r = -\cos \psi$, and

$$\frac{c}{\sin \theta_r} - U = \frac{c}{\sin \theta_i} = c_T, \tag{1.73}$$

and

$$\sin \theta_r = \sin \theta_i / (1 + \frac{U}{c} \sin \theta_i). \tag{1.74}$$

Equation 1.72 shows that when the sound wave and the wind have the same azimuth, the sound direction (or "ray") is refracted (bent) downwards since $\theta_r > \theta_i$ (see Figure 1.26a). Equation 1.74 shows that when the sound wave and the wind have the opposite direction, the sound is bent upwards, since $\theta_r < \theta_i$ (Figure 1.26b).

Of course we never exactly experience the idealized wind speed situations shown in Figure 1.26. However this theory shows us what happens in the more usual and complicated case shown in Figure 1.27. We normally find that the wind speed slowly increases with altitude.[10] If there is a temperature inversion (see Section 1.7.2), the increase with height is more rapid. If there is a temperature lapse the increase is more gradual. If the wind speed slowly increases with altitude as shown in Figure 1.27, consideration of the refraction processes shown in Figure 1.26 shows the manner in which the sound is refracted. Sound which propagates upwind (against the wind) is refracted upwards. This causes a sound "shadow" upwind. Sound which propagates downwind (with the wind) is refracted downwards which causes sound reinforcement. The shadowing or reinforcement effects can cause differences of as much as 20 or 30 dB from the expected inverse square law attenuation of a sound source. This refraction effect is caused by the increase in wind speed with altitude. The amount of refraction will depend upon the rate of wind speed change with altitude. The refraction does not occur if the wind speed is constant with altitude.

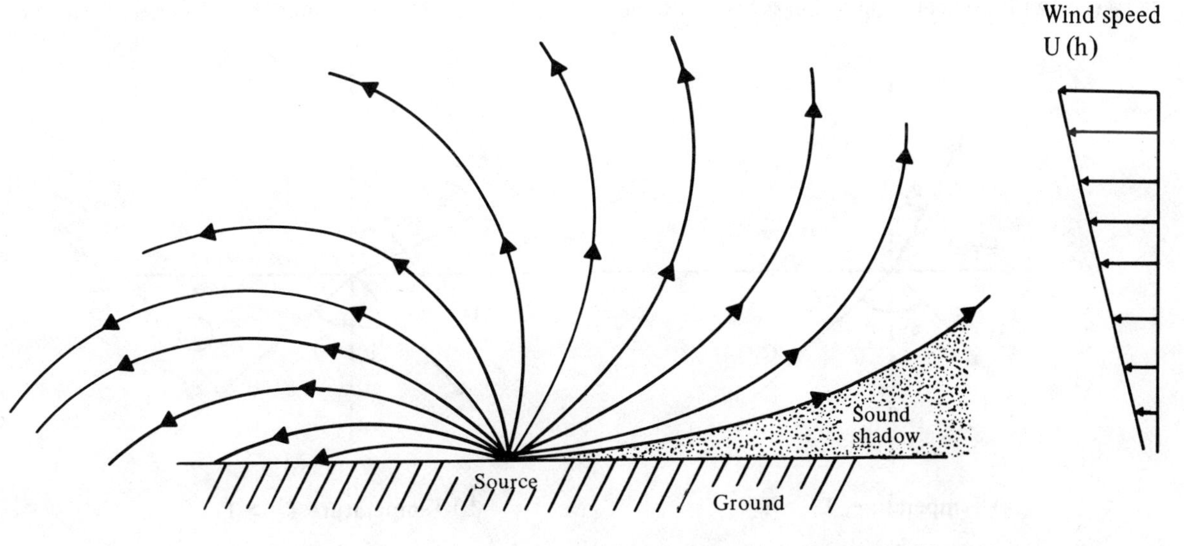

FIGURE 1.27. Refraction of sound in air with wind speed increasing with altitude.

1.7.2 Effects of Temperature Gradients

Temperature gradients in the atmosphere also cause sound waves to be refracted. Although the effects are similar to wind, there are differences. Again let us examine the idealized cases of sound propagating through a layer of atmosphere with one temperature into a layer with a different temperature (Figure 1.28). Again the condition which governs the angles of reflection and refraction is that the trace speed at the interface, c_T, must equal the trace speed of the incident, reflected and refracted waves. Hence:

$$c_T = \frac{c_1}{\sin\theta_i} = \frac{c_2}{\sin\theta_r}, \tag{1.75}$$

or

$$\sin\theta_r = (c_2/c_1)\sin\theta_i. \tag{1.76}$$

Equation 1.76 is similar to Snell's Law in optics. If the temperature $T_1 > T_2$, then $c_1 > c_2$ (from Equation 1.23) and $\theta_r < \theta_i$ as in Figure 1.28a. If $T_1 < T_2$, then $c_1 < c_2$ and $\theta_r > \theta_i$ as in Figure 1-28b.

Actually in practice we never encounter the idealized cases shown in Figure 1.28. If the atmosphere were assumed to be completely at rest, then the density and pressure would decrease exponentially with height, but the temperature would be constant throughout because of heat conduction. Continual agitation in the atmosphere negates this simplistic model. Some more insight may be gained by assuming that air moves from lower to higher altitudes by convection adiabatically (i.e., with no heat conduction). Assuming the perfect gas law, $p = \rho RT$, and also the adiabatic law, $p = \rho^\gamma \times$ constant, gives:[11]

$$dT/dh = -(g/R)(\gamma - 1)/\gamma, \tag{1.77}$$

where T is the temperature, h is the height, g the gravitational constant, R the gas constant, and γ the ratio of specific heats. Equation 1.77 gives the adiabatic lapse rate dT/dh. When inserting values for dT/dh, the result is dT/dh = −0.9°C/100 m.

Actually this theoretical lapse rate is inaccurate close to the ground where the ground heats the air in contact during the day and cools it during the night. Figure 1.29 shows typical examples of variation of atmospheric temperature with height. During a clear afternoon the temperature decreases steadily with altitude (Figure 1.29a) and refraction as shown in Figure 1.30a results. This result may be deduced by studying the simpler model in Figure 1.28a. During a clear night there is a temperature inversion at low altitude and the refraction shown in Figure 1.30b results. This result may be deduced from Figure 1.28b.

We note immediately that unlike wind gradient effects (Figure 1.27), the refraction pattern for temperature gradients is symmetrical about a vertical line drawn through the source. During a normal clear afternoon then, a sound shadow can exist all around a source (Figure 1.30a) due to the normal temperature lapse provided there is no wind. The sharpness of this shadow would be reduced at low frequency. Also at all frequencies, scattering due to atmospheric turbulence would diminish the shadow sharpness. In the presence of wind the shadow could be eliminated in the

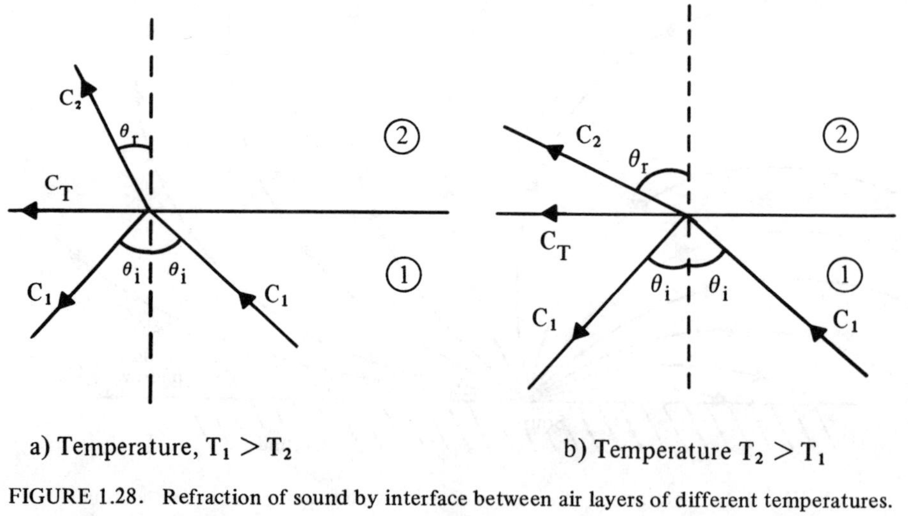

FIGURE 1.28. Refraction of sound by interface between air layers of different temperatures.

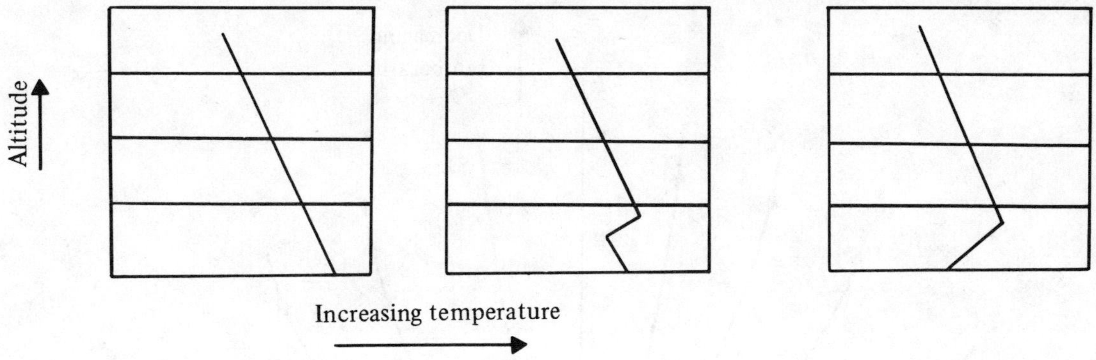

FIGURE 1.29. Typical examples of temperature variation with height.

downwind direction, but enhanced in the upwind direction (Figure 1.27).

During a clear night, due to the temperature inversion, no sound shadow would be observed, but instead an enhancement of sound would be heard (compared with the normal inverse square law attenuation) (see Section 1.5). During early morning at low altitudes a channelling of sound can occur with only a small decrease in level with distance. More detailed discussion on temperature and wind effects may be found in the literature.[12,13] The actual numerical values of attenuation caused by wind and temperature gradients can be calculated by ray acoustics and some experimental data available.[12,13]

1.7.3 Humidity

Even if the atmosphere were stationary and isotropic (same properties at all points) there would still be excess attenuation due to atmospheric absorption. Atmospheric absorption is caused by heat conduction and viscosity, and rotational and vibration relaxation of oxygen molecules in the air. The second effect is normally more important, except at very low temperatures, and is dependent on humidity and temperature. The excess attenuation at 20°C (68°F) is given by.[14,4]

$$A_e = 7.4\,(f^2 r/\phi)\,10^{-8}\ \text{dB} \tag{1.78}$$

where f is the frequency, Hz; ϕ is the relative humidity, %; and r is the source to receiver distance, m. Formulae are also available for excess attenuation at other temperatures. The aircraft industry[15] has also produced standard charts for calculating aircraft to ground excess attenuation (see Figures 1.31 and 1.32).

1.7.4 Ground Effects

Even in a stationary isotropic atmosphere, propagation near the ground does not follow the inverse square law and again excess attenuation is observed. This is thought to be because ground does not have an infinite impedance ($\rho c = \infty$) and most soils or materials are absorbing to a certain extent. If the ground had an infinite impedance we could use the idea of an *image* source the same distance under the ground as the source is above (as in optics). In practice, except for very hard surfaces, the reflected image is weaker than the original source. However, as with a perfect image source there will be positions of both destructive and constructive interference.

1.7.5 Absorption Due to Grass, Shrubs, and Trees

The excess absorption due to thick grass or shrubbery is much greater than that due to absorbent soils such as sand and can reach values of about 20 dB/100 m. Despite the exaggerated claims of some people, attenuation due to trees is disappointingly low. Measurements have been made of the excess attenuation for bare trees, jungles, and dense pine forests. Attenuations range from 3 dB/100m for bare trees to 15 or 20 dB/100 m for dense pine forests. An average value for trees can be assumed:

$$A = 0.01\,(f)^{1/3} r,\ \text{dB}, \tag{1.79}$$

where f is the frequency, Hz; and r is the distance, m. Unless there are other compelling reasons such

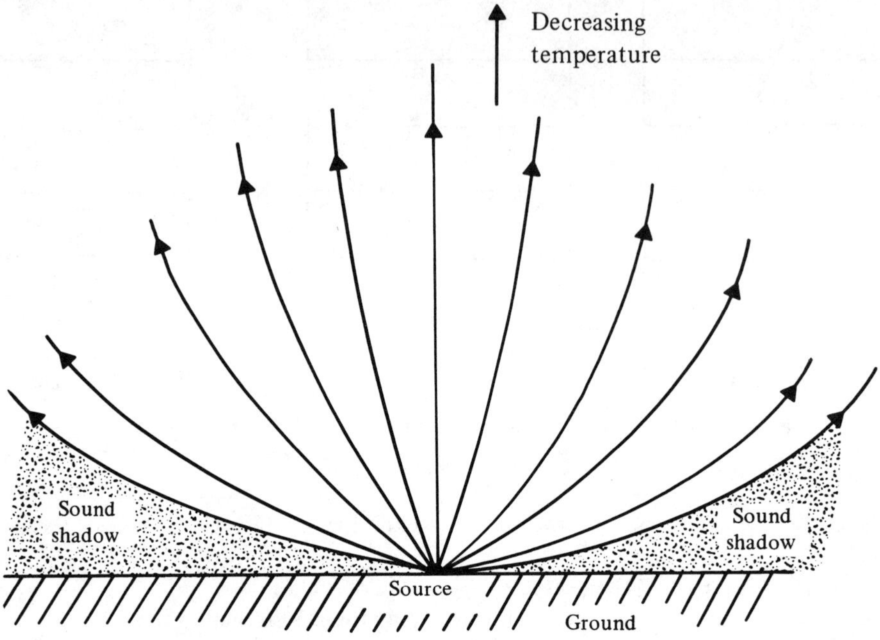

a) Normal temperature lapse (temperature decreases with altitude)

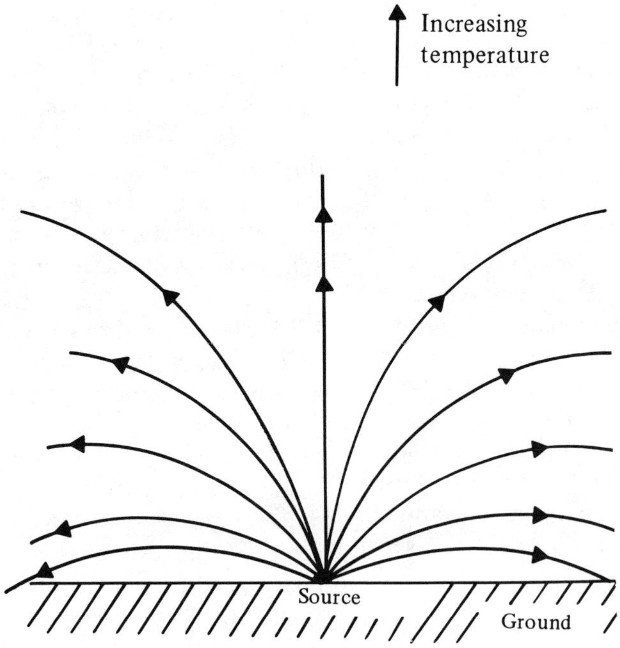

b) Temperature inversion

FIGURE 1.30. Refraction of sound in air with normal temperature lapse and with temperature inversion.

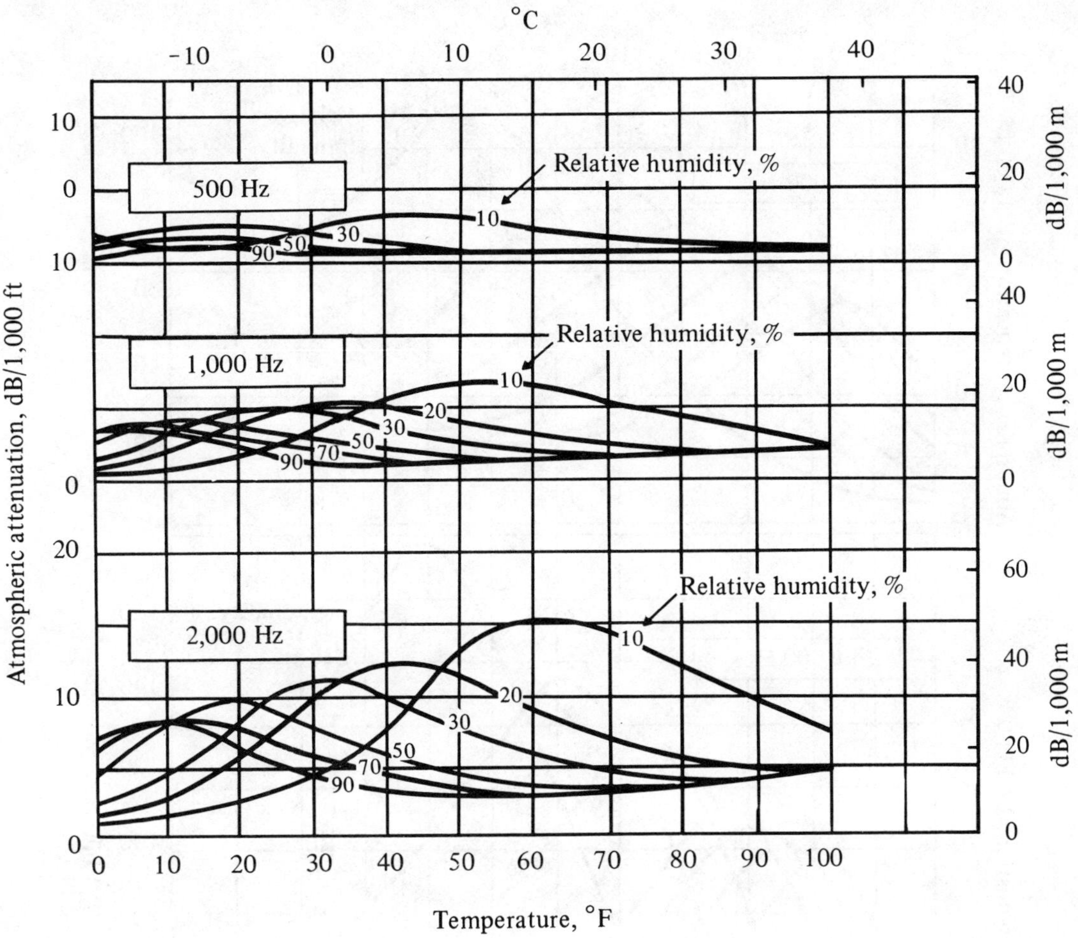

FIGURE 1.31. Atmospheric excess attenuation for aircraft to ground propagation for octave band center frequencies 500, 1,000, and 2,000 Hz.

as "cosmetic" appearance or public relations it seems that trees form a poor barrier against noise and their use for such should not be seriously considered. Impervious barriers constructed specifically for sound attenuation are preferred (Volume II).

1.7.6 Absorption Due to Rain, Fog, and Snow

Most experimental measurements seem to indicate that there is no change in excess attenuation due to the presence of light rain or fog. Although many people observe that sound seems to carry particularly well during light rain or fog, this is probably due to the decrease in human activity during such times and when there is snow on the ground. During such times it seems that both movements of surface vehicles and aircraft are reduced and that children's shouts and the use of appliances such as lawn mowers and chain saws, etc. decreases. This general decrease in activity reduces the community background level and hence makes any noise seem more prominent. This is probably why many people believe that sounds carry better in the presence of rain, fog, or snow.

1.8 FREQUENCY ANALYSIS

1.8.1 Periodic Waveforms

Sometimes in acoustics we encounter sounds which are *pure tones* (or very nearly so) e.g., the 120 Hz hum from an electric motor. In the case of a pure tone, the pressure-time history is simple harmonic and could be represented by the waveform in Figure 1.3, $p(t) = A \sin 2\pi f_1 t$. This has been redrawn in Figure 1.33a. The pure tone can

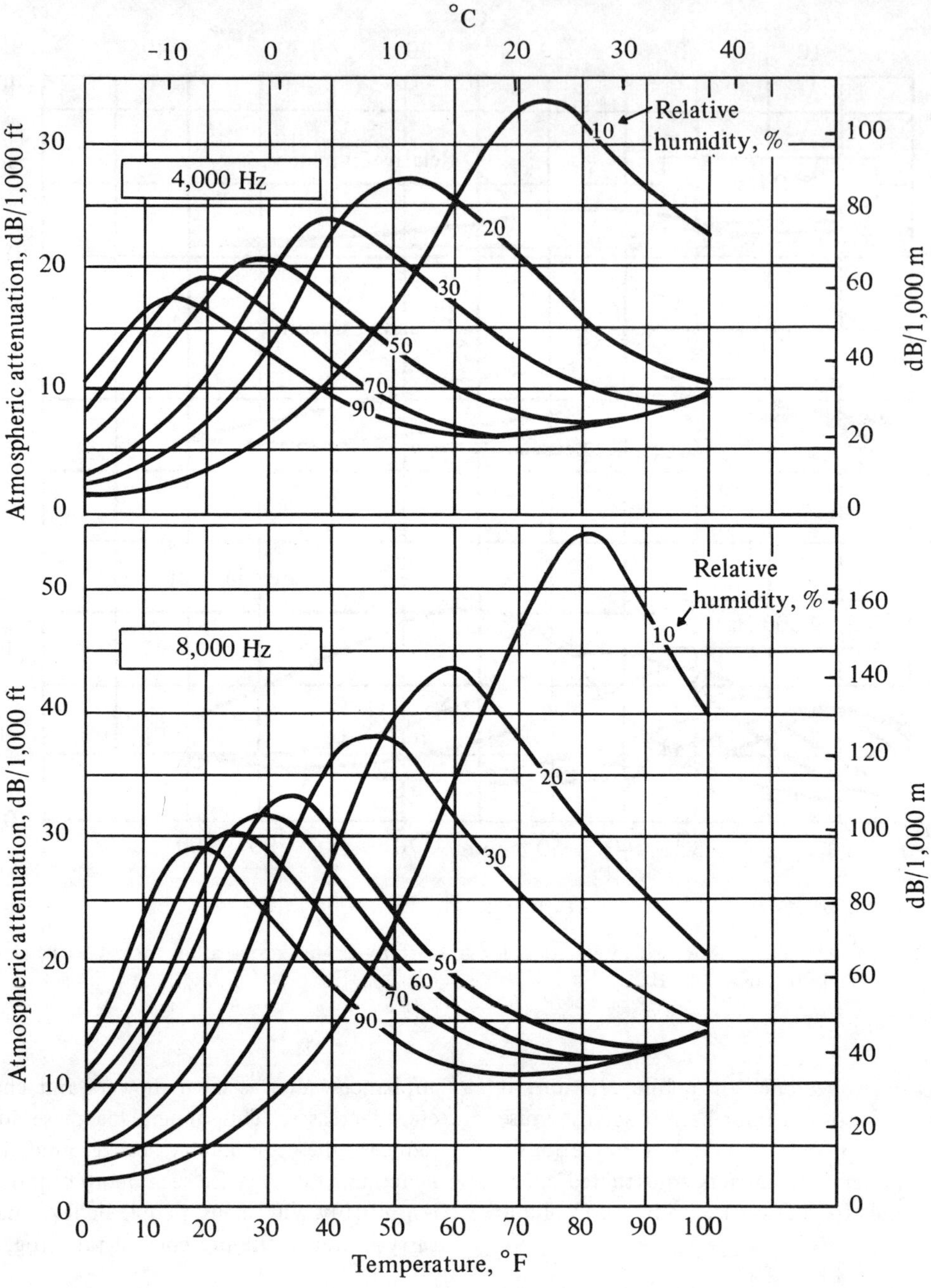

FIGURE 1.32. Atmospheric excess atenuation for aircraft to ground propagation for octave band center frequencies 4,000 and 8,000 Hz.

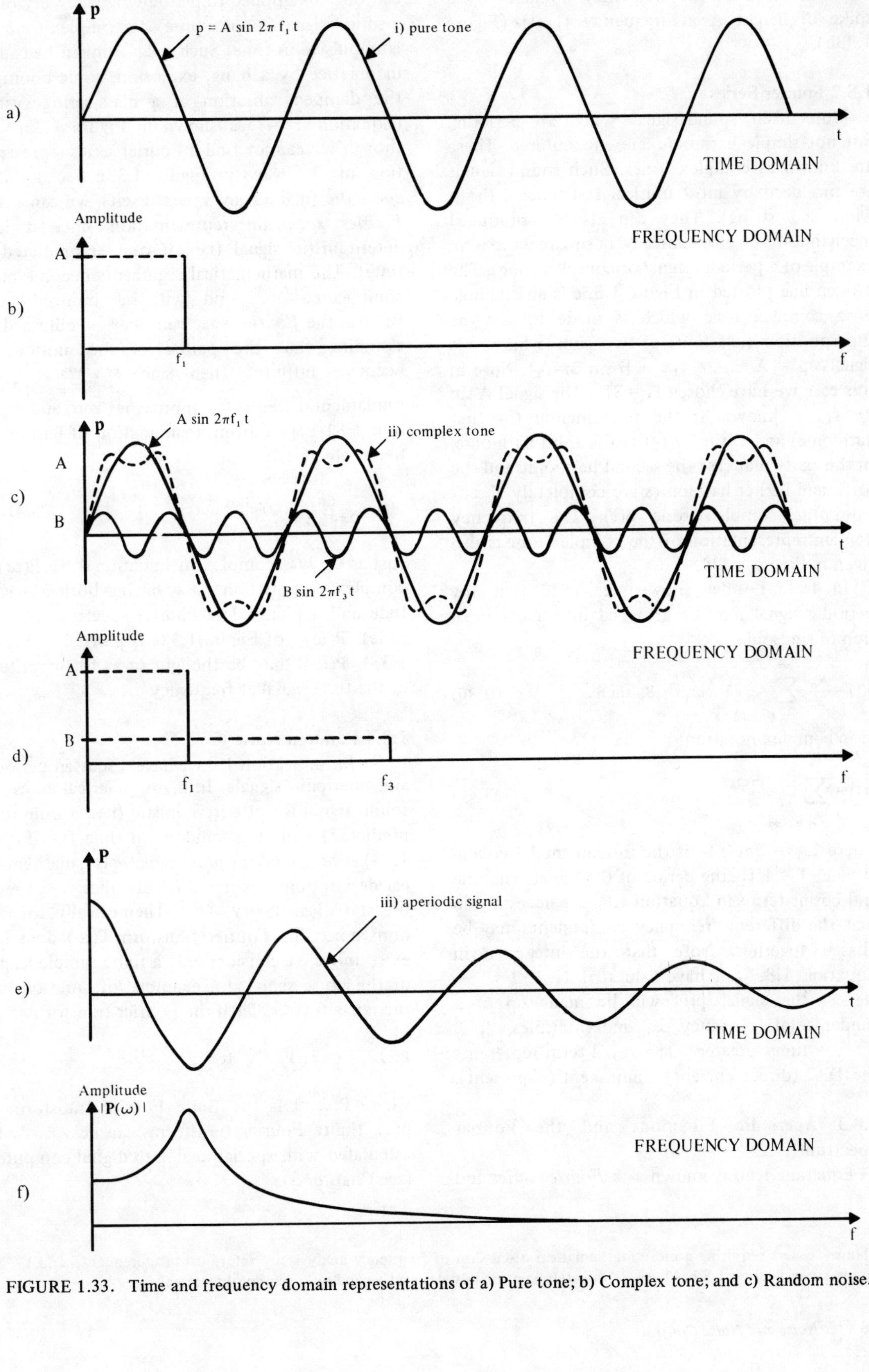

FIGURE 1.33. Time and frequency domain representations of a) Pure tone; b) Complex tone; and c) Random noise.

be represented in the frequency domain* by a spike of height A at frequency f_1, Hz (Figure 1.33b).

1.8.2 Fourier Series

More often, sound signals which are periodic, but not simple harmonic, are encountered. These are known as complex tones. Such sound signals are produced by most musical instruments (both wind and string). They can also be produced mechanically or electronically (a square wave is an example of a periodic signal or complex tone). The broken line plotted in Figure 1.33c is an example of a complex tone which is made up by the superposition (addition) of two simple harmonic signals, $p = A \sin 2\pi f_1 t + B \sin 2\pi f_3 t$. Note in this case we have chosen $f_3 = 3f_1$. The signal $A \sin 2\pi f_1 t$ is known as the fundamental (or first harmonic) and $B \sin 2\pi f_3 t$ is the third harmonic. In this particular case the second harmonic and the forth and higher harmonics are completely absent from the complex tone p(t). The frequency domain representation of the complex tone is also given in Figure 1.33d.

In fact, Fourier showed in 1822 that any periodic signal may be analyzed into a combination of sinusoids:

$$p(t) = \frac{A_0}{2} + \sum_{n=1,2,3\ldots}^{\infty} (A_n \cos n\omega_1 t + B_n \sin n\omega_1 t). \qquad (1.80)$$

or in complex notation:

$$p(t) = \sum_{n=-\infty}^{\infty} C_n e^{in\omega_0 t}, \qquad (1.81)$$

where $\omega_1 = 2\pi f_1$; f_1 is the fundamental frequency; and $T = 1/f_1$, the period of the signal. The sine and cosine terms in Equation 1.80 are necessary so that the different frequency components may be phased together. Note that the integer n in Equation 1.80 can have values of 1, 2, 3 . . . ∞. Hence, the signal p(t) will be made up of a fundamental frequency ω_1 and multiples, 2, 3, 4 . . . ∞ times greater. The $A_0/2$ term represents the D.C. (direct current) component (if present).

1.8.3 Aperiodic Functions and the Fourier Spectrum

Equation 1.80 is known as a *Fourier series* and can only be applied to periodic signals. Very often a sound signal is not a pure or a complex tone but is impulsive in time. Such a signal might be caused in practice by a bang, explosion, sonic boom, or the damped vibration of a mass-spring system (Equation 1.14) as shown in Figure 1.33e. Although we cannot find a Fourier series representation of the wave in Figure 1.33e because it is *aperiodic* (it does not repeat itself), we can find a *Fourier spectrum* representation since it is a deterministic signal (i.e., it can be predicted in time). The mathematical arguments become more complicated[16,17] and will be omitted here. Briefly, the *Fourier* spectrum may be obtained by assuming that the period of the motion, T, becomes infinite. Then since $f_1 = \frac{\omega_1}{2\pi} = \frac{1}{T}$, the fundamental frequency approaches zero and Equation 1.81 passes from a summation of harmonics to an integral:

$$p(t) = \frac{1}{2\pi}\int_{-\infty}^{\infty} \mathbf{P}(\omega)\, e^{i\omega t}\, d\omega. \qquad (1.82)$$

Just as $\mathbf{C}_n$ was complex in Equation 1.81, $\mathbf{P}(\omega)$ is complex in Equation 1.82, having both a magnitude and a phase. The Fourier spectrum (magnitude), $|\mathbf{P}(\omega)|$, of Figure 1.33e is plotted in Figure 1.33f. $|\mathbf{P}(\omega)|$ may be thought of as the amplitude of the time signal at frequency ω.

1.8.4 Random Noise

So far in Section 1.8 we have discussed periodic and aperiodic signals. In many practical cases the sound signal is not deterministic (i.e., it cannot be predicted) and it is random in time (see Figure 1.34). For a random noise signal, p(t), mathematical descriptions become difficult since we have to use statistical theory.[16,17] Theoretically, for random noise the Fourier transform $\mathbf{P}(\omega)$ does not exist unless we consider only a finite sample length of the noise signal, for example, of duration τ in the range $0<t<\tau$. Then the Fourier transform is

$$\mathbf{P}(\omega,\tau) = \int^{T} p(t)\, e^{-i\omega t}\, dt \qquad (1.83)$$

where $\mathbf{P}(\omega,\tau)$ is the finite Fourier transform of p(t). Finite Fourier transforms can now easily be calculated with special analog to digital computers (see Chapter 3).

*Those readers requiring a more mathematical discussion of frequency analysis are referred to References 16 and 17. The discussion in Section 1.8 has been deliberately kept mainly descriptive and only semimathematical.

1.8.5 Mean Square Values

In the case of the pure tone (Section 1.8.1) a useful quantity to determine is the *mean square* value, i.e., the time average of the signal squared, $\langle p^2(t)\rangle_t$:

$$\langle p^2(t)\rangle_t = \frac{1}{T}\int_0^T p^2(t)\,dt. \qquad (1.84)$$

We have already carried out just this sort of integration when we computed the time average of the instantaneous intensity in Equations 1.32 and 1.33. For the pure tone in Figure 1.33(a) then we obtain

$$\langle p^2(t)\rangle_t = A^2/2. \qquad (1.85)$$

The root mean square value is given by the square root of $\langle p^2(t)\rangle_t$ or

$$p_{rms} = A/\sqrt{2}. \qquad (1.86)$$

For the general case of the complex pure tone in Equation 1.80 or 1.81 we obtain:

$$\langle p^2(t)\rangle_t = \frac{A_0{}^2}{4} + \frac{1}{2}\sum_{n=1}^{\infty}(A_n{}^2 + B_n{}^2), \qquad (1.87)$$

or

$$\langle p^2(t)\rangle_t = C_0{}^2 + 2\sum_{n=1}^{\infty}|C_n|^2, \qquad (1.88)$$

since $|C_n|^2 = (A_n{}^2 + B_n{}^2)/4$. The mean square value then is the sum of the squares of all the harmonic components of the wave (weighted by a constant of 1/2 or 2).

1.8.6 Energy and Power Spectral Densities

In the case of aperiodic signals (Section 1.8.3), a quantity called the *energy density function* or equivalently the *energy spectral density*, S(f), is defined:

$$S(f) = |P(f)|^2. \qquad (1.89)$$

The energy spectral density S(f) is the "energy" of the noise signal in a band width of 1 Hz. Note that $S(\omega) = 2\pi\, S(f)$ where $S(\omega)$ is the "energy" in a one radian/sec bandwidth. We use the term "energy" because if p(t) were converted into a voltage signal, $S(\omega)$ would have the units of energy if the voltage were applied across a 1 ohm resistor. In the case of the pure tone, if p(t) is assumed to be a voltage, then the mean square value in Equation 1.85 represents the power in watts.

In the case of random noise we define a *power spectral density* $G_p(f)$. This may be derived through the filtering–squaring–averaging approach or the finite Fourier transform approach. We will consider both approaches in turn.

Suppose we filter the time signal through a filter of bandwidth Δf, then the mean square value

$$M_p{}^2(f,\Delta f) = \lim_{T\to\infty}\frac{1}{T}\int_0^T p^2(t,f,\Delta f)\,dt, \qquad (1.90)$$

where $p(t,f,\Delta f)$ is the filtered frequency component of the signal after passing through a filter of bandwidth Δf centered on frequency f. In the practical case the filter bandwidth, Δf, could be, for example, a one third octave or smaller. The *power spectral density* is defined as:

$$G_p(f) = \lim_{\Delta f\to 0}\left[\frac{M_p{}^2(f,\Delta f)}{\Delta f}\right]. \qquad (1.91)$$

The power spectral density may also be defined via the finite Fourier transform:

$$G_p(f) = \lim_{T\to\infty}\frac{1}{T}|P(f)|^2. \qquad (1.92)$$

Note the difference between Equations 1.92 and 1.89. We notice that G_p must be a *power* spectral density because of the division by time. For duscussion in greater detail on analysis of random signals the reader is referred to References 16 and 17. Often in practice we define a new spectral density $\overline{G}_p(f) = 2\,G_p(f)$ which does not contain energy at negative frequencies but only exists in the region $0<f<\infty$. The power spectral density of the random noise in Figure 1.34 is plotted in Figure 1.35. Measurements of $M_p{}^2(f,\Delta f)$ and $\overline{G}_p(f)$ are discussed in Chapter 3.

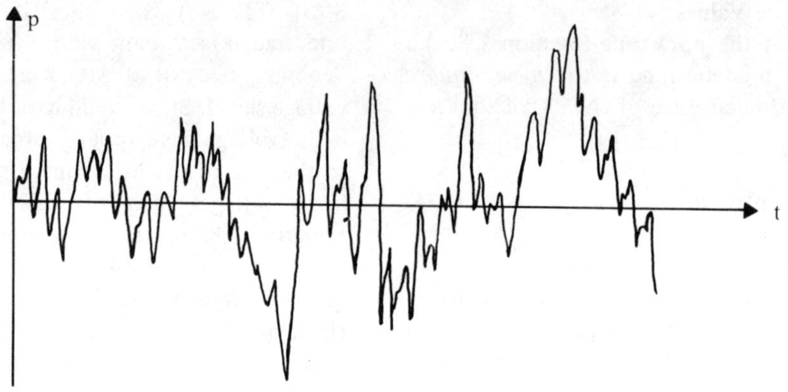

FIGURE 1.34. Random noise signal.

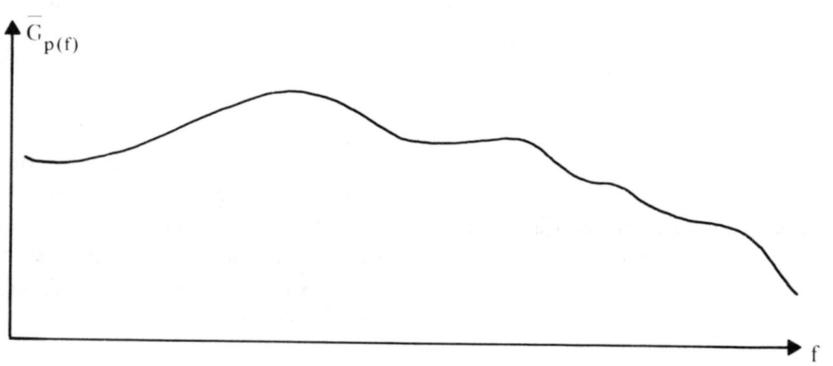

FIGURE 1.35. Power spectral density of random noise.

REFERENCES

1. **Wood, A.,** *Acoustics,* Blackie & Son, London, 1940, 56. Dover reprint, New York, 1960.
2. **Kinsler, L. E. and Frey, A. R.,** *Fundamentals of Acoustics,* 2nd ed., John Wiley & Sons, New York, 1962, 113.
3. **Kinsler, L. E. and Frey, A. R.,** *Fundamentals of Acoustics,* 2nd ed., John Wiley & Sons, New York, 1962, 117.
4. **Kurze, V. and Beranek L. L.,** Sound propagation outdoors, in *Noise and Vibration Control,* Beranek, L. L., Ed., McGraw-Hill, New York, 1971, 170.
5. **Wood, A.,** *Acoustics,* Blackie & Son , London, 1940, Dover Reprint, New York, 1960, 59.
6. **Kinsler, L. E. and Frey, A. R.,** *Fundamentals of Acoustics,* 2nd ed., John Wiley & Sons, New York, 1962, 112.
7. **Kinsler, L. E. and Frey, A. R.,** *Fundamentals of Acoustics,* 2nd ed., John Wiley & Sons, New York, 1962, 164.
8. **Sabine, W. C.,** *Collected Papers on Acoustics,* Harvard University Press, 1922. Dover reprint, 1964.
9. **Vaidya, P. G.,** Sound propagation outdoors, in *Reduction of Machinery Noise,* Crocker, M. J., Ed., W. Lafayette, Purdue University Press, 1974, 34.
10. **Rudnick, I.,** Propagation of sound in the open air, in *Handbook of Noise Control,* McGraw-Hill, New York, 1957, Chap. 3, p. 7.
11. **Rudnick, I.,** Propagation of sound in the open air, in *Handbook of Noise Control,* McGraw-Hill, New York, 1957, Chap. 3, p. 5.
12. **Rudnick, I.,** Propagation of sound in the open air, in *Handbook of Noise Control,* McGraw-Hill, New York, 1957, Chap. 3.
13. **Kurze, V. and Beranek, L. L.,** Sound propagation outdoors, in *Noise and Vibration Control,* Beranek, L. L., Ed., McGraw-Hill, New York, 1971, chap. 7.
14. **Cremer L.,** *Raumakustik,* Vol. 2, Springer-Verlag, Berlin, 1961, 27.
15. **Anon.,** Aerospace Recommended Practice ARP 866, Standard Values of Atmospheric Absorption as a Function of Temperature and Humidity for Use in Evaluating Aircraft Flyover Noise, Society of Automotive Engineers, New York, 1964.
16. **Magrab, E. B. and Blomquist, D. S.,** *The Measurement of Time-Varying Phenomena,* Interscience, New York, 1971.
17. **Bendat, J. S. and Piersol, A. G.,** *Random Data: Analysis and Measurement Procedures,* John Wiley & Sons, New York, 1971.

Chapter 2

HUMAN HEARING AND SUBJECTIVE RESPONSE TO SOUND

TABLE OF CONTENTS

2.1 INTRODUCTION

The human ear is a marvelous mechanism for detecting sound. It is very sensitive. If it were only slightly more so, we would be able to hear the Brownian (random) motion of the air molecules and we would have a perpetual buzz in our ears! The ear has a wide frequency response from about 15 or 20 Hz to about 16,000 Hz. Also the ear has a large dynamic range; the ratio of the loudest sound pressure we can tolerate to the faintest we can hear is about ten million (10^7).

There are three essential reasons to consider the ear in this volume. Sound levels are now so high in industrialized societies that many individuals are exposed to intense noise and *permanent damage* results. Large numbers of other individuals are exposed to noise from aircraft, surface traffic, construction equipment or machines and appliances, and disturbance and *annoyance* results. Lastly are *subjective reasons*. An understanding of people's subjective response to noise allows environmentalists and engineers to reduce noise in more effective ways. For example, noise should be reduced in the frequency range in which the ear is most sensitive. Noise reduction should be by a magnitude which is subjectively significant. There are several other subjective parameters which are important in hearing.

2.2 CONSTRUCTION OF EAR AND ITS WORKING

The ear can be divided into three main parts (Figure 2.1): the *outer, middle,* and *inner* ear. The outer ear consisting of the fleshy pinna and ear canal conducts the sound waves onto the eardrum. The middle ear converts the sound waves into mechanical motion of the auditory ossicles, and the inner ear converts the mechanical motion into neural impulses which travel along the auditory nerves to the brain. The anatomy and functioning of the ear are described more completely in various other references and textbooks[1-7] and will only be discussed briefly here.

2.2.1 Construction of the Ear

The fleshy appendage on the side of the head (the pinna) is not as well developed in man as in some animals. Its function is to focus sound into the ear canal. It helps us to localize the source of sound, particularly in the vertical direction, and is more effective at higher frequencies. The ear canal is about one in. long and ends at the tympanic membrane (eardrum) which is under tension and has the thickness of a sheet of paper.

The eardrum is connected to the malleus, the first of the three small bones known as the auditory ossicles (see Figure 2.2). The inner ear air cavity is connected to the back of the mouth by the Eustachian tube. The smallest of the ossicles, the stapes, which is about half the size of a grain of rice, is connected to a small oval window in the cochlea. The cochlea consists of spiral fluid-filled cavities inside the bone of the skull. The cochlea is comprised of a passageway which makes two and one half turns rather like a snail shell. Connected to the cochlea are the semicircular canals which are the balance mechanism and unrelated to hearing.

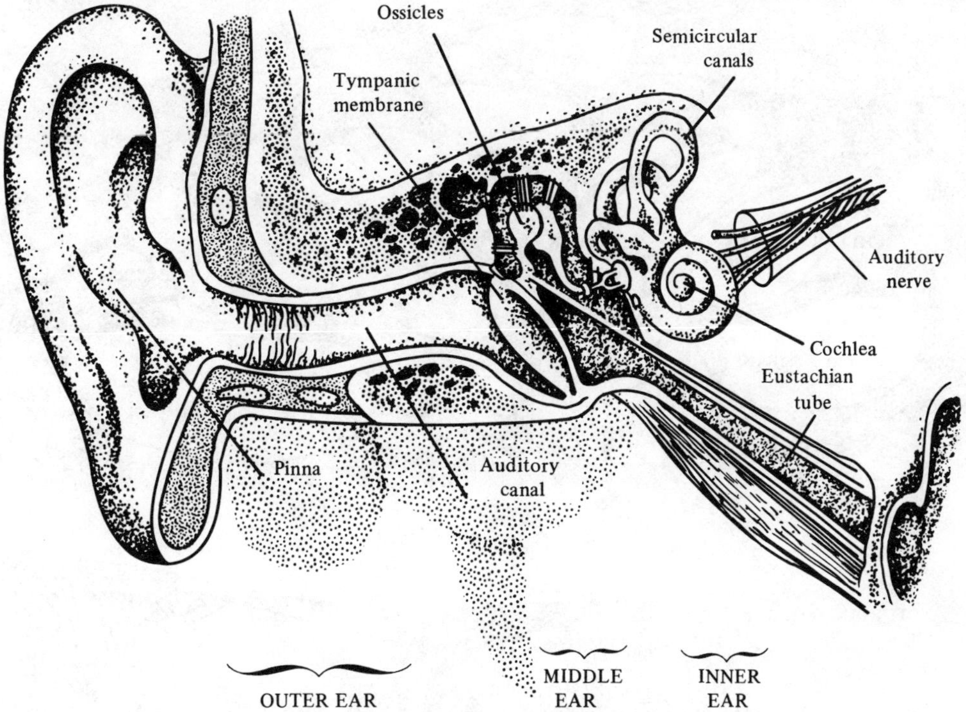

FIGURE 2.1 Simplified cross section through the human ear. (From Palmer, J. W., *Anatomy for Speech and Hearing,* Harper & Row, New York, 1972. With permission.)

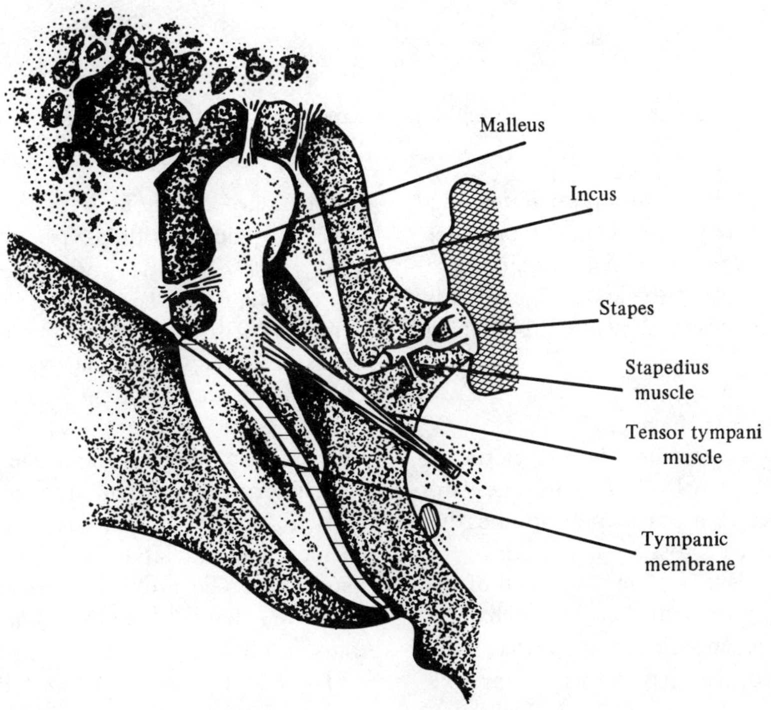

FIGURE 2.2 Eardrum and three auditory ossicles. (From Palmer, J. W., *Anatomy for Speech and Hearing,* Harper & Row, New York, 1972. With permission.)

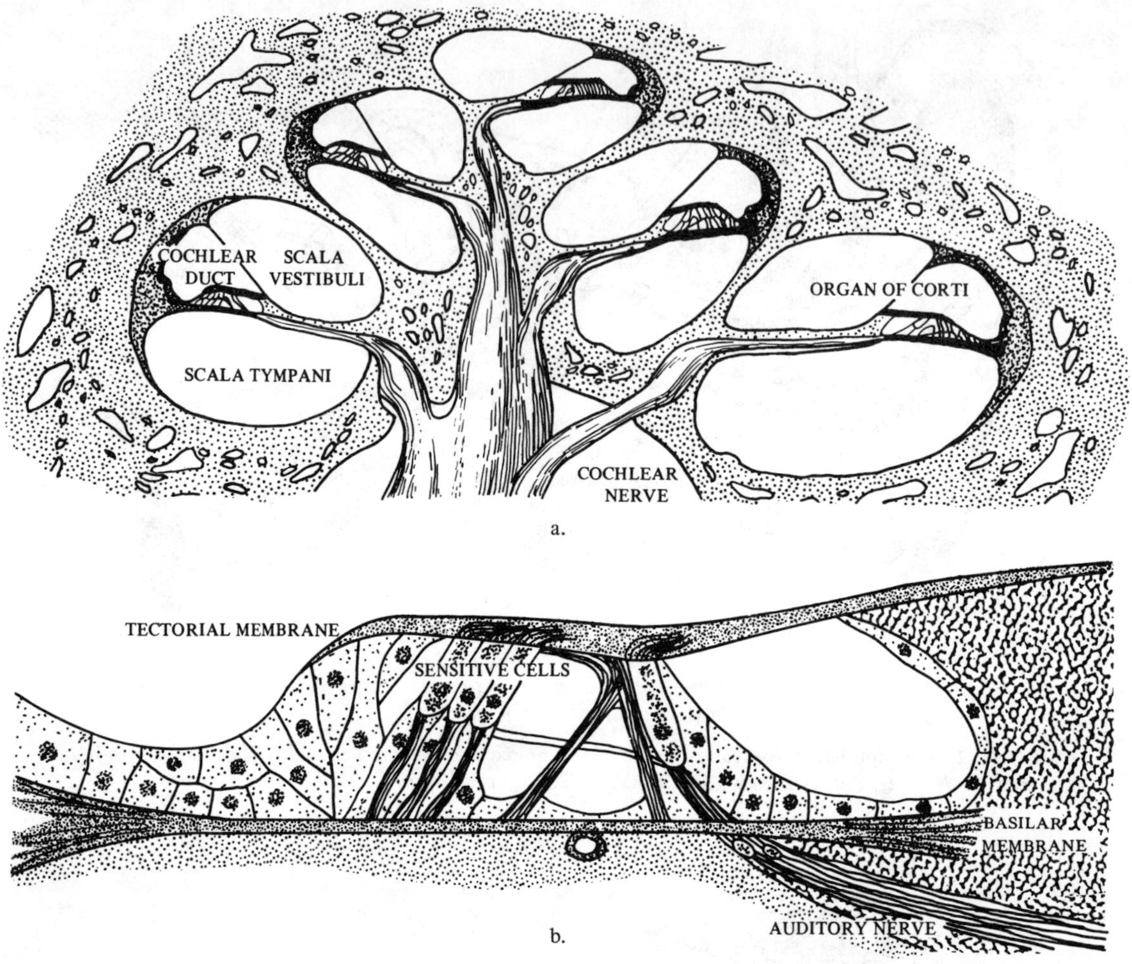

FIGURE 2.3 Cochlea. (a) Section through the cochlea. (b) Enlargement of Organ of Corti. (From von Békésy, G., The ear, *Sci. Am.*, 197, 69, 1957. With permission.)

The passageway of the cochlea is separated into a lower and upper gallery by a membraneous duct (Figure 2.3). The upper and lower galleries are connected together only at the apex. Figure 2.4 is a schematic representation showing the cochlea "unrolled."

2.2.2 Working of the Ear Mechanism

When a sound wave reaches the ear, it travels down the auditory canal until it reaches the eardrum. It sets the eardrum in motion and this vibration is transmitted across the 2 mm gap to the oval window by the lever system comprised of the auditory ossicles. It is thought that this mechanical system is an impedance-matching device. The characteristic impedance, ρc, of air is approximately one two-thousandth of the impedance of the cochlea fluid (see Equation 1.30). The area of the tympanic membrane is 20 or 30 times larger than that of the oval window.* Also the pivot of the ossicle system may be assumed closer to the oval window than the eardrum, hence providing a mechanical advantage of 2 or 3 times. The net result is that low pressure, high particle velocity amplitude air waves at the eardrum are converted into high pressure low particle velocity amplitude fluid waves in the cochlea, approximately matching the air to fluid impedances. We probably remember from electrical theory that in order to obtain maximum power transfer, impedances must be matched.

There is a "safety" device built into the inner

*Some believe that it is not the area of the oval window which is important, but rather the area of the footplate of the stapes. The tympanic membrane is about 20 times greater in area than the footplate area. However, not all of the tympanic membrane vibrates because it is firmly attached at its periphery. The ratio of the part of the tympanic membrane that moves to the footplate area is about 14 to 1.

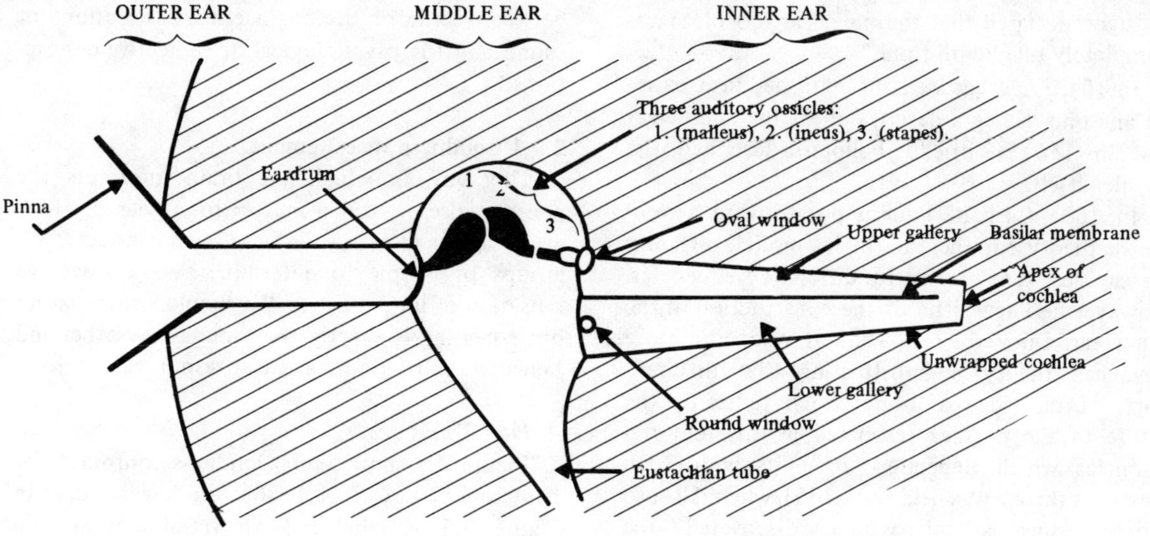

FIGURE 2.4 Cochlea "unwrapped" to show schematically working of the ear.

ear mechanism. Attached to the malleus and stapes are two muscles: the tensor tympani and the stapedius. If continuous intense sounds are experienced, the muscles contract and rotate the ossicles so that the force passed onto the oval window does not increase correspondingly to the sound pressure. It seems that this safety device may be somewhat fortuitous, however, since such man-made intense sounds are not experienced in nature. There seems to be some evidence that the muscles also contract when we speak, to prevent us hearing so much of our own speech.

The Eustachian tube is used to equalize the pressure across the eardrum by swallowing. This explains why our ears "pop" in airplanes when we ascend and the atmospheric pressure changes. We may experience some pain in an airplane when landing again if we have a cold; mucus, blocking the Eustachian tube, can prevent us from equalizing the pressure by swallowing.

Movement of the footplate of the stapes which is connected to the oval window causes pressure waves in the fluid of the upper gallery of the cochlea (Figures 2.3 and 2.4). The fluid in the lower gallery is separated from that in the upper gallery by the cochlear duct containing the organ of Corti. The organ has about 35,000 sensitive hair cells distributed along its length which are connected in a complicated way to about 18,000 nerve fibers which combine into the auditory nerve which runs into the brain. The pressure waves cause the basilar membrane to deflect and a shearing motion occurs between the basilar and tectorial membranes. The hair cells sense the shearing motion and if the stimulus is great enough the neuron to which each hair cell is attached sends an impulse along the nerve fiber to the brain cortex. Each neuron takes about 1/1,000th of a second to recharge and so individual neurons are limited to "firing" no more than 1,000 times/sec. With the neurons, a triggering level must be reached before they "fire" and so they have an all-or-nothing response. The brain must interpret the neural impulses to give us the sensation of hearing and, as we can imagine, the way in which this is done is not well understood.

2.2.3 Theories of Hearing

Phythagoras* in the 6th century BC was perhaps the first to recognize that sound was an airborne vibration.[8] Hippocrates in the 4th century BC recognized that the air vibrations were picked up by the eardrum but thought that the vibrations were transmitted directly to the brain by bones. In 175 AD, Galen of Pergamum, a Greek physician, realized that it was nerves that transmitted the sound sensations to the brain. Galen and most other early scientists and philosophers proposed, mistakenly, however, that somewhere deep in the head was a sealed pocket of implanted air which was the "seat" of hearing. This view was popularly held until 1760 when Domenico

*Professor S. S. Stevens has given a fascinating semi-technical account of the mechanism of hearing and the early theories of hearing in Reference 8.

Cotugno declared that the inner ear (cochlea) was completely filled with fluid.[8]

In 1543, Andreas Vesalius published his treatise on anatomy giving a description of the middle ear and in 1561 Gabriello Fallopio described the cochlea itself.

In 1605 Gaspard Bauhin put forward a resonance theory for the ear. In his model, different air cavities were excited by different frequencies. However he knew little of the construction of the inner ear. Du Verney, in 1683, developed a more advanced theory by postulating that different parts of the ridge of bone which twists up the inside of the cochlea resonated at different frequencies which depended upon its width. Du Verney's theory was held until 1851 when Alfonso Corti, using a microscope, discovered the thousands of hair cells on the basilar membrane attached to the ridge of bone in the cochlea.

A few years later, Hermann von Helmholtz used Corti's findings to suggest a new theory of hearing. In Helmholtz's theory, as it became refined, different parts of the basilar membrane resonated at different frequencies. Later workers showed that Helmholtz was not exactly right (the basilar membrane is not under tension). However, in 1928 Georg von Békésy did show that waves do travel along the basilar membrane and different sections of the basilar membrane do respond more than others to a certain sound. The region of maximum response is frequency-dependent and as Helmholtz had predicted, von Békésy found that the high frequency sound is detected nearer to the oval window and the low frequency sound, nearer to the apex (Figures 2.3 and 2.4).

2.3 SUBJECTIVE RESPONSE

So far we have traced the sound signal down the ear canal to the eardrum, through the auditory ossicles, through the oval window to the cochlear fluid to the basilar membrane and the hair cells, and finally to the neural impulses sent to the brain. How does the brain interpret these signals? Our study now enters the realm of psychology. While the physicist or engineer talks about sound pressure level (Equation 1.36) and frequency (Equation 1.1), the psychologist talks about loudness (Equation 2.1) and pitch, respectively. In Section 2.3 we shall discuss the relationships between some of the engineering descriptions of sound and the psychological or subjective descriptions.

2.3.1 Loudness Measurement

The way in which the brain interprets the neural pulses is still a matter for research. However, various experiments have been conducted on groups of people to determine people's average sensation of loudness, etc. We should stress that no one's hearing is exactly the same as any other and hence we must find statistical responses.

A. Pure Tones

Figure 2.5 shows equal loudness contours* for pure tone sounds.[9] Note that the lowest curve in Figure 2.5 is labeled MAF (minimum audible field). This is the hearing threshold, the quietest sounds, on average, at any frequency that average young people can hear.

We should note that there are two ways that we can measure the hearing threshold and the equal loudness contours. The first way is to present the listener with a free progressive wave field at a discrete frequency. This is the method which was used to obtain the results in Figure 2.5. Such measurements are normally made with the listener facing the source in an anechoic room (Section 4.6) where there are no reflections. The second way (which is used more frequently) is to present the listener with sounds played through earphones. There are some small differences in the results obtained by the two methods.

The equal loudness contours are determined as follows. For the 60-Phon curve, the listener is first presented with a pure tone at 1,000 Hz at a sound pressure level of 60 dB. Then he is presented with a pure tone at, for example, 500 Hz. The pure tones are presented alternately to the listener every few seconds. The level of the 500-Hz tone is adjustable and the listener is asked to adjust the level of the 500-Hz tone until the two tones sound equally as loud. The procedure is repeated with pure tones at other frequencies which are continually adjusted to sound equal in loudness to the 1,000-Hz pure tones at 60 dB. The curve drawn through these points (when the result is averaged for a large number of people) is called the 60-Phon equal loudness contour. The 70-Phon contour is obtained by finding the sound pressure level of pure

*These equal loudness contours measured by Robinson and Dadson[9] at the National Physical Laboratory at Teddington in England have now been adopted in ISO Recommendation ISO/R 226–1961.

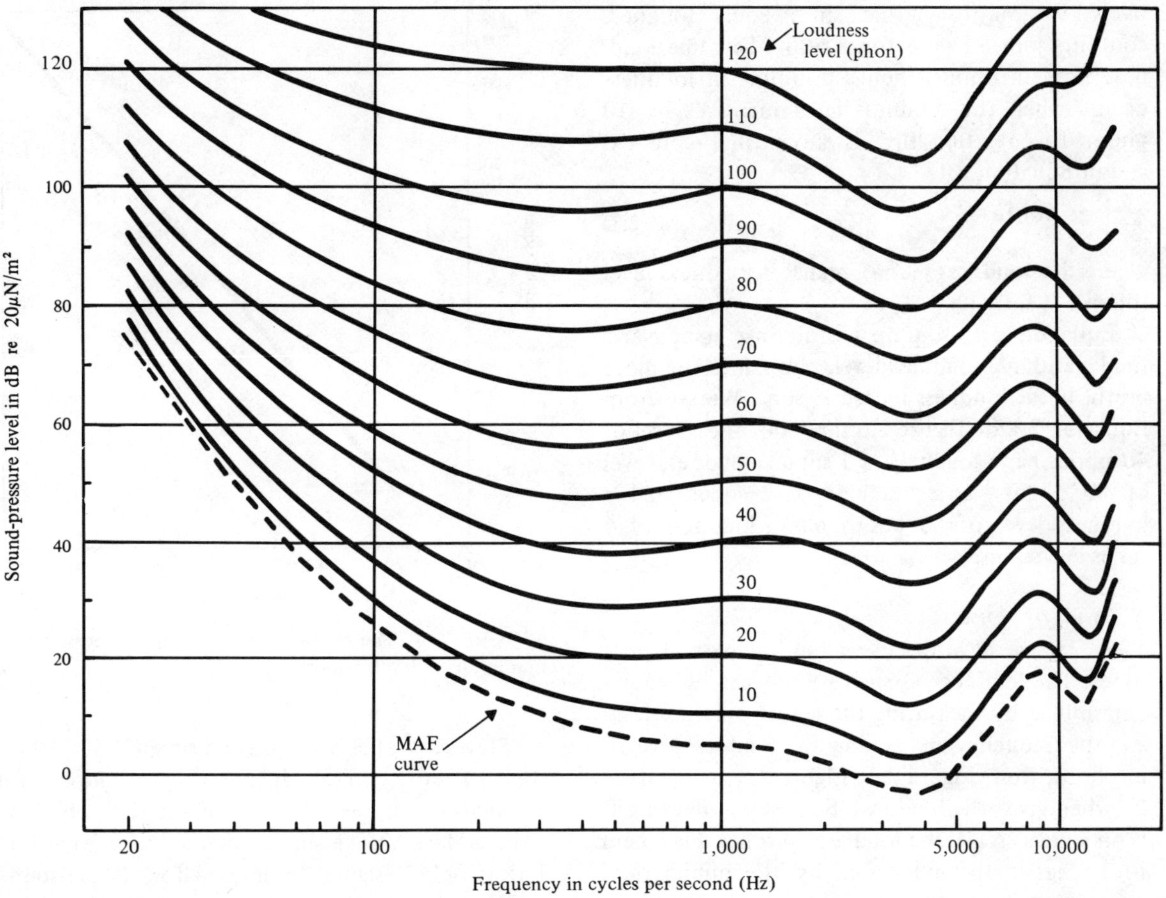

FIGURE 2.5 Equal loudness contours for pure tones. (From Robinson, D. W. and Dadson, R. S., A re-determination of the equal loudness relations for pure tones, *Br. J. Appl. Phys.*, 7, 166, 1956. With permission.)

tones which seem equally as loud as a 1,000-Hz pure tone at 70 dB and so on. Finally, the minimum audible field, MAF, is the contour joining all pure tones which are just audible. Figure 2.5 was determined for a group of young people with healthy ears in the age group 18 to 25. Each person listening with both ears, facing the source, in a free progressive wave acoustic field.

We see from the equal loudness contours that the ear is most sensitive to sound at about 4,000 Hz. This is mainly due to a quarter wavelength resonance in the ear canal at this frequency.* The increase in sensitivity at 12,000 Hz is mainly caused by diffraction effects around the head.

The equal loudness contours are not flat as frequency is changed. Notice at 1,000 Hz we can just detect sounds of 4 dB, while at 100 Hz they must be 25 dB. Thus the intensity must be 21 dB higher at 100 Hz than 1,000 Hz for us to detect a pure tone. This represents a 100-fold increase in intensity or a 10-fold increase in sound pressure amplitude (see Section 1.4). As the sound pressure level is increased, the equal loudness contours become flatter. A 40 dB tone at 1,000 Hz appears equally as loud as a 51 dB tone at 100 Hz, an 11 dB increase. A 70 dB tone at 1,000 Hz is as loud as a 75-dB tone at 100 Hz, only a 5 dB increase.

A further experiment has been performed on the loudness of pure tones. This experiment is to determine the increase in sound pressure level for a pure tone apparently to double in loudness. It has been found that for pure tones at 1,000 Hz, when the sound pressure level is increased by about 10 dB, the sound appears twice as loud.** Alterna-

*Assume the average length of the ear canal is 1 in. Then $\lambda/4 = 1/12$ ft, and $\lambda = 1/3$ ft. Thus $1{,}120/f = 1/3$, giving $f \approx 3{,}400$ Hz.

**Note that Fletcher and Munson first commented on this in 1933.[10]

tively we may say that since equal loudness contours join all pure tones with the same loudness level in phons, then a doubling in loudness occurs when the loudness level increases by 10 phon. This relationship* is shown in Figure 2.6 and in Equation 2.1.

$$S = 2^{(P-40)/10} \tag{2.1}$$

where S = loudness (sone), and P = loudness level (phon).

Equation 2.1 is quite useful since it converts the logarithmic loudness level, phon, into the more useful linear loudness measure, sone. We see from Equation 2.1 or Figure 2.6 that a loudness level of 40 phons has a loudness of 1 sone, a loudness level of 50 phons has a loudness of 2 sones, and a loudness level of 60 phons has a loudness of 4 sones and so on.

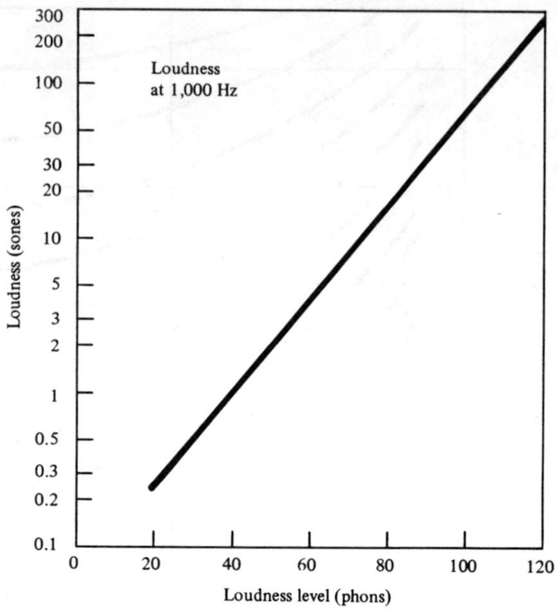

FIGURE 2.6 The relationship between loudness, sones, and loudness level, phons.

B. Bands of Noise

So far we have discussed the loudness of pure tones. The loudness of such sounds can be readily determined by measuring the sound pressure level and the frequency and finding the loudness level, P in phons, from Figure 2.5. Figure 2.6 or Equation 2.1 then gives the loudness, S, in sones. Because it is on a linear scale, the loudness in sones does seem to be easier to understand by the public than either the sound pressure level or the loudness level of a sound. We may compare an unknown sound in sones with some other well-known sound and say it is equally as loud, twice as loud, or half as loud, and so on. However, many of the noises we experience, although they may contain pure tones, are predominantly broad-band (Section 1.8). Similar loudness rating schemes have been worked out for broad-band noise. In 1958–1960, Zwicker,[11] taking into account masking effects (see later discussion in Section 2.3.2), devised a graphical method to compute the loudness of a broad-band noise. It should be noted that Zwicker's method can be used both for diffuse and free field conditions and for broad-band noise even when pronounced pure tones are present. However, the method is somewhat time-consuming and involves measuring the area under a curve. It has been well described elsewhere.[12]

Stevens[13] at about the same time (1957–1961) developed quite different procedures for calculating loudness. Stevens' method[13] which he named Mark VI** is simpler than Zwicker's method but is only designed to be used for diffuse sound fields and when the spectrum is relatively flat and does not contain pure tones. Both Zwicker's method and Stevens' Mark VI method have been standardized*** by the International Organization for Standardization (ISO). ISO has recommended Stevens' Mark VI method to be used when the noise is measured with octave band filters and Zwicker's method when the noise is measured with third octave band filters.

We should mention, in passing, that the late S. S. Stevens reviewed the several existing methods of rating loudness, noisiness, and annoyance of noise and produced a new scheme which he designated his Mark VII method.[14] This new scheme is somewhat different from his Mark VI and specifies a doubling of loudness (sones) for each 9-Phon increase in loudness level, instead of the 10-Phon increase. In view of the fact that the new Mark VII method appears a little more complicated to use,

*Both S. S. Stevens and D. W. Robinson independently suggested this relationship which has now been adopted in ISO Recommendation ISO/R 131-1959.

**Stevens' Mark VI procedure for calculating loudness has been standardized in the U.S. in ANSI S3.4–1968.

***"Method for Calculating Loudness" and "Method for Calculating Loudness Level" ASA S3–1965 ISO Recommendation R532–1967. American National Standards Institute, New York.

and that the previous Mark VI method has been standardized by ANSI and ISO, Stevens' more recent Mark VII method will not be discussed further. Readers interested in this Mark VII method should consult References 14 and 15.

Stevens' Mark VI method of loudness computation for broad-band noise[13] will now be discussed in more detail. Figure 2.7 shows contours of equal loudness index. Stevens conducted subjective experiments similar to those used to produce Figure 2.5 on a group of people, asking them to compare the loudness of octave bands of noise with a pure tone at 1,000 Hz. The experimental results, together with some curve-smoothing, masking information, and empiricism produced Figure 2.7.

The loudness of a broad-band noise source such as a machine may be computed as follows:

1. The machine noise is measured in octave bands normally between 31.5 and 8,000 Hz. (See Section 3.6.2)

2. The octave band levels are plotted on Figure 2.7 or else for more accuracy compared with tabulated values of Loudness Index.[15]

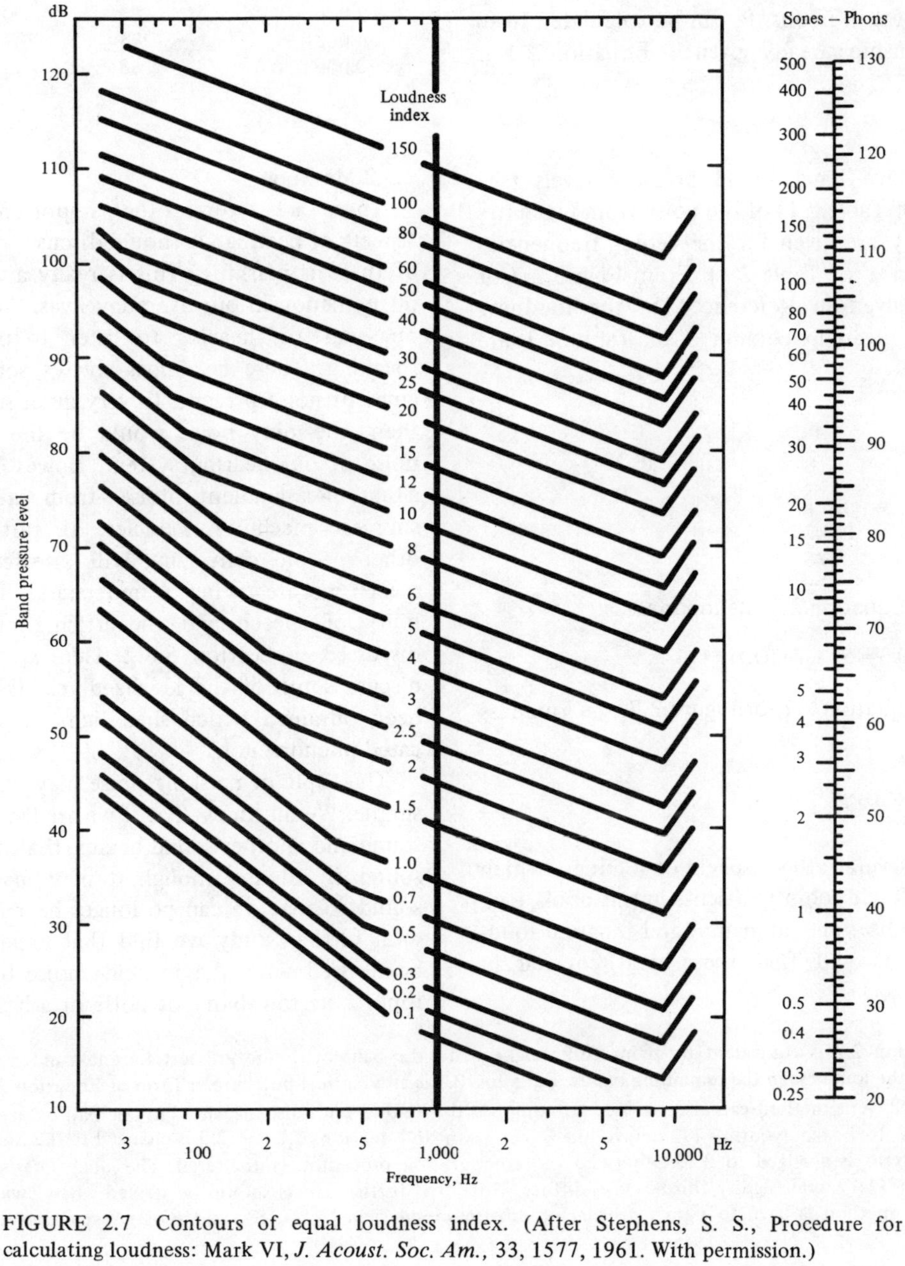

FIGURE 2.7 Contours of equal loudness index. (After Stephens, S. S., Procedure for calculating loudness: Mark VI, *J. Acoust. Soc. Am.*, 33, 1577, 1961. With permission.)

3. The Loudness Index for each octave band is obtained from Figure 2.7 or a table.

The loudness in sones, S, is then given by:*

$$S = S_{max} + 0.3\ (\Sigma S - S_{max}) \qquad (2.2)$$

or

$$S = 0.7\ S_{max} + 0.3\ \Sigma S, \qquad (2.3)$$

where S_{max} is the maximum loudness index and ΣS is the sum of all the loudness indices. The loudness level in phons, P, can be calculated from the relationship already given in Equation 2.1 if desired.

Example 2.1:

The octave band sound pressure levels (see column 2 of Table 2.1) of the noise from machines in a factory are given for the center frequencies (see column 1 of Table 2.1). From Figure 2.7 or the table given in Reference 15, the loudness indices, S, given in column 3 of Table 2.1 are found.

We obtain:

S_{max} = 26.5
$0.7\ S_{max}$ = 18.6
ΣS = 134.2
$0.3\ \Sigma S$ = 40.3

Thus from Equation 2.3 the loudness is:

S = 18.6 + 40.3 = 59 sone (OD),**

and from Equation 2.1, or Figure 2.7, the loudness level is:

P = 99 phon (OD)**.

There are several other aspects of loudness which we have not had room to discuss in this book, e.g., impulsive noise, and monaural and binaural loudness. Readers will find these well covered in Kryter's book.[16]

TABLE 2.1

Octave Band Levels of Factory Noise and Stephens' Loudness Indices

Octave band center frequency Hz	Octave band level dB	Band loudness index S
31.5	75	3.0
63	79	6.2
125	82	10.5
250	85	15.3
500	85	18.7
1,000	87	26.5
2,000	82	23.0
4,000	75	17.5
8,000	68	13.5

2.3.2 Masking

There are several other important subjective aspects of hearing we should discuss here. The first of these is masking. This is really a very familiar phenomenon in our everyday lives. We mainly use our sense of hearing to listen to other people speaking. There is almost always some form of ambient noise present. In very quiet surroundings, then, the only noise would be the background noise of our hearing system. However, in normal situations, ambient noise from a variety of sources: machines, vehicles, aircraft, and even other people conversing will interfere with the speech we are trying to understand. The intelligibility of speech and the articulation index are discussed in Section 5.8.2. Here we will briefly discuss some of the idealized results which have been obtained which shed light on this complicated phenomenon.

The topic of masking noise may at first appear simple. We all know that if we are listening to one sound and another sound begins, that if the second sound is intense enough it may mask the first sound so that we can no longer hear it. However, with further study we find that experiments can be devised with the masking noise being a pure tone, a narrow band of noise or a broad band of

*Note Equation 2.2 is equivalent to giving full weighting to the band with the greatest loudness index and only 30% weighting to the loudness in the remaining bands. Equation 2.3 is an identical but simpler form of Equation 2.2.

**Note that Stevens' method can also be used to compute the loudness of noise measured in one half octave and one third octave bands. If the bandwidth is 1/2 octave, the 0.3 constant in Equations 2.2 and 2.3 is changed to 0.2 and if 1/3 octave, the 0.3 constant is changed to 0.1. Otherwise the computation procedure is unaltered. The label (OD) signifies octave diffuse, while (TD) would signify third-octave diffuse. Note that further labels should be affixed when Zwicker's loudness computation method is used to signify frontal or diffuse sound fields[12], (G-F), and (G-D) respectively; (G stands for gruppen).

noise and being either steady or intermittent. Likewise the sound we are trying to hear can be a pure tone, a narrow or broad band of noise, and either steady or intermittent. This gives rise to a large number of variables in controlled experiments that we can devise. Results from different experiments, although showing some variation, do agree on the following:

1. The masking effect is different when the masking noise is a pure tone from when it is a band of noise (of the same level). The band of noise causes more masking around its center frequency than does the pure tone.

2. At low levels of masking noise, the masking effect is limited to a narrow frequency range around the center masking frequency. As the level of the masking noise is increased, the frequency range of the masking effect increases.

3. The masking effect is asymmetrical in frequency. Frequencies above the masking noise or tone are more easily masked than those below.

Result 1 above may be seen easily by studying the measurements of Ehmer[17,18] given in Figure 2.8. Results 2 and 3 above may be seen in Figure 2.8 and even more clearly in Figure 2.9. In Figure 2.9, after Spieth,[19] we clearly see the asymmetry of the masking effect with frequency and the way the frequency range of the masking effect rapidly increases as the level of the masking noise is increased.

Fletcher and Munson[20] were perhaps the first to postulate an explanation of the masking effect. They suggested that the sensation of loudness is proportional to the number of nerve impulses per second reaching the brain. Two tones or noises only separated by a small frequency interval would tend to "compete" with each other in exciting the same neuron or nerve fiber and the loudness sensation would not necessarily increase compared with only one tone or noise. Fletcher and Munson suggested that it is necessary to group all noise components into small frequency bands as far as loudness is concerned. The width of these grouping bands was estimated by Fletcher and Munson to be 100 Hz for frequencies below 2,000 Hz, 200 Hz for frequencies between 2,000 and 4,000 Hz, and 400 Hz for frequencies between 4,000 and 8,000 Hz.

Zwicker, Flottorp, and Stevens[21] also studied these frequency groupings (or "frequenzgruppen") and decided these bands were somewhat wider.[22] These frequency groupings are also sometimes known as critical bands.[16,22] Zwicker and his colleagues[21] suggested that one critical band corresponds to a distance of 1.3 mm along the basilar membrane (Section 2.2) and defined it as one Bark. Table 2.2 gives the critical bands. It is interesting to note that Zwicker, Flottorp, and Stevens' results showed that in much of the frequency range (particularly above several hundred Hz), the critical bandwidth was not much different from the one third octave bandwidth of 23% (see Section 3.1). This is sometimes given as a justification for making so many noise measurements in one third octaves.

Kryter[23] has given a good recent review of knowledge on masking. So far we have mainly discussed what is known as direct masking. This may be defined as masking that occurs when the neurons which are normally most excited at a certain frequency in the cochlea become unavailable at that frequency. They may already be excited by a masking noise at the same frequency or by a masking noise at a frequency sufficiently close so that the band of neurons on the basilar membrane excited by the masking noise encompasses those neurons most excited by the sound frequency to be received.

Some interesting but slightly less important aspects of masking have been discussed by Kryter.[23] Remote masking is the phenomenon discovered by Bilger and Hirsh[24] in which high frequency bands of noise mask pure tones of low frequency. Figure 2.9 clearly shows this effect.

Central masking is masking which occurs in one ear when noise is presented to the other.[25] Such masking cannot be attributed to the noise being transmitted through the cranium from one ear to the other by vibration, nor to the action of aural reflex or binaural phase interactions;[23] it is a minor phenomenon.

Temporal masking is the phenomenon whereby a sound may be masked by a short burst of masking noise either immediately proceeding or following the sound (normally known as forward and backward masking, respectively). Such masking has been investigated since about 1958 by several workers.[26-28]

Forward masking is easy to explain, while backward masking is more difficult.[23] Forward masking has some potential application in noise control problems. It is possible that forward

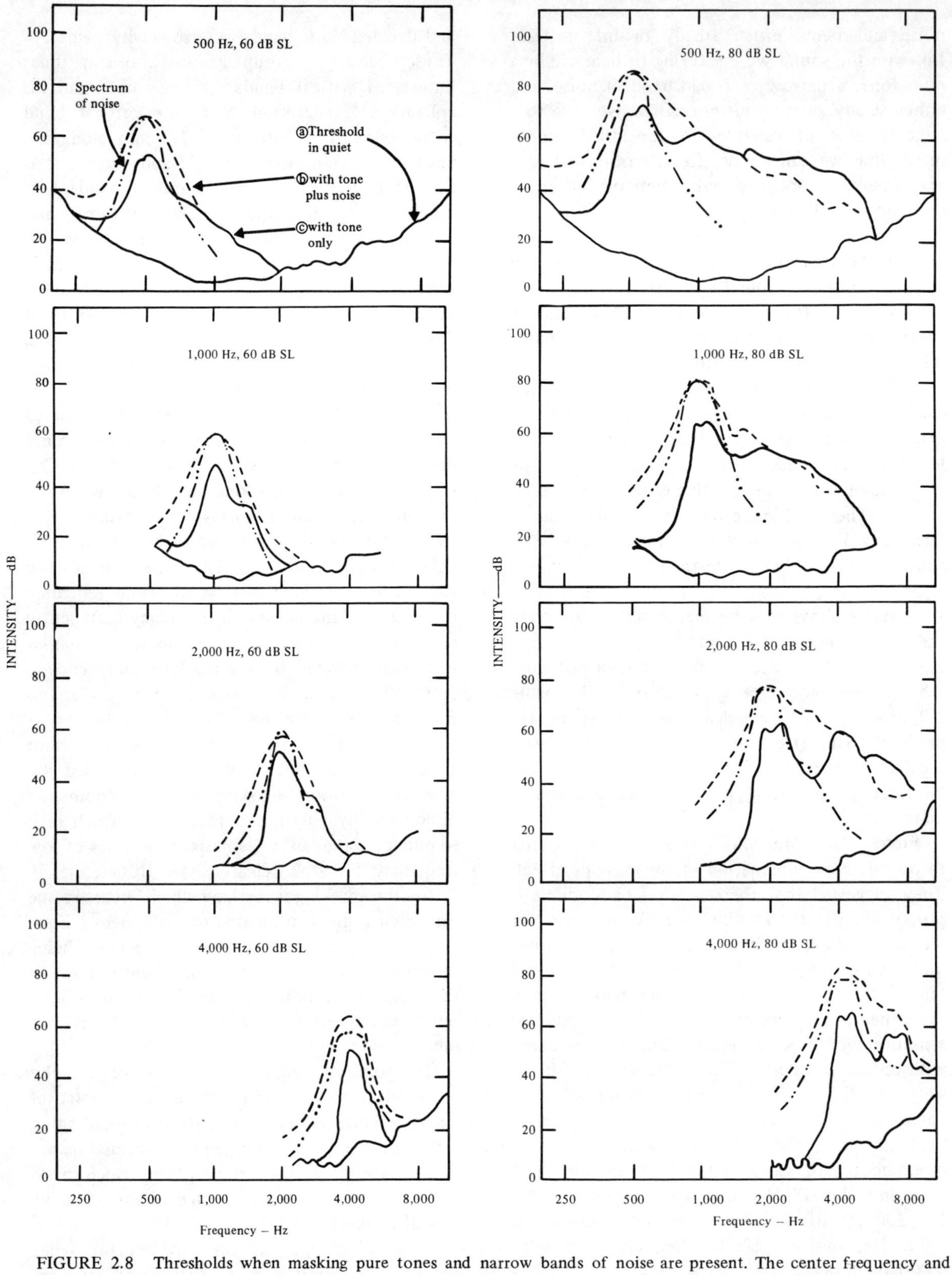

FIGURE 2.8 Thresholds when masking pure tones and narrow bands of noise are present. The center frequency and sensation level (SL) of the masking pure tone and noise band are the variables. (After Ehmer, R. H., Masking by tones vs. noise bands, *J. Acoust. Soc. Am.*, 31, 1253, 1959. With permission.)

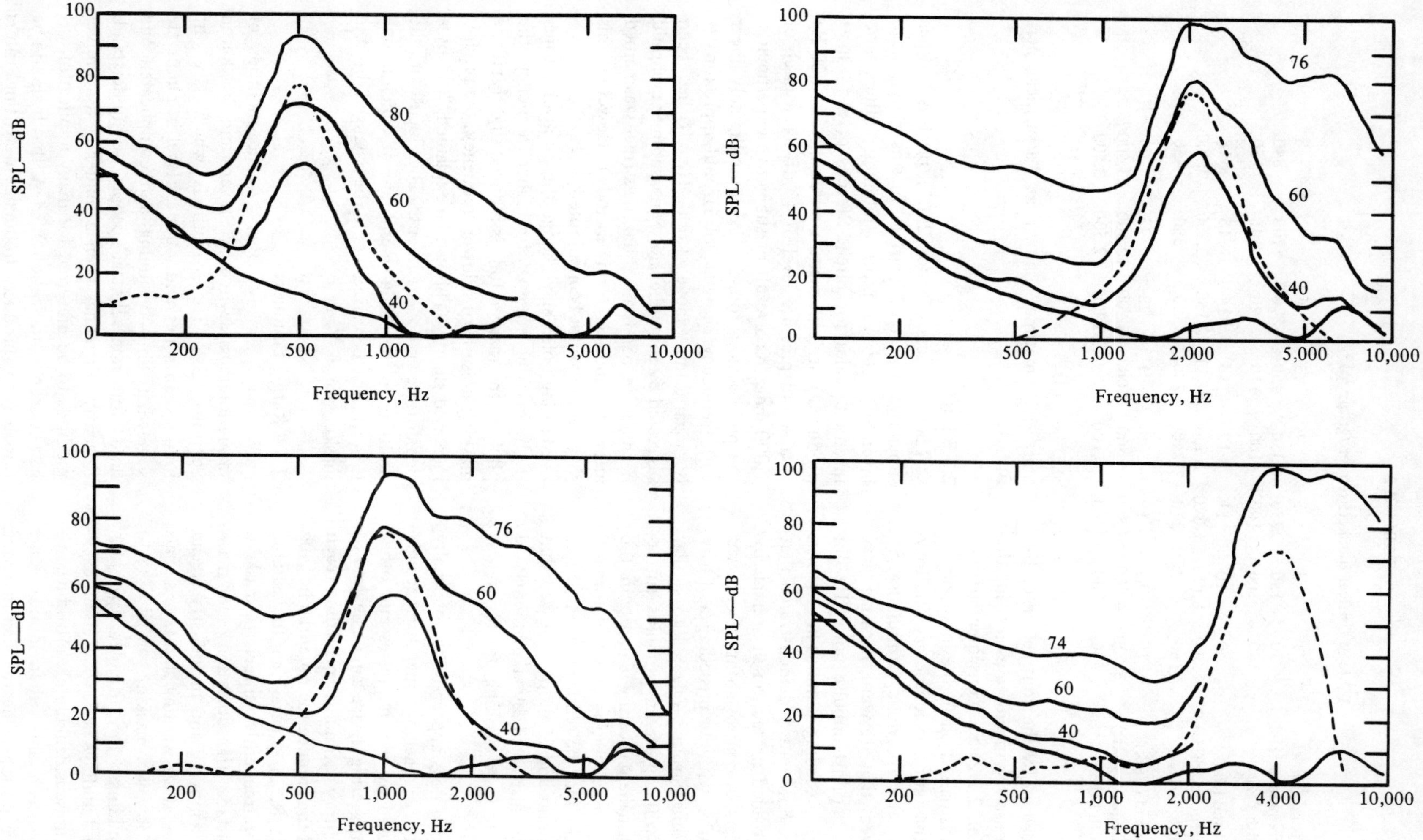

FIGURE 2.9 Remote masking. The solid curves represent the pure tone thresholds in the presence of bands of noise centered at 500, 1,000, 2,000, and 4,000 Hz. The spectrum of each noise band (at its maximum level used) is shown in the dotted lines. The levels on each solid curve represent the masking noise levels. (After Spieth, W., Downward spread of masking, *J. Acoust. Soc. Am.*, 29, 502, 1957. With permission.)

TABLE 2.2

Table of Critical Bands (Frequenzgruppen).[21]

Critical band (Bark)	1	2	3	4	5	6	7	8
Center frequency (Hz)	50	150	250	350	450	570	700	840
Bandwidth f (Hz)	100	100	100	100	110	120	140	150
Critical band (Bark)	9	10	11	12	13	14	15	16
Center frequency (Hz)	1,000	1,170	1,370	1,600	1,850	2,150	2,500	2,900
Bandwidth f (Hz)	160	190	210	240	280	320	380	450
Critical band (Bark)	17	18	19	20	21	22	23	24
Center frequency (Hz)	3,400	4,000	4,800	5,800	7,000	8,500	10,500	13,500
Bandwidth f (Hz)	550	700	900	1,100	1,300	1,800	2,500	3,500

masking may be caused by the contraction of the stapedius and tensor tympani muscles in the middle ear (Section 2.2.2) and consequent reduction in the stimulus passed on to the inner ear.[29] An alternative explanation is temporary auditory fatigue due to the previous stimulation.[23] Fletcher[30] showed that forward masking gave some "protection" against impulse .noise. He did this by providing a preceding stimulus in the form of noise or clicks. Simmons[31] demonstrated that this technique provided protection in animals by experiments on cats. A control group of cats was subjected to gunfire without preceding auditory stimuli, while another group subjected to similar gun reports each with a preceding stimulus, suffered much less permanent hearing damage in the cochlea. The noise ameliorating effect of forward masking in humans may be reduced when people are distracted. Durant and Shallop[32] showed that subjects could be distracted by a mathematical task, so that this forward masking protection was reduced.

Forward masking does appear to be potentially useful in protecting machine operators against impact noise caused by punch presses and other impact tools. A triggering device can be devised to produce an auditory stimulus preceding each impact noise and thus reduce the potential hearing damage. Masking is, in general, an undesirable effect, since very frequently, speech is masked. Aircraft and surface transportation can mask speech which is annoying and in industry it can even be dangerous, since warning shouts or signals can be masked. However, masking can be used to advantage in buildings: in particular open-plan offices. In such buildings a low level of background masking noise is deliberately introduced so that unwanted speech and other sounds some distance away are no longer heard. Such masking noise is sometimes known as "acoustical perfume".

2.3.3 Pitch

Like loudness (Section 2.3.1), pitch is another subjective aspect of hearing. Just as people have invented scales to express loudness, others have invented scales for pitch. Stevens et al.[33] have produced a scale in mels. A pure tone of 1,000 Hz at a sound pressure level of 40 dB has a pitch of 1,000 mels. As a result of subjective experiments, the pitch scale is found to be approximately linear with frequency below 1,000 Hz, but approximately logarithmic above 1,000 Hz. It has been suggested by some that noise measurements should be made in bands of equal mels (mel was named after the musical term, melody). However, this suggestion has not been adopted.

Masking noise may change the pitch of a tone. If the masking noise is of a higher frequency, the pitch of the masked tone is reduced slightly; if the masking noise is of a lower frequency, the pitch is increased slightly. This can be explained[34,35] by a signal/noise ratio argument. The locus of the position on the basilar membrane at which a tone is normally perceived could be changed by the masking noise.[34]

2.3.4 Rating Loudness

We have already discussed the methods of rating loudness devised by Stevens and Zwicker in Section 2.3.1. These methods which give the loudness in sones and loudness level in phons are somewhat time-consuming and complicated and hard for nontechnical people to understand. Recently, there has been a tendency for simpler methods to be preferred for many applications. It can be seen from Section 2.3.1.A. that it should be possible to produce frequency-weighting filters

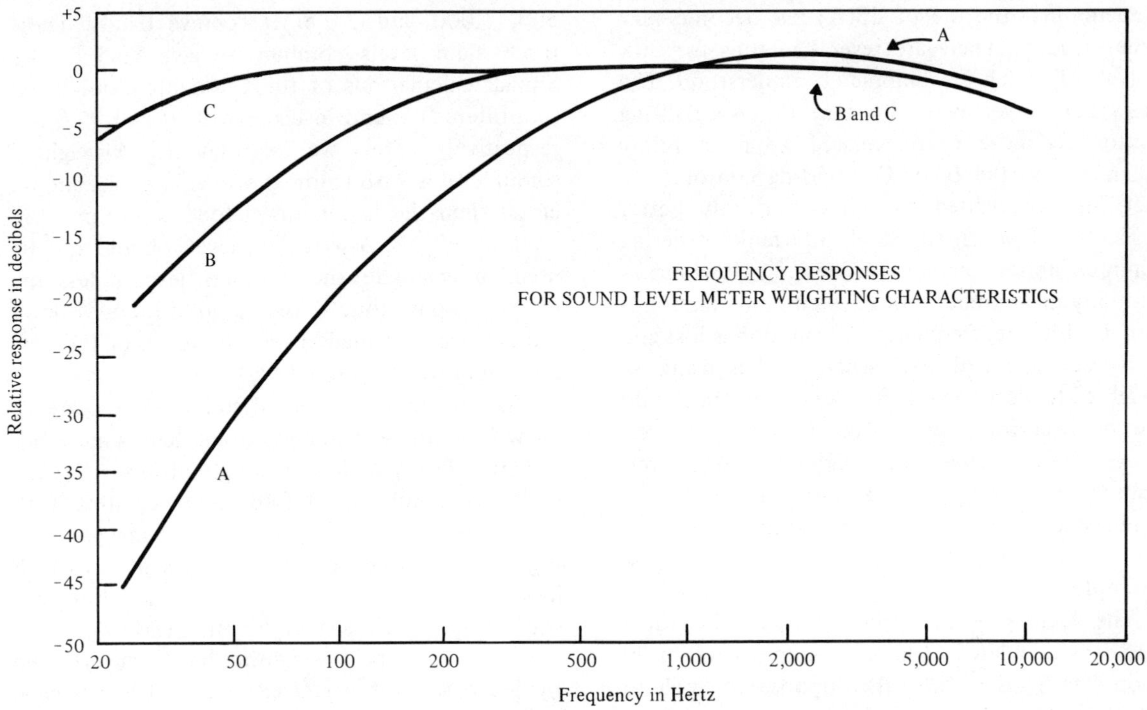

FIGURE 2.10 A-, B-, and C-frequency weightings used with sound level meters.

which would, if inserted after a microphone and amplifier, give a direct reading of loudness. However, it should now be obvious from the discussion in Sections 2.3.1.A. and B. that such simple frequency-weighting filters will only successfully give the loudness of pure tones, and that they cannot possibly be used to give the loudness of broad-band noise. It has, in fact, been internationally agreed[36] that sound pressure level measurements made with simple frequency-weighting filters should not be accepted as measurements of loudness level. Such measurements are extremely useful, however, when used for comparison purposes for rating like noises in rank order. For instance if we wish to rank several truck noises (which all have similar frequency spectra), a weighting filter is very useful. It is normally found in such cases that a weighting filter will rank order the truck noises in the same order as the more complicated Stevens or Zwicker loudness methods, although the loudness levels will not be accurately predicted. See, for example, experiments done on ranking the noise from electric clocks by Viebrock, Crocker, and Cooper.[37]

The weighting filters which are commonly used date back to the 1930s and to the early Fletcher-Munson equal loudness contours.[10] The A, B, and C frequency-weightings (shown in Figure 2.10)* approximately correspond to the inverse of the Fletcher-Munson, 40-, 70-, and 100- Phon contours after some frequency smoothing is applied (see, e.g., Figure 2.5).

The sound pressure level measurements made with the frequency-weighting filters are described as sound levels. The readings then are known as A-weighted sound levels, B-weighted sound levels, and C-weighted sound levels, or L_A, L_B, and L_C. Since the A-weighting filter is the one most commonly used, if the description sound level is given without qualification, it is assumed to be A-weighted. The most complete description would be to say that the A–weighted sound level was 50 dB (for example). A common abbreviation is to say the sound level is 50 dB(A). This is preferable to the use of dBA.

Although it was originally intended to use dB(C) for high level noise measurement, dB(B) for medium level noise, and dB(C) for low level noise,

*The frequency response characteristics shown in Figure 2.10 have been standardized in The American National Standard Specification for Sound Level Meters (ANSI S1.4–1971), in the International Standard (IEC-123) and in most other national standards.

it seems that the use of dB(A) has become very widespread.[38] There are several reasons for this, not all of which are completely understood. The main reason seems to be that the A-weighting contour is more nearly related to most rating schemes than the B- or C-weighting contours. In fact, the A-weighted level is consistently better than the C-weighted level in rank ordering common noises for their loudness.[38] However, as with any one number rating system, information is lost. In this case, frequency information is lost and it is recommended that when studies using A-weighted levels are made, frequency spectra should also be taken for noise control and other purposes.

As already remarked, weighted sound levels cannot be used directly as an estimate of the loudness level, as the following example shows:

Example 2.2:

The factory noise spectrum (given in Table 2.1), was calculated to have a loudness level of 99 Phon (see Section 2.3.1 B). Suppose we wish to calculate the approximate A-weighted sound level from the octave band levels given in Table 2.1. Since we do not have the directly measured A-weighted sound level, we can calculate this approximately using Figure 2.10. The A-weighting corrections at the octave band center frequencies have been read off Figure 2.10 and entered in Column 3 of Table 2.3. The so-called A-weighted octave band sound pressure levels have been calculated in Column 4 and these have been combined two–at–a–time (using Figure 1.16) to give the A-weighted sound level of 89.8, i.e., 90 dB(A). Note this A-weighted sound level is dominated by the A-weighted band levels in the 500, 1,000 and 2,000 Hz octave bands. These three band levels combine to give 89.5 dB(A). Similar calculations of the C-weighted and linear (nonfiltered) sound levels give 91.9 and 92.0 dB, respectively. Thus we see that the A-weighted sound level is 9 dB below the level in phons and no closer than the linear unweighted sound pressure level of 92 dB. A-weighted levels should not be used to calculate the loudness level unless the noise is a pure tone – then a good loudness level estimate can be made using the A, B, or C filters (depending on the noise level).

The fact that A-weighted sound levels correlated with loudness rankings better than some other weighted levels such as C-weighted levels has led some to search for a frequency-weighting with better correlation than the A-weighting contour. Figure 2.11 shows several other weightings which have been recently proposed. The D-weighting has been proposed by Kryter for the measurement of aircraft noise. The E-weighting has been suggested by Stevens[14] for perceived level and SI has been proposed by Webster[39] for speech interference studies.

Further methods of rating loudness, noisiness, and annoyance of noise are found in other relevant parts of Volumes 1 and 2 of this book where noys, noise pollution level, traffic noise index, and other units and rating schemes are discussed.

2.4 HEARING LOSS AND DISEASES (DISORDERS)

2.4.1 Introduction

Some are born with the severe handicap of deafness. Others suffer sudden hearing loss later in

TABLE 2.3

Combination of Octave Band Sound Pressure Levels of Factory Noise to Give the A-weighted Sound Level

Octave band center frequency Hz	Octave band level dB	A-weighting correction dB	A-weighted octave band levels dB
31.5	75	−42	33
63	79	−28	51
125	82	−18	64
250	85	−9	76
500	85	−3	82
1,000	87	0	87
2,000	82	+1.5	83.5
4,000	75	+0.5	75.5
8,000	68	−2.0	66.0

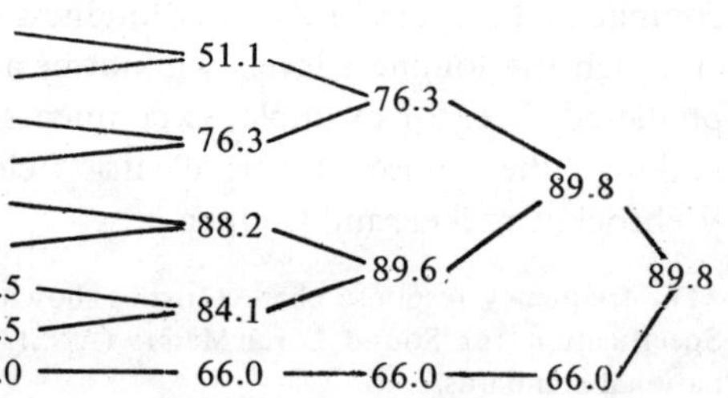

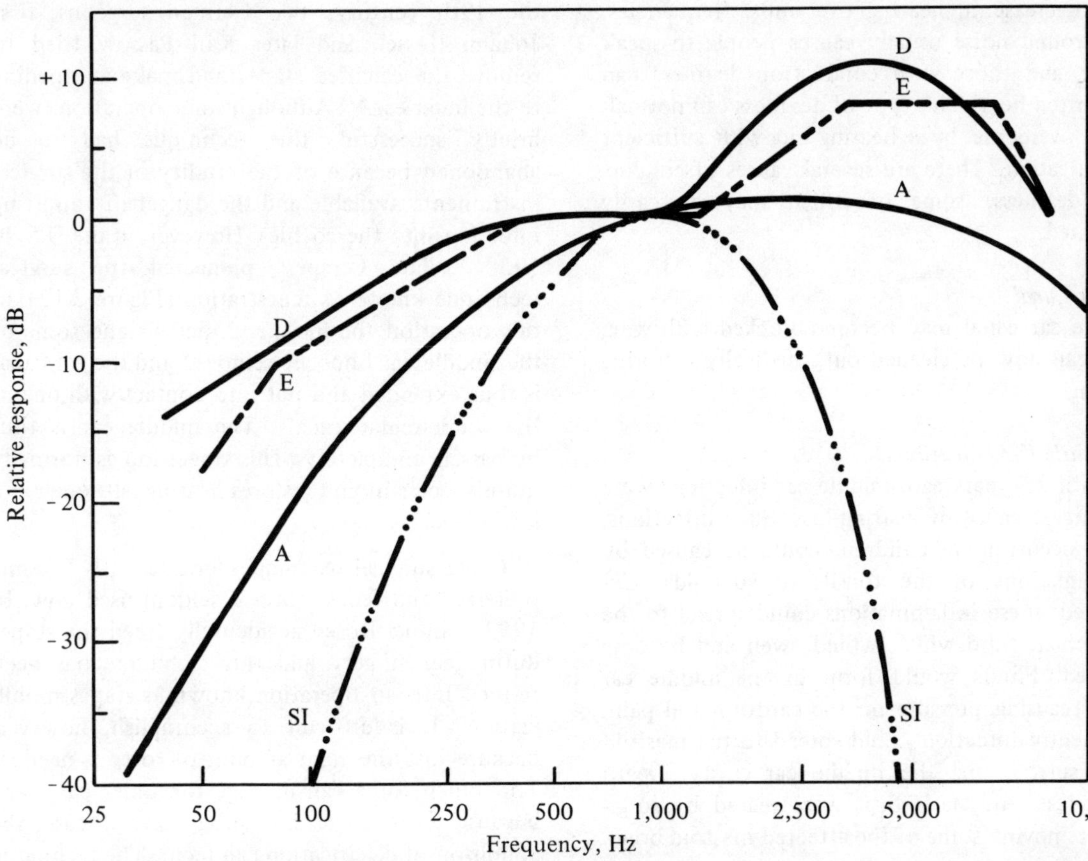

FIGURE 2.11 The A-weighting contour compared to several other recently proposed frequency weightings.

life or (much more commonly) gradually lose their hearing over a period of time. For all these persons, deafness is a crippling loss of one of the most important senses. Deafness can be very slight or complete. There are different legal and medical definitions of deafness. For our purpose we will define the onset of impaired hearing when a person's hearing level is 25 dB, averaged at 500, 1,000 and 2,000 Hz. We will define total deafness when a person's hearing level is 92 dB averaged at the same important speech frequencies[40] (see Section 2.4.5). Someone who is hard-of-hearing, then, may have a hearing level anywhere between these two extremes of 25 and 92 dB. The average hearing level of a nonimpaired person is 0 dB.

Deafness not only causes a lowering in the level at which sound is heard, but in many cases also causes a loss in hearing quality so that consonants cannot be distinguished and words are confused. In addition, some deaf people also suffer from tinnitus or a ringing in the ears which is disturbing and competes with other sounds.

Because of their affliction, deaf people tend to withdraw from society. This is sometimes reinforced by society itself. In ancient Rome, the deaf were classed with the mentally incompetent. In modern societies the deaf are treated with more compassion, although some people still tend to equate deafness with stupidity. Education, advances in medicine, and use of hearing aids are now improving the plight of the deaf. Strict control of noise levels in industry will also reduce the number of people receiving needless hearing damage and consequent deafness.

Hearing disorders (deafness) can be divided into two main types: conduction deafness and sensory-neural deafness. Conduction deafness is related to disorders in the outer and middle ear, while sensory-neural deafness occurs in the inner ear in the organ of corti, the auditory nerve, or the auditory cortex in the brain.

2.4.2 Conduction Hearing Loss

This normally manifests itself as a fairly uniform decrease in hearing over most frequencies. Background noise usually causes people to speak louder and those with conduction deafness can then often hear. This type of deafness can normally be overcome by a hearing aid with sufficient amplification. There are several causes of conduction deafness, some of which may be easily corrected.

A. Ear Canal

The ear canal may become blocked with wax. This can now be cleaned out, medically restoring hearing.

B. Middle Ear Infections

Until 25 years ago, middle ear infections were the largest cause of hearing loss. Such infections, often occurring in children, could be caused by inflammations of the tonsils or adenoids. Unchecked, these inflammations could spread to the Eustachian tube which would swell and become blocked. Fluids would form in the middle ear cavity causing pressure on the eardrum and pain. Frequently infection could spread to the mastoid bone surrounding the middle ear cavity. Death often resulted. Mastoiditis was treated by surgically removing some of the affected mastoid bone. With the advent of penicillin and other antibiotics, there is no need for middle ear infections to progress so far and now they rapidly can be cleared up with drugs in almost every case.

C. Calcification of Stapes

Another common cause of hearing loss occurs when bony material forms around the footplate of the stapes preventing it moving and transmitting vibrations through the oval window into the cochlea. This condition known as Otosclerosis normally causes a severe loss in hearing, at most frequencies of 50 dB or more. Otosclerosis can often be helped quite successfully with a hearing aid. By providing appreciable amplification of the sound pressure at the eardrum, sufficient vibration amplitude can be transmitted to the oval window despite the calification.

However, calcification can now be successfully treated by surgical methods. Towards the end of the 19th century, two German surgeons, first Johann Kessel, and later Karl Passow tried to remove the calcified stapes and make an opening to the inner ear.[40] Although some operations were briefly successful, this technique had to be abandoned because of the crudity of the surgical instruments available and the danger of admitting infection into the cochlea. However, in the U.S. in 1938, Julius Lempert pioneered the surgical technique known as fenestration (Figure 2.12). In this operation the incus and malleus and some of the middle ear bone are removed and the ear canal is thus extended and put into contact with one of the semicircular canals. The middle ear is thus by-passed completely. This operation is normally quite successful and restores hearing, although at a lower level.*

Other surgical techniques have recently become preferred and fenestration is seldom used now. In 1952, Samuel Rosen accidentally freed the stapes during ear surgery and this technique has been refined into an operation known as stapes mobilization. It is difficult to accomplish, however, because just the right amount of force is needed. Too much force can damage the inner ear. Also because Otosclerosis is a progressive disease, the condition of calcification can recur. The technique now most preferred is stapedectomy: the complete removal of the stapes and its replacement with a plastic part. Stevens[8] gives a detailed account of one such operation. The operation must be conducted using a microscope and can be accomplished in less than 30 min. After the stapes is chipped away it is replaced with an artificial one which is attached both to the incus and the oval window and almost perfect hearing is restored.

2.4.3 Sensory-Neural Hearing Loss

There are several causes of this and all are associated with disorders of the inner ear, the auditory nerve fibers, the auditory cortex in the brain, or combinations of all three. Unlike conductive deafness, sensory-neural deafness is often most severe at higher frequencies. Background noise can thus mask the consonants in speech and make it unintelligible. Although a hearing aid

*The *impedance-matching* of the middle ear discussed in Section 2.2.2 is confirmed by the results of fenestration surgery. The result of fenestration is absence of the intervening middle ear structure. Thus, the sound waves meet the fluid of the inner ear at the new fenestra. If a tympanic membrane to flootplate area ratio of 14:1, and a mechanical advantage of 2 for the ossicular chain are assumed, then the hearing loss remaining after fenestration might be expected to be about 10 log $(28)^2$ or about 30 dB. This agrees quite well with the surgical results where it is found that a hearing level of about 25 or 30 dB is the best that can be obtained from fenestration.

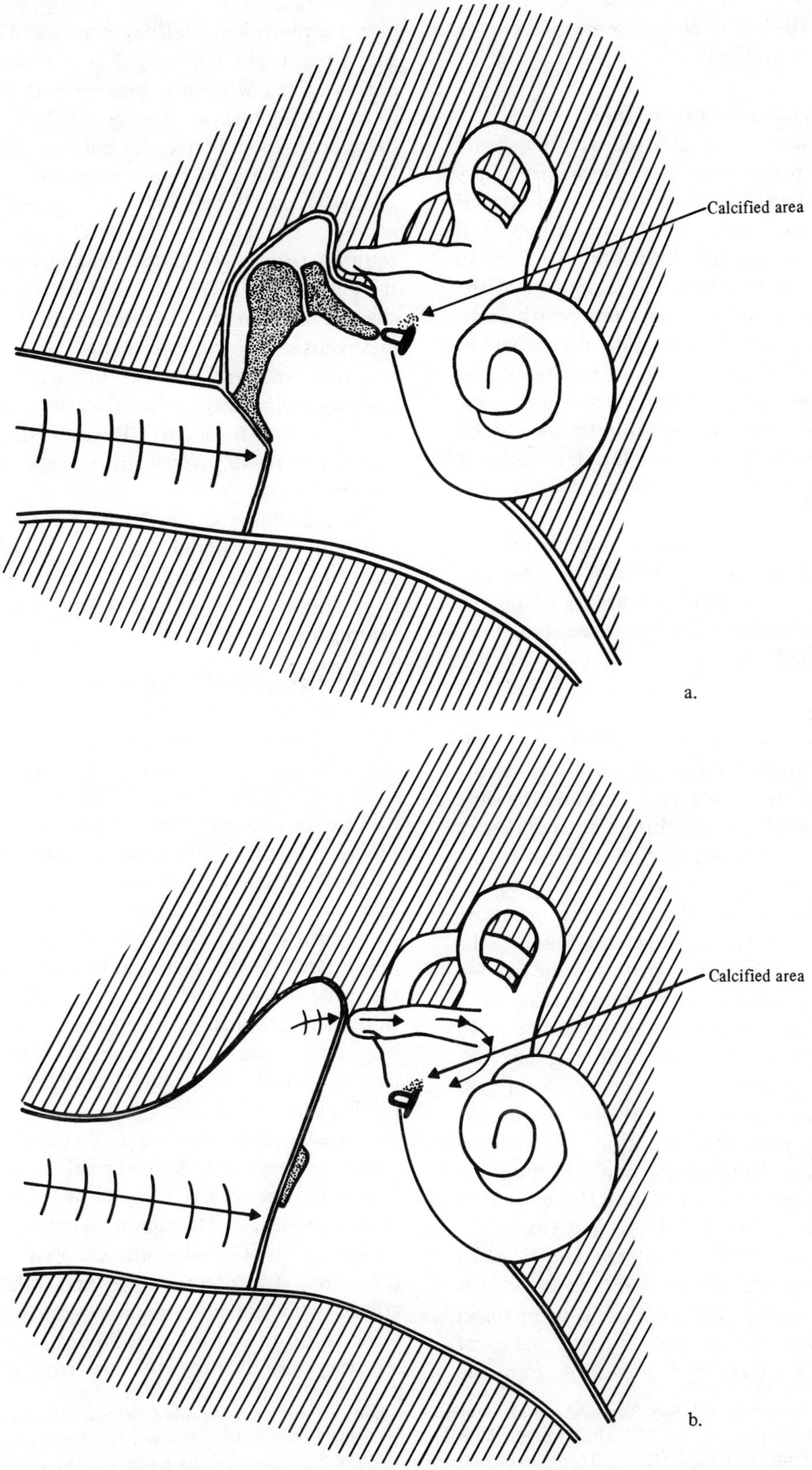

FIGURE 2.12 Fenestration: An operation designed to relieve Otosclerosis. (a) Diseased ear before surgery. (b) Ear after surgery.

brings some relief, it is perhaps not so useful as with conductive deafness.

A. Causes of Sensory-Neural Deafness

Congenital deafness is often a cause. Here genes producing imperfect hearing mechanisms are inherited. During the first three months of pregnancy, if a mother contracts certain viral diseases such as mumps, influenza, or German measles these can impair the development of the hearing mechanism in the fetus. During birth itself, brain damage can affect the hearing mechanism. In early childhood, diseases such as meningitis can damage the inner ear or auditory nerve, producing deafness. Perforations in the eardrum can also be caused. However, the eardrum has a remarkable ability to heal itself, even if ruptured by intense sound.

All these forms of sensory-neural hearing loss mentioned are presently medically untreatable. One exception is Ménière's disease in which the inner ear becomes distended. Drugs can sometimes improve this condition.

2.4.4 Presbycusis

Another form of sensory-neural hearing loss is presbycusis. This loss in hearing sensitivity occurs in all western societies. Figure 2.13 shows the shift in hearing threshold at different frequencies against age as determined by Spoor.[41] As can be seen, presbycusis mainly effects the high frequencies, above 2,000 or 3,000 Hz. It effects men more than women. The curves shown in Figure 2.13 show the average hearing loss (with average hearing at the age of 25 assumed to be zero hearing loss). The group of people tested was subject to both presbycusis (the so called natural hearing loss with age), and sociocusis (a term coined by Glorig[42] to include all the effects of noise damage in our everyday lives with the exception of excessive industrial noise). Although the control group was deliberately chosen to screen out the effects of industrial noise, as von Schulthess and Huelsen[43] and von Schulthess[44] have pointed out, it is not possible to separate presbycusis and sociocusis through audiological means. For a comparison with Spoor's data, the British data of Hinchcliffe[45] are given in Figure 2.14. Here the data are plotted in a different way and the curves show the median hearing loss for nonindustrial noise-exposed women compared with the median hearing threshold at the age of 21.5. Although there are minor differences between Spoor's and Hinchliffe's data, the trends are similar.

Presbycusis is believed to be caused by central nervous system deterioration with aging, in addition to changes in the hearing mechanism in the inner ear.[4,45] Hinchcliffe[47] believes that these changes explain the features normally found in presbycusis: loss of discrimination of normal or distorted speech, increase in susceptibility to masking particularly by low frequency noise, and a general loss in sensitivity. Rosen[48] has suggested that degenerative arterial disease is a major cause of presbycusis.

Rosen, Olin, and Rose[49] have suggested that a diet low in saturated fat may, in addition to protecting a person against coronary heart disease, also protect him or her against sensory-neural hearing loss (presbycusis).

2.4.5 Hearing Loss Caused by Intense Noise

If very intense noise levels of the order of 135 dB or above at any frequency in the hearing range are experienced, immediate hearing damage is likely to result. However, permanent hearing damage is also produced at much lower sound pressure levels if the noise is experienced over much longer periods (weeks, months, or years). This reminds us of the similar phenomenon of metal fatigue where failure can occur at very much lower stress levels than the breaking stress, provided the stress is produced over a sufficiently long time. The problem is that noise-induced hearing loss, or to give it its technical name – noise-induced permanent threshold shift (NIPTS) is hard to distinguish from presbycusis.

Since we all suffer from presbycusis* in western societies, we should like to be able to subtract the effects of presbycusis from those of NIPTS. The only way we can attempt to do this is to study groups of people who are exposed throughout their lives to intense noise and other control groups who are not. Some attempts have been made to do this in the past 20 years. Such studies are very useful but still are troubled by the

*A study by S. Rosen et al. (See Rosen, S. et al., Presbycusis study of a relatively noise-free population in the Sudan, *Ann. Otol. Rhinol. Laryngol.*, 71, 727, 1962.) of the Mabaan tribe in the Sudan in 1962 showed that these people suffered much less presbycusis (see Figure 2.15). This may be because of the low background noise levels, but is more likely caused by the different diet or exercise habits of these people.

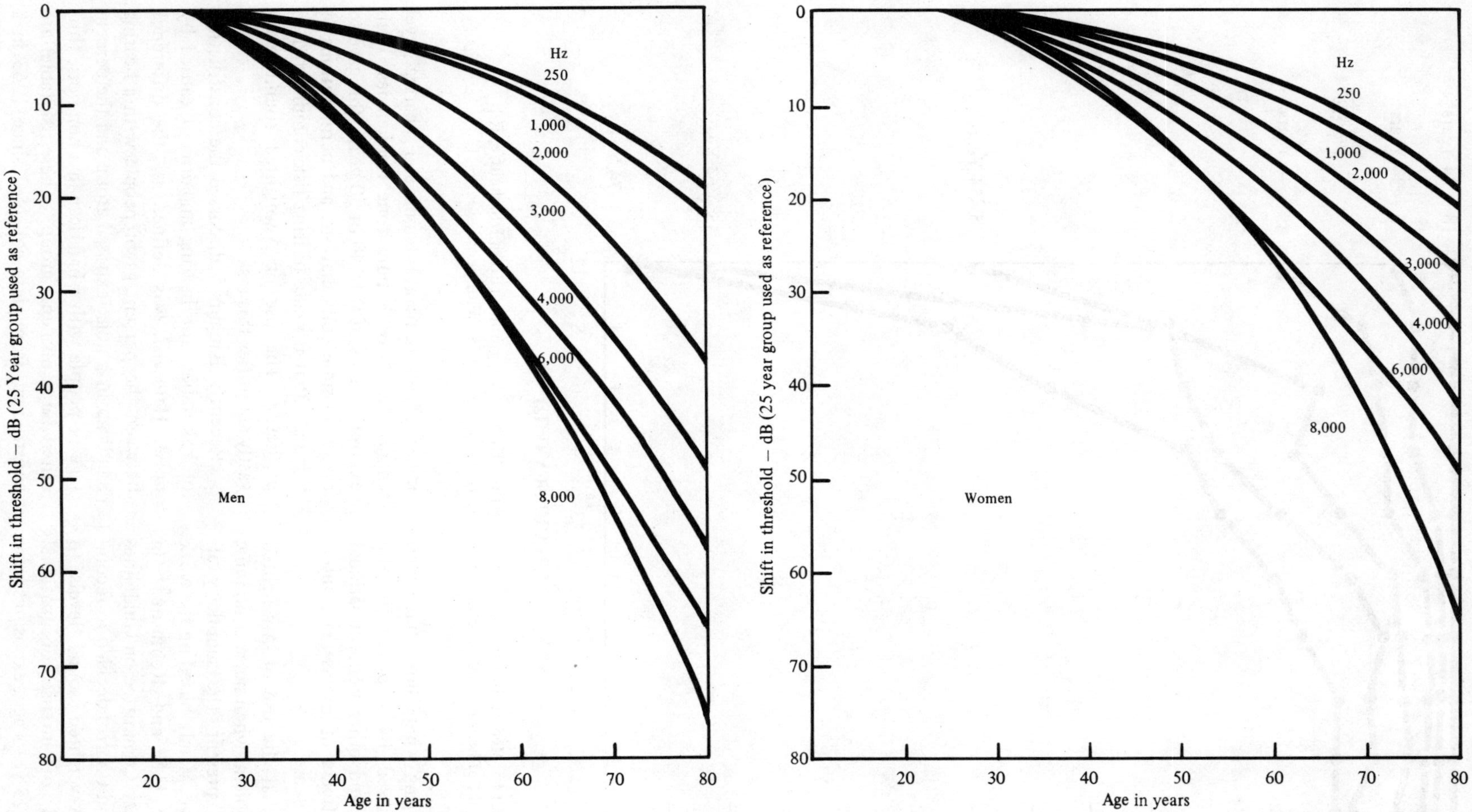

FIGURE 2.13 Shift in hearing threshold at different frequencies for men and women. (From Spoor, A., Presbycusis values in relation to noise-induced hearing loss, *Int. Audiol.*, 6, 48, 1967. With permission.)

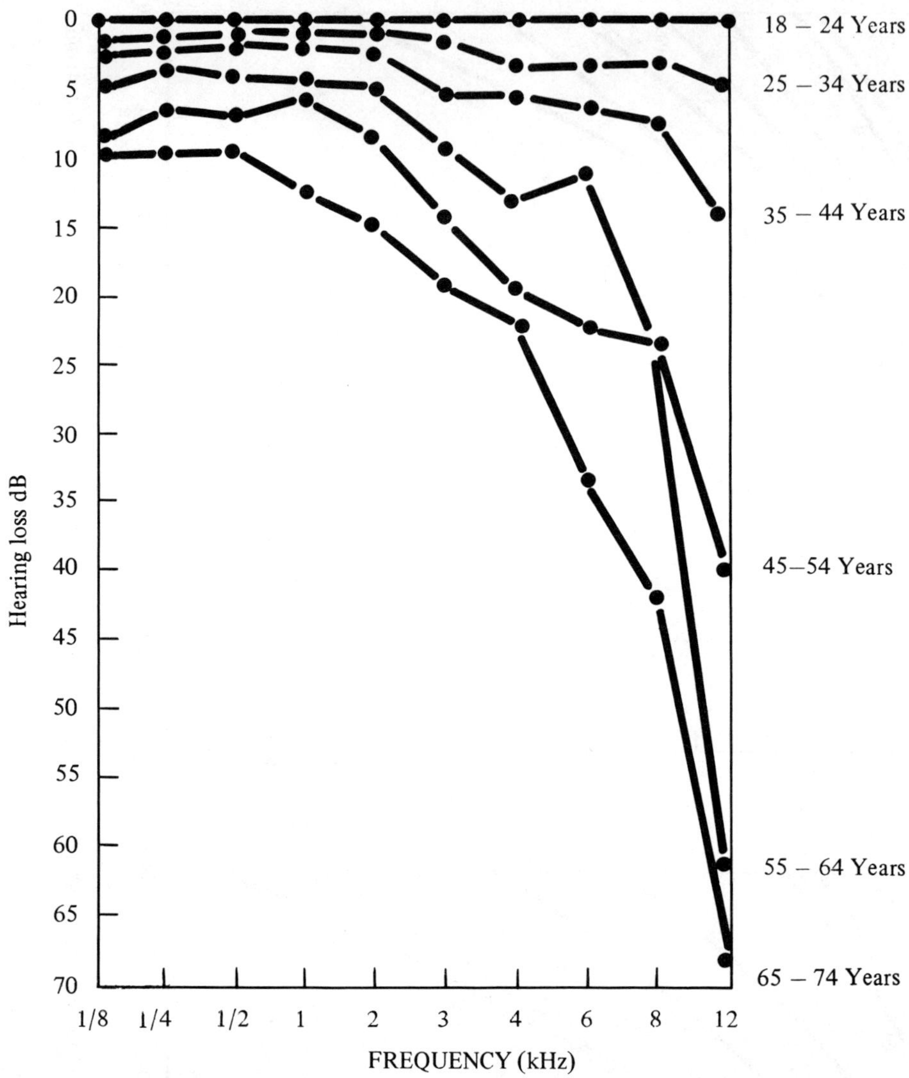

FIGURE 2.14 Median hearing loss for women in different age groups. (From Hinchcliffe, R., The threshold of hearing as a function of age, *Acustica,* 9, 303, 1959. With permission.)

difficulties in eliminating what Glorig has called *Sociocusis* (losses caused by exposure in the military, recreational, and other nonoccupational activities) and the effects of disease and intermittency in jobs and disease.

Despite the fact that the use of A-weighted sound levels cause problems when noise containing strong pure tones is present, the simplicity afforded is very attractive. Mainly based on the work of Baughn[50] in the U.S. and Robinson[51] in Britain, the Intersociety Committee on Guidelines for Noise Exposure and Control decided to recommend the use of A-weighted sound levels to indicate the hazard of intense occupational noise.[52] About 20 or 25 years ago, exposures were measured in 8 octave bands and hearing level changes at 6 or 8 pure tone frequencies. With exposure times of 0 to 40 or 50 years, this gave so many possible combinations and permutations of this data that it made hearing risk complicated to predict.[53] The use of A-weighted sound levels vastly simplifies this task.

Recently Baughn[53] discussed the construction of risk tables for hearing impairment caused by noise. Here *risk* was defined as "the difference between the percentage of people with a hearing handicap in a noise-exposed group and the percentage of people with a handicap in a non-noise (but otherwise equivalent) group." These risk tables are constructed from data collected since 1963 in a

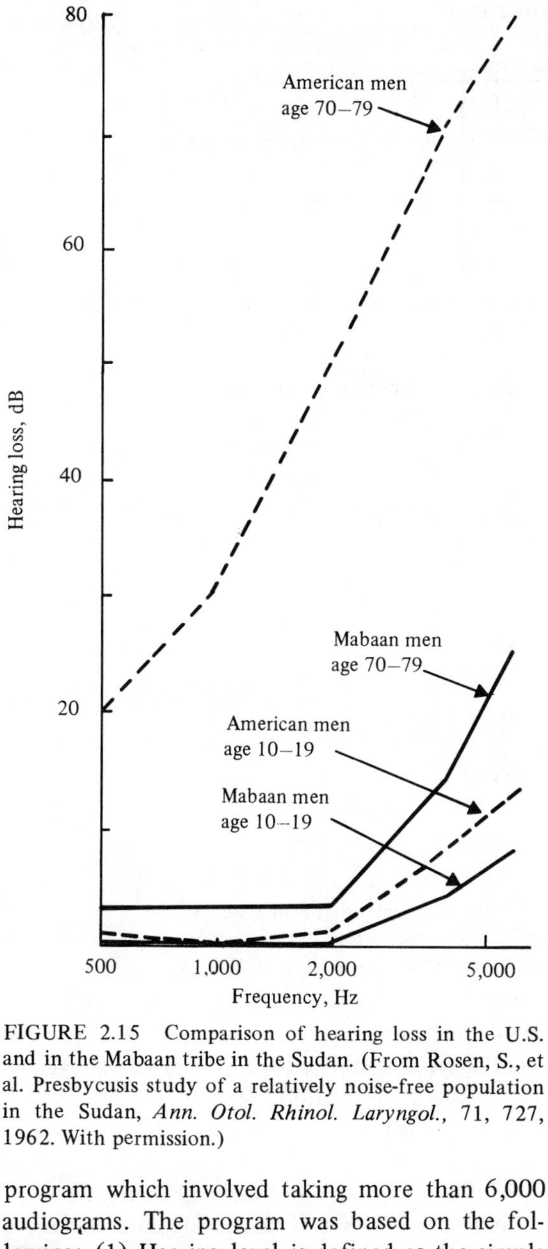

FIGURE 2.15 Comparison of hearing loss in the U.S. and in the Mabaan tribe in the Sudan. (From Rosen, S., et al. Presbycusis study of a relatively noise-free population in the Sudan, *Ann. Otol. Rhinol. Laryngol.,* 71, 727, 1962. With permission.)

program which involved taking more than 6,000 audiograms. The program was based on the following: (1) Hearing level is defined as the simple arithmetic average of audiometric hearing levels* at 500, 1,000 and 2,000 Hz, (2) Exposures were measured with sound level meters set to measure the A-weighted sound level (on the slow dynamic setting) with the microphone located at the subject's ear position, and (3) Exposure duration was defined as the years on the job (this was approximated by subtracting 18 years from the center year of each age group).

The data were subjected to regression analysis and risk tables constructed. The American Academy of Opthalmology and Otorhinolaryngology (AAOO) defined the onset of hearing impairment as occurring when the arithmetic mean of hearing levels at the 3 speech frequencies of 500, 1,000 and 2,000 Hz exceeds 25 dB (when measured with an audiometer conforming to current standards of the American National Standards Institute (ANSI). AAOO has also defined[54] the percentage of impairment (see Figure 2.16). Note total impairment occurs when the average of the three hearing levels exceeds 92 dB (on the same ANSI or ISO scale).

A typical risk table is given in Table 2.4.[53] Baughn[53] warned that several slightly different risk tables have been constructed using the same audiometric data. It is beyond the scope of this book to explain all of these reasons. Briefly, however, Baughn[53] explained that all of the tables are "correct." One reason is that it is necessary to refer all the audiometric data to some base-line. The base-line of the mean hearing level of a group of 18-year-old men tested at the time of their preemployment examination was first used. The hearing level was found to be 14.7 dB (averaged over the 3 frequencies). Later an ASA** working group (now ANSI) recommended the examinations should not be conducted in the plant audiometric department, but in a private otologist's office and after an off-the-job period of at least six months. This changed the hearing level to 10 dB and thus the base-line was changed and a new risk table emerged. Later when the base age of 18 was changed to 20 years, further risk tables were produced. Finally, when new audiometric data were included with the original data the tables changed again. All these changes have caused some confusion.

Baughn[53] recommended the present base-line of 10 dB hearing level (ANSI) for 80 dB(A) exposure (which is claimed to produce no risk of NIPTS) at the age of 20 as a good choice.

Baughn further explained that the risk due to noise at higher levels (e.g., 105 dB(A)) increases up to about 30-years exposure and then decreases. This can be explained by Figure 2.17 where it can be seen that the noise exposed curve OP is convex, while the nonnoise exposed curve OG (mainly

*The audiometric hearing level at a certain frequency is defined as the increase in the hearing threshold level compared with an agreed standard hearing threshold level at that same frequency. Different hearing levels will result if a different standard hearing threshold level is used.

**American Standards Association (now defunct).

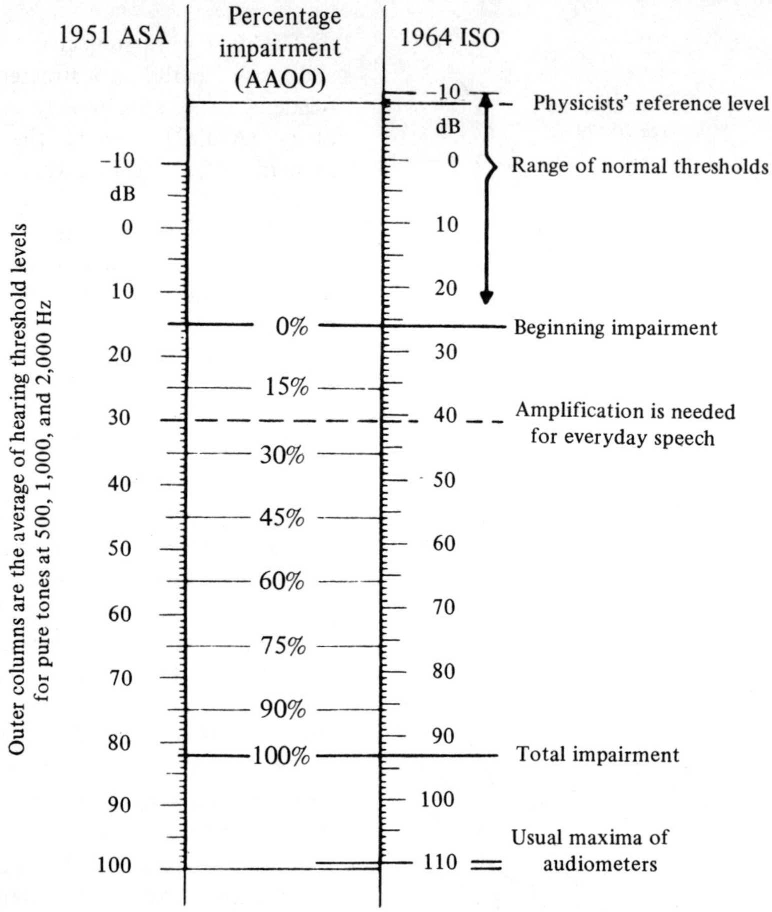

FIGURE 2.16 Average hearing levels and hearing impairment. (From Davis, H. and Kranz, F. W., The international standard reference zero for pure tone audiometers and its relation to the evaluation of impairment of hearing, *J. Speech Hear. Res.*, 7, 7, 1964. With permission.)

presbycusis) is concave. The difference between these two curves which may be attributed to noise (percentage due to noise in Table 2.4) reaches a peak and then decreases.

Baughn stated that the Occupational Safety and Health Administration (OSHA) 90 dB(A) limit for an 8-hr work day seems reasonable since over a working lifetime (45 years), about 10% of the exposed population would experience about 10% hearing impairment (see Figure 2.16). In recent congressional records, Kryter,[55] on the other hand, feels that the OSHA limits are inadequate and more hearing damage will result from the 90 dB(A), 8-hr limit than Table 2.4.

It is difficult to reconcile Kryter's and Baughn's results.* According to Kryter's results (Figure 2.18), after 42 years of exposure at the age of 62, 65% of noise exposed workers would suffer hearing loss compared with 15% in the general nonexposed population (which experiences presbycusis alone).

2.4.6 Theories on Noise-Induced Hearing Loss

Since World War II, a considerable amount of information has been gathered on hearing loss connected with intense noise. Most of these data have been collected in industry, but there have been other sources (e.g., aircraft pilots). Because in most instances the data have been collected for individuals working 8-hr days, little information is available for people working shorter or longer shifts each day and some empiricism is necessary to extrapolate results for use in noise regulations. Two main theories have emerged which are used to

*The probable explanation of this anomoly is the different assumptions for nonexposed populations. Baughn assumes workers in an 80 dB(A) environment at age 20 have normal hearing, while Kryter assumes a random sample from the 1960–1962 U.S. Public Health Survey by Glorig & Roberts for H.E.W. For further discussion see the six papers in *J. Acoust. Soc. Am.*, 53(3), 1211, 1973.

TABLE 2.4

Percentage of Population Exhibiting Average Hearing Threshold Level in Excess of 15 dB ASA 1951* as a Function of Years of Exposure and Noise Level, dB(A).

(All data have been adjusted to a median hearing level of 0 dB for 80 dB(A) exposure and 20 years of age.)

Age		18	23	28	33	38	43	48	53	58	63
Exp. years (Age − 18)		0	5	10	15	20	25	30	35	40	45
Exp. level	Total % expected	0.7	1.0	1.3	2.0	3.1	4.9	7.7	13.5	24.0	40.0
	% due to noise	0.0	0.0	0.0	0.0	0.0	0.0	0.0	0.0	0.0	0.0
80 dB(A)	% due to other	0.7	1.0	1.3	2.0	3.1	4.9	7.7	13.5	24.0	40.0
Exp. level	Total %	0.7	2.0	3.9	6.0	8.1	11.0	14.2	21.5	32.0	46.5
	% due to noise	0.0	1.0	2.6	4.0	5.0	6.1	6.5	8.0	8.0	6.5
85 dB(A)	% due to other	0.7	1.0	1.3	2.0	3.1	4.9	7.7	13.5	24.0	40.0
Exp. level	Total %	0.7	4.0	7.9	12.0	15.0	18.3	23.3	31.0	42.0	54.5
	% due to noise	0.0	3.0	6.6	10.0	11.9	13.4	15.6	17.5	18.0	14.5
90 dB(A)	% due to other	0.7	1.0	1.3	2.0	3.1	4.9	7.7	13.5	24.0	40.0
Exp. Level	Total %	0.7	6.7	13.6	20.2	24.5	29.0	34.4	41.8	52.0	64.0
	% due to noise	0.0	5.7	12.3	18.2	21.4	24.1	26.7	28.3	28.0	24.0
95 dB(A)	% due to other	0.7	1.0	1.3	2.0	3.1	4.9	7.7	13.5	24.0	40.0
Exp. level	Total %	0.7	10.0	22.0	32.0	39.0	43.0	48.5	55.0	64.0	75.0
	% due to noise	0.0	9.0	20.7	30.0	35.9	38.1	40.8	41.5	40.0	35.0
100 dB(A)	% due to other	0.7	1.0	1.3	2.0	3.1	4.9	7.7	13.5	24.0	40.0
Exp. level	Total %	0.7	14.2	33.0	46.0	53.0	59.0	65.5	71.0	78.0	84.5
	% due to noise	0.0	13.2	31.7	44.0	49.9	54.1	57.8	57.5	54.0	44.5
105 dB(A)	% due to other	0.7	1.0	1.3	2.0	3.1	4.9	7.7	13.5	24.0	40.0
Exp. level	Total %	0.7	20.0	47.5	63.0	71.5	78.0	81.5	85.0	88.0	91.5
	% due to noise	0.0	19.0	46.2	61.0	68.4	73.1	73.8	71.5	64.0	51.5
110 dB(A)	% due to other	0.7	1.0	1.3	2.0	3.1	4.9	7.7	13.5	24.0	40.0
Exp. level	Total %	0.7	27.0	62.5	81.0	87.0	91.0	92.0	93.0	94.0	95.0
	% due to noise	0.0	26.0	61.2	79.0	83.9	86.1	84.3	79.5	70.0	55.0
115 dB(A)	% due to other	0.7	1.0	1.3	2.0	3.1	4.9	7.7	13.5	24.0	40.0

*25 dB (ANSI) or 26 dB ISO.

From Baughn, W. L., The risk of hearing impairment as a function of noise exposure, in *Noise and Vibration Control Engineering,* Crocker, M. J., Ed., Purdue University, West Lafayette, Ind., 1972, 133. With permisison.

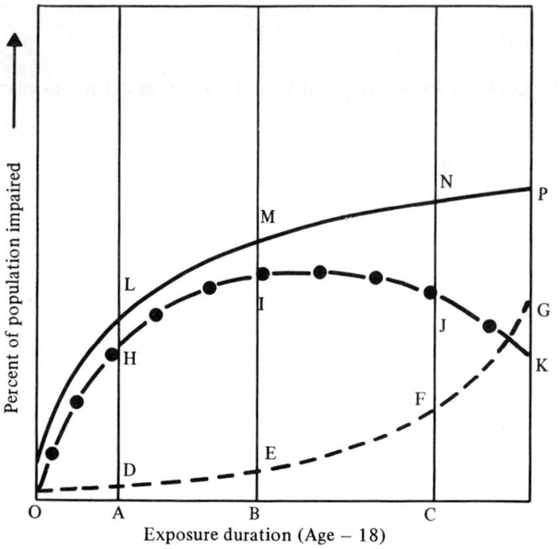

FIGURE 2.17 Schematic representation of noise-exposed group of people ——— OP, and non-noise-exposed group - - - - - OG. The difference —•—•—• OK represents the % noise = risk. (From Baughn, W. L., The Risk of Hearing Impairment as a Function of Noise Exposure, in *Noise and Vibration Control Engineering*, Crocker, M. J., Ed., Purdue University, West Lafayette, Ind., 1972, 133. With permission.)

aid in these extrapolations: (1) the *equal energy hypothesis,* and (2) the *equal temporary effect hypothesis.*

The equal energy hypothesis assumes that hearing damage is directly proportional to the acoustic energy an individual has experienced in his or her lifetime. Since acoustic energy is intensity multiplied by the time duration, it is also proportional to the sound level times the duration (see Section 1.4.2). This equal energy hypothesis suggests that the sound level may be increased by 3 dB for each halving in exposure time per day (see e.g., Reference 56). This concept lead to the adoption of the 1956 U.S. Air Force Regulation AF 160-3[57] and is widely used in European regulations or noise codes. Robinson[56] several other European acousticians, and the U.S. Environmental Protection Agency support this hypothesis. However, several U.S. acousticians, e.g., Botsford[58] have argued against this hypothesis suggesting that there has been little experimental verification. This hypothesis does allow a simple, attractive approach to the regulation problem which also seems thoroughly reasonable.

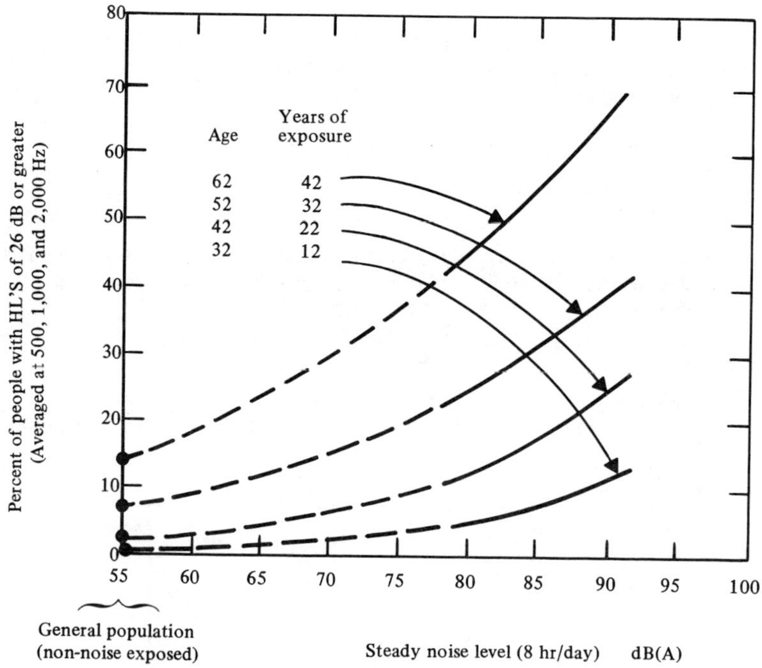

FIGURE 2.18 Percent of population with hearing loss as a function of steady noise level, dB(A), for eight-hr workdays. (From Kryter, K. D., Setting tolerable limits of noise exposure with respect to hearing impairment, *J. Acoust. Soc. Am.,* 51, 142(A), 1972: in Noise Pollution, Hearing before the Subcommittee on Air and Water Pollution, 92nd Congr., U.S. Gov. Print. Off., Washington, D.C., 1972, 154. With permission.)

The equal temporary effect hypothesis, on the other hand, assumes that the permanent hearing hazard of noise (NIPTS) can be related to the noise induced temporary threshold shift (NITTS) caused by the same levels of noise exposure. This hypothesis is largely based on the work of Ward, Glorig, and Sklar[59,60] and is supported philosophically by Kryter[61] and Botsford.[58]

Botsford argues that this hypothesis is more plausible because it can be related to an observable physiological effect in the ear. Also there seems to be recent evidence that both the noise induced permanent threshold shift (NIPTS) and the temporary threshold shift (NITTS) can be related to the same metabolic process ocurring in the inner ear (perhaps fatigue through a depletion of nutrients and a build-up of waste products). This hypothesis leads to the conclusion that noise which is intense enough to produce temporary loss (NITTS) will also be intense enough to produce permanent loss (NIPTS).

2.5 EFFECTS OF NOISE AND VIBRATION ON MAN

2.5.1 Introduction

Chapter 2 has, so far, been concerned with the auditory effects of noise on people: the loudness of sounds (Section 2.3), and noise-induced hearing loss (Section 2.4.5). However, noise is known to have other nonauditory effects on people. These other effects may be classed as psychological (e.g., annoyance, fear, effects of noise on actions, mental performance and work performance in industry) and physiological (e.g., startle, stress, sleep, vision, pulse rate, blood pressure, and muscle tension). There is concern that some of these effects mentioned above could be dangerous to health. However, it is difficult to isolate such effects and research and experimental data have been inconclusive to date. Most U.S. acousticians trained in psychoacoustics hold the view that noise-induced hearing loss is the major effect of intense noise and that all other observed psychological and physiological effects are of a minor nature, at least as far as the health of people is concerned. In the summary of his book, Kryter[65] makes the following statement "... other than as a damaging agent to the ear and as a masker of auditory information, noise will not directly harm people or interfere with psycho-motor performance. Man should be able, according to this concept, to adapt physiologically to his noise environment, with only transitory interference effects of physiological and mental and motor behavioral activities during this period of adaption. This concept, or its converse, is difficult to substantiate by scientific research and must be recognized as being hypothetical at this time."

Kryter's view is not universally shared, and in many other countries in Europe, nonauditory effects of noise are considered to be more serious. In the U.S.S.R. it is believed that noise can lead to increased accidents and absenteeism at work.[66] Since this subject is controversial and of considerable interest and importance, some of the nonauditory effects of noise are discussed in the remainder of Section 2.5.

2.5.2 Annoyance and Sleep Loss

A. Annoyance

Cohen[67] recently reviewed noise annoyance. Perhaps the major noise two annoyance problems in the U.S. are caused by surface transportation and aircraft noise.

In 1968, Wakstein[68] estimated that trucks and cars probably expose about 60 million people in the U.S. to noise levels exceeding those recommended by the International Organization for Standardization (ISO). He estimated that about 8 or 10 million people living near major airports were also seriously disturbed by noise. The actual number disturbed near all airports in the U.S. might be nearer to 15 million. It is interesting to compare these results with other estimates and results from other countries. In Britain, the London noise surveys[69] in 1948 and 1961 show that an increasing number of people are being disturbed by noise sources external to the home (see Table 2.5). Table 2.6 shows that in 1961, both in and out of the home, about four times as many people were disturbed by road traffic as by aircraft noise. Road traffic was the major source of noise disturbance both in- and outdoors. This 4:1 ratio agrees approximately with Wakstein's 6:1 ratio in the U.S. discussed above. Close[70] reviewed several noise surveys in the U.S. Some agree with the 4:1 ratio in the London surveys. Those which are in disagreement were probably taken in cities and close to airports so that aircraft noise received increased emphasis.

As early as 1958, Borsky[71] attempted to relate the annoyance reactions of people to the time that aircraft noise exceeded certain speech

TABLE 2.5

Comparison of Noise at Home in 1948 and 1961 Surveys in London, England[80]

Individuals' reaction to noise	External noise %		Internal noise %	
	1948	1961	1948	1961
Those disturbed by noise (a)	23	50	19	14
Those who notice but are not disturbed (b)	19	41	21	14
Total of people who notice noise (a + b = c)	42	91	40	28
Those who do not notice noise (d)	58	9	60	72
Total = c + d	100	100	100	100

TABLE 2.6

Comparison of Most Disturbing Sources of Noise from Survey* in London, England in 1961[80]

Type of noise	Number of people disturbed/100 questioned		
	Home	Outdoors	Work
Road traffic	36	20	7
Aircraft	9	4	1
Trains	5	1	–
Industry or construction	7	3	10
Domestic/light appliances	4	–	4
Neighbors	6	–	–
Children	9	3	–
Adult voices	10	2	2
Radio/television	7	1	1
Bells/alarms	3	1	1
Pets	3	–	–
Other noise	–	–	–

*1,400 people questioned

interference levels (SIL).* Borsky found that more than 50% of people interviewed complained of frequent disturbance and annoyance when an SIL value (measured outdoors) of 60 dB was exceeded for at least 80 sec an hour. However only 15% of the people were annoyed when an SIL of 60 dB was exceeded for less than 20 sec an hour.

Noise annoyance is, of course, difficult to measure and is possibly subject to biasing effects. Miller[63] has reviewed this topic recently. Whenever a team of people is introduced into a neighborhood to make a social survey, there is a danger that their mere presence may affect people's attitudes. The design of the questionnaire and the way in which it is conducted can also seriously affect the results. McKennel[72] has suggested, however, that annoyance caused by noise may not be affected so severely by biasing effects. Figure 2.19 shows average annoyance scores plotted against the average peak loudness level, PNdB. Figure 2.20 shows the mean annoyance index for people exposed to traffic noise.[73] The average annoyance scales in Figures 2.19 and 2.20 cannot, of course, be directly compared. Such annoyance is caused by interference with people's activities such as television or radio listening (21%), conversation (15%), telephone use (14%), resting outdoors (13%), resting indoors (11%), listening to records or tapes (9%), sleeping (8%), reading (6%), and eating (4%). The percentages shown are for people disturbed by aircraft noise found in a Tracor study in 1971.[74]

Attempts have been made to relate community responses to noise levels or other noise rating schemes. Figure 2.21 shows an early attempt by

*The speech interference level (SIL) of a noise is the average of the levels in the three octave bands: 600 to 1200, 1200 to 2400, and 2400 to 4800 Hz. SIL is no longer used, instead the preferred speech interference level (PSIL) which is the average of the levels in octave bands centered at 500, 1000 and 2000 Hz is used.

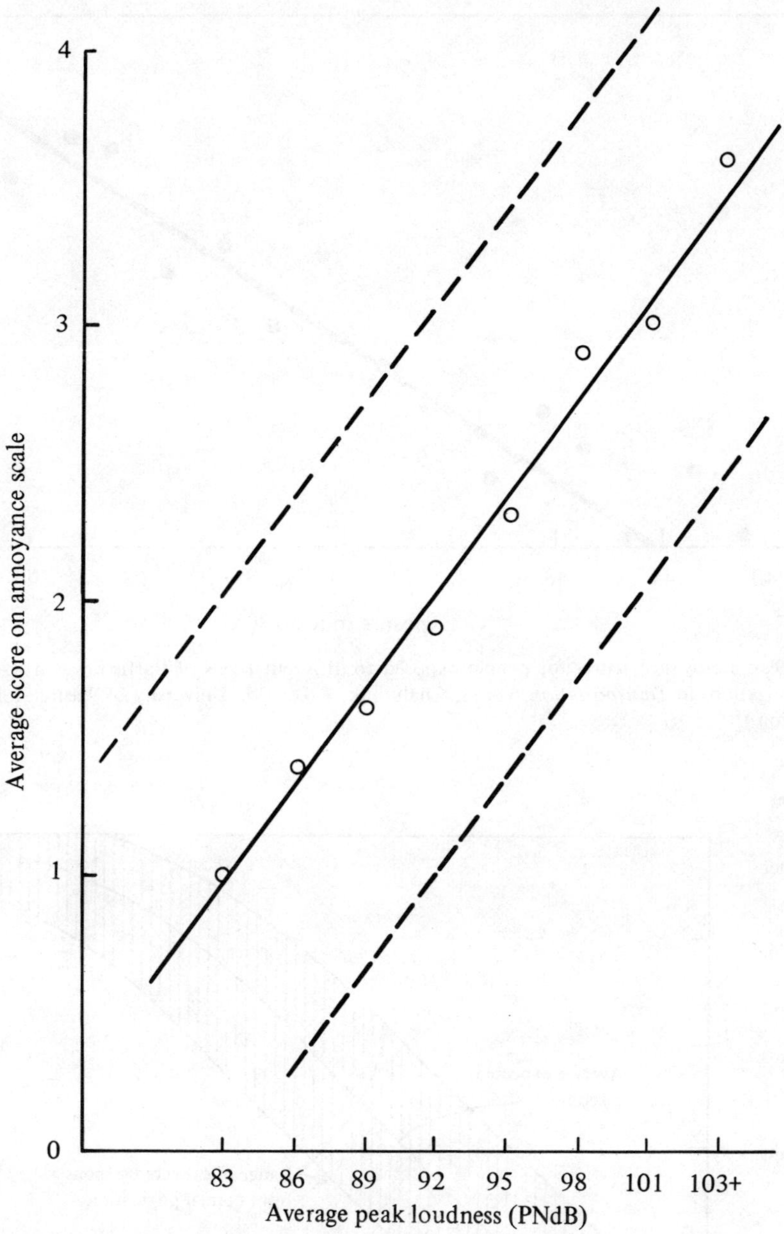

FIGURE 2.19 Average annoyance scores for people exposed to aircraft noise. The dashed lines include two thirds of the people interviewed. (From McKennell, A. C., Noise complaints and community action, in *Transportation Noises,* Chalupnik, J. D., Ed., University of Washington Press, Seattle, 1970, 228. With permission.)

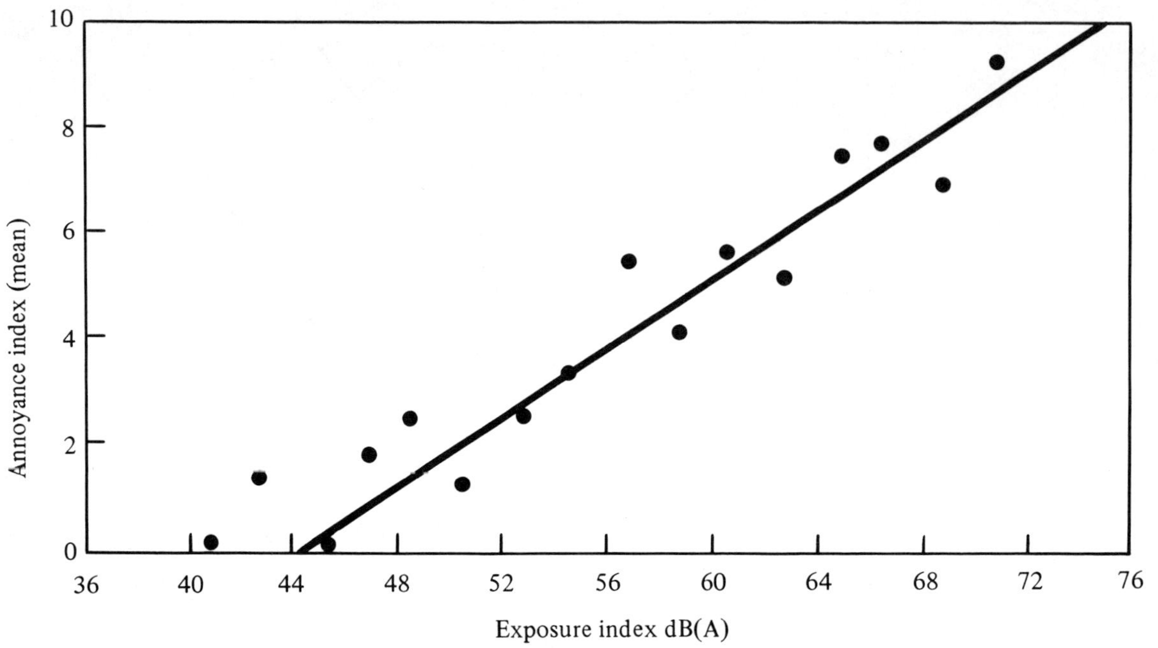

FIGURE 2.20 Mean annoyance scores for people exposed to different levels of traffic noise, dB(A). (From Kajland, A. Traffic noise investigation, in *Transportation Noises,* Chalupnik, J. D., Ed., University of Washington Press, Seattle, 1970 327. With permission.)

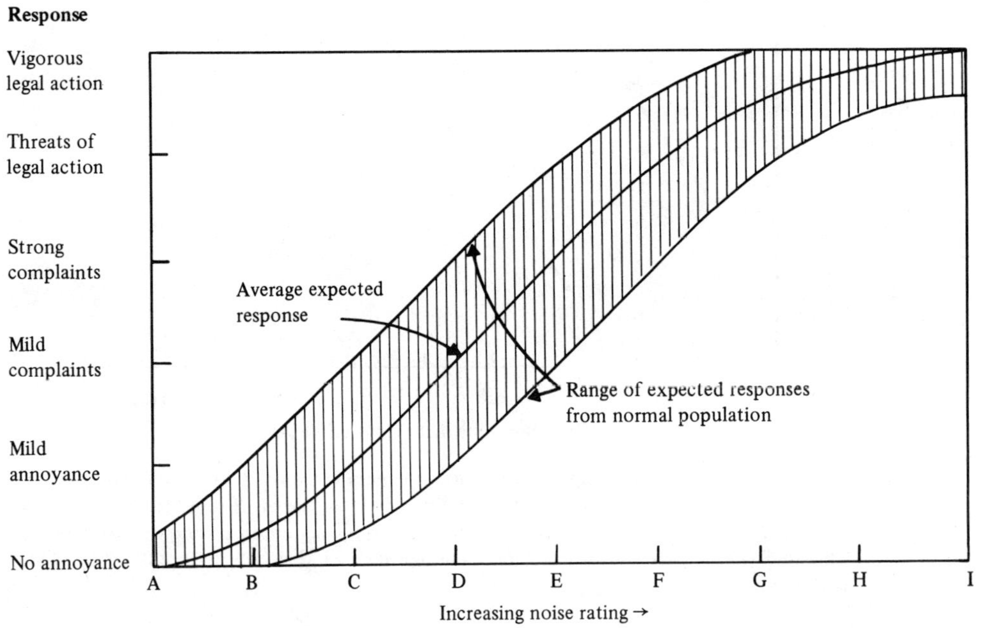

FIGURE 2.21 Relationship between community response and noise rating. (From Rosenblith, W. A., Stevens, K. N., and the staff of Bolt Beranek & Newman, *Handbook of Acoustic Noise Control,* Vol. II, *Noise and Man,* Wright–Patterson Air Force Documentation Center Tech. Rep. 52–204, 1953, 181.)

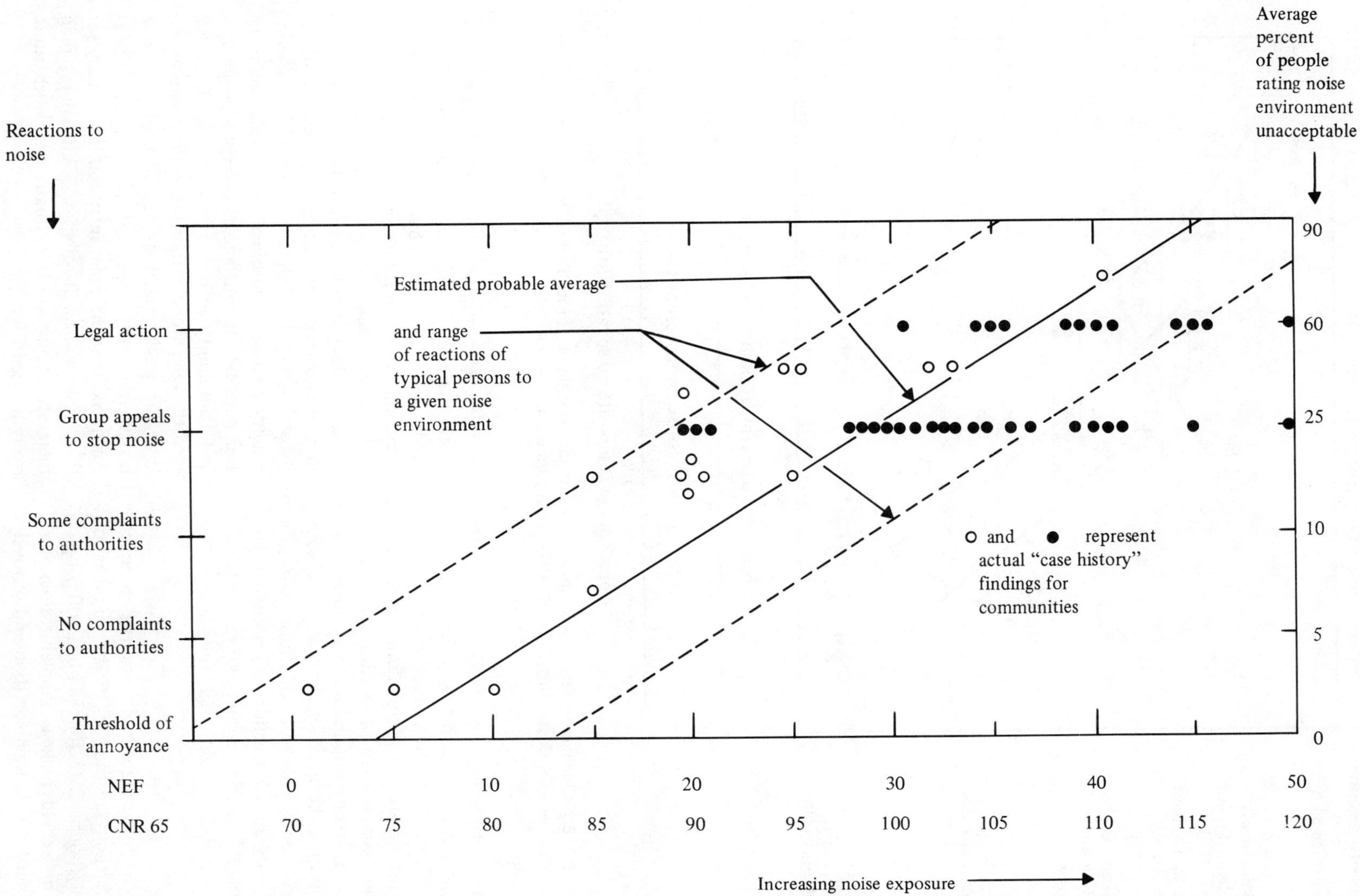

FIGURE 2.22A. Community response and relationship to Noise Exposure Forecast (NEF) and Composite Noise Rating (CNR). (From Kryter, K. D., *The Effects of Noise on Man*, Academic Press, New York, 1970. With permission.)

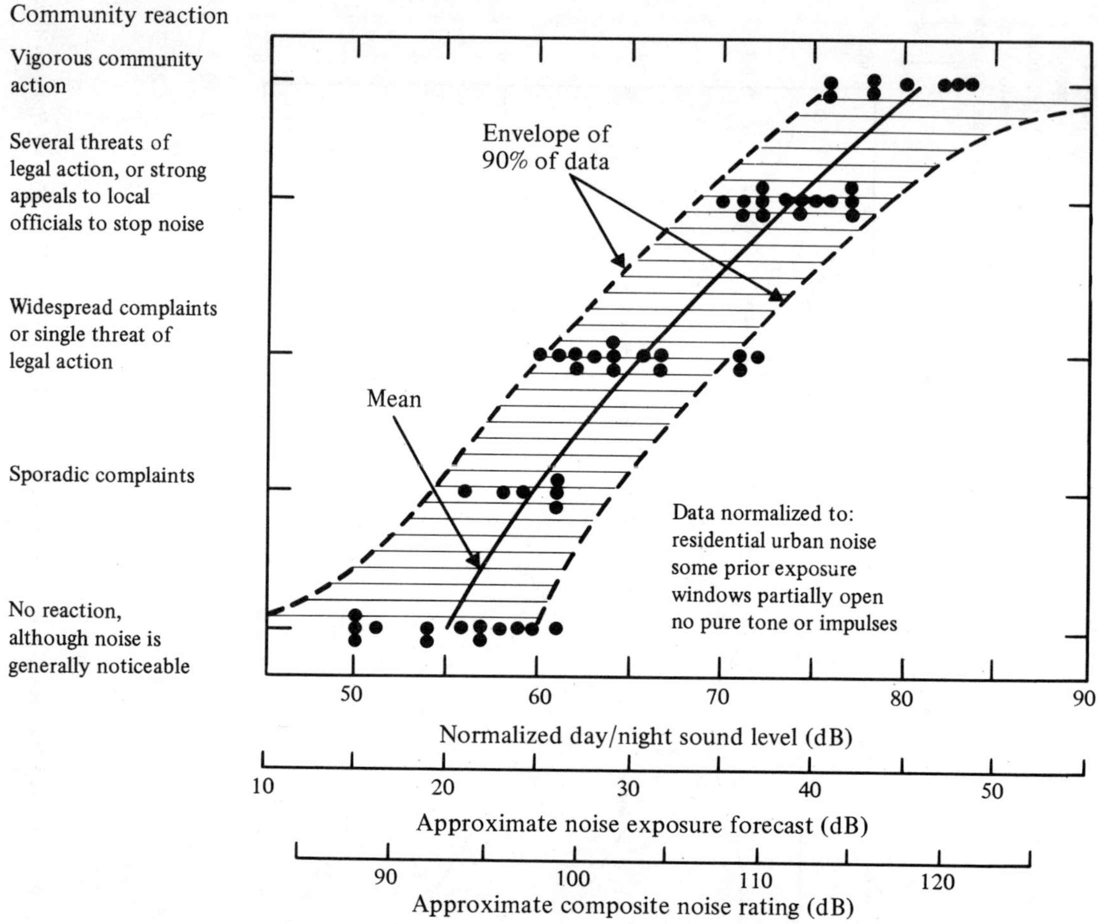

FIGURE 2.22B. Community reaction to many types of intrusive noise. (From Eldred, K. M., Assessment of community noise, *Noise Contr. Eng.*, 3(2), 88, 1974. With permission.)

Rosenblith and Stevens[75] in 1953. Figure 2.22a shows a more recent attempt by Kryter et al.[76] in 1971 and Figure 2.22b shows another by Eldred in 1974.* The response of a community to noise is even more complicated than the response of individuals. It has been found that individuals who are driven to make a complaint are not always those most annoyed. Also of those seriously disturbed only a small fraction will complain. The Tracor study[74] produced the empirical relationship for neighborhoods with high noise levels:

$$h = 196 + 2c,$$

where h is the number of seriously annoyed households/thousand and c is the number of complaints/thousand. If, for example, 100 people/thousand households make a complaint, then there will be almost 400 households/thousand seriously annoyed. This empirical relationship suggests that when the number of seriously disturbed households decreases to about 200 households/thousand, complaints will drop to zero.

With individuals, noise annoyance is complicated by other annoyance factors, e.g., whether a man or woman is annoyed by other factors such as marital quarrels. When concerned with community response additional factors such as social and political circumstances, noise action groups, and community attitudes towards the noise and even the regulatory authorities may be important. Here the public relations activities of the noise maker are clearly important. If the community is convinced that the noise intrusion is necessary for their economic well-being or their safety, then their acceptance of the situation is much more likely to occur.

*See Eldred, K. M., Assessment of community noise, *Noise Contr. Eng.*, 3(2), 88, 1974.

B. Sleep Loss

It is common knowledge that noise will interfere with sleep. We are all awakened by intense noises and most will easily be awakened by an alarm clock or radio. However, the phenomenon is more complicated then it might appear at first sight. Sleep has been studied extensively in the laboratory in the last few years. It has been found that there are several stages of sleep. A person passes from one stage to another during a night's sleep. There are several ways of labeling these sleep stages or states. One common way is: REM, 1, 2, 3, and 4. The rapid eye movement stage (REM) has aroused much interest. Figure 2.23 shows a typical sleep pattern for a young adult. It is in the REM stage that most dreams occur. Here the person has internalized his attention but is not in a deep stage of unconsciousness. However, paradoxically in the REM stage a person is particularly insensitive to sound.[78] The different stages of sleep can be detected by attaching electrodes to the body and monitoring brain waves and other behavior. Miller[63] gives a series of curves (Figure 2.24) which show the sound levels which will awaken a given percentage of people. The solid circles in the figure represent the percentage of people that awake to a 3-min noise at each A-weighted sound level. These were obtained by Steinicke in 1957 on 350 people each tested in their own bedrooms. It is reasonable to assume that most people were aroused from stages 2 or REM. The solid curves *noise wakes me up* and *noise keeps me from going to sleep* are derived from the Wilson Report.[80] It appears from Figure 2.24 that sleep in Stage 2 is most easily disturbed while deep sleep in Stages 3 and 4 is least easily disturbed.

Some researchers believe that there is a direct connection between noise-disturbed sleep and a person's health. Richter[81] has suggested that since noise prevents a person from the recovery and strengthening process of sleep that this will have an effect on health. However, this hypothesis is as yet unproved. Miller[63] has suggested that while noise undoubtedly produces sleep patterns in people similar to those of a poor sleeper, a person compensates by spending more time in deep sleep, by becoming less responsive to stimuli and by napping. Thus, noise may not deprive a person of efficient sleep to produce an adverse effect on health. Kryter[82] shares this view. We must conclude that without further experimental evidence, we still cannot ascertain for certain that noise-disturbed sleep has an adverse effect on health. However, we do know that sleep disturbance caused by noise will reduce a person's feeling of well-being and that if this disturbance is sufficient it could conceivably have an adverse effect on health.

2.5.3 Other Physiological Responses to Noise

Kryter,[82] Miller,[63] and Reference 64 discuss several other physiological responses to noise. The most important of these will be discussed briefly here. Although, it has been undoubtedly established that people have a number of further responses to noise, it is still debatable whether these further responses are potentially hazardous

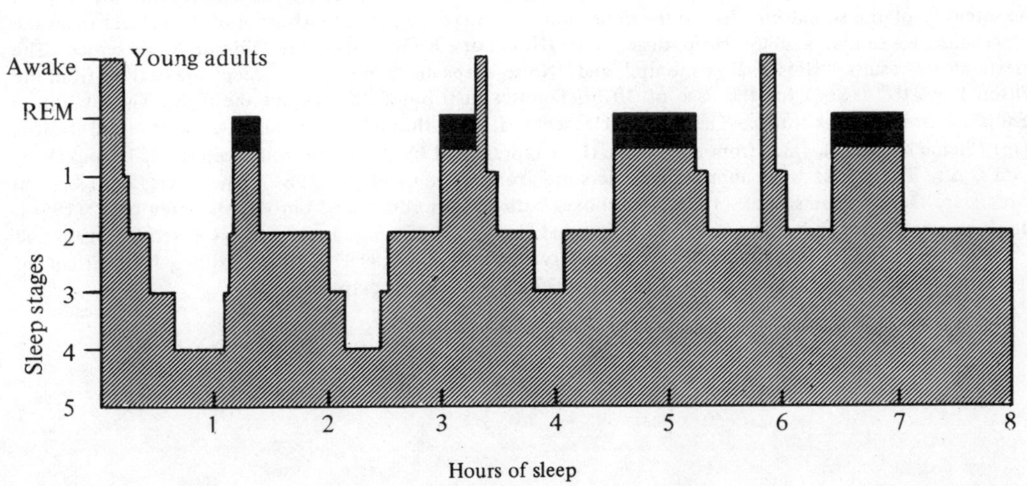

FIGURE 2.23 The sleep pattern for a young adult. (From Berger, R. J., The sleep and dream cycle, in *Sleep: Physiology and Pathology, A Symposium*, Kales, A., Ed., Lippincott, Philadelphia, 1969, 17. With permission.)

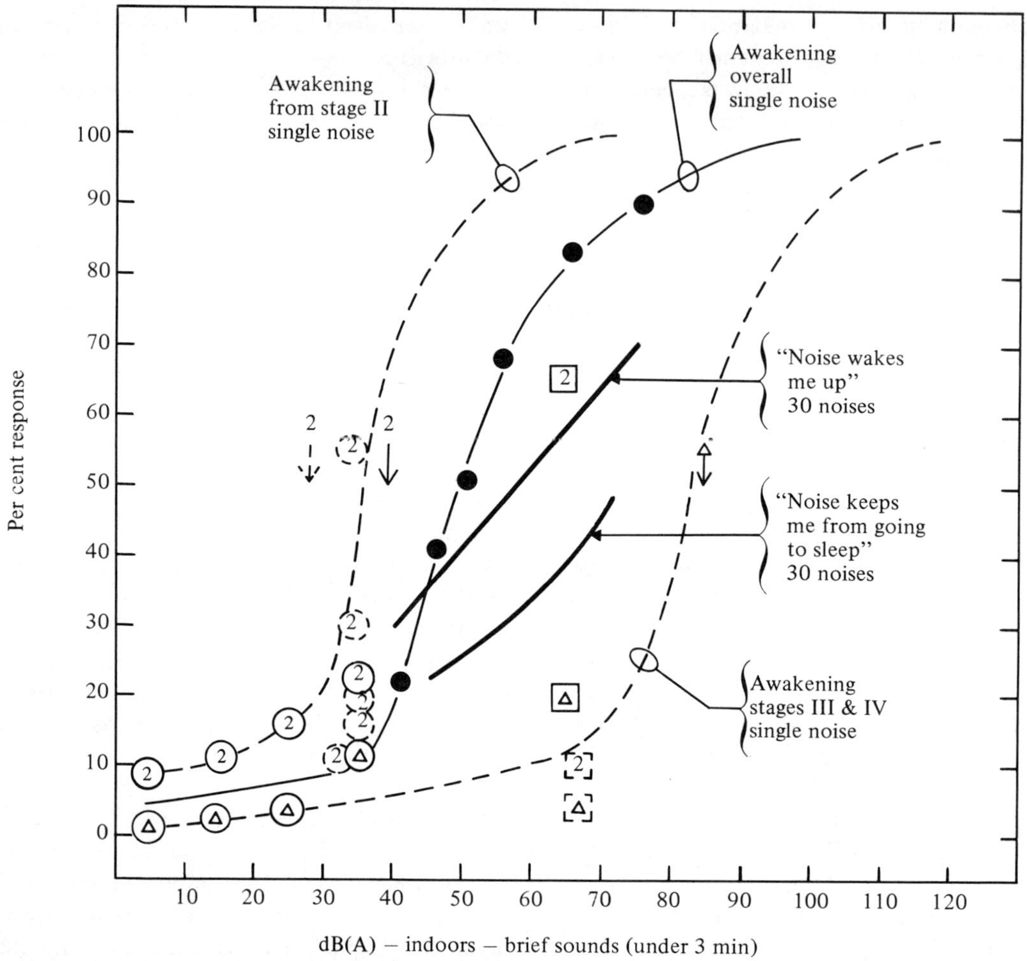

FIGURE 2.24 Awakenings to sound from various laboratory and questionnaire studies are shown. The horizontal axis gives the approximate A-weighted sound level dB(A) of the noise. The curves labeled "awakening" are from normally rested young adults who were sleeping in a laboratory and were moderately motivated to awake in response to sound. The percentage of awakening responses will depend not only on the intensity of the sound but also on the definition of "awakening," the motivation of the subject to awake in response to sound, and the sleep stage (I, II, III, IV, or I-REM) when the stimulus is presented. The questionnaire results, "Noise wakes me up," and "Noise keeps me from going to sleep" are derived from the Wilson Report[80] (1963) for the case of 30 brief noises distributed throughout the night. The laboratory results are from various studies. The filled circles were gathered throughout the night without regard to sleep stage (Steinicke, 1957). Data from sleep stage II are represented by *2*'s; those from sleep stages III and IV by delta's, Δ's. The circles with unbroken borders are from Williams et al. (1964). The circles with broken borders are from Williams et al. (1965). The boxes with solid borders are from Rechtshaffen et al. (1966). The boxes with broken borders are from Lukas and Kryter (1970). The broken arrow is from Watson and Rechtshaffen (1969). The solid arrows are from Kryter and Williams (1970). (From Miller, J. D., Effects of Noise on People, U.S. Environ. Protect. Rep. NTID 300.7, 1971. With permission.)

to health or whether people adapt to noise over a long period and these responses decrease. Perhaps the most important response yet to be considered is noise-induced stress.

We may define stress as a state of arousal which can eventually have adverse health effects. The general response to any form of stress is called the *adaptation syndrome.* This consists of three states: alarm, a stage of resistance, and a stage of exhaustion. It is suggested by some medical authorities that some nonauditory diseases can be produced in the resistance stage of stress such as ulcers and high blood pressure. Such diseases are called diseases of adaption. However, many authorities do not accept this hypothesis.

Kryter[82] supports the view that people adapt to noise-induced stress. Certainly it appears that laboratory experiments are inconclusive at present. Although it appears that short-term changes in blood circulation and the secretion of chemicals in the urine are produced, it is uncertain if any long-term effects are produced. Miller[63] suggests that it is unlikely that any long-term stress effects are produced unless noise levels are sufficient to produce hearing damage and loss. However, some studies have been published which suggest that in some industrial situations noise-induced stress can produce adverse health effects. Figure 2.25 shows the result of a study by Jansen[83] on 1,005 German industrial workers. Figure 2.26 shows the incidence of hypertension in male and female workers in noisy workshops after Andrukin.[84] As Kryter[82] pointed out, however, these results could be misleading. In the noisy environment other factors may be important such as poor ventilation, heat and light, danger from accidents, anxiety over job security, and the possible selection of people with lower general health and economic or social status for intense-noise jobs.

Kryter,[82] Miller[63] and Reference 64 also discuss the following effects of noise. Because they are probably somewhat less important than annoyance, stress, or sleep loss, they will only be discussed briefly here.

A. Balance

As long ago as 1926, Tullio[85] reported that pigeons could be made to swerve off course if exposed to intense low frequency noise. From this experiment some have concluded that noise has a direct effect on the balance. More recent experiments both at work and in the laboratory show that intense noise can produce complaints of vertigo (dizziness), nystagmus (rapid involuntary side-to-side eye movements), and balance problems in people. However, the levels to produce these

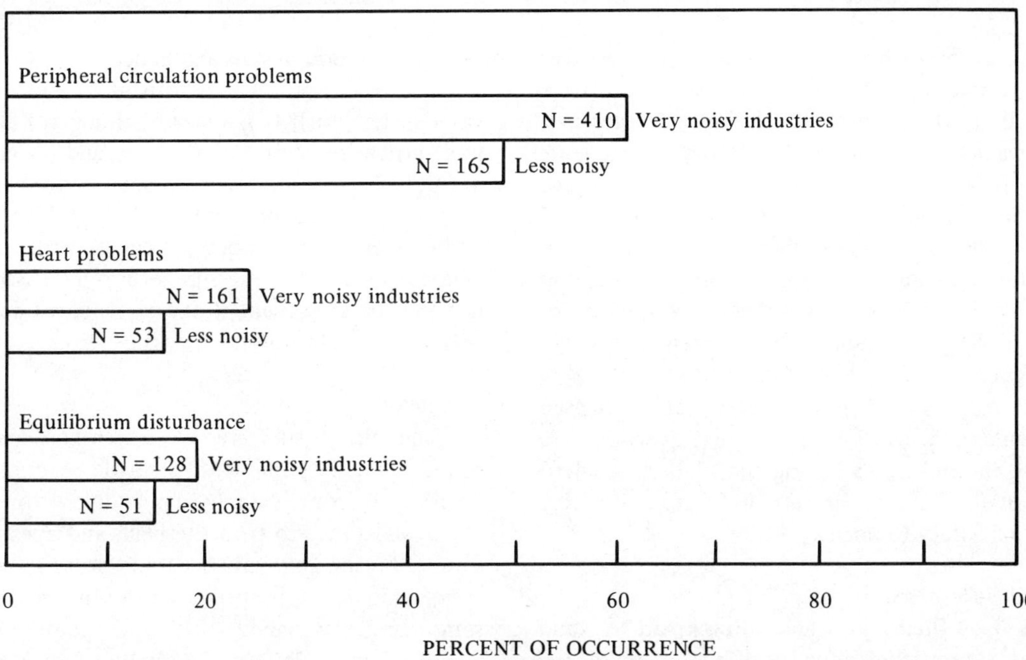

FIGURE 2.25 Differences in percentages of occurrence of various physiological problems in two different noise levels. Data compiled from 1,005 German industrial workers. (From Jansen, G., Adverse effects of noise on iron and steel workers, *Stahl. Eisen.*, 81, 217, 1961. With permission.)

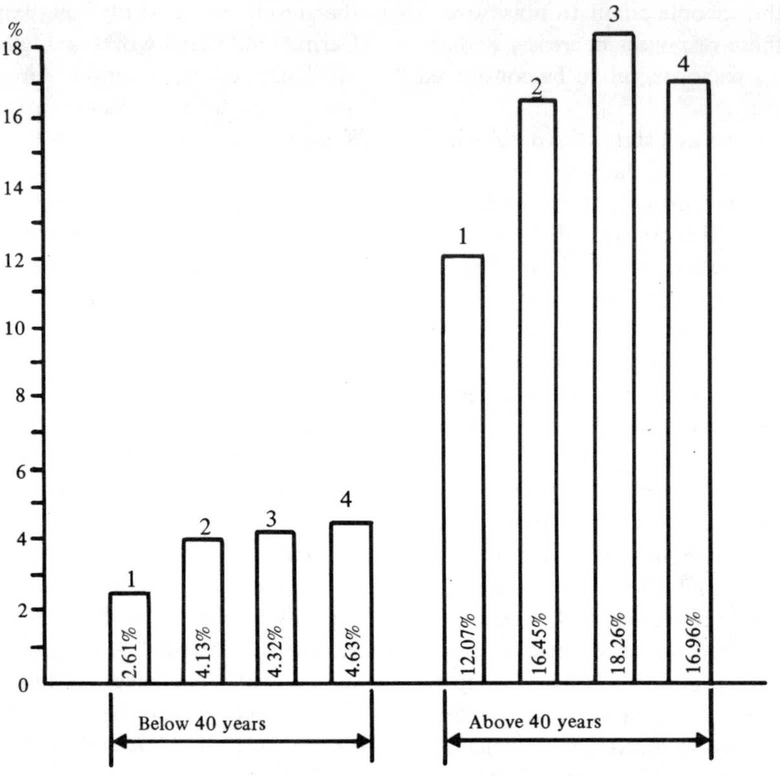

FIGURE 2.26 Incidence of hypertensive men and women workers (above and below the age of 40) in noisy workshops: (1) Tool making; (2) Sorting workshop; (3) Workshop with automatic lathes; (4) Workshop producing ball bearings (see Andriukin, A. A., Influence of sound stimulation on the development of hypertension: Clinical and experimental results, *Cor. Vasa,* 3, 285, 1961.)

effects are quite high – about 130 dB.[86] Lower levels, about 120 dB, can produce these effects particularly if the levels are different in each ear. This was demonstrated in laboratory experiments where people had to balance themselves on rails of different widths.[87] Since the levels discussed above are so intense, they are perhaps not encountered often in work environments except in such intensely noisy environments as quarry or rock drilling. In such environments, vertigo, nystagmus, and balance problems have been mentioned as possible causes of increased accidents, and certainly mining and quarrying are among the industries having the highest accident rates in the U.S. per man-hour (according to the National Safety Council).

B. The N-Response

Davis, Buchwall, and Frankman,[88] and Kryter[82] term the following effects of noise, the *N-response:* (a) blood circulatory changes caused by vasoconstriction (temporary shrinkage) of the peripheral blood vessels and hence increased blood flow to the brain (since this does not exhibit vasoconstriction); (b) slow breathing; (c) changes in electrical resistance of the skin, and (d) changes in muscle tension. Such responses do begin at quite low sound pressure levels (e.g., 70 dB at 1,000 Hz). The response becomes progressively greater as the sound pressure level is increased. It is not known at present if these responses decrease with adaption to noise.

C. Startle

If people are subjected to sudden impulsive-like sounds they respond with a startle reaction. The reaction is controlled by an auditory–muscular reflex and tends to turn the head and eyes to the source. Presumably this occurs in order to prepare the person for appropriate action in case there is some danger at hand.[89] This reflex action begins at quite low levels[64] and normally causes a change in the rate of heart beat and blood pressure. However, it appears that there is considerable adaption to startle. Kryter[82] reported several experiments which show adaption. We will quote two here. Bartoshuk[90] reported that the acceleration in the heart beat rate of unborn babies to bursts of acoustic clicks adapted out after 40 trials. Ando and Hattori[91] reported that babies of mothers who were pregnant in neighborhoods exposed to aircraft noise were much less aroused from sleep by aircraft noise than were babies of mothers who were pregnant in nonnoise exposed neighborhoods.

D. Vision

Kryter[82] reviews several effects of noise on vision including visual acuity and field, color vision, critical flicker frequency, and also a narrowing of the visual field during noise. Jansen[92] reported that noise caused a dilation of the pupil during exposure. Figure 2.27 shows the pupil size Jansen measured for different levels of white noise. Adaptation with time apparently does not occur and apparently quite low levels such as 70 dB are sufficient to cause dilation which increases until at least 110 dB. The significance of this reaction is not yet understood. It is another of the many physiological responses to noise which need further study.

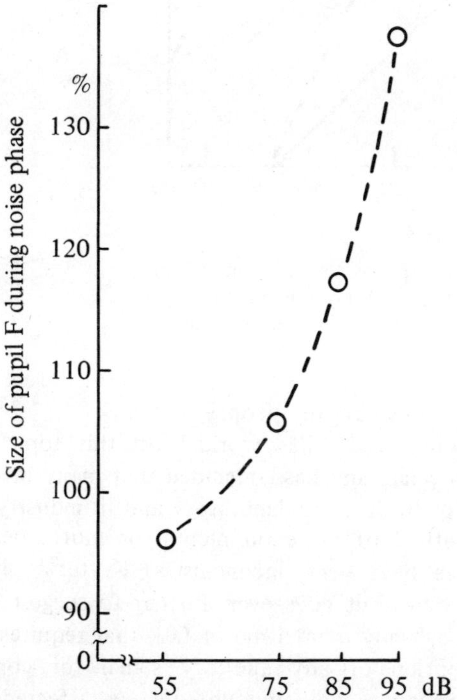

FIGURE 2.27 Relationship between pupil size (percentage) and sound pressure level (of white noise). (From Jansen, G., Effects of noise on physiological state, in *Noise as a Public Health Hazard,* ASHA Rep. No. 4, American Speech and Hearing Assoc., Washington, D.C., 1969. With permission.)

E. Infrasound and Ultrasound

Infrasound below 20 Hz can excite some of the body organs into resonance as can low frequency sound in the range 20 to 75 Hz. The eyeball has a resonance frequency of about 5 Hz. Kryter[82] gave some qualitative results obtained by the U.S.A.F. and NASA when people were exposed to intense (above 130 dB) low frequency (below 100 Hz) pure tones and noise. Some effects noticed included chest wall, abdominal, and nasal cavity vibration. Additional effects included nausea, giddiness, and coughing. However, these effects did not appear unless the levels were above about 130 dB. Such low frequency levels are not experienced frequently except in a very few environments. Such physiological effects can be related to the resonance of organs in the body. This will be discussed further in the next section.

Although ultrasound is inaudible, it can produce effects on people. Examples of ultrasonic sources include dentist drills, ultrasonic cleaners*, air jets, and jet and rocket engines. Probably the tinnitus, dizziness, nausea, and headaches sometimes attributed to ultrasound[82] are really caused by any parts of the noise which are in the audible range (below 20,000 Hz). However, von Gierke has suggested[82] that ultrasound can become "audible" because subharmonics can be generated in the middle ear by ultrasound. The inner ear will then respond to the subharmonics.

F. Effects of Vibration on People

A discussion of the effects of vibration on people is given by Goldman.[93] Several of the human organs have resonances which can produce sensations of pain if exited by vibration or noise at their resonance frequencies. These resonance frequencies are approximately in the range of 5 to 100 Hz. People seem most responsive to vertical vibration. Figure 2.28 shows the Reiher – Meister chart showing human response to vibration.[94] This was produced in 1931 and later measurements with people produced somewhat similar subjective results. It is rather interesting to note that for small vibrations the subjective responses follow constant velocity contours as frequency is changed. For large vibration, the responses change to constant acceleration contours with frequency.

*Note that ultrasonic cleaners also make some audible sound in the 2,000 to 10,000 Hz range.

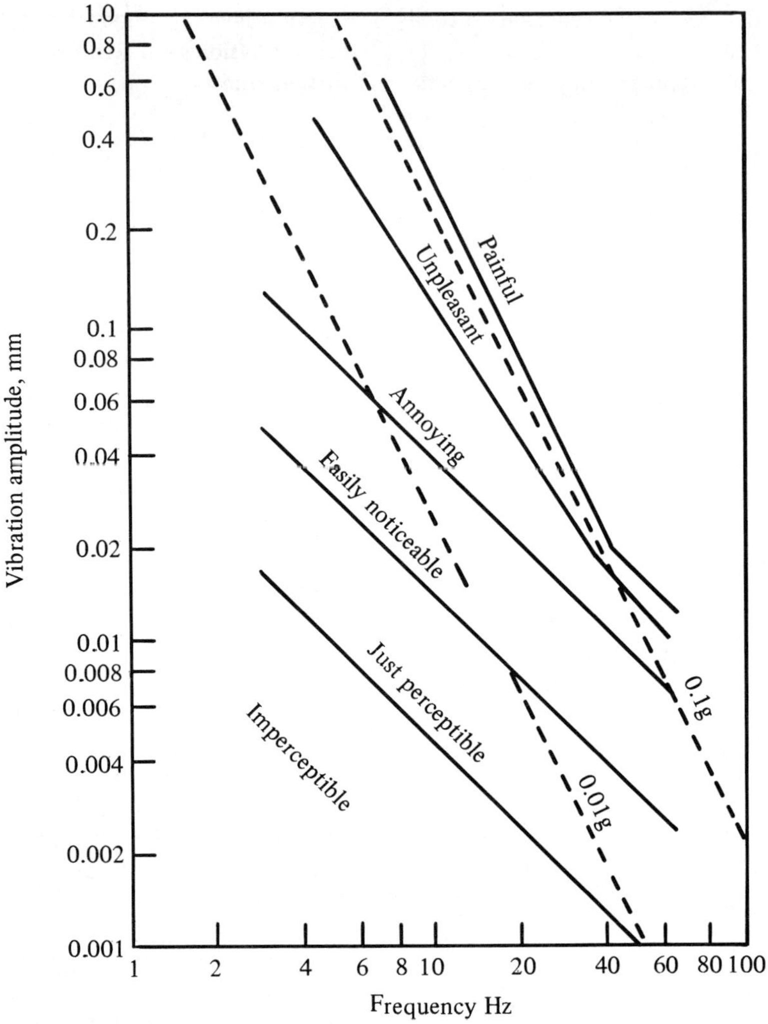

FIGURE 2.28 Reiher-Meister chart of human response to vibration. (From Reiher, H. J. and Meister, F. J., Human Sensitivity to Vibration, *Forsch. Geb. Ingenieurmes.*, Stuttgart, Inst. Tech., Nov. 1931; Abstract in *Eng. News Rec.*, March 31, 1932, 470. With permission.)

Interestingly, building response to vibration is somewhere between the constant velocity and constant acceleration contours according to Monk[95] (see Figure 2.29). Generally, vibration will be insufficient to cause cracks in plaster finishes or in walls until it becomes unpleasant or painful. Vibration which is only just perceptible to people is unlikely to cause any damage to buildings unless they are very old or highly stressed. In such cases the damage normally appears first as cracks in old ceilings.

2.5.4 Effects of Noise on Performance

Several reviews have recently been written on the possible effects of noise on the performance and efficiency of people at work.[63,64,67,96] Broadbent[97-99] has worked on this topic for many years and has concluded that many investigations both in the laboratory and in industry on the effect of noise on mental or motor performance have been inconclusive. Kryter[96] shares this view, but goes even further to suggest that although any mental motor task that requires the perception of any auditory signal for correct performance will undoubtedly be affected by noise, that in his opinion, experimental evidence shows noise has no adverse effects on nonauditory mental or motor tasks. He suggests that adaption to noise is responsible for this. Most researchers are not as dogmatic as Kryter on this point and

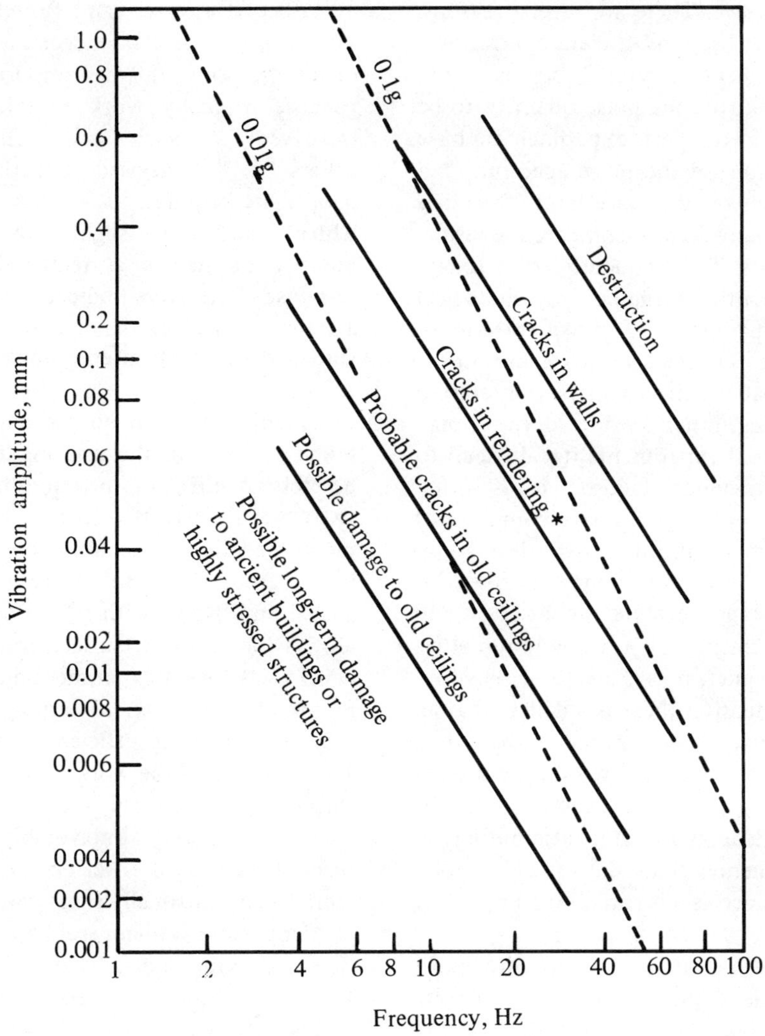

FIGURE 2.29 Response of buildings in good condition to vibration. (From Monk, R. G., Mechanically Induced Vibration in Buildings, Environ. Eng., No. 51, Soc. Environ. Eng. Lond., 1971. With permission.)
*Rendering is plaster put on stone or brick walls.

suggest although many experiments are inconclusive, there is sufficient evidence to conclude that noise does have an adverse effect on work performance. We may summarize some of the generally agreed findings[63,64] as follows:

a. Steady noise (without meaning) does not interfere with human performance unless the sound level exceeds 90 dB(A) and even then not consistently.[97]

b. Intermittent or impulsive noises are more disrupting to performance than steady noises.[67] Bursts of such noise can be disruptive even for levels below 90 dB(A).

c. High frequency noise (above 2,000 Hz) appears more disrupting than low frequency noise (below 2,000 Hz).[97]

d. Noise does not seem to change the rate of work but rather increases the variability of the work; there may be work pauses compensated by increases in work rate.[100]

e. Noise increases the number of errors made at work rather than the total amount of work.[97]

f. Complex tasks are more affected by noise than simple tasks.[101]

The reasons are not difficult to see why both laboratory and industrial experiments designed to test the effects of noise on performance are sometimes inconclusive. There are several reasons. In the past there has been insufficient control over the noise environment. Kryter[96] suggests that loudness level should be the noise quantity to be monitored. Often in the past, experimenters have given no information on the noise spectrum and sometimes not even on the noise level. Thus it is difficult or even impossible to compare the results of such experiments. Too frequently, even today, it is difficult to control extraneous psychological factors in the experiments. For example, motivation is an important fact in work output. If workers believe that the management is interested in their well-being, then work output may increase. In industrial experiments it is difficult to control such extraneous factors. It is often difficult to obtain control groups (for comparison) which will perform identically to the test group when both groups are in the absence of noise. For this reason some experimenters sometimes try to monitor the performance of just one group alternately in a noisy, quiet, noisy, then quiet environment. In such a situation there is a danger that an average performance is monitored rather than two different performances in two different environments. Despite all these experimental difficulties, experimenters are still attempting to study this phenomenon since the effect of noise on work performance is potentially so important and controversial.

Cohen[67] suggested that one of the most positive conclusions on the adverse effects of noise on task performance came from laboratory studies of *vigilance*. If vigilance decreased in the presence of intense noise, this had important consequences because in many work situations, people were involved in such tasks. Broadbent[98] and others[102,103] devised experiments in which people are required to watch several dials, any of which could show slight movements. Both the number of signals correctly detected and the response time were indicators of such performance in noise. Both of these indicators were shown to suffer in intense noise levels (see Figure 2.30).

Broadbent[97] also quoted an early noise study taken in 1935 in the weaving industry[104] which he felt could not be questioned because of extraneous effects (see Figure 2.31). He mentioned that data on the ten individual workers in the group showed that every person's performance improved with the use of earplugs. Motivation to work harder because the group was being used for an experiment did not seem to be a factor. In fact, when earplugs were used, an average increase in efficiency of 12% was found. The opinions of the workers on the benefits of the earplugs were also recorded. Even those workers who reacted very unfavorably to the earplugs recorded increased efficiency. These results would seem to be statistically significant.

Many more well-controlled experiments of sufficient duration must be made in industrial work situations before the effects of noise on work

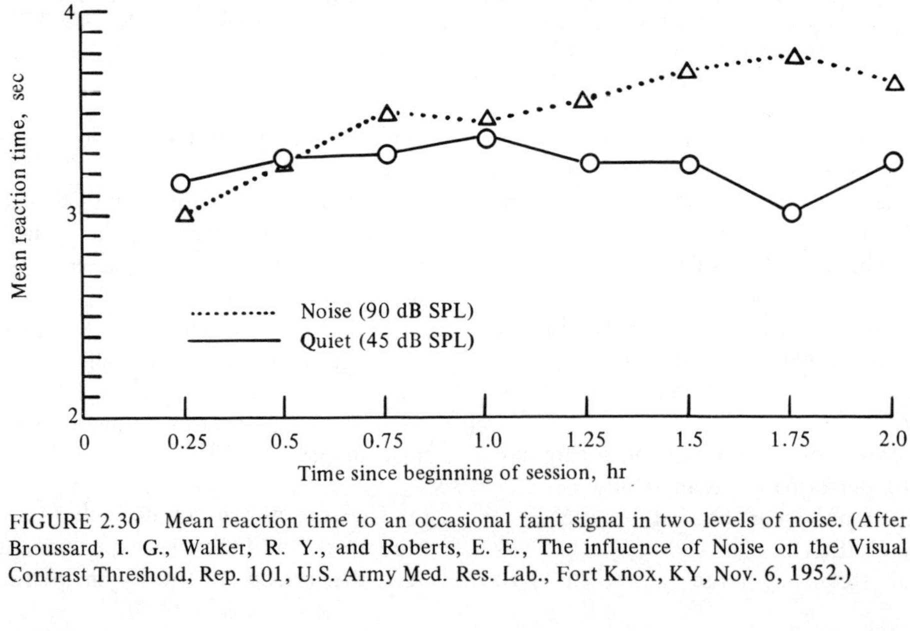

FIGURE 2.30 Mean reaction time to an occasional faint signal in two levels of noise. (After Broussard, I. G., Walker, R. Y., and Roberts, E. E., The influence of Noise on the Visual Contrast Threshold, Rep. 101, U.S. Army Med. Res. Lab., Fort Knox, KY, Nov. 6, 1952.)

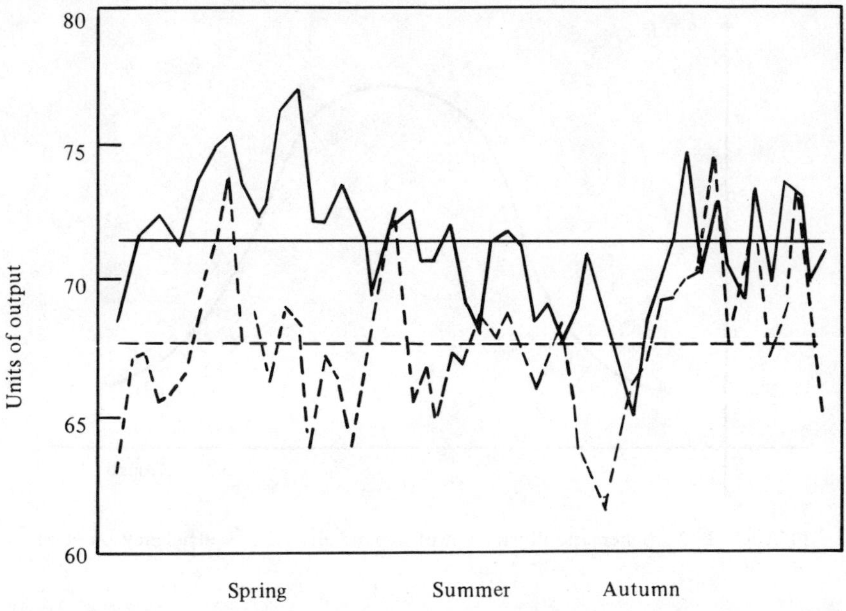

FIGURE 2.31 Output of two groups of weavers, both exposed to noise but one group, indicated by the solid line, wearing earplugs. (From Weston, H. C. and Adams, S., Rep. 70, Industrial Health Research Board, Her Majesty's Stationary Off., London, 1935. With permission.)

performance can be reliably established. Laboratory experiments are more controllable, however, they are normally of short duration and work motivation and other psychological factors are different. Perhaps as the U.S. Department of Labor Noise Regulations brings a lowering of noise levels in some industrial situations, statistical information may become available as to changes in efficiency, number of accidents, absenteeism and job turnover, etc. In some situations it may be possible to say that nonnoise environmental and psychological factors have remained constant while the noise has been reduced. However, in many cases it may still not be possible to eliminate other contaminating influences.

Kryter[96] discussed several theories which were proposed to explain changes in work performance caused by noise, although he personally did not subscribe to them. The most common theory advanced[63,64] is the *distraction–arousal* theory. It is well known that a state of *arousal* is needed to perform a task efficiently. Too much arousal causes a decrease in efficiency while too little arousal also causes a low efficiency. The right amount of arousal produces optimum efficiency as is shown schematically in Figure 2.32. Noise definitely increases a person's arousal, although intense noise (110 dB if it is expected) is needed before the arousal thus produced is equivalent to the arousal from the task itself.[96] Figure 2.32 suggests that if a person is sleepy, then noise could actually increase efficiency because the noise could increase arousal until nearer peak efficiency is reached. However, if a person is already sufficiently awake, and somewhat fatigued, and in a state of arousal caused by the task itself, and perhaps already at peak task efficiency, noise could increase the arousal sufficiently to bring a person past peak efficiency. This hypothesis is in agreement with some observed experimental results.[64]

Broadbent[99] has also recently discussed how noise can act as a *distracting* stimulus. He suggests that people have a limit to the number of features of their surroundings that they can observe in a certain time interval. All the stimuli which are received compete for attention, and the human sensory system receiving all these stimuli containing both useful and useless information has to filter out useless information such as noise. It is supposed that this human mental filter functions as follows:

a. It rejects or ignores signals which do not change with time, even though these signals may be important.

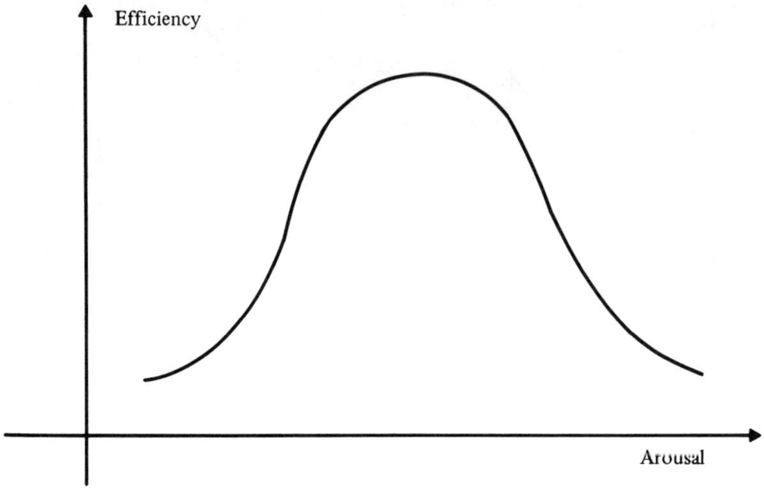

FIGURE 2.32 Schematic illustration of dependence of task efficiency on state of arousal.

b. A person's state of arousal, stress, or fatigue may alter the mental filter's ability to discriminate.

c. The filter can be overridden or overloaded by useless information which demands attention such as new, intense, or unexpected stimuli. The overloading of the mental filter can cause momentary lack of attention and other specific effects such as auditory masking and muscular activation.

It is hoped that sufficient research in the laboratory and actual work environments will be devoted in the future to determining the effects of noise on work performance and efficiency more conclusively. It is indisputable that noise has short-term effects, however, some argue that people are likely to adapt to a noise environment and long-term effects on work performance and efficiency may not exist. Until this hypothesis is conclusively proved, it would seem wise to assume that some long-term effects of noise on work performance and efficiency may exist.

2.6 PREVENTION OF HEARING LOSS IN INDUSTRY

2.6.1 Introduction

The preceding discussions in this chapter, particularly in Sections 2.4.5, 2.4.6, 2.5.3, and 2.5.4, suggest that noise-induced hearing loss is the only physiological effect of occupational noise on people which has been conclusively proved. Thus, several countries have recently produced regulations, codes of practice, or guidelines on permissible industrial noise exposure. It is, indeed, surprising that such regulations are so recent, since noise-induced hearing loss has been recognized for centuries. For example, Chadwick[105] cites references going back to the 12th Century. With the advent of the Industrial Revolution in Britain in the last century, noise-induced hearing loss problems inevitably became more severe. In 1831, Fosbroke[106] described the gradual onset of deafness in blacksmiths which insidiously increased and became most evident at the age of 40 or 50 when a ringing and noise in the ears was also often experienced. Later in 1886, Barr[107] measured the deafness among boilermakers, ironmolders, and mailmen. Walker and Martin describe Barr's measurements in a chapter on hearing conservation in a recent book.[108] Barr measured the distance in inches from a person's ear at which a ticking watch was just audible, and concluded that the boilermakers' hearing had suffered the most. At the age of 35 they could only hear the ticking watch at an average distance of 7 in., compared with 70 in. for normal nonnoise exposed people!

Now, of course, we have more scientific methods for monitoring hearing loss. These will be described in Section 2.6 together with the use of noise surveys and management and engineering controls to protect hearing. Also described in Section 2.6 are the use and performance of hearing protectors. Engineering controls to reduce industrial noise will be discussed in more detail in the chapter on industrial noise in Volume 2.

2.6.2 Industrial Noise Regulations

The U.S. was the first western country to introduce industrial noise regulations. In 1969, new standards for industrial noise became effective under a revision of the Walsh–Healey Act.[109] These standards only applied to firms which had federal contracts of $10,000 or more during the course of 1 year. These standards set the following permissible noise exposures in Table 2.7. These standards originally affected only 28 million out of the 40 million U.S. nongovernmental and nonagricultural work force. However, in 1971 these standards were extended to apply to the employees of all U.S. companies engaged in interstate commerce when the Occupational Safety and Health Act became law.* These noise standards were not strictly enforced to begin with. However, the Occupational Safety and Health Administration (OSHA) of the U.S. Department of Labor increased inspections in subsequent years which encouraged compliance from companies. In 1972 and 1973 OSHA did not classify violations under noise, air, or water, but total OSHA inspections in 1973 averaged approximately 1,695 per week, compared to 919 per week a year earlier. In 1973 and 1974, a standards advisory committee reporting to OSHA considered the proposal by the National Institute for Occupational Safety and Health (NIOSH) that the noise exposure of 90 dB(A) for an eight-hour day should be reduced to 85 dB(A) in the OSHA regulations. The committee did not recommend a reduction to 85 dB(A). However, it is likely that this lower level will eventually be adopted. The limit of 85 dB(A) for an 8-hour day was originally first proposed for use in the Walsh – Healey Act in 1969, but was then raised to 90 dB(A).

TABLE 2.7
Permissible Noise Exposures†

Duration per day, hours	Sound level dB(A) slow response
8	90
6	92
4	95
3	97
2	100
1–1/2	102
1	105
1/2	110
1/4 or less	115

†When the daily noise exposure is composed of two or more periods of noise exposure at different levels, their combined effect should be considered, rather than the individual effect of each. If the sum of the following fractions: $C_1/T_1 + C_2/T_2 \ldots C_n/T_n$ exceeds unity, then, the mixed exposure should be considered to exceed the limit value. C_n indicates the total time of exposure at a specified noise level, and T_n indicates the total time of exposure permitted at that level.

Exposure to impulsive or impact noise should not exceed 140dB(C) peak sound pressure level, fast response.

It should be noted that in other countries, industrial noise regulations or codes of practice similar to those enforced by OSHA now exist. However, in most cases these regulations are based on the equal energy principle (3 dB increase allowed for halving of exposure time) rather than the 5 dB increase for halving of exposure time allowed by OSHA. Both the British Code of Practice[110] published in 1972, and the ISO Recommendation R 1999[111] published in 1971, for instance, were based on the equal energy principle. In 1974, the British Code of Practice was still used only in an advisory nature although the H.M. Factory Inspectorate used the code to decide whether working conditions did or did not pose a hearing hazard. The ISO recommendation does not set levels which must not be exceeded. Instead, it gives a relationship between occupational noise exposure and risk of hearing impairment. The British Code, like the OSHA regulations, starts with an allowed level of 90 dB(A) for an 8-hr workday. A nomogram is also provided for the calculation of the equivalent continuous sound level L_{eq}. Exposures at different levels for a person are totaled for a workday by the equal energy principle.

In a 1974 study commissioned by OSHA, the estimated cost to selected U.S. industries most directly affected, of complying with the OSHA noise regulations (see Table 2.7) was $13.5 billion. If the noise exposure for an eight-hour workday were reduced from 90 to 85 dB(A), the compliance cost would increase to $31.6 billion.

Perhaps it would be appropriate at this stage to discuss how a plant can comply with the OSHA noise regulations. Bulletin 334, "Guidelines To The Department of Labor's Occupational Noise

*The Occupational Safety and Health Act of 1970, S.2193 was reported out of a Senate-House Conference Committee on December 14, 1970, approved by the Senate on December 16 and by the House on December 17 and signed by the President on December 29. The Act took effect on April 28, 1971.

Standards" (1971) is very useful to consult. These guidelines suggest the following approach:

A. OSHA noise surveys – "A noise survey of each area in the plant in which sound levels exceed 90 dB(A) shall be made at least once each year to ensure that sound levels have not increased above those originally existing. The survey may also establish that noise levels in some areas have been reduced to levels below 90 dB(A) and thereby justify discontinuing application of requirements for administrative controls, ear protection, and audiometric tests of individuals in such areas. A noise survey of an area is recommended whenever a change is made in either equipment or type of operations so that significant changes in noise level will be acted upon immediately."

Whenever a noise survey shows noise levels in excess of those listed inTable 2.7, the Bulletin 334 Guidelines state that "necessary steps to reduce the noise exposure to or below those levels shall be ascertained and a detailed plan, with completion dates for individual steps, shall be prepared. Following the original survey which shows existence of overexposures, the steps in a typical compliance plan might include the following, not necessarily in this order and some usually going on simultaneously."

B. OSHA compliance plan –

1. A detailed survey of sound levels and sound spectra to determine the sources of excessive sound levels.
2. Initiation of engineering studies to determine methods for reducing sound levels at their sources.
3. Planning and initiation of feasible administrative controls, such as modifying production schedules to divide noisy jobs among enough people to bring each below the permissible limit or spreading part-time noisy operations.
4. Initial audiograms for personnel excessively exposed.
5. Installation of a personal protective equipment program.
6. Followup audiograms at appropriate intervals to assess effectiveness of the personal protective equipment program and administrative controls.
7. Installation of engineering controls, or process changes to reduce noises at their sources.
8. Repeated noise surveys to measure effectiveness of the engineering changes."

The OSHA guidelines state that the orderly completion, on schedule, of the various phases of the compliance plan, together with other components of the hearing conservation program, will be considered as compliance with the regulations.

2.6.3 Audiometry and Monitoring of Hearing

If employees are working in areas with intense noise levels, then it is very desirable to monitor their hearing regularly. These hearing tests will determine whether or not the administrative and engineering controls and ear protection devices are being adhered to or effectively used. The 90 dB(A) eight-hour workday limit only protects a certain percentage of workers (about 93%). Since people's ears are different, some workers (about 77%) will still be suffering appreciable hearing damage in these conditions. For these reasons monitoring of hearing is very desirable; with monitoring, it may be possible to identify those persons with "weak" ears who may still be susceptible to hearing damage at the levels set by OSHA.

In fact, the U.S. Department of Labor's Occupational Noise Standards (Bulletin 334) make it mandatory to conduct audiometric tests periodically on all individuals who are working either regularly or infrequently in places where noise levels exceed 90 dB(A). It is suggested that the tests should be conducted as frequently as suggested by the plant's regular or consulting physician, but only in exceptional circumstances less than once a year. Several books have recently described audiometry and monitoring of hearing[108,112–115] and these subjects will be described briefly in the remainder of Section 2.6.

A. Types of Audiometers

Any audiometers used in a hearing monitoring program should conform with the current U.S. Standard ANSI S.3.6 – 1969.*

There are three main classes of audiometers: (a) *Monitoring audiometers* which are used to determine whether people have hearing loss; (b) *Diagnostic audiometers* used by professionals to diagnose causes of hearing loss and possible corrective action; and (c) *Research audiometers,* used to conduct basic research into speech and hearing defects.

We shall only concentrate on the first type here – monitoring audiometers. These again can be subdivided into two types, *manual,* and *automatic.*

*The current U.S. Standard ANSI S3.6–1969 "Standard Reference Threshold Hearing Levels for Audiometers for General Diagnostic Purposes" is issued by the American National Standard Institute and is based on the 1964 Recommendation of the International Organization for Standardization (ISO/R389–1964).

Manual types are cheaper and simpler but require more skilled operators or technicians. Automatic types may cost three times as much, but need less skilled operators. Although the accuracy and results are similar with manual and automatic audiometers, it is easier to detect malingerers with the manual type. However, results from the automatic type may be less open to criticism on grounds of interpretive judgment or errors.

B. Manual Audiometers

A block diagram of a typical manual audiometer is shown in Figure 2.33. Comprehensive discussions on the methods of conducting such hearing tests and obtaining audiograms are given by other writers and will only be briefly discussed here. It is of considerable importance that the operator, nurse, or technician not give the subject any cues during the examination. For that reason it is best that the subjects sit with their faces at right angles to the operator so that they cannot observe the operator's actions or see the audiometer, but at the same time, their facial expressions can be seen by the operator. The subjects' facial expressions can often reveal whether they have heard the sound signal. Alternatively, the subjects should sit with their backs to the operator. It is important that instructions given to the subject be concise, unambiguous and consistent. Glorig[116] suggests a procedure similar to the following: The subjects should be told that they will hear both soft and loud sounds, also high and low tones, and that first one ear will be tested and then the other. The subjects should be told to raise their fingers (or press a button) each time they hear a sound, no matter how faint and to keep their fingers raised (or buttons pressed) as long as they hear the sound. The earphones should then be placed over the subjects' heads, a snug fit ensured, and the audiometric booth door closed. A few test tones should be tried at different levels and frequencies to ascertain that the subjects have understood the instructions and that everything is in order. Then the test is ready to begin.

There are several ways of conducting manual audiometric tests. It is normally recommended that the test starts with a 1,000 Hz signal at 40 dB which is decreased in level successively in 10-dB steps by the operator until the subjects indicate they can no longer hear the tone. Normally the tone is played for several 1-sec durations with breaks. At any one level, at least 50% of the 1-sec bursts should produce a response from the subject, otherwise this level should be assumed below the hearing threshold. For instance, if four 1-sec bursts are produced at a certain level, either 2, 3, or 4 responses would indicate audibility. When a level is reached that elicits no response whatsoever, this level is assumed below the hearing threshold. The minimum audible level is then selected again. The level is now decreased in 5-dB steps until a level of inaudibility is reached again. The levels are then increased in 5-dB steps again. The threshold levels which produced at least two out of four responses (i.e., 50%) are noted. If they are different they may be averaged in the ascending and descending tests. The test is then repeated at 2,000, 4,000, 6,000, 8,000, 250, 500 and finally at 1,000 Hz again. In the repeat at 1,000 Hz, a check can be made to see if the subjects' thresholds are repeatable or if they have "learned" anything during the test.

Figure 2.34 shows a photograph of a manual audiometer. It is noted that the attenuator (Control 5) is marked HTL, dB (hearing threshold level). The pure tones in each earphone can be produced at levels from −10 dB less to 90 dB above what has now been standarized as the "normal" threshold of hearing at each test frequency (ISO 1964 and ANSI 1969) (see Figure 2.16 and Section 2.4.5).

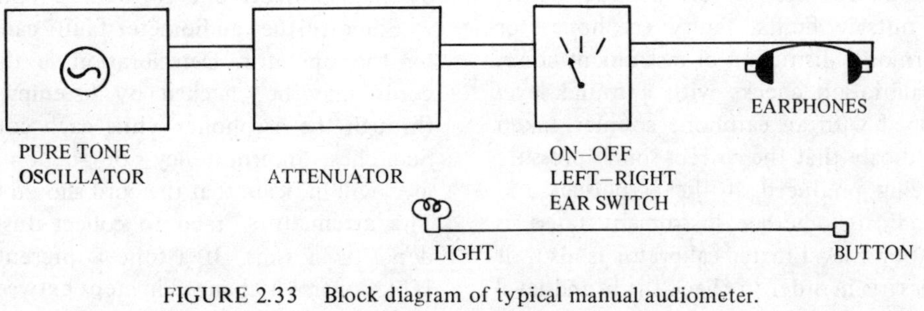

FIGURE 2.33 Block diagram of typical manual audiometer.

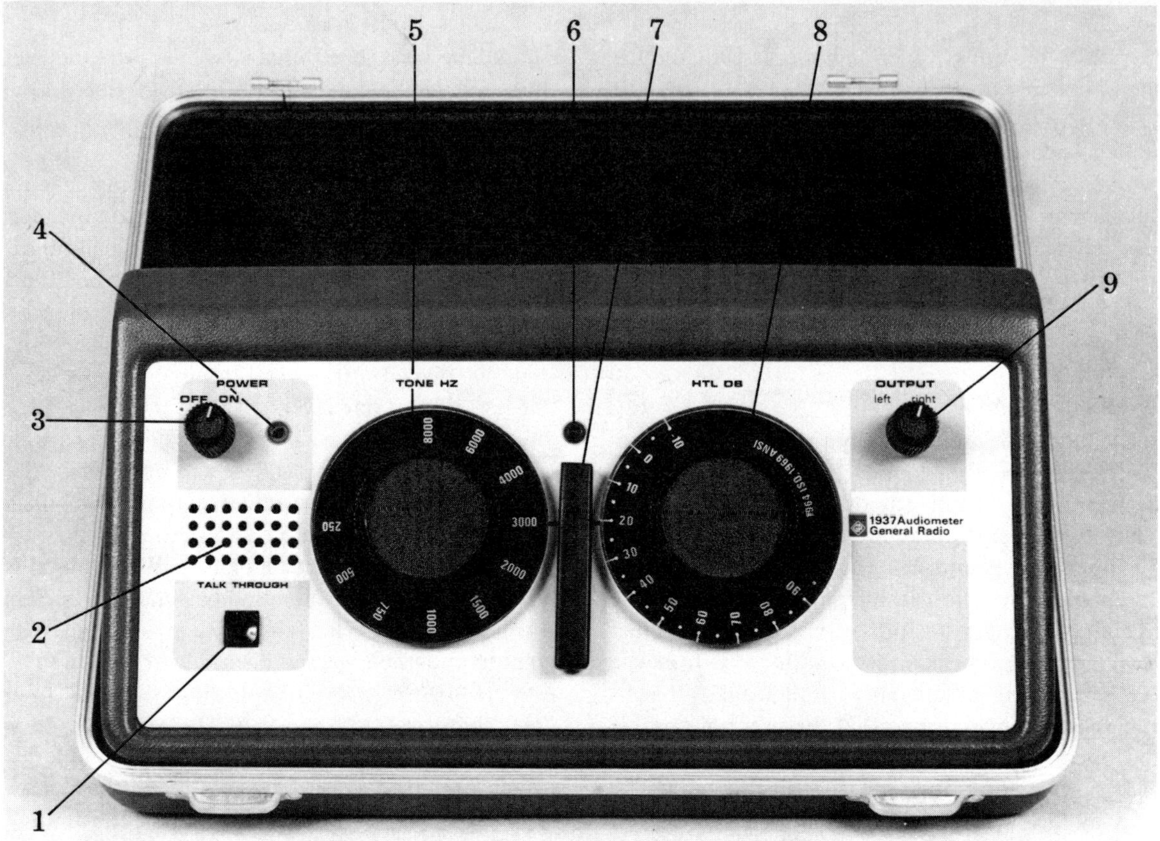

FIGURE 2.34 A typical manual audiometer with explanation of controls. 1. Intercom. control, 2. Intercom. microphone, 3. Power on-off switch, 4. Power on pilot lamp, 5. Ten position pure tone frequency selector switch, 5. Indicator light (glows when bar [7] is depressed), 7. Bar switch, presents test tone to subject when depressed, 8. HTL control, determines level of test tone presented to subject, 9. Left-right earphone selector switch. (With permission of General Radio Co., Concord, Mass.)

There are several practical things to note about the use of the audiometer. First, good sealing between the earphones and the head is most important. All obstructions such as hair, frames of eyeglasses, and earrings should be removed. The headband should be adjusted so that it exerts equal pressure firmly on each earphone to maintain the seal. Second, the calibration of the audiometer should be checked daily. In 1963, a study[117] showed that 50% of audiometers in use were inaccurate. Errors were caused by sound pressures produced with levels outside limits, frequencies outside limits, faulty earphones, or excessive harmonic distortion or instrument noise.

Simple calibration checks with a sound level meter equipped with an earphone coupler, taken daily, will indicate that the correct sound pressure levels are being produced at the earphones and may also indicate whether instrument noise is present. A sound level meter calibrator is also, of course, necessary in order to check the sound level meter itself. These instruments are discussed in more detail in Chapter 3.

Variations of ±5 dB may be allowed in the audiometer output, since a person's hearing threshold level may change by this much from day to day. However, variations in audiometer output greater than ±5 dB may indicate more serious problems in the audiometer performance and it should be returned to the manufacturer for calibration. In any case, it is advisable to have the audiometer calibration checked comprehensively by the manufacturer every 6 to 12 months.

Some further audiometer faults can be checked by the operator. Deterioration in the earphone cords may be checked by listening to a tone through the earphones while each cord is flexed. Scratches, intermittency, or sudden changes in level will indicate that the cord should be replaced. The attenuators tend to collect dust and other deposits in time. If a tone is presented and the HTL control is changed in steps between 0 and 60,

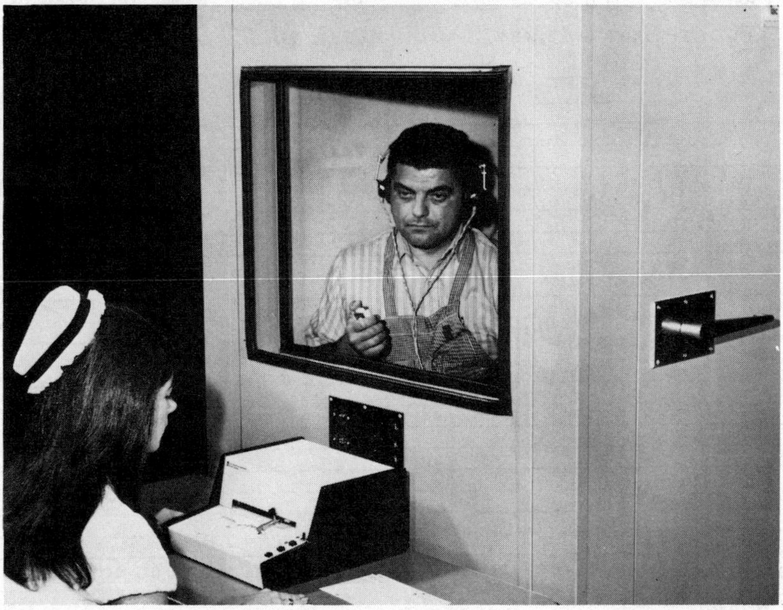

FIGURE 2.35 Typical automatic recording audiometer. (With permission of General Radio Co., Concord, Mass.)

then if scratching noises are present, these will again suggest trouble. Hum and random noise may also be observed as the HTL control is changed from 0 to 60. This would suggest maintainence is required. Some random noise above 60 HTL is allowable. Finally, frequency distortion can be detected by listening to the audiometer output through the earphones with the HTL set at 70 or 80. A harsh or rasping quality to the sound will indicate distortion.

C. Automatic Audiometers

Von Bekesy[118] is often credited with first designing an automatic audiometer. With this type of audiometer, the pure tone frequency supplied to the earphones is gradually increased (often from 100 to 10,000 Hz). The results are plotted automatically on an X-Y recorder with the deflection of the pen in the X-direction made proportional to the frequency. The HTL is plotted in the Y-direction. The signal is continually increased in level until it becomes audible to the subject who then pushes a button. While the sound remains audible and the subject is pushing the button, the level is continually decreased. When the sound becomes inaudible and the subject releases the button the signal level is increased again and so on until the maximum frequency is reached. In this way the subject's HTL against frequency is automatically plotted out.

Recently automatic audiometers which work on a slightly modified principle appear to be more popular. The procedure is similar except that the frequency is not varied continuously, but the test is conducted at seven sequentially presented pure tone frequencies. The audiometer shown in Figure 2.35 works on this principle. The signal is presented sequentially at 500, 1,000, 2,000, 3,000, 4,000, 6,000, and 8,000 Hz first to the left and then to the right ear followed by a right ear retest at 1,000 Hz (see Figure 2.36).

Each signal is presented for 30 sec and the intensity is increased initially at +10 dB/sec in the HTL range from −10 to +90 until the first subject response. While the subject holds the button depressed, the intensity is changed by −5 dB/sec. The intensity is then varied at + or −5 dB/sec between each response until the 4th response and then at 2.5 dB/sec* until the end of the 30-sec period when the frequency is changed.

This procedure allows the hearing threshold level to be approached rapidly at each frequency and then to be defined accurately. A typical audiogram is shown in Figure 2.36 which indicates a

*However, if the subject fails to respond in 5 sec, the intensity rate is returned to 5 dB/sec.

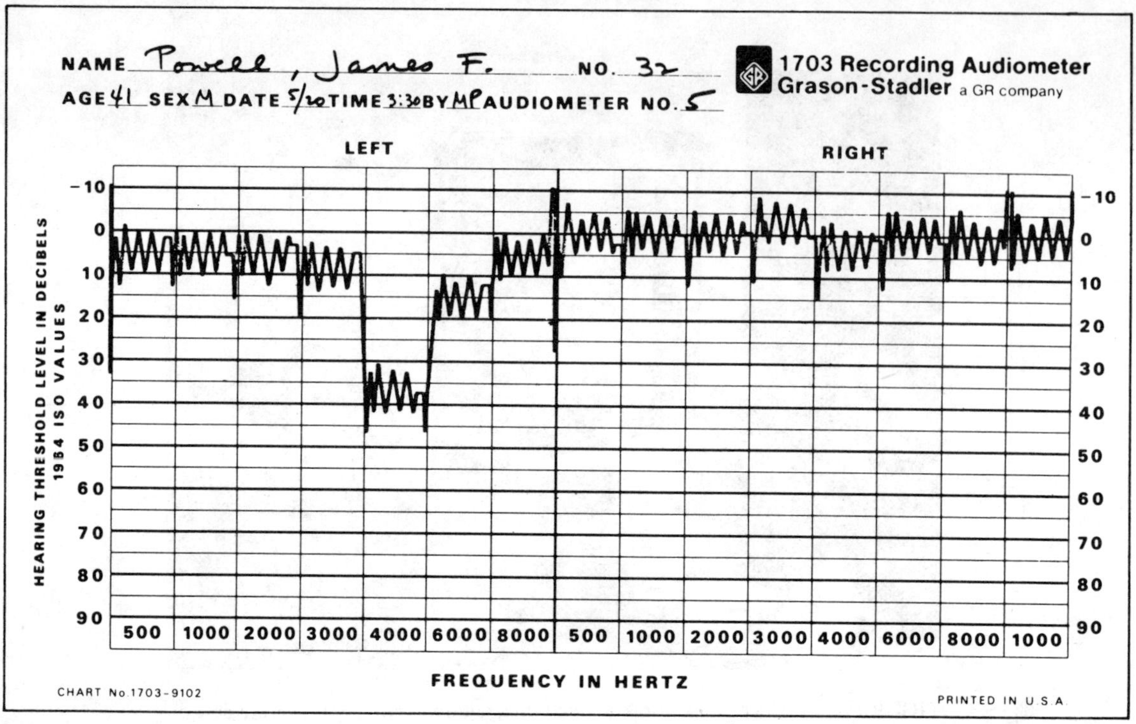

FIGURE 2.36 Typical audiogram. Note hearing loss in left ear. (With permission of General Radio Co., Concord, Mass.)

considerable hearing loss in one ear. Note that hearing loss is often characterized by a notch (loss) at 4,000 Hz. This seems to be true whether it is caused by acoustic trauma (e.g., explosions or gunfire) or by chronic exposure to less intense noise in industrial environments. The reason is presumably because the ear is most sensitive in this frequency range (see Section 2.3.1).

D. Audiometric Rooms

An audiometric testing program must often be carried out in an industrial plant or a hospital where the background noise levels are high and may interfere with the results. In high background levels, incorrect hearing threshold levels (HTL) may be obtained because the audiometer signals are masked by the background noise. In such instances, the values of HTL will be artificially raised by the background noise and will be almost useless.

Unfortunately, it is not always possible to choose the most quiet location for audiometric testing because often other factors such as proximity to medical or personnel departments in a plant may be given prime consideration. For these reasons it is usually necessary to build a special audiometric room or booth in order to guarantee low enough background noise levels which will not interfere with the audiometric test program. Such rooms or booths may be built in-house, provided plant personnel have sufficient acoustical knowledge. However, if this knowledge is not available, it may be more economical (unless several rooms or booths are needed) to purchase a prefabricated room from one of several suppliers. In this way, development costs may be avoided and the room or booth will probably be sufficiently portable to make relocation easy. Figure 2.37 shows a typical audiometric testing booth.

Whether the audiometric room or booth is made in-house or purchased, the background noise levels in the room should not exceed those recommended by ANSI S3.1–1960 (R1971).* These background noise levels are given in line two

*This standard issued by the American National Standards Institute (ANSI) is based on the American Standards Association (ASA) Standard ASA S3.1–1960. It was revised in 1971, and reissued by ANSI. The 1960 standard itself was based on the earlier ASA Standard ASA Z24.5–1951, Standard Reference Threshold Hearing Levels for Audiometers for General Diagnostic Purposes. Note that the American Standards Association changed to become the American National Standards Institute in 1969.

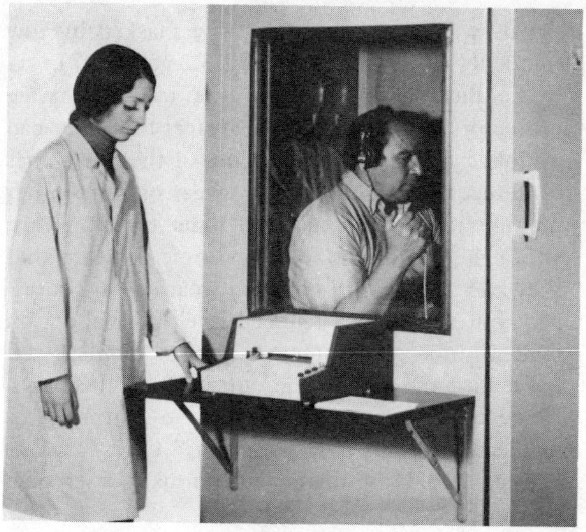

FIGURE 2.37 Small audiometric testing booth. (With permission of Industrial Acoustics Inc., Bronx, NY.)

of Table 2.8. Above 1,000 Hz the permissible octave band levels increase. Below 1,000 Hz they are constant at 40 dB. Line three gives the noise reduction of the room[119] shown in Figure 2.37. However, because such measurements are made in ideal laboratory conditions where there are no reflecting walls or nonuniform sound fields, a safety factor of 10 dB is given in line four.[119] This safety factor also allows for the possibility of background noise levels increasing gradually with time and unexpected interruptions such as doors slamming, loud voices, etc. Line five of Table 2.8 shows the maximum outside octave band noise levels which can be allowed with this room.

Hirschorn and Singer suggest that in most cases the 500 Hz octave band level is critical. This, presumably, is because of the shape of most background noise spectra. If background noise levels in excess of those given in row five exist, then an audiometric room or booth providing a greater noise reduction will be needed. Hirschorn and Singer[119] describe such rooms. By using thicker walls and other details, about 10 dB more noise reduction throughout the frequency range 125 to 8,000 Hz can be achieved. By using a double wall construction consisting of 4-in. walls with a 4-in. air space, a double door, and a quiet ventilation system, 30 dB more noise reduction than in row three of Table 2.8 throughout the same frequency range can be obtained. Such a room should be usable even in the most adverse industrial noise environments.

2.6.4 Personal Hearing Protection

In some countries (e.g., Japan) it is considered that the most effective method of controlling hearing damage caused by intense industrial noise is the issuance and strict enforcement of the use of ear protective devices. In Japan, community noise is considered to be a more serious problem.

Reliance on ear protective devices is economically attractive. As stated earlier, in the U.S. it is estimated that the cost of complying with present OSHA noise regulations is \$13.5 billion. If the allowable noise exposure for an 8-hr workday were reduced from 90 to 85 dB(A), the compliance cost would increase to \$31.6 billion!

In the U.S., Britain, and several other countries, the use of ear protectors is considered to be an *interim* measure and must only be used until engineering methods successfully bring noise levels below specified limits. Both the U.S. Guide-

TABLE 2.8

Allowable Inside and Outside Permissible Noise Levels for a Typical Audiometric Room

Octave band center frequencies, Hz	125	250	500	1,000	2,000	4,000	6,000	8,000
Permissible octave band levels ANSI S3.1 – 1960 (R1971)	40	40	40	40	47	57	62	67
Room noise reduction, dB	18	32	38	44	51	52	52	50
Safety factor, dB	–10	–10	–10	–10	–10	–10	–10	–10
Outside allowed noise levels, dB	48	62	68	74	88	99	104	107

lines[120] and the British Code of Practice[110] embrace this philosophy. The U.S. Guidelines state that . . .

"The use of personal protective equipment is considered by the Department (of Labor) to be an interim measure while engineering and administrative controls are being perfected. There will be very few cases in which the use of this equipment will be acceptable as a permanent solution to noise problems."

The British Code of Practice states that . . .

"Ear protectors should not be used as a substitute for effective noise control. They should normally be regarded as an interim measure while control of noise exposure by other means is being perfected."

A. Disadvantages of Wearing Ear Protectors

The reasons why personal ear protection is not a completely satisfactory solution to the industrial noise problem are not hard to find. Coles[121] has recently discussed these. Perhaps the overriding reason is *discomfort.* Most types of ear protectors are uncomfortable to wear. They may cause the ears to get hot and sweaty or the tightness of some types of ear protectors can cause pain or headaches.

Another reason is *communication* difficulties. Many people claim that ear protectors impair communication which to a certain extent is true. However, in continuous intense noise conditions, hearing protectors actually do not reduce intelligibility of speech for people with normal hearing because noise and speech signals are both reduced equally and the speech signal-to-noise ratio remains the same. Thus, in continuous noise, the Articulation Index should remain constant (see Section 5.8.2) and speech intelligibility should be unimpaired. Of course, in intermittent or impulsive noise situations, conversation can normally be carried on in the "quiet" intervals and ear protectors will interfere with communication in such instances.

Another communication-related disadvantage of ear protectors is the difficulty in hearing *warning signals.* In continuous noise, hearing protectors may not offer a disadvantage in this regard except for people with high frequency hearing loss. Such people, who are often found working in noisy environments, may well have increased difficulty in hearing warning signals and even normal conversation, because ear protectors attenuate high frequencies more and high frequency signals will then be more easily masked by low frequency noise (see Sections 2.3.2 and 5.8.7).

Another disadvantage of the use of hearing protectors is that workers may feel that they can no longer judge the *performance* of their machines or detect the noise caused by wear or some other malfunction. There may be some truth in this, although it is important for people to learn that they can still hear noise when wearing ear protectors, but simply at a reduced level. A final reason is that some ear protectors such as earplugs can exacerbate diseases of the outer ear such as otitis. However, in these cases the use of earmuffs or medications such as Cortisporin® Otic drops or Corticosteroid ointment will normally overcome this problem.

B. Educational Programs

As already discussed, there are some legitimate reasons for personnel refusing to use hearing protection; however, there are also some reasons which are incorrect. Many of these objections are based on ignorance or misinformation concerning occupational hearing loss. Some men feel that it is "unmanly" to wear ear protectors. Others think their ears have become "hardened" to noise. It is extremely helpful to use an educational program to teach management at all levels and all shop personnel in an industrial plant the facts of occupational hearing loss and ear protection devices.[122] With the commitment and example of management, after such a program the majority of shop personnel will cooperate with a hearing conservation program. It is useful for the plant manager, an otologist or medical director, and the plant nurse to take part in the program. Films on hearing loss can be shown and educational literature distributed. After consultation with the medical director or nurse, shop personnel should be allowed to choose the type of ear protectors they prefer, wherever possible, and to change types after a few weeks if they desire. Personnel may need encouragement to continue the wearing of ear protectors if they initially find them uncomfortable. Often if the protectors are worn for short periods to begin with, after a few days or weeks, eight-hr periods can be tolerated with minimal discomfort.

C. Types of Ear Protectors

Each type of ear protector has advantages and disadvantages.[108,110,121]

TABLE 2.9

Noise Protection of Dry Cotton

Frequency, Hz	Noise protection, dB at frequencies shown 250	500	1,000	2,000	4,000	8,000
Cotton: old type: coarse fibers	3	3	4	7	10	19
Cotton: new type: finer fibers	5	9	10	15	16	25

From Coles, R. R. A., in *Noise as a Public Health Hazard,* ASHA Rep. No. 4, American Speech and Hearing Assoc., Washington, D.C., 1969. With permission.

Cotton – Dry cotton placed in the ears has little value. The noise protection gained is only a few dB at low frequency (see Table 2.9) and only 10 or 15 dB at high frequency. Although finer fibers seem to be becoming available for medical use, dry cotton as an ear protector is still not recommended. The noise protection is minimal and it becomes loose with jaw movement after a short time, thus providing even less protection. Moreover, it provides a false sense of security.

If the cotton is mixed with petroleum jelly or wax, the noise protection increases (Figure 2.38). However, petroleum jelly is messy, and waxed cotton, although better, still becomes loose with jaw movement and must be periodically removed and reinserted. This process is hygienically undesirable with dirty hands. However, this type of earplug is disposable.

Glass-wool – In the last few years, very fine glass fibers (of diameter of about 1 micron) have been developed. These fibers seem to cause no irritation to the ear canal. The material is normally supplied in pieces which must be folded according to manufacturers' instructions. The material is more resilient than cotton and maintains a better fit. The noise protection is quite acceptable (Figure 2.38) and disposable plugs formed from this material provide one of the most acceptable forms of ear protection.

Permanent earplugs – Unlike cotton or glass-wool which are disposable, this type of plug is reusable provided it is kept clean. Such plugs are

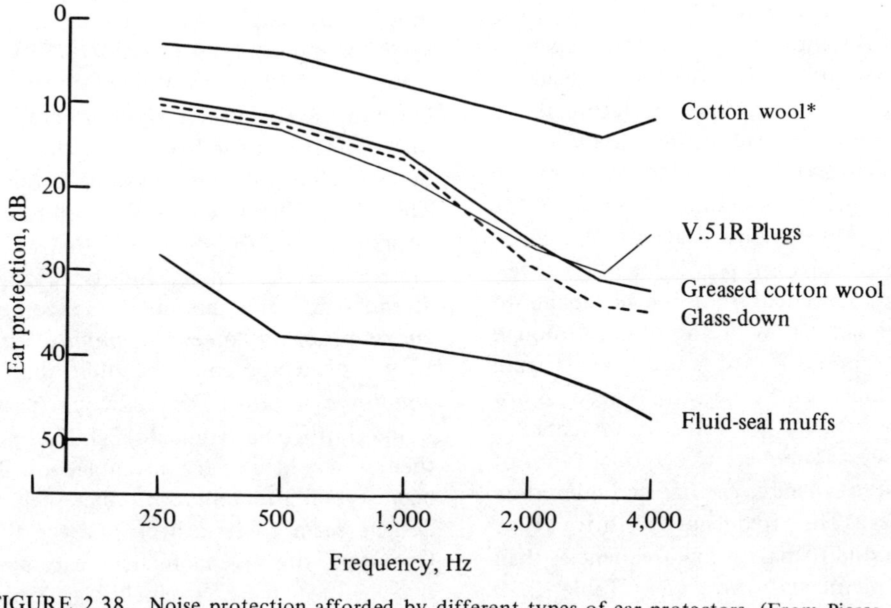

FIGURE 2.38 Noise protection afforded by different types of ear protectors. (From Piesse, R. A., Ear Protectors, ASHA Rep. No. 4, American Speech and Hearing Assoc., Washington, D.C., 1969. With permission.)
*Cotton wool is dry cotton.

made of plastic or rubber and are designed to fit tightly into the ear canal. The plugs are manufactured either in three or five different sizes and also some manufacturers make a "universal fit" type. The "universal fit" type is undesirable because it is difficult to obtain a good fit for many people. All such plugs should initially be fitted by trained personnel. Some people require a different size plug in each ear. Because of the close fit required, these plugs tend to be uncomfortable, but they do not cause the sweating which earmuffs do. The noise protection is quite good and similar to that provided by glass-wool plugs (Figure 2.38). Unfortunately they do tend to harden with repeated washing. It is also difficult to tell if they are being worn unless a marker is attached to them.

Semi-insert types – These are normally made from two soft rubber caps attached to a headband which holds them in the ear canal under some pressure. They have the advantage that one size will fit most people, they cannot be lost easily (which may be important in the food industry), and they can be slipped off and placed with the band around the neck when not needed. However, many people find the pressure applied by the headband on the ear caps unpleasant or intolerable.

Earmuffs – These consist of plastic shells which fit over the entire ear and are held in place with a headband (Figure 2.39). The inside of the shell is lined with acoustically absorbent material and a seal is provided around the edge of each shell. The best type of seal is the fluid seal first developed by Shaw and Thiessen.[124] The fluid seal usually is filled with glycerine, and with moderate pressure from the headband the head contours are followed accurately which prevents much sound leakage.* Unfortunately, such seals can be damaged and the fluid will leak. More robust seals are also made which are filled with a soft plastic foam. However, more pressure is needed from the headband to maintain good contact with the head. Although the noise protection afforded is very good (Figure 2.38), the earmuff causes the ears to become hot and sweaty, particularly in warm environments.

Special types – There are two special types of ear protectors: *frequency sensitive* and *amplitude sensitive* types. The frequency sensitive type allows more sound to pass at low frequencies than conventional *permanent ear plugs.* Table 2.10 shows a comparison between a frequency sensitive

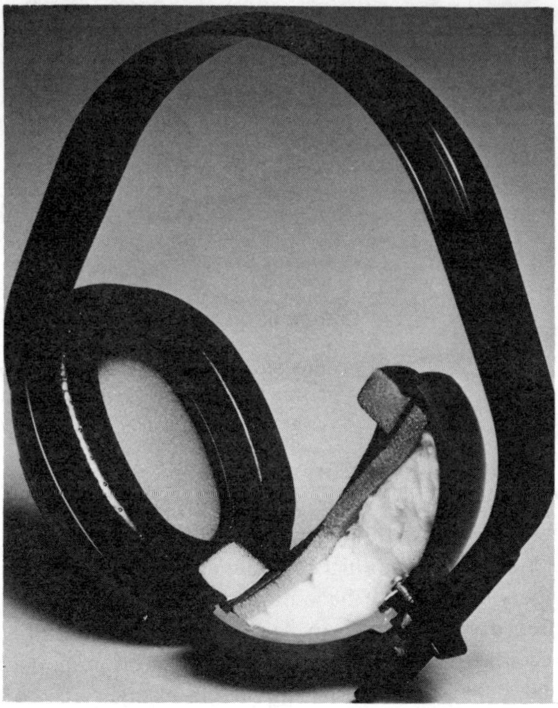

FIGURE 2.39 Modern type of earmuff. (With permission of Bilsom International, Inc., Reston, Virginia).

type (Selectone-K) and a conventional type (V.51R). In use in impulsive noise environments, the Selectone-K plug has an advantage because it allows the passage of more low frequency speech components. In continuous noise environments, the Selectone-K plugs would normally be worse because the low frequency noise components pass more easily and tend to mask the high frequency speech components (see Section 2.3.2). In environments with high levels of low frequency noise, the Selectone-K should not be used because it affords low protection. The Selectone-K has several small holes to allow the passage of low frequency sound. These holes tend to have a nonlinear resistance because of the turbulence created at high sound pressure levels. Thus, this type of plug may attenuate higher amplitude sounds somewhat more and is probably somewhat *amplitude sensitive.* If an ear protector could be built that were truly *amplitude sensitive* (i.e., it would pass low level sounds but would attenuate high level noise more), then it would be very advantageous. There have been several attempts to produce such plugs. Most designs seem to have been unsuccessful to date. One type, the "Gundefender" earplug,[110] is designed to attenuate extremely high explosive levels by the nonlinear resistance effect in small holes,

*A disadvantage of earmuffs is that sound leakage past the seal can be increased by frames of eyeglasses, earrings and long hair.

TABLE 2.10

Noise Protection of Frequency Sensitive Plug Compared with a Standard Plug

	Noise protection, dB at frequencies shown					
Frequency, Hz	500	1,000	2,000	3,000	4,000	6,000
Frequency sensitive plug (Selectone-K)	6	10	21	31	31	34
Conventional plug (V.51R)	13	17	26	28	27	25

From Coles, R. R. A., in *Noise as a Public Health Hazard,* ASHA Rep. No. 4, American Speech and Hearing Assoc., Washington, D.C., 1969. With permission.

already discussed. However, although it may be effective in gunfire, it appears that the attenuation of impulsive noise is insufficient at levels such as 110 dB, e.g., from punch presses in industrial situations. Such plugs are not recommended for industrial use.[110] Some earmuffs have been built which incorporate an electronic system to attenuate high levels of noise but not low levels. These earmuffs are successful but expensive; however, they are very useful in impulsive noise environments since communication in the "quiet" periods becomes easy.

Helmets – All the ear defenders discussed so far have an upper limit of noise protection. They produce an *air-conduction* hearing loss by blocking the air path. Thus, assuming good sealing, the use of earplugs *together* with earmuffs normally does not produce much additional noise protection (at the most 6 dB, see Reference 108). However, if a helmet which covers the whole head and seals well to the face is worn, up to an additional 10 dB noise protection can be gained if the helmet is worn in conjunction with earmuffs or earplugs. This is because the helmet will attenuate the bone conduction path through the skull directly to the cochlea avoiding the ossicles (which normally becomes dominant when only earplugs or muffs are used on their own). [If the bone conduction path is also attenuated by a helmet, then an upper limit of noise protection has probably been reached because the dominant path for noise becomes conduction through the body directly to the cochlea[121] (see Section 2.2).]

Built-in communication systems – Earmuffs have been produced with built-in communication systems such as telephone receivers in the earmuff and noise-canceling voice microphones.[121] These systems are quite successful but rather expensive.

D. Tests for Ear Protectors

In the U.S., a measurement standard is available* for determining the *noise protection* produced by ear protectors. Noise protection data should be demanded from a manufacturer when purchasing hearing protectors. Standard ASA Z.22.4–1957 requires a comparison in a low background noise room between the hearing threshold with and without the ear protectors. The difference in the threshold may be termed the *noise protection* of the ear protectors. The experiment is carried out at several frequencies with pure tones and each person is normally asked to remove and replace the ear protectors twice so that three results are obtained at each frequency. Normally ten people are tested. Both the mean *noise protection* at each frequency and the *standard deviation*** of the measurements are normally reported (see Table 2.11).

Although this standard method is useful for determining the noise protection of ear protectors and is used in a number of other countries, there are several objections to and difficulties in its use:[110] (1) The noise protection is determined at very low levels and the performance at high levels may be slightly different (the test will, of course, be completely unsuitable for amplitude-sensitive

*ASA Z.22.4 – 1957, Measurement of Real Ear Attenuation of Ear Protectors at Threshold.

**The standard deviation gives a statistical measure of the scatter of the results. The noise protection of the ear protectors may be expected to be less than the mean value for 50% of people, worse than the mean value minus the standard deviation for 16% of people and worse than the mean minus twice the standard deviation for 3% of people.

TABLE 2.11

Calculation on Reduction of Sound Level of Factory Noise by Typical Fluid Seal Ear Muffs

Frequency, Hz	125	250	500	1,000	2,000	4,000	8,000
Octave band factory noise levels, dB	102	105	105	107	102	95	88
A-weighting correction	–18	–9	–3	0	+1.5	+0.5	–2.0
A-weighted octave band levels, dB	84	96	102	107	103.5	95.5	86
Mean noise protection for typical fluid seal ear muff, dB	13	20	33	35	38	47	41
Standard deviation, dB	6	6	6	6	7	8	8
Assumed noise protection, dB	7	14	27	29	31	39	33
Assumed octave band levels at ear, dB	77	82	75	78	72.5	56.5	53

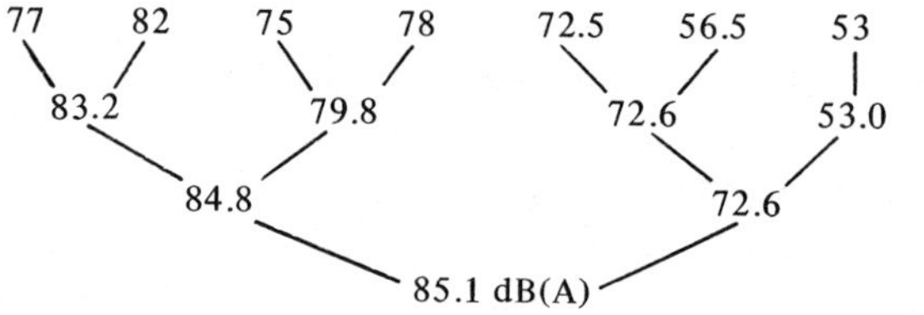

ear protectors). (2) A low background noise room may not be available in a plant so field checks may be difficult. (3) The standard deviation results partly from variations in fit of the ear protectors, but also partly because of the experimental difficulty in determining a person's hearing threshold. (4) The standard deviation should probably be increased for industrial situations since the fit obtained in ideal laboratory tests is likely to be better.

Although the standard method has been criticized on these grounds and other methods have been devised, none is in general use and probably ASA Z.22.4–1957 remains the best method available to test ear protectors.*

E. Calculation of Reduction in Sound Levels

In order to determine whether ear protectors are giving sufficient noise protection to reduce sound levels below those required in some criteria, it is necessary to perform a frequency analysis of the noise. The U.S. Department of Labor Bulletin 334[120] states that . . .

"If such (engineering or administrative) controls fail to reduce sound levels within the levels of the table (Table 2.7), personal protective equipment shall be provided and used to reduce sound levels within the levels of the table."

In Table 2.11 we will make a calculation to see whether the levels of factory noise at the ear have been reduced to 90 dB(A). Line two shows typical octave band sound pressure levels (actually these are the factory noise levels used in Table 2.3 but increased uniformly by 20 dB). Lines five and six show the mean noise protection and standard deviation of a typical fluid seal earmuff.[110] Line seven shows the assumed noise protection (mean minus one standard deviation); we can assume that only16% of people would have less noise protection than the values given. Line eight shows the assumed octave band levels at the ear (row four minus row seven). These levels are combined

*As this book was going to press, a new standard was issued. This is titled "Method for the Measurement of Real-Ear Protection of Hearing Protectors and Physical Attenuation of Earmuffs." It is the first standard to be issued by the Acoustical Society of America, Number 1–1975. It was adopted by ANSI on August 14, 1974 as ANSI S3.19–1974. The main differences from the old 1957 standard are the use of one-third octave bands of noise instead of pure tones, a pseudo-diffuse sound field environment instead of an almost anechoic one, and an extra provision for measurement of hearing protector performance on an artificial head.

(using Figure 1.16) as shown to give 85.1 dB(A). This compares with a value of 110 dB(A) obtained by summing the A-weighted levels in line four – a 25 dB(A) reduction.

It is seen that 84% of people would be expected to experience levels of less than 85 dB(A) in such an environment. We also see that with the earmuffs the 250-Hz octave band now dominates in determining the A-weighted level instead of the 1,000-Hz octave band.

If we use an assumed noise protection of the mean minus two standard deviations in line seven, we find that the 250-Hz band again dominates with this example with a value of 96–(20 – 12) = 88. Repeating this procedure at each frequency and combining the A-weighted levels obtained gives a sound level of 91.2 dB(A) instead of 85 dB(A). We may assume that 97% of people would experience levels less than 91dB(A) in this environment.

REFERENCES

1. **Palmer, J. W.,** *Anatomy for Speech and Hearing,* Harper & Row, New York, 1972.
2. **Cooper, W. A.,** The ear, hearing, loudness and hearing damage, in *Reduction of Machinery Noise,* Crocker, M. J., Eds., Purdue University, West Lafayette, Ind., 1974, 43.
3. **Crocker, M. J.,** The ear, hearing, loudness and units for rating noise, in *Tutorial Papers on Noise Control,* INTER-NOISE 72, Crocker, M. J., Ed., INCE, Poughkeepsie, N.Y., 1972, 124.
4. **Davis, H. and Silverman, S. R.,** *Hearing and Deafness,* Holt Rinehart & Winston, New York, 1970.
5. **Gulick, W. L.,** *Hearing: Physiology and Psychophysics,* Oxford University Press, New York, 1971.
6. **Rose, D. E.,** *Audiological Assessment,* Prentice-Hall, Englewood Cliffs, N.J., 1971.
7. **Welch, B. L. and Welch, A. S.,** *Physiological Effects of Noise,* Plenum Press, New York, 1970.
8. **Stevens, S. S. and Warshofsky, F.,** *Sound and Hearing,* Time-Life Books, New York, 1965.
9. **Robinson, D. W. and Dadson, R. S.,** A re-determination of the equal loudness relations for pure tones, *Br. J. Appl. Phys.,* 7, 166, 1956.
10. **Fletcher H. and Munson, W. A.,** Loudness, its definition, measurement and calculation, *J. Acoust. Soc. Am.,* 5, 1933.
11. **Zwicker, E.,** Ein Verfahren zur Berechnung der Lautstarke, (A means for calculating loudness), *Acustica,* 10, 304, 1960.
12. **Broch, J. T.,** Acoustic Noise Measurements, Brüel & Kjaer Co., Naerum, Denmark, January 1971, 174.
13. **Stevens, S. S.,** Procedure for calculating loudness: Mark VI, *J. Acoust. Soc. Am.,* 33, 1577, 1961.
14. **Stevens, S. S.,** Perceived level of noise by Mark VII and decibels (E), *J. Acoust. Soc. Am.,* 51, 575, 1972.
15. **Peterson, A. P. G. and Gross, E. E., Eds.,** *Handbook of Noise Measurement,* 7th ed., General Radio Co., Concord, Mass., 1972, 28.
16. **Kryter, K. D.,** *The Effects of Noise on Man,* Academic Press, New York, 1970, chap. 7.
17. **Ehmer, R. H.,** Masking patterns of tones, *J. Acoust. Soc. Am.,* 31, 1115, 1959.
18. **Ehmer, R. H.,** Masking by tones vs. noise bands, *J. Acoust. Soc. Am.,* 31, 1253, 1959.
19. **Spieth, W.,** Downward spread of masking, *J. Acoust. Soc. Am.,* 29, 502, 1957.
20. **Fletcher, H. and Munson, W. A.,** Relation between loudness and masking, *J. Acoust. Soc. Am.,* 9, 1, 1937.
21. **Zwicker, E., Flottrop, G., and Stevens, S. S.,** Critical bandwith in loudness summation, *J. Acoust. Soc. Am.,* 29, 548, 1957.
22. **Broch, J. T.,** Acoustic Noise Measurements, Brüel & Kjaer Co., Naerum, Denmark, January 1971, 25.
23. **Kryter, K. D.,** *The Effects of Noise on Man,* Academic Press, New York, 1970, chap. 2.
24. **Bilger, R. C. and Hirsh, I. J.,** Masking of tones by bands of noise, *J. Acoust. Soc. Am.,* 28, 623, 1956.
25. **Ward, W. D.,** Studies of the aural reflex. I. Contralateral remote masking as an indicator of reflex activity, *J. Acoust. Soc. Am.,* 33, 1034, 1961.
26. **Samojlova, I. K.,** The masking effect of short signals as a function of the time between the masked and masking sound, *Biofizika,* 4, 550, 1959.
27. **Pickett, J. M.,** Backward masking, *J. Acoust. Soc. Am.,* 30, 1613, 1958.
28. **Chistovich, L. A. and Ivanova, V. A.,** Backward masking by short sound pulses, *Biofizika,* 4, 170, 1959.
29. **Anon.,** Public health and welfare criteria for noise, July 27, 1973, U.S. Environ. Protect. Agency, Office of Noise Abatement and Control, Rep. 550/9-73-002, sect. 4.
30. **Fletcher, J. L.,** Protection from high intensities of impulse noise by way of preceeding noise and click stimuli, *J. Audit. Res.,* 5, 145, 1965.

31. **Simmons, F. B.,** Individual sound damage susceptibility: Role of middle ear muscles, *Ann. Otol. Rhinol. Laryngol.,* 72, 528, 1963.
32. **Durant, J. D. and Shallop, J. K.,** Effects of differing states of attention on acoustic reflex activity and temporary threshold shift, *J. Acoust. Soc. Am.,* 46, 907, 1969.
33. **Stevens, S. S., Volkmann, J., and Newman, E. B.,** A scale for the measurement of the psychological magnitude pitch, *J. Acoust. Soc. Am.,* 8, 185, 1937.
34. **Egan, J. P. and Meyer, D. R.,** Changes in pitch of tones of low frequency as a function of the pattern of excitation produced by a band of noise, *J. Acoust. Soc. Am.,* 22, 827, 1950.
35. **de Boer, E.,** Note on the critical bandwidth, *J. Acoust. Soc. Am.,* 34, 985, 1962.
36. **Broch, J. T.,** Acoustic Noise Measurements, Brüel & Kjaer Co., Naerum, Denmark, January 1971, 29.
37. **Viebrock, W. M., Crocker, M. J., and Cooper, W. A.,** *J. Appl. Acoust.,* in press, 1975.
38. **Peterson, A. P. G. and Gross, E. E.,** *Handbook of Noise Measurement,* 7th ed., General Radio Co., Concord, Mass., 1972, 42.
39. **Webster, J. C.,** Speech communications as limited by ambient noise, *J. Acoust. Soc. Am.,* 37, 692, 1965.
40. **Eldridge, D. H. and Miller, J. D.,** Acceptable noise exposures – damage risk criteria, in *Noise as a Public Health Hazard,* ASHA Rep. 4, American Speech and Hearing Association, Washington, D.C., 1969, 110.
41. **Spoor, A.,** Presbycusis values in relation to noise induced hearing loss, *Int. Audiol.,* 6, 48, 1967.
42. **Glorig, A.,** *Noise and Your Ear,* Grune & Stratton, New York, 1958.
43. **von Schulthess, G. and Huelsen, E.,** Statistical evaluation of hearing losses in military pilots, *Acta Oto-Laryngol.,* 65, 137, 1968.
44. **von Schulthess, G.,** Statistical evaluation of hearing losses in military pilots (second rep.) *Acta Oto-Laryngol.,* 68, 250, 1969.
45. **Hinchcliffe, R.,** The threshold of hearing as a function of age, *Acustica,* 9, 303, 1959.
46. **Konig, E.,** *C. R. Seances, 11 Congr. Ord. Soc. Int. Audiol.,* 87, 1955.
47. **Hinchcliffe, R.,** Report from Wales: Some relations between aging noise exposure and permanent hearing level changes, Committee S3-W-40, American National Standards Institute, August 1969.
48. **Rosen, S.,** Noise, hearing and cardiovascular function, in *Physiological Effects of Noise,* Welch, B. L. and Welch, A. S., Eds., Plenum Press, New York, 1970.
49. **Rosen, S., Olin, P., and Rosen, H. V.,** Dietary prevention of hearing loss, *Acta Oto-Laryngol.,* 70, 242, 1970.
50. **Baughn, W. L.,** Noise control – Percent of population protected, *Int. Audiol.,* 5, 331, 1966.
51. **Robinson, D. W.,** Estimating the risk of hearing loss due to continuous noise, in *Occupational Hearing Loss,* Robinson, D. W., Ed., Academic Press, New York, 1971.
52. **Mercer, D. M. A.,** Noise damage criterion using A-weighting levels, *J. Acoust. Soc. Am.,* 43, 636, 1967.
53. **Baughn, W. L.,** The risk of hearing impairment as a function of noise exposure, in *Noise and Vibration Control Engineering,* Crocker, M. J., Ed., Purdue University, West Lafayette, Ind., 1972, 133.
54. **Davis, H. and Kranz, F. W.,** The international standard reference zero for pure tone audiometers and its relation to the evaluation of impairment of hearing, *J. Speech Hear. Res.,* 7, 7, 1964.
55. **Kryter, K. D.,** Setting tolerable limits of noise exposure with respect to hearing impairment, *J. Acoust. Soc. Am.,* 51, 142(A), 1972; in Noise Pollution, Hearings before the Subcommittee on Air and Water Pollution, 92nd Congr., U.S. Gov. Print. Off., Washington D.C., 1972, 154.
56. **Burns, W. and Robinson, D. W.,** Hearing and noise in industry, Her Majesty's Stationary Office, London, 1970.
57. **Anon.,** Hazardous noise exposure Air Force Regulation 160-3, Department of the Air Force, 1956.
58. **Botsford, J. H.,** Damage risk in, *Transportation Noises: A Symposium on Acceptability Criteria,* Chalupnik, J. D., Ed., University of Washington Press, Seattle, 1970.
59. **Ward, W. D., Glorig, A., and Sklar, D. L.,** Temporary threshold shift produced by intermittent exposure to noise, *J. Acoust. Soc. Am.,* 31, 791, 1959.
60. **Ward, W. D., Glorig, A., and Sklar, D. L.,** Dependence of temporary threshold shift at 4 kc on intensity and time, *J. Acoust. Soc. Am.,* 30, 944, 1958.
61. **Kryter, K. D.,** *The Effects of Noise on Man,* Academic Press, New York, 1970, chap. 5.
62. **Kryter, K. D.,** *The Effects of Noise on Man,* Academic Press, New York, 1970, chap. 12, 13.
63. **Miller, J. D.,** Effects of Noise on People, U.S. Environ. Protect. Agency Rep. NTID 300.7, 1971; also reprinted in *J. Acoust. Soc. Am.,* 56, 729, 1974.
64. **Anon.,** Public Health and Welfare Criteria for Noise, July 27, 1973, U.S. Environ. Protect. Agency Rep. 550/9-73-002, 1973.
65. **Kryter, K. D.,** *The Effects of Noise on Man,* Academic Press, New York, 1970, 587.
66. **Filippov, V. and Ilyashuk, Y.,** Approach to the Problem of Noise Control in the Soviet Union, in INTER-NOISE 72, Crocker, M. J., Ed., INCE, Poughkeepsie, N.Y., 1972, 32.
67. **Cohen, A.,** Effects of noise on psychological state, in *Noise as a Public Health Hazard,* ASHA Rep. 4, American Speech and Hearing Association, Washington D.C., 1969.
68. **Wakstein, C.,** The Noise Problem in the United States, 5th International Congress for Noise Abatement, London, England, May 1968.

69. **Wilson A.,** (Chairman), Noise, Final report, Her Majesty's Stationary Off., London, England, 1963, 24.
70. **Close, W. H.,** Surface transportation noise, in *Noise and Vibration Control Engineering,* Crocker, M. J., Ed., Purdue University, West Lafayette, Ind., 1972, 20.
71. **Borsky, P.,** Some of the Human Factors Underlying Community Reactions to Air Force Noise, Sixth Annual Meeting of Armed Forces, Washington, D.C., Natl. Res. Counc. Comm. Hear. and Bioacoust. Biomech. (CHABA), 1958.
72. **McKennell, A. C.,** Noise complaints and community action, in *Transportation Noises: A Symposium on Acceptability Criteria,* Chalupnik, J. D., Ed., University of Washington Press, Seattle, 1970, 228.
73. **Kajland, A.,** Traffic noise investigation, in *Transportation Noises: A Symposium on Acceptability Criteria,* Chalupnik, J. D., Ed., University of Washington Press, Seattle, 1970, 327.
74. Tracor Staff; Community Reaction to Airport Noise, Vol. 1, NASA CR-1671, 1971.
75. **Rosenblith, W. A., Stevens, K. N., and the staff of Bolt Beranek & Newman,** *Handbook of Acoustic Noise Control,* Vol. II, *Noise and Man,* Wright-Patterson Air Force Documentation Center Tech. Rep. 52–204, 1953, 181.
76. **Kryter, K. D., Jansen, G., Parker, D., Parrack, H. O., Thiessen, G., and Williams, H. L.,** Non-Auditory Effects of Noise Rep. WG–53, NAS-NRC Comm. Hear. Bioacoust. Biomech. (CHABA), Washington, D.C., 1971.
77. **Berger, R. J.,** The sleep and dream cycle, in *Sleep Physiology and Pathology, A Symposium,* Kales, A., Ed., Lippincott, Philadelphia, 1969, 17.
78. **Williams, H. L., Hammack, J. T., Kaly, R. L., Demert, W. C., and Lubin, A.,** Responses to auditory stimulation, sleep loss and the EEG stages of sleep, *Electroencephalogr. Clin. Neurophysiol.,* 16, 209, 1964.
80. **Wilson, A.,** (Chairman), Noise, Final Report, Her Majesty's Stationary Off., London, England, 1963.
81. **Richter, R.,** Sleep Disturbances Which We Are Not Aware of Caused by Traffic Noise, EEG Station of the Neurological University Station, Basel, Switzerland, (undated).
82. **Kryter, K. D.,** *The Effects of Noise on Man,* Academic Press, New York, 1970, chap. 12.
83. **Jansen, G.,** Adverse effects of noise on iron and steel workers, *Stahl. Eisen.,* 81, 217, 1961.
84. **Andriukin, A. A.,** Influence of sound stimulation on the development of hypertension: Clinical and experimental results, *Cor. Vasa,* 3, 285, 1961.
85. **Tullio, P.,** *Sulla Funzione Delle Varie Parti Dell-Orrechio Interno,* 1926.
86. **Dickson, E. D. D. and Chadwick, D. L.,** Observations of disturbances of equilibrium and other symptoms induced by jet engine noise, *J. Laryngol. Otol.,* 65, 154, 1965.
87. **Nixon, C. W., Harris, C., and von Gierke, H. E.,** Rail Test to Evaluate Equilibrium in Low-Level Wideband Noise, Report AMRL-TR-66-85, U.S. Aerospace Med. Res. Lab., Wright-Patterson Air Force Base, Dayton, Ohio 1966.
88. **Davis, R. C., Buchwald, A. M., and Frankman, R. W.,** Autonomic and muscular responses and their relation to simple stimuli, Psychol. Monogr., 69, No. 405, 1955.
89. **Moline, J.,** Reflexes oculaines d'origine auditive, *Rev. Laryngol.,* 1, 385, 1916.
90. **Bartoshuk, A. K.,** Response decrement with repeated elicitation of human neonatal cardiac acceleration to sound, *J. Comp. Physiol. Psychol.,* 55, 1962.
91. **Ando, Y. and Hattori, H.,** Effects of intense noise during fetal life upon postnatal adaptability (Statistical study of the reactions of babies to aircraft noise), *J. Acoust. Soc. Am.,* 47, 1128, 1970.
92. **Jansen, G.,** Effects of noise on physiological state, in *Noise as a Public Health Hazard,* ASHA Rep. No. 4, American Speech and Hearing Association, Washington, D.C., 1969.
93. **Goldman, D. E.,** Effects of vibration on man, in *Handbook of Noise Control,* Harris, C. M., Ed., McGraw-Hill, New York, 1957.
94. **Reiher, H. J. and Meister, F. J.,** Human Sensitivity to Vibration, Forsch. Geb. des Ingeniermes., Stuttgart, Inst. Tech., Nov. 1931; Abstract in *Eng. News Rec.,* March 31, 1932, 470.
95. **Monk, R. G.,** Mechanically Induced Vibration in Buildings, Environ. Eng., No. 51, Soc. Environ. Eng., Lond., 1971.
96. **Kryter, K. D.,** *The Effects of Noise on Man,* Academic Press, New York, 1970, chap. 13.
97. **Broadbent, D. E.,** Effects of noise on behavior, in *Handbook of Noise Control,* Harris, C. M., Ed., McGraw-Hill, New York, 1957.
98. **Broadbent, D. E.,** *Perception and Communication,* Pergamon Press, London, 1958.
99. **Broadbent D. E.,** Factors increasing and decreasing the effects of noise, paper presented at the Int. Congr. on Noise as a Public Health Problem, Dubrovnik, Yugoslavia, 1973.
100. **Carpenter, A.,** Effects of noise on performance and productivity in control of noise, Symp. No.12., Her Majesty's Stationary Off., London, 1962, 297.
101. **Boggs, D. H. and Simon, J. R.,** Differential effect of noise on tasks of varying complexity, *J. Appl. Psychol.,* 52, 148, 1968.
102. **Jerison, H. J. and Wing, S.,** Effects of noise and fatigue on a complex vigilance task, Wright-Patterson Air Force Documentation Center Tech. Rep., TR-57-14, 1957.
103. **Broussard, I. G., Walker, R. Y., and Roberts, E. E.,** The Influence of Noise on the Visual Contrast Threshold, Rep. 101, U.S. Army Medical Research Laboratories, Fort Knox, KY, Nov. 6, 1952.
104. **Weston, H. C. and Adams, S.,** Rep. 70, Industrial Health Research Board, Her Majesty's Stationary Off., London, 1935.

105. **Chadwick, D. L.,** Noise-Induced Deafness, *Practitioner,* 191(1146), 733, 1963.
106. **Fosbroke, J.,** Practical observations on the pathology and treatment of deafness. II, *Lancet,* I, 1830, 645.
107. **Barr, T.,** Enquiry into effects of loud sounds on the hearing of boilermakers and others who work amid noisy surroundings, *Proc. Glasgow Phil. Soc.,* 17, 223, 1886.
108. **Walker, J. G. and Martin, A. M.,** Hearing conservation, in *Noise and Vibration Control for Industrialists,* Petrusewicz, S. A. and Longmore, D. K., Eds., American Elsevier, New York, 1974.
109. **Federal Register,** 34(96), Part 2, May 20, 1969, 7948.
110. Code of Practice for Reducing the Exposure of Employed Persons to Noise, Her Majesty's Stationary Off. London, 1972.
111. International Organization for Standardization, Recommendation R 1999, Assessment of Occupational Noise Exposure for Hearing Conservation Purposes, 1971.
112. **Kryter, K. D.,** *The Effects of Noise on Man,* Academic Press, New York, 1970, chap. 4.
113. **Grason, R. L. and Hetzel, C.,** Hearing conservation programs in industry, in *Handbook of Noise Measurement,* Peterson, A. P. G. and Gross, E. E., Eds., General Radio Co., Concord, Mass., 1972.
114. **Newby, H. A.,** *Audiology,* 3rd ed., Appleton–Century Crofts, New York, 1972.
115. **Price L. L.,** Pure-tone audiometry, in *Audiological Assessment,* Rose, D. E., Ed., Prentice-Hall, Engelwood Cliffs, N. J., 1971.
116. **Glorig A.,** Audiometric testing in industry, in *Handbook of Noise Control,* Harris, C. M., Ed. McGraw-Hill, New York, 1957, chap. 6.
117. **Eagles, E. L., Wishik, S. M., Doefler, L. G., Meinick, W., and Levine, H. S.,** Hearing Sensitivity and Related Factors in Children, Laryngoscope, St. Louis, Mo, 1963.
118. **von Bekesy, G.,** A new audiometer, *Acta Oto-Laryngol.,* 35, 411, 1947.
119. **Hirschorn, M., and Singer, E.,** The effect of ambient noise on audiometric room selection, *Sound Vib.,* 7(2), 18, 1973.
120. **Anon.** Guidelines to the Department of Labor's Occupational Noise Standards, Bull. 334, revised, U.S. Dep. of Labor, Washington, D.C., 1971.
121. **Coles, R. R. A.,** in *Noise as a Public Health Hazard,* ASHA Rep. No. 4, American Speech and Hearing Assoc., Washington, DC, 1969.
122. **Imbus, H. R.,** Medical Aspects of Hearing Conservation, in INTER-NOISE 72, Crocker, M. J., Ed., INCE, Poughkeepsie, N.Y., 1972, 8.
123. **Piesse, R. A.,** Ear Protectors, Commonwealth Acoust. Lab. Rep. No. 21, Sydney, Australia, 1962.
124. **Shaw, E. A. G. and Thiessen, G. J.,** Improved cushion for ear defenders, *J. Acoust. Soc. Am.,* 30, 24, 1958.

Chapter 3

INSTRUMENTATION FOR SOUND AND VIBRATION MEASUREMENT

TABLE OF CONTENTS

3.1 INTRODUCTION

There is a bewildering variety of acoustics and vibration measurement equipment now available to the noise control engineer. The variety of equipment produced, the sophistication, and the price all seem to increase each year with rapid technical changes in electronics taking place, it is always difficult to be current and not to make statements which become dated. It is not the purpose of this chapter to discuss in great detail the acoustical or design theory or construction of this equipment; this has been done elsewhere.[1-6] Rather, it is the intent here to classify the equipment available and to describe briefly the most important instruments for noise work. Suggestions will be made in other chapters of Volumes I and II on which instrumentation systems are most suitable for different noise measurement problems.

In Section 3.2 we will review briefly some electrical theory and some of the basic principles of measurement systems and transducers in order to understand better the discussion to follow in the rest of the chapter. Sections 3.3 and 3.4 will discuss the basic characteristics and types of microphones available for noise measurements. In the remaining Sections 3.5 through 3.8, the use of vibration transducers, spectrum analyzers, portable instruments such as sound level meters and data recording equipment will be discussed.

It is not possible to describe every type of instrument which will be used by the noise control engineer in Chapter 3. Some instruments such as voltage amplifier, moving coil galvanometer, rms detector, and others are familiar to most people and are omitted. They have been described elsewhere in any case.[1,2]

3.2 TRANSDUCERS AND MEASUREMENT SYSTEMS

3.2.1 Electrical Circuit Theory

The fundamental work by Ohm, Faraday, and Coulomb in the last 200 years has shown that the basic elements of an electric circuit are *resistors, inductors,* and *capacitors.* We will consider the simple circuit (Figure 3.1) built from elements with resistance, R, inductance, L, and capacitance, C, respectively. The instantaneous electric current I which will flow is defined as the rate of change of the charge passing through a specified area.*

*The area may be the cross-sectional area of the wire or some other area where the charge is in motion. Note the charge of one electron is 1.602×10^{-19} *coulomb.*

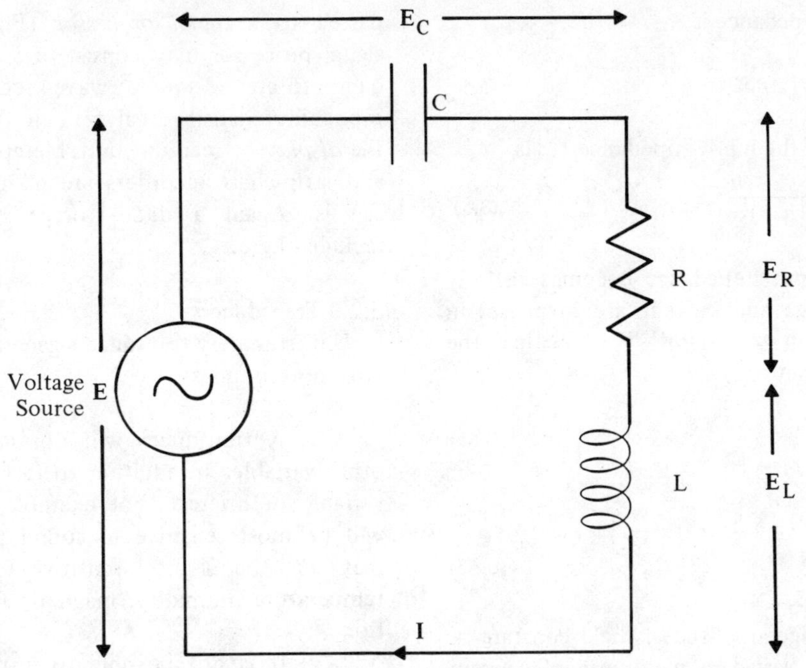

FIGURE 3.1 Simple series R-L-C circuit.

If a steady time-invariant potential difference, **E**, is applied as a voltage source in the circuit of Figure 3.1, no current will flow. However, if a time-variant voltage **E** is applied as a voltage source, elementary electrical considerations give the result that

$$E_L + E_R + E_C = E, \tag{3.1}$$

or

$$L\frac{dI}{dt} + RI + \frac{q}{C} = E, \tag{3.2}$$

where the current **I** = dq/dt = $\dot{q}$. Note also that dI/dt = $\ddot{q}$. On making these substitutions, Equation 3.2 becomes

$$L\ddot{q} + R\dot{q} + \frac{q}{C} = E. \tag{3.3a}$$

Equation 3.3 is identical in form to Equation 1.16b. In this case, the inductance, resistance, and capacitance in the electrical system are equivalent, respectively, to the mass, damping coefficient, and inverse of stiffness* in the mechanical system. The charge, current, and voltage are equivalent to the displacement, velocity and force, respectively. The analogies between electrical, mechanical, and acoustical systems are termed *analogues* and are extremely useful in analysis and design. They have been discussed by many authors.[6–10] It is often useful to make an electrical analogue of a mechanical or acoustical system in order to simplify the analysis. This is particularly true in transducer design where acoustical, mechanical, and electrical systems are interconnected.

The analogue we have just described is commonly known as the *impedance analogue.* However, note that a different analogue also exists known as the *mobility analogue.*[6,10] In the mobility analogue, current and voltage are equivalent to the force and velocity, respectively, and inductance, resistance, and capacitance are equivalent to mechanical compliance, inverse resistance, and mass respectively. Similar impedance and mobility analogues exist with acoustical systems.

In most cases of interest we will assume that the voltage, **E**, applied in Figure 3.1 is *simple harmonic* (see Section 1.2) and we will assume that the charge, **q** and hence current **I** will also be simple harmonic. Using the complex exponential notation as we did in Section 1.2.3 C gives from Equation 3.3a

$$(i\omega L)I + RI + I/(i\omega C) = E. \tag{3.3b}$$

*The inverse of stiffness is known as mechanical *compliance,* C_m. Note $C_m = 1/K$.

Thus the input impedance is

$$|Z| = E/I = R + i(\omega L - 1/\omega C). \quad (3.4)$$

The magnitude of the input impedance $|Z|$ is

$$|Z| = \sqrt{R^2 + (\omega L - 1/\omega C)^2}. \quad (3.5)$$

Note that the input impedance becomes entirely real (when voltage and current are in phase) in Equation 3.4 when $\omega L = 1/\omega C$. This occurs at the resonance frequency

$$\omega_R = 1/\sqrt{LC}, \quad (3.6a)$$

or

$$f_R = 1/(2\pi\sqrt{LC}). \quad (3.6b)$$

If the input voltage amplitude is held constant as the frequency is varied, a maximum of current flows at the resonance frequency.

3.2.2 Typical Measurement System

When measurements are made of noise or vibration it is necessary to combine several different types of instruments into one measurement system. A typical system is shown in Figure 3.2.

In order to make useful measurements, it is desirable to have a good understanding of the phenomena being investigated, and to have a reasonable understanding of the functioning of the instrumentation. This chapter is designed to help with the latter requirement.

Since about 1920, most sound and vibration measurement systems have made extensive use of electrical networks. Electrical amplification of signals made possible several measurement techniques previously impossible. Prior to this time, sound and vibration measurements either were not made at all or had to be made using mechanical or acoustical-mechanical systems.

The first item in any measuring system is the *transducer.* As its name implies, this instrument converts a signal in one physical form into another, e.g., a sound pressure signal into an electrical signal. Normally the electrical signal from a transducer is not suitable for direct read-out and a *signal conditioner* is then used to either amplify, attenuate, or transform the signal.

It is optional at this stage to include a *data storage* item in the system before the signal is passed to a *signal processor* (Figure 3.2). The signal processor may consist of a narrow or wide band filter, a square wave rectifier (rms), a probability density analyzer, etc. The last item is the *display* or read-out unit. Meters, oscilloscopes, and strip-chart recorders are all used to display signals. Again a data storage item could be included here.

3.2.3 Transducers

Unfortunately transducers generally suffer from two shortcomings:

1. A transducer will normally respond to other variables in addition to its response to the variable of interest. For example, a microphone will be most sensitive to sound pressure, but it may also be slightly sensitive to variations in temperature, humidity, magnetic field, and vibration.
2. It is not possible to introduce a transducer into the measurement medium without disturbing the medium in some way. The transducer will extract some energy from the medium. In addition, other disturbances will be caused. An accelerometer will add mass to the system. A Pitot tube will disturb the flow. A microphone will cause reflections.

3.3 MICROPHONE CHARACTERISTICS

An ideal microphone should have the following characteristics:[11]

1. It should cause negligible diffraction of the sound field (i.e., its dimensions should be small compared with the smallest wavelength of interest).
2. It should have a high acoustic (driving point) impedance compared with the fluid medium so that little energy is extracted.
3. It should have low electrical noise.
4. Its output should be independent of temperature, humidity, magnetic fields, static pressure, and wind velocity, and it should be rugged and stable with time.
5. Its sensitivity should be independent of sound pressure level.
6. Its frequency response should be flat.
7. It should introduce a zero phase shift between the sound pressure and the electrical output signal.

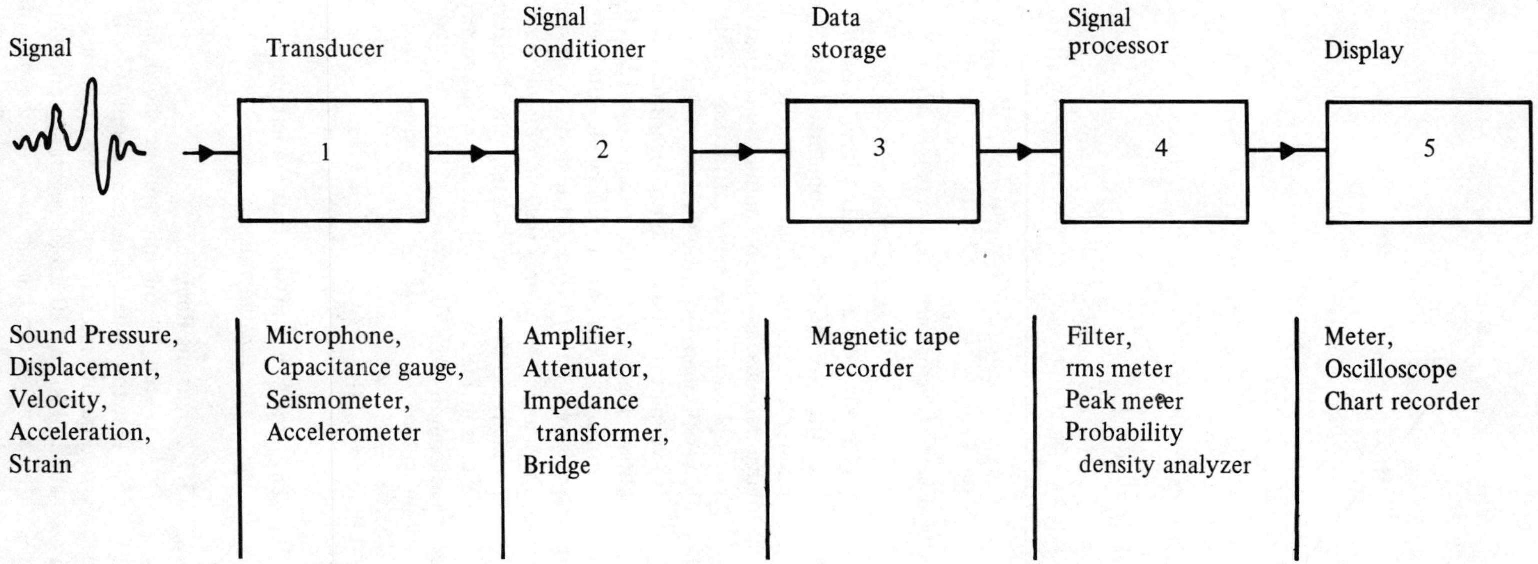

FIGURE 3.2 Idealized measurement system.

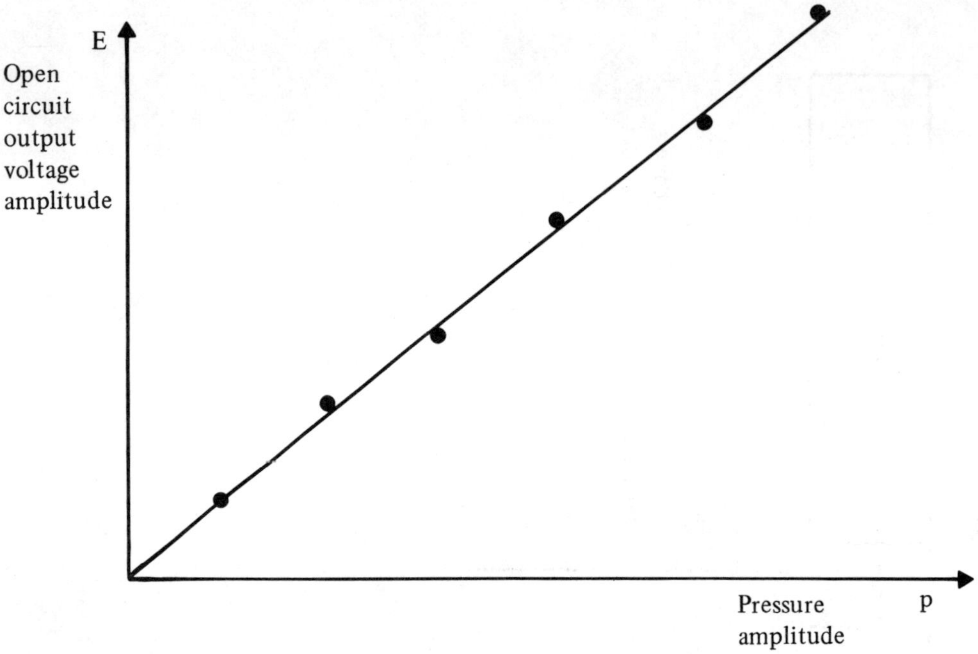

FIGURE 3.3 Sensitivity of an ideal microphone.

No microphone can meet all of the above criteria, and thus different types of microphones are preferred for different measurements.

3.3.1 Sensitivity

An ideal microphone should have an output voltage amplitude E proportional to the exciting pressure amplitude p (see Figure 3.3). The ratio of open circuit output voltage to input pressure is normally called the sensitivity M_p.

$$M_p = E/p. \tag{3.7a}$$

The sensitivity depends on the type of microphone (different types are discussed later in Section 3.4) and the size of the microphone. Larger microphones are normally more sensitive and this is advantageous in many applications (but larger microphones experience diffraction problems at lower frequencies as we shall see in Section 3.3.5). If we plot Figure 3.3 on a logarithmic scale we obtain Figure 3.4. The reference voltage E_{ref} is normally taken as 1 V, while the reference pressure is either taken as 1 μbar or 1 N/m^2 (1 Pa). Thus, at any frequency the *microphone response* is normally given by subtracting the X- from the Y-axis in Figure 3.4 and the microphone response is given in terms of

$$R = 20 \log [(E/p)/E_{ref}/p_{ref})] \text{ or } 20 \log [E/E_{ref})/(p/p_{ref})], \tag{3.76}$$

i.e., decibels re 1V/N/m^2 or else re 1V/μbar. For very low sound pressure amplitudes, electrical noise will exceed the voltage generated by the microphone and this sets a lower limit for measurements. At very high sound pressure amplitudes, the diaphragm displacement may become so large that the voltage generated is no longer proportional to the displacement. Nonlinearity thus sets an upper use limit. At pressures a little above the nonlinear limit, physical damage to the microphone can occur. Thus between the upper and lower limits we have the range in which the microphone can be used – the *dynamic range.*

3.3.2 Dynamic Range

It is seen for the microphone shown in Figure 3.4 that the dynamic range is about 100 dB. Most good quality microphones have a dynamic range of about 100 to 120 dB (interestingly enough about the same as the human ear). As the microphone diaphragm diameter is increased, the sensitivity is increased so that electrical noise is less of a

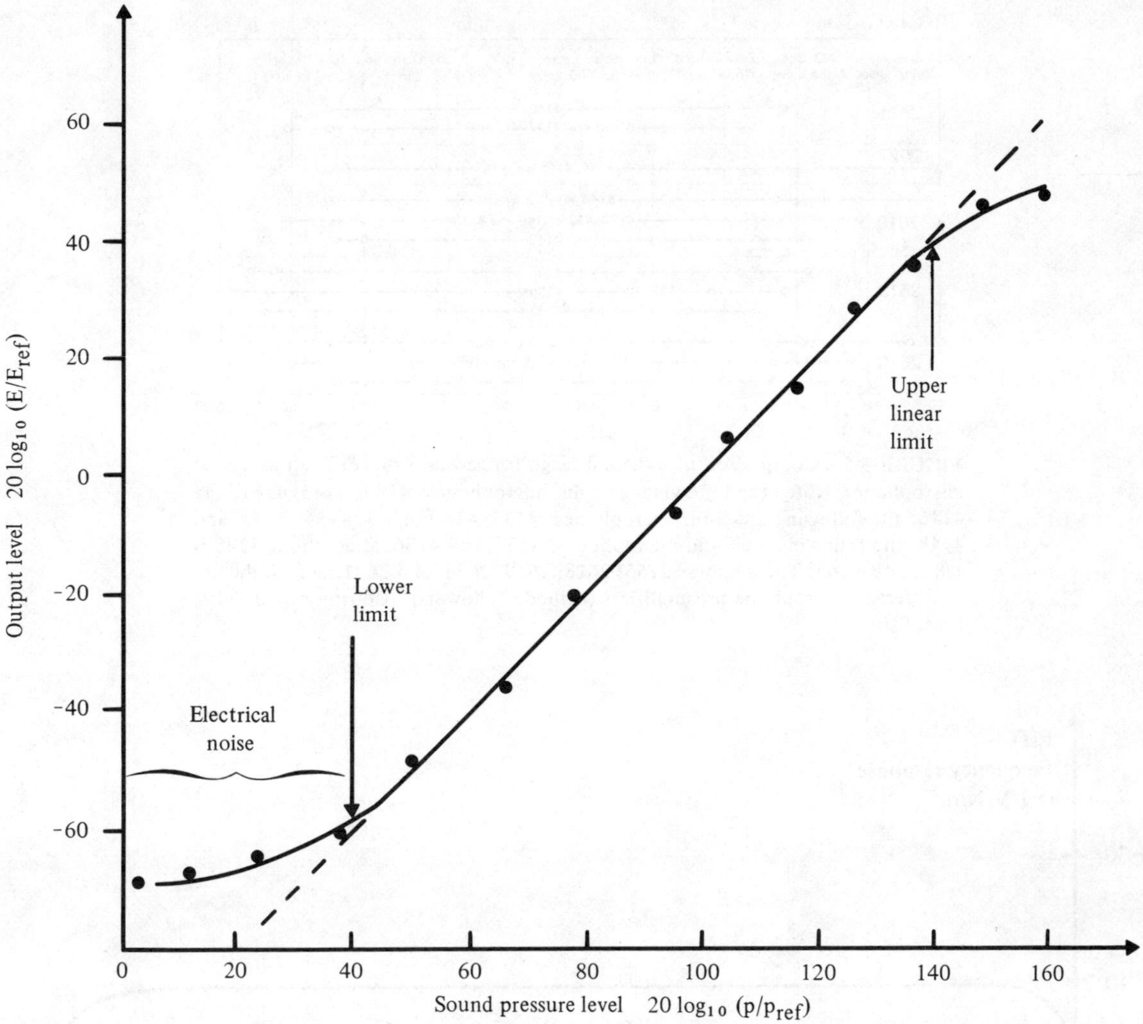

FIGURE 3.4 Sensitivity of ideal microphone showing upper and lower limits.

problem and the lower limit for measurements is decreased. However, a larger diaphragm diameter means larger deflections for a given pressure and a reduced upper limit because of nonlinearity problems. Figure 3.5 shows the dynamic range of several different diameter capacitor microphones. Because of the dynamic range problems, the small diameter microphones cannot be used for "quiet" sounds and the large diameter microphones cannot be used for intense noise.

3.3.3 Frequency Response

The *frequency response,* R(f), of the microphone is given by Equation 3.7b as the frequency of the sound pressure amplitude p is changed. Figure 3.6 shows the frequency response of an ideal microphone used for audible sound measurements. Since the human hearing range extends from about 20 to 20,000 Hz, it is desirable that the frequency response be "flat" between these limits. For certain types of measurements (e.g., for measurement of sonic booms or explosive blasts), it may be necessary to measure to frequencies lower than about 20 Hz. For model studies or for the measurement of noise environments of structures subject to fatigue, it may be necessary to measure to frequencies higher than 20,000 Hz. At low frequencies, e.g., below about 1,000 Hz, the frequency response of a microphone is independent of the angle of incidence of the sound waves. However, at higher frequencies as the microphone dimensions and the wavelength of the sound become comparable, *diffraction* effects are important and the frequency response of a microphone is strongly dependent on the angle of incidence of

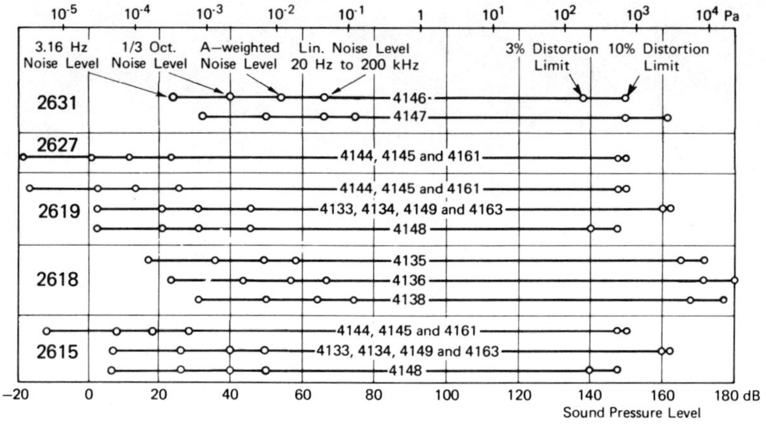

FIGURE 3.5 Comparison of dynamic range for several Brüel & Kjaer capacitor microphones. Note: the following are 1-in. microphones: 4144, 4145, 4161, and 4146; the following are ½-in. microphones: 4133, 4149, 4163, 4134, 4147, and 4148; the following are ¼-in. microphones: 4135, and 4136. Microphone 4138 is 1/8 in. diameter. The numbers 2165, 2618, 2619, 2627, and 2631 refer to the use of different microphone preamplifiers (cathode followers). (Courtesy of Brüel & Kjaer Co.)

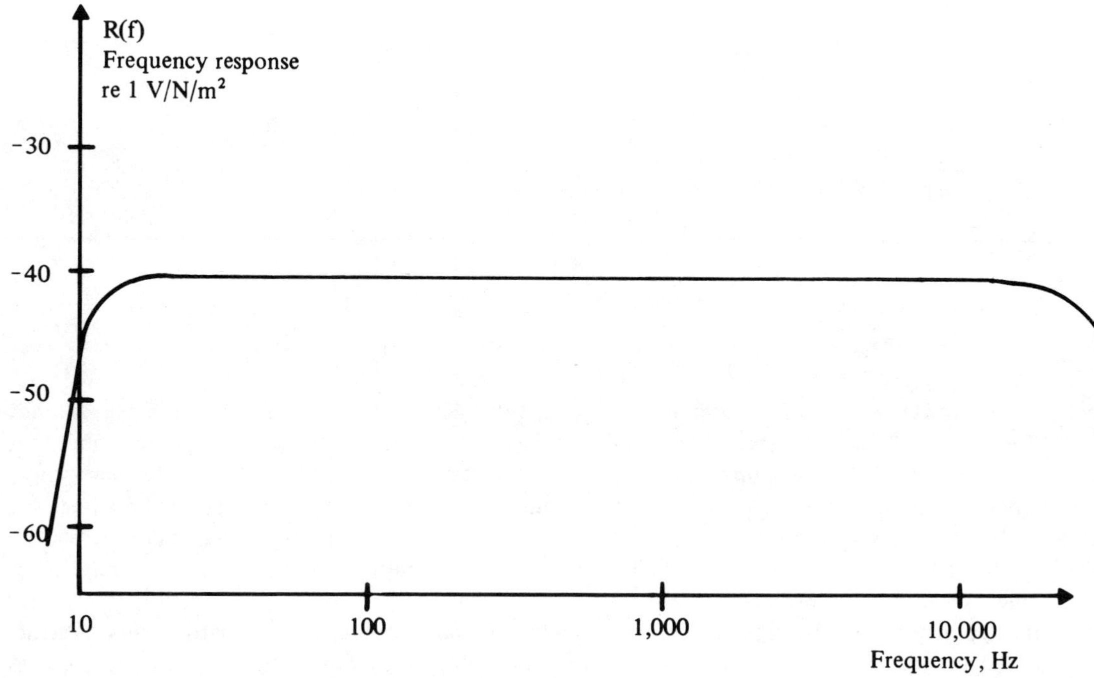

FIGURE 3.6 Frequency response R(f) of ideal microphone for use in audible range.

the sound. Before discussing *directivity* in detail in Section 3.3.5, we will first discuss methods by which microphones are calibrated.

3.3.4 Microphone Calibration

When a microphone is introduced into a sound field, it causes undesirable reflections and hence changes the sound field. This effect is more pronounced at high frequency. Ideally we should like to measure the sound pressure in the unperturbed or "free field" but this is difficult. Brüel and Rasmussen[12] have discussed the problems of calibrating microphones. It is extremely important that manufacturers and users calibrate microphones accurately. Unfortunately, some users do not calibrate microphones sufficiently frequently to ensure the accuracy of sound pressure level readings. There are several calibration procedures which will only be described briefly in the following. We shall define a microphone as having two major responses: *pressure* response and *free-field* response.

The pressure response of a microphone may be defined as the response of the microphone to a uniform pressure applied over its diaphragm. This is the response of an ideal microphone of *zero size* introduced into a free progressive plane wave field. However, when an actual finite size microphone is introduced into such a field, reflection and diffraction are caused which give a different microphone response called the *free-field* response. Because the microphone can be oriented at any arbitrary angle in the plane wave field, perhaps the pressure response is more fundamental. There are several ways of measuring the pressure response of a microphone:[12-14] (A) Pistonphone, (B) "Loudspeaker" Calibrators, (C) Electrostatic Actuator, (D) Reciprocity Method, and (E) Substitution Method. The only absolute method of calibration is the Reciprocity Method. This is somewhat complicated and time consuming and the other four methods although not giving absolute calibration are more convenient and normally sufficiently accurate.

A. Pistonphone

This is a very accurate, reliable, and simple device for calibrating a microphone which is convenient for use in the field. The principle of operation is quite simple. Figure 3.7 shows a photograph of a commercial pistonphone and Figure 3.8 shows two cross-sectional drawings which explain its operation. A small battery-powered electric motor drives a shaft on which is mounted a cam disc (Figure 3.8b). The cam disc drives two pistons (each of area A_p) symmetrically. The cam gives the pistons a sinusoidal motion at four times the shaft rotational speed. The stroke of the pistons (or peak amplitude from mean position), ℓ, is thus 1/4 of the difference in maximum and minimum diameters of the cam. The pistons vary the cavity volume V sinusoidally

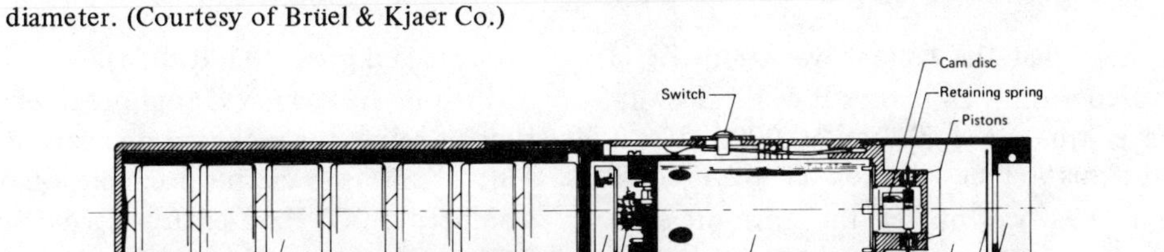

FIGURE 3.7 Photograph of a pistonphone. The dimensions are 230 mm long and 36 mm in diameter. (Courtesy of Brüel & Kjaer Co.)

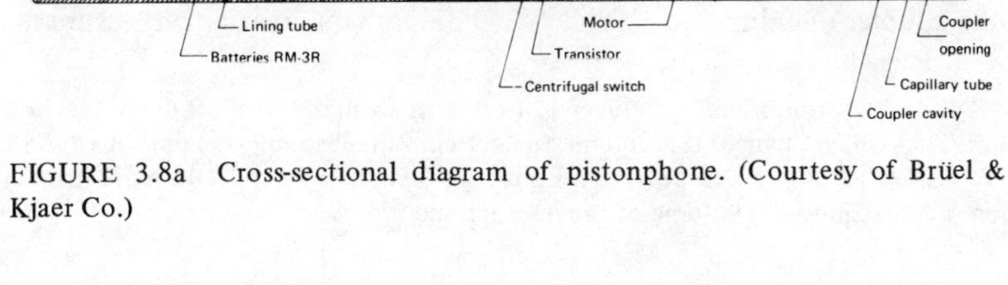

FIGURE 3.8a Cross-sectional diagram of pistonphone. (Courtesy of Brüel & Kjaer Co.)

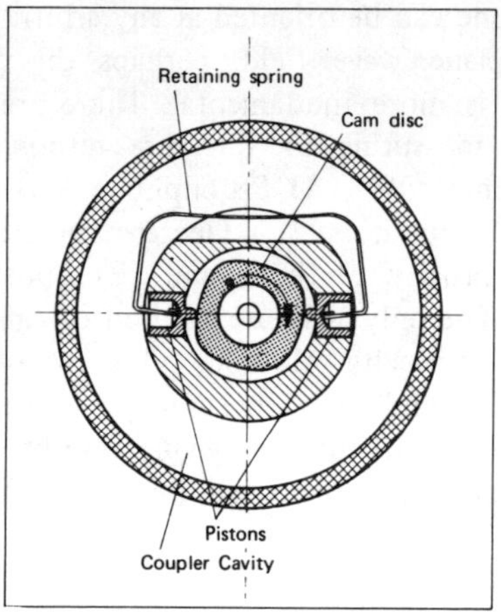

FIGURE 3.8b Cross section of cam and piston region of pistonphone showing the principle of operation. (Courtesy of Brüel & Kjaer Co.)

in time and since for pistonphones a low frequency (e.g., 250 Hz) is normally chosen (for mechanical reasons), a corresponding sinusoidal variation in pressure occurs. It is accurate to assume the adiabatic law:

$$pV^{\gamma} = \text{constant}; \tag{3.8}$$

where

p is the total pressure
V the cavity volume
γ = ratio of specific heats = 1.4 (for air).

By differentiation of Equation 3.8 we obtain

$$dpV^{\gamma} + p\gamma V^{\gamma-1}\, dV = 0,$$

or

$$dp = -\gamma \frac{p\, dV}{V}. \tag{3.9a}$$

We assume here that the volume variation dV is small compared with V. For the B & K® Pistonphone, this is true since $dV/V < 0.0003$. If we substitute the rms volume variation $dV = 2A_p\ell/\sqrt{2}$ in Equation 3.9a we obtain the rms pressure fluctuation dp experienced by a microphone flush-mounted in the coupler opening

$$dp = \gamma p\sqrt{2}\, A_p\ell/V, \tag{3.9b}$$

where p is the atmospheric pressure. We note that all the terms on the right side of Equation 3.9b are easily measured. Provided the piston stroke ℓ is carefully controlled, the sound pressure level,

$$L_p = 20 \log (dp/p_{ref}), \tag{3.10}$$

can be accurately predicted. Good sealing should be maintained when the microphone is fitted into the coupler opening. In principle, this type of pistonphone can be used to calibrate any microphone provided that good sealing is maintained. If microphones of different diameters are to be calibrated with the pistonphone, and different coupler connections are used to keep the volume V unchanged, the sound pressure level L_p (Equations 3.9 and 3.10) is unchanged. If a different volume results, the change in sound pressure level can be predicted from Equations 3.9 and 3.10.* A change in atmospheric pressure p will alter the calibration (Equation 3.9) but a correction is simply made by the use of a barometer provided by the manufacturer.

B. "Loudspeaker" Calibrators

Somewhat simpler, lower cost calibrators are also available which work on the loudspeaker principle. Figure 3.9 shows a photograph of a commercially available type which produces 114 dB at 1,000 Hz. Figure 3.10a shows a block diagram of this calibrator and Figure 3.10b shows details of the calibrator to microphone coupling. The loudspeaker is driven by an oscillator which is powered by a 9-V battery. The same company makes a similar calibrator (Type 1562®) which generates 5 selected frequencies: 125, 250, 500, 1,000 and 2,000 Hz. With this calibrator, corrections must be made for changes in ambient temperature and pressure (altitude).

Figure 3.11 shows a photograph of another type of calibrator mounted on a sound level meter. Figure 3.12 shows the principle of operation. A stabilized 1,000-Hz oscillator feeds a piezoelectric driven element which vibrates the metallic diaphragm creating a pressure in the front coupling

*For example, the B & K 4220 Pistonphone® produces 124 dB with a volume V of 19.6 cm^3. If a special microphone with a coupler is to be calibrated which changes this volume to 24.1 cm^3 the Pistonphone output would be changed to 124.0 + $20 \log_{10} (19.6/24.1)$ = 124.0 − 1.8 = 122.2 dB (from Equations 3.9a, 3.9b, and 3.10). Note that it may also be necessary to include a correction for the equivalent volume of the microphone.[14]

a. b.

FIGURE 3.9 Photograph of "loudspeaker" type microphone calibrator (General Radio Co. Type 1567). (a) Control end; (b) Microphone end.

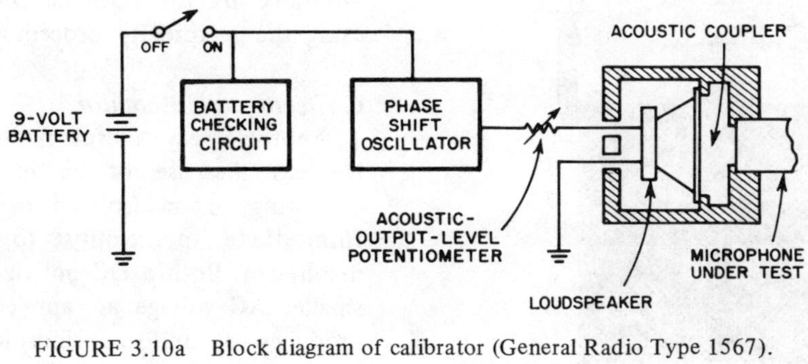

FIGURE 3.10a Block diagram of calibrator (General Radio Type 1567).

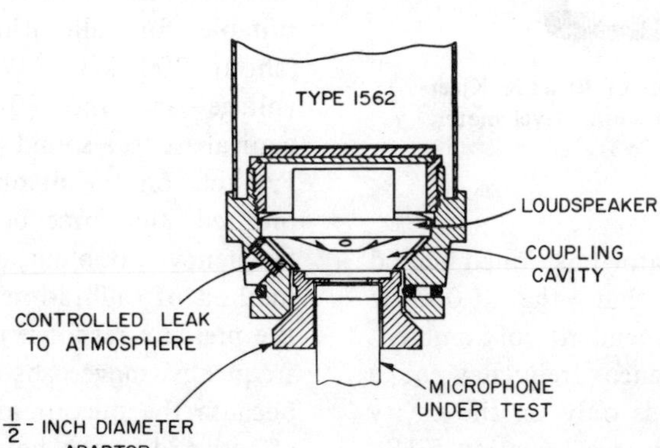

FIGURE 3.10b Details of calibrator-microphone coupling (General Radio Type 1567).

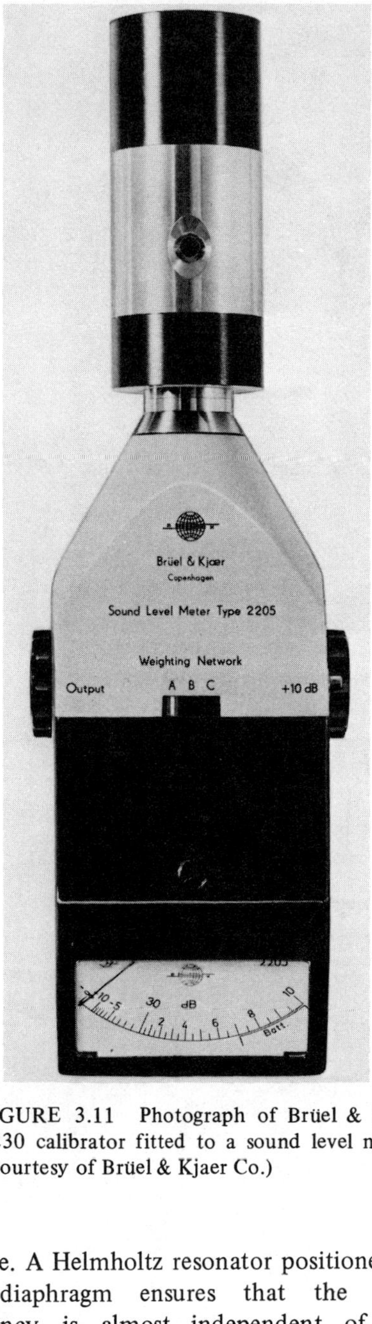

FIGURE 3.11 Photograph of Brüel & Kjaer 4230 calibrator fitted to a sound level meter. (Courtesy of Brüel & Kjaer Co.)

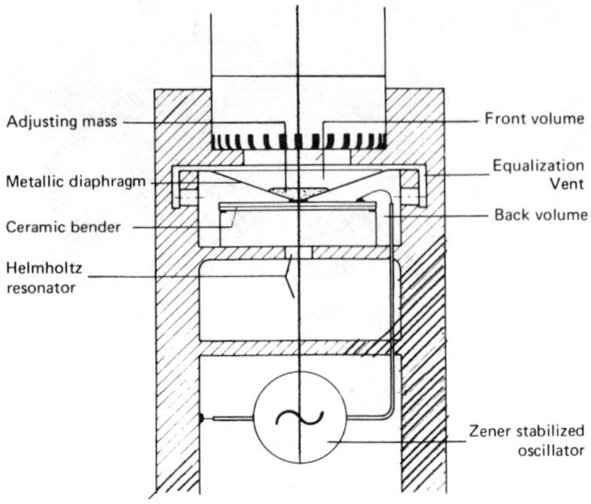

FIGURE 3.12 Principle of operation of Brüel & Kjaer Acoustic Calibrator 4230. (Courtesy of Brüel & Kjaer Co.)

volume. A Helmholtz resonator positioned behind the diaphragm ensures that the 1,000-Hz frequency is almost independent of ambient pressure (since the resonance frequency of a Helmholtz resonator depends only on the cavity volume and neck dimensions, see Equation 5.18). Because of this design, the manufacturer claims that the sound pressure level produced: 94 dB re 2×10^{-5} N/m^2, or 2×10^{-5} Pa is independent of the microphone's volume and the care taken in sealing. These loudspeaker calibrators are normally adjusted by the manufacturers who calibrate them against a laboratory standard microphone (e.g., Western Electric type 640-AA®) which has been calibrated by the National Bureau of Standards using the reciprocity procedure discussed below.

C. Electrostatic Actuator

Another very useful method of calibration involves the use of an *electrostatic actuator* consisting of an isolated metal grid placed in immediate proximity to the microphone diaphragm. Both a DC polarization voltage and a smaller AC voltage are applied to the grid. This produces a uniform sinusoidal force on the diaphragm simulating a uniform acoustic pressure. Figure 3.13 shows a measurement arrangement using an electrostatic actuator. This method is suitable for calibration up to high frequencies (about 200 kHz). With the 800-DC polarizing voltage and the 30-volt AC voltage, a force equivalent to a sound pressure of about 1 N/m^2 (1 Pa) acts on the diaphragm. If the DC voltage is omitted, the force becomes nonsinusoidal and a frequency doubling occurs.[11,13] Although this method of calibration is very useful for obtaining the pressure response of a microphone over a wide frequency range, absolute calibration is difficult because the measurement is very sensitive to the effective distance between the actuator and the microphone diaphragm.

D. Reciprocity Method

The most widely used method of absolute microphone calibration is the *reciprocity method.* This method is described in detail in U.S. Standard ANSI S1.10–1966.[14] The method can be used to

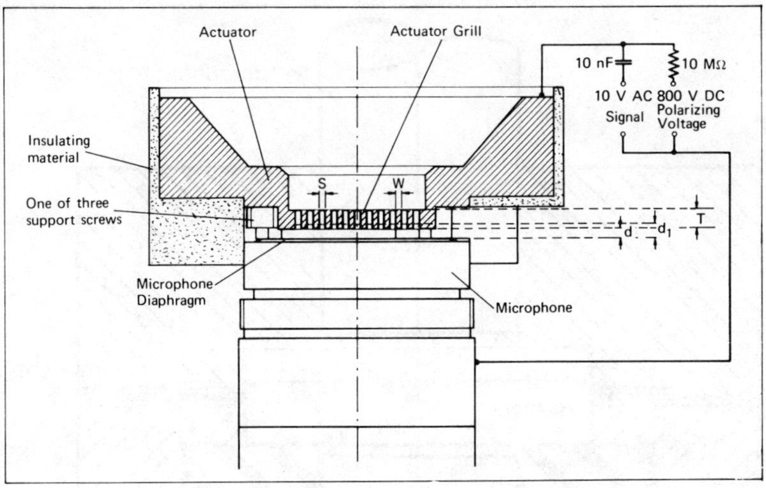

FIGURE 3.13 Measuring arrangement for determining the pressure response of a microphone using an electrostatic activator. (Courtesy of Brüel & Kjaer Co.)

measure both the *pressure* response and the *free-field* response of a microphone. In order to measure the pressure response, two microphones (which we will assume here are of the same type) and also an independent sound source are used. The first step in the calibration procedure is to measure the open-circuit voltage amplitude E of each microphone when subjected to the independent sound source producing a pressure amplitude p (say from the pistonphone). The sensitivity of each microphone M = E/p is thus obtained and the ratio of sensitivities eliminates the unknown pressure amplitude p

$$\frac{M_{p1}}{M_{p2}} = \frac{E_1/p}{E_2/p} = \frac{E_1}{E_2}. \tag{3.11}$$

The two microphones are then acoustically coupled closely together by means of a small cavity. Figure 3.14 shows a typical arrangement. One microphone is used as a transmitter (i.e., a "loudspeaker" by supplying it with a current amplitude i) and the other is used as a receiver. If the second microphone is the transmitter, then it may be shown that

$$M_{p1}\, M_{p2} = \frac{E_1'}{i_2}\, \frac{\omega V_c}{\gamma P_o}, \tag{3.12}$$

where

V_c is the volume of the cavity (usually chosen to be between 10 and 20 cm^3

γ is the ratio of specific heats for the gas in the cavity (for air $\gamma = 1.4$)

ω is the angular frequency of the sound signal ($\omega = 2\pi f$)

P_o is the atmospheric pressure.

Note here the impedance of the cavity is given by $Z = \gamma P_o/\omega V_c$. The sensitivity of microphone 2 used as the transmitter is p/i_2 and the sensitivity of the microphone 1 as a receiver is E_1'/p. In this method, the pressure amplitude p is eliminated in Equation 3.12. By combining Equations 3.11 and 3.12, the microphone sensitivity M_{p2} may be eliminated giving:[11]

$$M_{pl} = \left(\frac{E_1}{E_2} \cdot \frac{E_1'}{i_2} \cdot \frac{\omega V_c}{\gamma P_o} \right)^{1/2}. \tag{3.13}$$

All these quantities are measureable and the only restriction with this method is that the cavity is small enough so that wave effects are negligible in the cavity (λ is much greater than a typical cavity dimension), and the pressure fluctuations may be assumed uniform throughout the cavity. In the cavity shown in Figure 3.14, the upper frequency[11] is 2,600 Hz for wave effects to produce inaccuracies no greater than 0.1 dB. If hydrogen is used in the cavity, this upper frequency may be extended to 10,000 Hz. Microphone pressure response reciprocity methods are normally combined with the electrostatic actuator method to extend the pressure response calibration to much higher frequencies (e.g., 200,000 Hz).

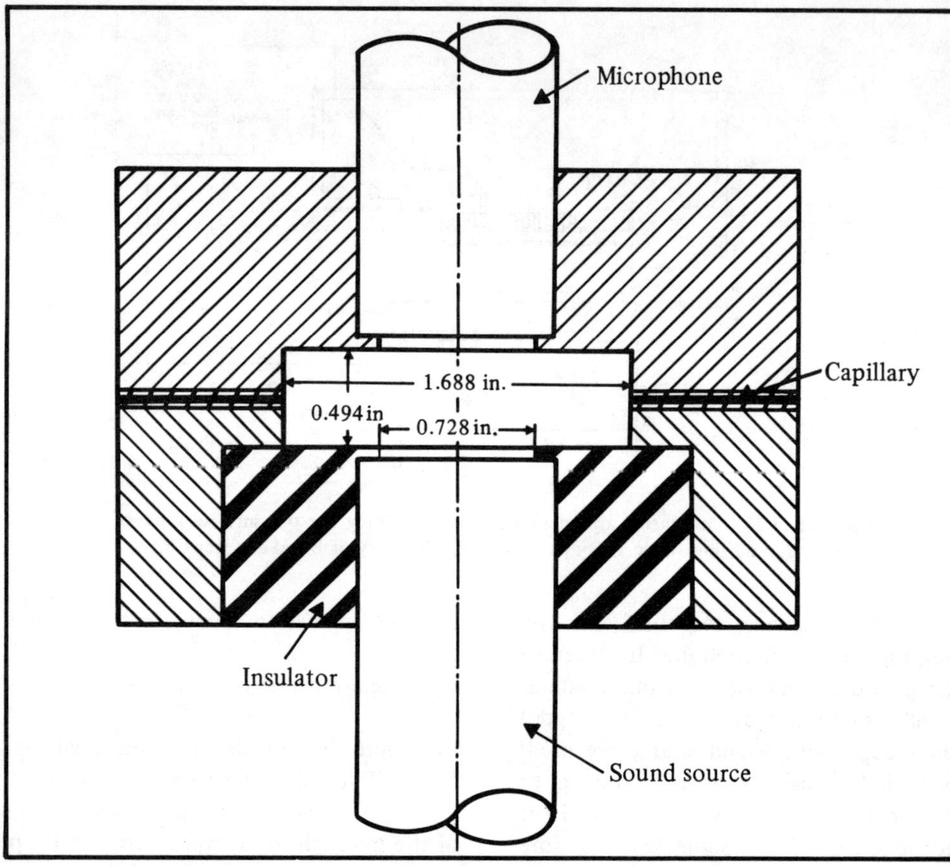

FIGURE 3.14 Typical arrangement for reciprocity calibration of a microphone.

The reciprocity method can also be used to measure the *free-field* response of a microphone. The method is very similar to that used to measure the *pressure* response. Again two microphones (again assumed here to be of the same type) and an independent sound source are used. The angle between the plane wave sound field (from the source usually produced in an anechoic room) and the microphone axis can be arbitrarily set, although 0° is normally chosen. First the ratio of sensitivities is obtained

$$\frac{M_{f1}}{M_{f2}} = \frac{E_1}{E_2} \tag{3.14}$$

by exposing both microphones to the same sound field. Note the equation is identical in form to Equation 3.11. As before, microphone 2 is operated as a loudspeaker, this time giving:[11,15-17]

$$M_{f1}\, M_{f2} = \frac{E_1'}{i_2} \cdot \frac{2d\lambda}{p\,c} \tag{3.15}$$

where

d is the distance between the microphones (d must be large compared with the microphone dimensions)

λ is the sound wavelength c/f, f is the frequency Hz

$p\,c$ is the characteristic impedance of air, where p and c are the air density and speed of sound, respectively.

Combining Equations 3.14 and 3.15 gives

$$M_{f1} = \left(\frac{E_1}{E_2}\, \frac{E_1'}{i_2}\, \frac{2d\lambda}{p\,c} \right)^{1/2}, \tag{3.16}$$

a similar result to Equation 3.13.

E. Substitution Method

The substitution method is often used to obtain the *pressure* or the *free-field* response of a microphone. First the sound pressure against frequency (either in a cavity or a free field) is measured with a standard microphone whose characteristics are very well known (the Western Electric Type 640 AA® has become widely accepted as a standard). Then the microphone under test is substituted and the sound pressure against frequency is measured again. Finally sound pressure may be measured again with the standard microphone to check that nothing has been inadvertently changed during the experiment. From the differences in response between the standard microphone and the microphone under test, the frequency response of the microphone under test can be calculated. This method is of course not so basic as the reciprocity method and is subject to some error.

3.3.5 Diffraction and Directivity

Suppose a microphone is designed to have a *pressure* response similar to that shown in Figure 3.6. When the microphone is introduced into a progressive wave field because of reflections from the diaphragm and microphone body, there will be an increase in pressure. At low frequency (where the microphone diameter D is small compared with the sound wavelength λ) there will be no increase in pressure; at high frequency, where D >> λ there will be about a 6 dB increase (for waves approaching along the microphone axis).

Since microphones are most conveniently made in cylindrical shapes, this pressure increase effect can be examined both theoretically and experimentally without too much difficulty. The pressure increase at the center of a circular area has been examined theoretically by several authors.[18,19] Figure 3.15 from Brüel and Rasmussen[12] shows a comparison between the theoretical results of Müller, Black, and Davis[19] and some further experimental results.

The pressure increase shown in Figure 3.15 agrees qualitatively with the difference between the *free-field* frequency response and *pressure* response curves for a typical microphone. There are two main reasons why there is not complete agreement: 1. The pressure increase is not uniform all across the microphone diaphragm, and 2. the microphone diaphragm is more sensitive to the pressure at its center than near its circumference. Brüel and Rasmussen[12] have made measurements of both of these effects and Figure 3.16 shows the results of these and other very clever and carefully controlled experiments with a specially designed concentric ring actuator. The different voltages, supplied to the concentric rings of the actuator, applied forces to the diaphragm proportional to the measured pressure distribution across the microphone diaphragm in a free field. For complete details the reader should consult Reference 12. It is seen from the dashed line in Figure 3.16 that the average pressure increase for 0° incidence is smoother than that at the center of the microphone.[12] The sharp notch at (D/λ) = 2 is smoothed out considerably. These experiments have been repeated for sound waves incident on the microphone diagram at other angles, and Brüel and Rasmussen give some of these results.[12]

It is interesting to compare the free-field pressure increase (compared with the *pressure* response) determined by the reciprocity method for a typical microphone (Figure 3.17) with that predicted from Figure 3.16. It is seen that the agreement for 0° is quite good. Brüel and Kjaer

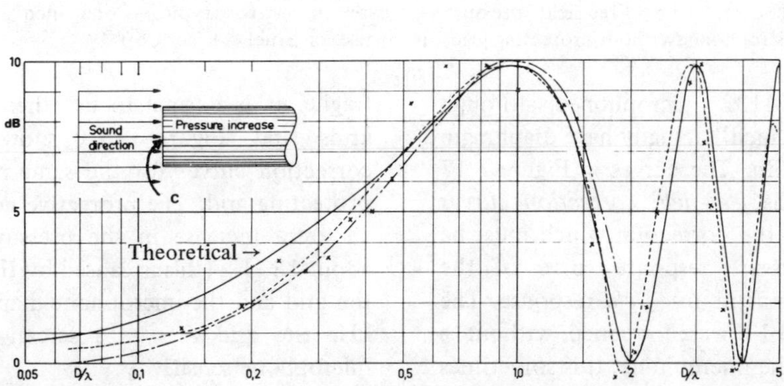

FIGURE 3.15 Pressure increase on axis at center of cylindrical surface (C). xxxxx Measured by Müller, Black, and Davis (1937); – • – • – Measured by Danish Technical University (1948); - - - - - Measured by Brüel & Kjaer (1958). (From Brüel, P. V. and Rasmussen, G., Free field response of condenser microphones, Brüel & Kjaer Tech. Rev., 1, 12, 1959. With permission).

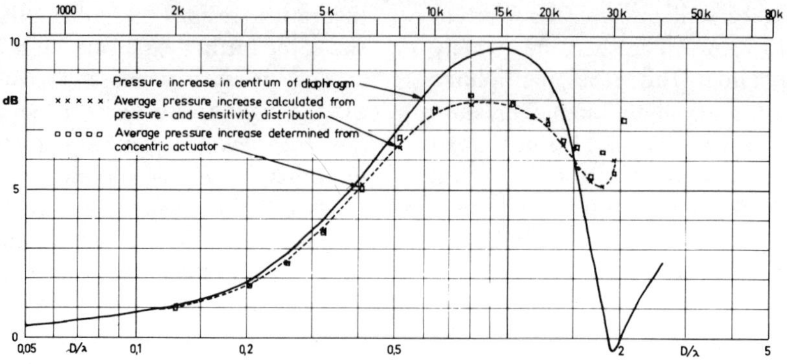

FIGURE 3.16 Pressure increase[1 2] at the center of a model microphone (solid curve). xx Average pressure increase calculated from the measured pressure and sensitivity distribution. □ □ Obtained from measurements with concentric actuator. All measurements on the model microphone were taken without protection grid. D = Microphone diameter; λ = Sound wavelength. (Courtesy of Brüel & Kjaer Co.)

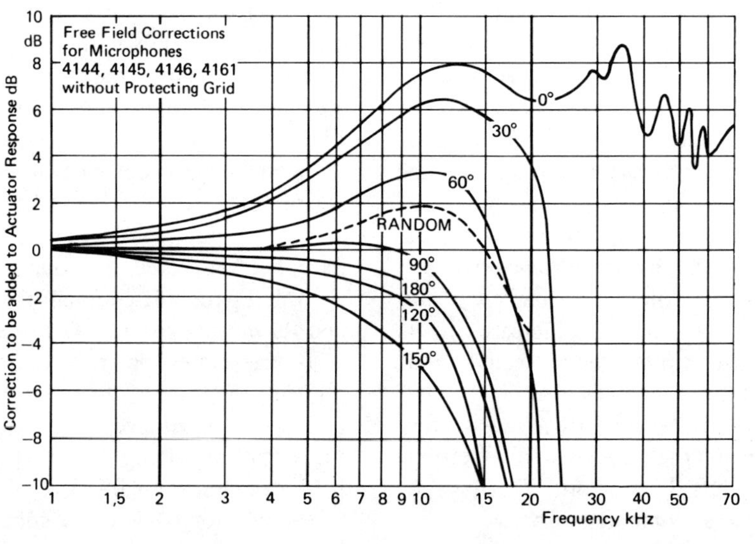

FIGURE 3.17 Free-field pressure increase curves for typical "one inch" microphone (without protecting grid). (Courtesy of Brüel & Kjaer Co.)

Type 4131 and 4132 microphones although referred to as "one inch" actually have diaphragm diameters of 0.910 in. The curves in Figure 3.17 are often known as *free-field correction curves* because they show the correction which must be added to the *pressure* response curve of the microphone to obtain the *free-field* response. The curves in Figure 3.17 were obtained without a protective grid on the microphone. It is sometimes desirable to use a microphone in this condition (e.g., if it is flush-mounted in the wall of a duct carrying an airflow, so that the diaphragm-protective grid does not produce turbulent noise). However, because microphone diaphragms are very fragile, it is normal to use them with protective grids and Figure 3.18 shows the free-field correction curves for the same microphones with protecting grid. The protective grid is seen to give an extra increase in the pressure increase up to about 18 kHz (this is caused by the cavity between the grid and the microphone diaphram). Above 18 kHz the grid acts as a screen and reduces the microphone sensitivity.

The directional characteristics of a microphone are perhaps more easily visualized on a polar plot (Figure 3.19). Here the same information as shown in Figure 3.18 is plotted in polar form (but normalized to give 0 dB at 0° incidence). The

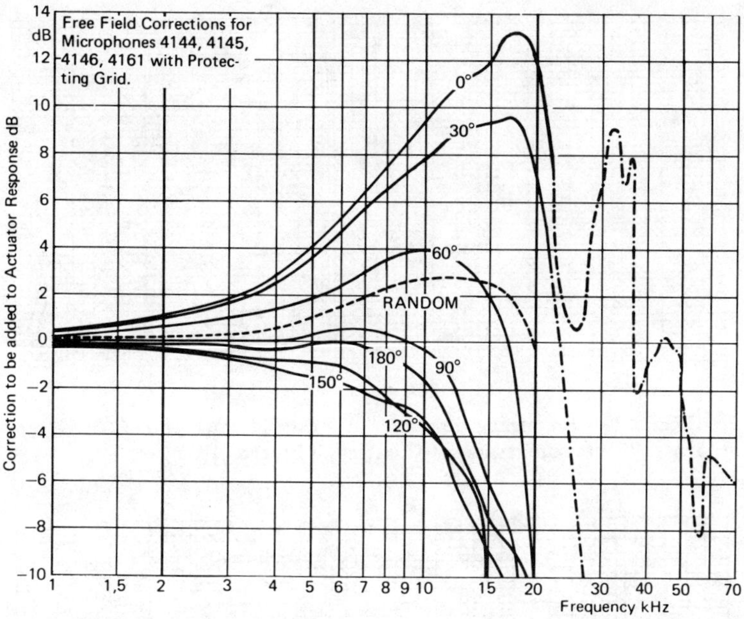

FIGURE 3.18 Free-field pressure increase curves for a typical "one inch" microphone (with protection grid). Angles are measured from microphone axis. (Courtesy of Brüel & Kjaer Co.)

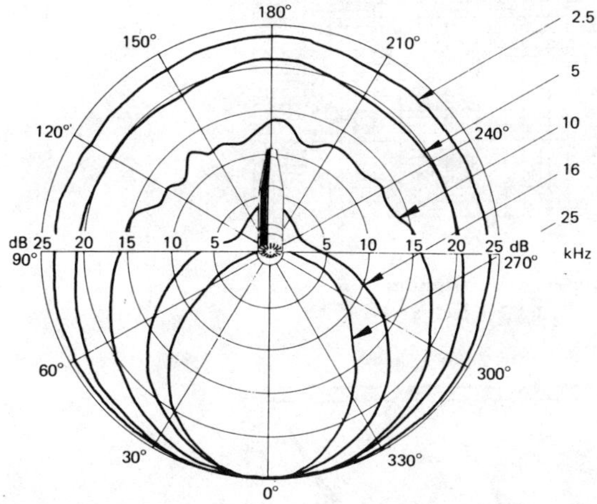

FIGURE 3.19 Directivity curves for "one inch" Brüel & Kjaer microphones (4131 and 4132) with protection grid.

curves are seen to become quite *directional* at high frequency where the sound wavelength starts to become comparable to the microphone dimensions. The high frequency directivity curves can be made less directional by fitting the microphone with a *random incidence corrector* or a *nose cone* supplied by the manufacturer.

3.3.6 Pressure and Free-Field Microphones

In some texts, for example those by Beranek[21] and Kinsler and Frey,[22] the authors denote microphones whose output is proportional to the sound pressure as *pressure microphones.* If the output is proportional to the pressure gradient, these are denoted as *pressure-gradient* microphones. For noise control work the *pressure* microphone is most commonly used. However, we should note that some authors and companies use a different terminology. Rasmussen[20] and the Brüel and Kjaer Company refer to a microphone with a pressure response which is designed to be flat as a *pressure microphone.* They refer to a microphone with a 0° incidence free-field response which is designed to be flat as a *free-field microphone.* The difference between these two types of microphones is that the diaphragm of the free-field microphone is damped in order to compensate for the pressure increase which occurs due to diffraction at high frequencies.

Figures 3.20 and 3.21 show the frequency response of Brüel and Kjaer *free-field* and *pressure* response capacitor microphones of different nominal diameters. Rasmussen[20] argues that the use of a *free-field microphone* pointed at a noise source in a building is much more desirable than using a *pressure microphone* or *random incidence corrected microphone.* Figure 3.22 illustrates this point well. A pressure microphone such as shown in Figure 3.21 must be placed at right angles to the source, while the free-field microphone must be

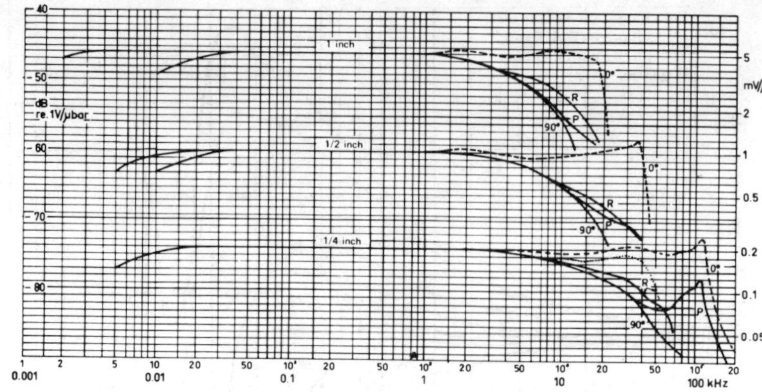

FIGURE 3.20 Frequency response of free-field microphones. R = random incidence, P = pressure response. (Courtesy of Brüel & Kjaer Co.)

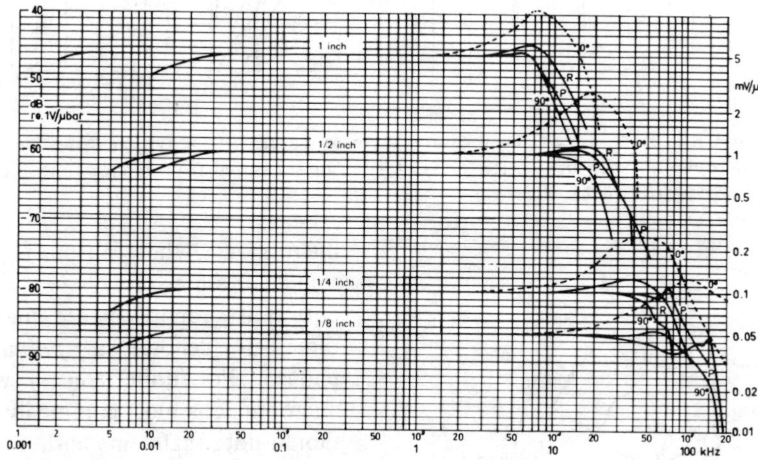

FIGURE 3.21 Frequency response of pressure microphones. R = random incidence, P = pressure response. (Courtesy of Brüel & Kjaer Co.)

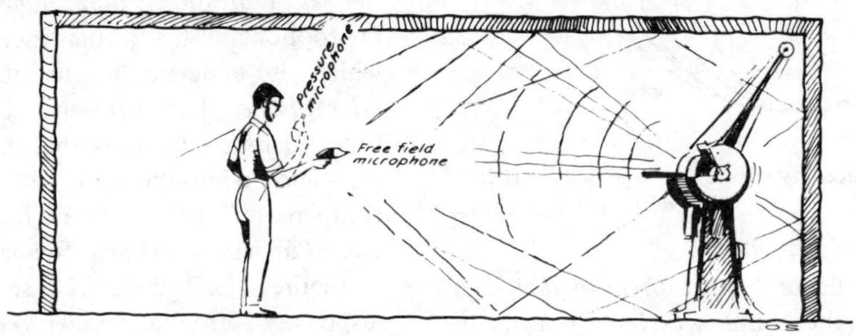

FIGURE 3.22 Noise measurement using free-field microphone or pressure microphone. (From Rasmussen, G., Measurement Microphones, in Inter-Noise 72 Proceedings, Crocker, M. J., Ed., Washington, D.C., October 4–6, 1972, 387. With permission.)

pointed at the source. In the higher frequency range, a free-field microphone will always give a "correct" reading with a value approximately equal to and never higher than the sound pressure level existing without the microphone. On the other hand, the pressure microphone may give a correct reading if no reflections are present, but it may severely overestimate the true reading if reflections are important. This is because reflections are measured at an angle of incidence of about 0° and the microphone response can be up to 8 or 10 dB too high in the higher frequency range.

3.4 TYPES OF MICROPHONES FOR NOISE MEASUREMENTS

3.4.1 Capacitor Microphones

Wood[23] related that in 1863 in a letter to Tait, Kelvin referred to the sounds that a condenser made when charged or discharged. Soon after this, numerous investigators began to study the use of a "condenser"* as a microphone. Wente is credited with developing the device to a form almost the same as is used today. In 1917[24] he described a microphone with almost uniform sensitivity from 0 to 10,000 Hz. With its uniform sensitivity, low distortion, and portability, the capacitor microphone revolutionized electroacoustics and soon became an integral part of any high quality sound system.[25] However, although the capacitor microphone has the above advantages and also can be used as either a microphone or a loudspeaker, it also suffers from some disadvantages: principally its rather low sensitivity, high internal impedance and the need for a polarizing voltage. These shortcomings lead to the development of other types of microphones such as piezoelectric** and electrodynamic which will be discussed shortly. The very large, almost distortionless electronic amplification that can now be achieved has made the lack of sensitivity of the capacitor microphone unimportant and because of its smooth sensitivity over a very wide frequency range and its well-defined geometry the capacitor microphone is still used in many applications.

Figure 3.23 shows a drawing of a modern 1-in. capacitor microphone. Figure 3.24 gives a simplified cross section of the same microphone. The capacitor microphone consists of a thin stretched metal diaphragm under tension spaced a short distance from an insulated back plate. The diaphragm and back plate constitute the electrodes of the capacitor. Holes are drilled in a back plate to provide the required air damping for the diaphragm. A small hole is provided to equalize the

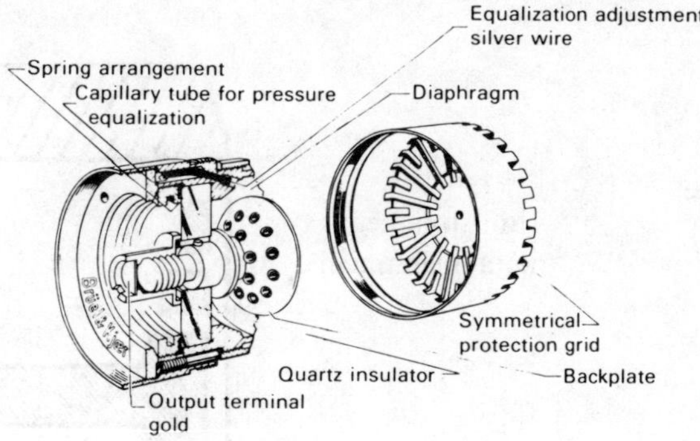

FIGURE 3.23 Modern 1-in. capacitor microphone. (Courtesy of Brüel & Kjaer Co.)

static pressure across the diaphragm, provided that it changes slowly. This hole may determine the lower cut-off frequency of the microphone. Some discussion of the electrical circuit of the system is now desirable in order to study further the capacitor microphone. Some authors[23,26,27] examine the electrical theory in some detail. We will limit ourselves to a simplified analysis.

Figure 3.25 gives a simplified electrical circuit for a capacitor microphone. A constant charge Q_0 is applied to the capacitor C by using a large DC (direct current) polarizing voltage E_0 of several hundred volts (usually between 100 and 300 V) and a high resistance R of several hundred megohms. The charge Q_0 will remain constant, provided the time constant RC is large in the frequency range of interest

$$RC >> (1/\omega). \tag{3.17}$$

This means that either R or C, or both, should be large in order to maintain constant charge at low frequency.

The circuit in Figure 3.25 may be replaced with the equivalent circuit shown in Figure 3.26. The equivalent circuit is composed of an equivalent capacitor C_e which is increased by the stray capacitance C_s of the connection with the amplifier and the input capacitance of the amplifier C_i. Hence $C_e = C + C_s + C_i$. The polarization resistor R is changed by the parallel input resistance of the amplifier R_i, so that the equivalent "load" resistance $R_e = R + R_i$.

*Such microphones, known as *condenser* microphones for many years are now more commonly called *capacitor* microphones.

***Piezoelectric* microphones are also sometimes known as crystal microphones.

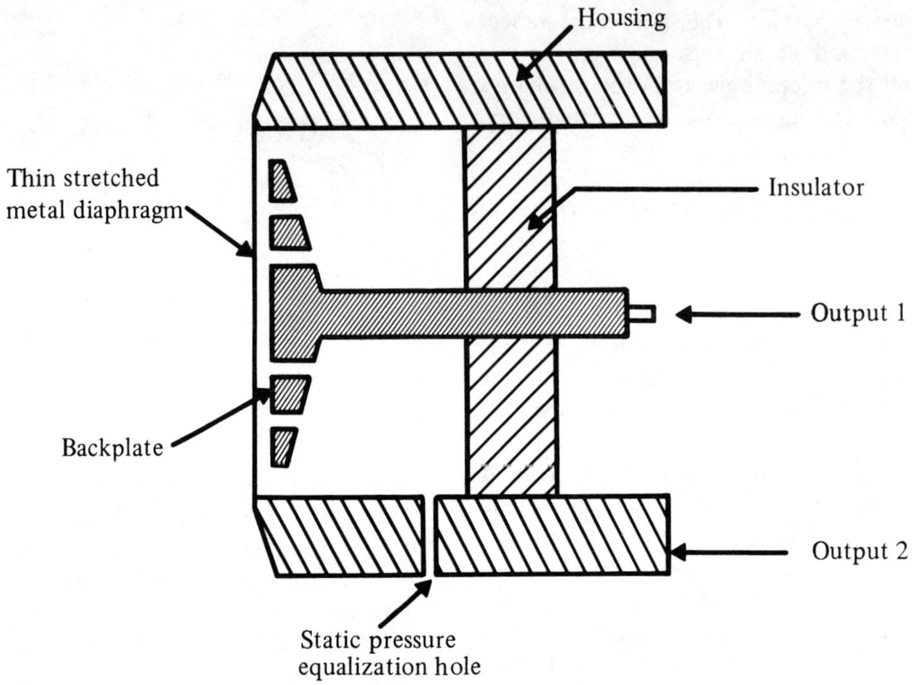

FIGURE 3.24 Simplified cross-sectional view of a capacitor microphone.

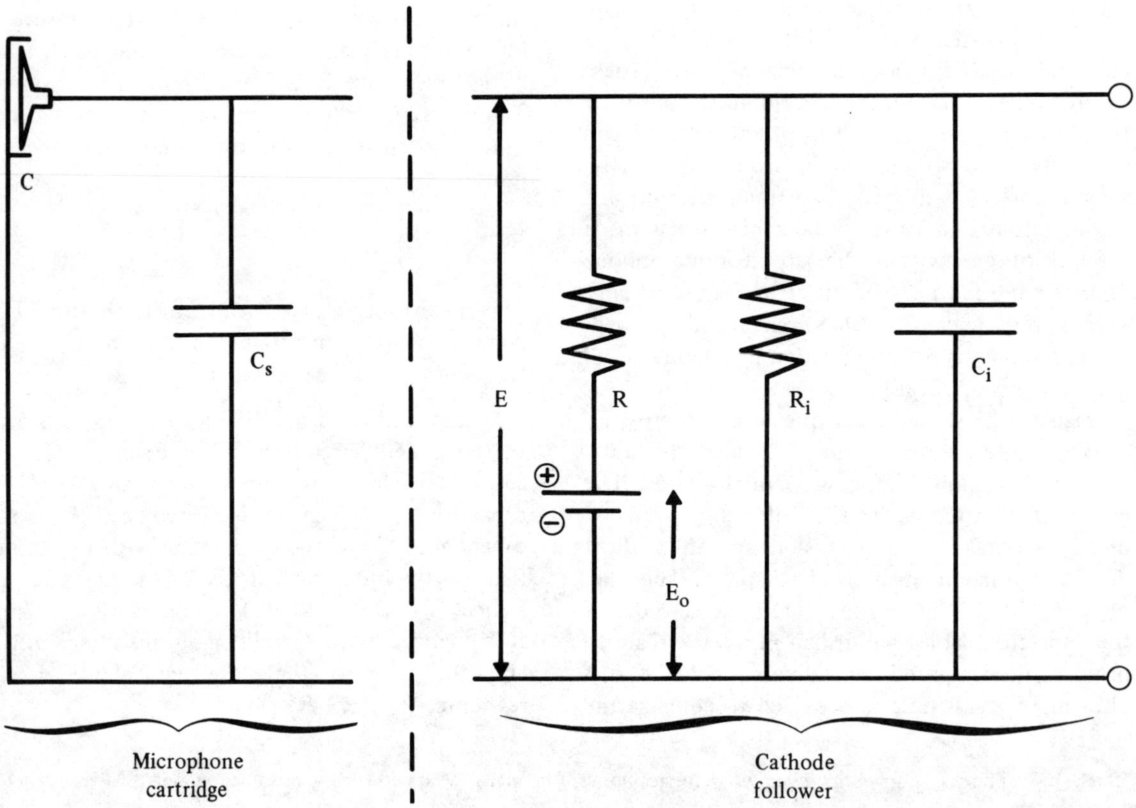

FIGURE 3.25 Simplified electrical circuit for a capacitor microphone.

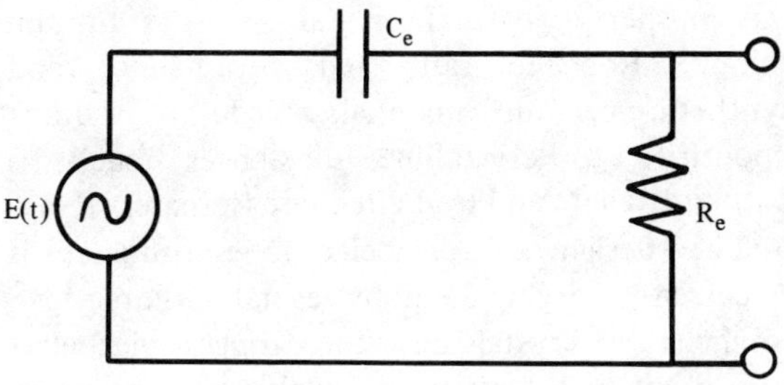

FIGURE 3.26 Simplified equivalent circuit for a capacitor microphone.

The circuit in Figure 3.26 has the appearance of a high-pass filter (see Section 3.6.1). Its low cut-off frequency f_c is given by (see Section 3.6.1)

$$f_c = \frac{1}{2\pi R_e C_e}. \tag{3.18}$$

This assumes that there is no vent hole to the atmosphere. Thus again we see that for a low cut-off frequency, R_e, C_e or both should be large. Now there is a reason why C_e cannot be made too large in a capacitor microphone; for a circular capacitor the capacitance C is given by

$$C = \frac{\epsilon_0 \pi a^2}{d}, \tag{3.19}$$

where

a is the microphone radius
d is the separation of the plates
ϵ_0 is the permittivity.

(For free space, $\epsilon_0 = 8.85 \times 10^{-12}$ farad/m.) Thus for steady state conditions, if d_0 is the static separation, then the static capacitance C_0 is

$$C_0 = 27.8(a^2/d_0)\ \text{pF}. \tag{3.20}^*$$

For geometric and constructional reasons a pF cannot be made too large or d_0 too small. The capacitance, C_0, of typical microphones varies between 10 and 200 pF (e.g., if $d_0 = 0.00004$m and a = 2 cm = 0.02m, then C_0 = 278 pF). Thus, in order to satisfy Equations 3.17 and 3.18, R_e must be made as large as possible. In typical microphones, R is of the order of several hundred megohms. This will also require a large amplifier input resistance R_i to prevent R_e from becoming small. Now the charge on a capacitor Q_0 is

$$Q_0 = E \cdot C. \tag{3.21}$$

For steady state conditions**

$$Q_0 = E_0 \cdot C_0. \tag{3.22}$$

From Equations 3.19 and 3.20, if the separation d_0 is increased to $d_0 + \Delta d$, then the capacitance,

$$C = C_0/(1 + \Delta d/d_0). \tag{3.23}$$

Using the binomial expansion,*** providing the displacement Δd is small compared with d_0,

$$C = C_0(1 - \Delta d/d_0); \tag{3.24}$$

or equivalently,

$$C = C_0 - \Delta C. \tag{3.25}$$

From Equations 3.21, 3.22, and 3.23, the voltage across the capacitance is

$$E = \frac{Q_0}{C} = \frac{E_0 C_0}{C_0 - \Delta C} = \frac{E_0}{1 - \frac{\Delta C}{C_0}} \approx E_0 \left(1 + \frac{\Delta C}{C_0}\right) \tag{3.26}$$

by the binomial expansion.*** We see that the fluctuating part of the output voltage is proportional to $\Delta C/C_0$. Hence, we have an additional reason for keeping C_0 small to ensure high sensitivity. If C_0 is increased too much by the addition of the stray capacitance C_s and the input capacitance C_i, then the sensitivity is decreased. Since the capacitance of screened cable is about 20 pfarad/ft, even just a few feet of cable connected from the microphone would decrease the microphone output voltage E. The normal practice in overcoming this problem is to place a noninverting amplifier right at the microphone electrode. This amplifier is known as a *cathode follower*[1,2] and converts the high internal impedance of the microphone $(1/i\omega C)$ (see Section 3.2.1) to a low impedance. Substantial lengths of cable can then

*1 picofarad = 1 pF = 10^{-12} farad.
**Ignoring for the moment the capacitances C_s and C_i.
***The binomial expansion states that $(1 + x)^n = 1 + nx + n(n - 1)x^2/1{\cdot}2 + n(n - 1)(n - 2)x^3/1{\cdot}2{\cdot}3 + \ldots$ provided $x < 1$. Thus $(1 + x)^{-1} \approx 1 - x$, if $x \ll 1$.

be connected to the cathode follower without reducing the microphone output.*

Figures 3.20 and 3.21 show the frequency responses of *free-field* and *pressure* capacitor microphones. The mechanical construction of the microphones is basically the same. The only difference is in the damping of the diaphragm. In the *free-field* microphone, the microphone resonance (at about 10,000 Hz for the 1-in. microphone) is overcritically damped (see Section 1.2.3A and B). In this way, the free field pressure increase (Figures 3.17 and 3.18) is compensated for by the decrease in pressure sensitivity to give an almost flat 0° free-field response up to 18,000 Hz. In the case of the pressure microphone, the resonance (at about 8,000 Hz for the 1-in. microphone) is approximately critically damped to obtain an almost flat pressure frequency response up to 6,000 or 7,000 Hz. The fundamental resonance frequency for a circular stretched diaphragm is:[29]

$$f_r = \frac{2.405}{2\pi a}\sqrt{\frac{T}{\sigma}}, \qquad (3.27)$$

where

a is the radius
T is the tension per unit length
σ is the surface mass density of the diaphragm.

Note above the resonance frequency the microphone response is mass controlled (see Figure 1.9) and falls off rapidly. The damping is controlled by the size of the holes in the stationary back plate (Figure 3.23) and the tension in the diaphragm.

3.4.2 Piezoelectric Microphones

The piezoelectric effect was discovered by J. and P. Curie[30,31] in the late 19th Century. If a slice is cut from a quartz crystal and pressures applied to the opposite faces of the slice causing a deformation, then equal and opposite charges are produced on the opposite faces of the slice with an electric potential developed between the faces (directly obeying Equation 3.21). Many other materials have since been discovered which have this property (natural crystals such as lithium sulfate, Rochelle salt, and tourmaline) and synthetic ceramic materials made by adding impurities to crystalline substances (such as barium titanate and lead zirconate titanate.)

The design of piezoelectric transducers is discussed in varying detail by several authors.[32–37] Rochelle salt crystals have the strongest piezoelectric effect and have been widely used for many years in transducer design. Unfortunately they deteriorate with moisture and are damaged by temperatures above 115°F. Ceramic materials such as barium titanate are more stable. They can withstand higher temperatures and are relatively unaffected by moisture or high humidity although their sensitivity is about 10 to 20 dB less than Rochelle salt.[35,37]

Piezoelectric crystals may be cut and used in many different orientations. Crystals are often directly exposed to liquids, (e.g., as hydrophones) where the high mechanical impedance of the liquid is not a disadvantage. In gases, the large acoustical/mechanical impedance mismatch is a disadvantage and piezoelectric materials are usually used in conjunction with a diaphragm.

Figure 3.27 shows a cross-sectional view of a commercial piezoelectric microphone. Figure 3.28 shows a simplified view of the microphone. Much thicker diaphragms are normally used with piezoelectric than capacitor microphones (usually about 50 times greater). This inevitably leads to a lower resonance frequency for the diaphragm (assuming the density of the diaphragm material is the same) (consider Equations 1.10 and 3.27). In order to obtain a flat free-field response, it is necessary to damp this resonance overcritically. Hence the upper frequency response is poorer than with a capacitor microphone because the mass controlled region is entered at a lower frequency (see Figure 1.9).

The diaphragm is connected to a ceramic bender element. In Figures 3.27 and 3.28 a bimorph simply supported** bender element is shown, although sometimes cantilever** benders are used.[16,38] The force needed to produce a

*The above elementary discussion, on the circuit theory of a capacitor microphone may be supplemented with more in-depth treatments given in References 23, and 26 to 28.

**A simply supported beam is one with no displacement or bending moment at each end. A cantilever beam is one with no displacement or slope at one end, but is completely free at the other. In microphone use, the free end of the cantilever is connected to the center of the diaphram.

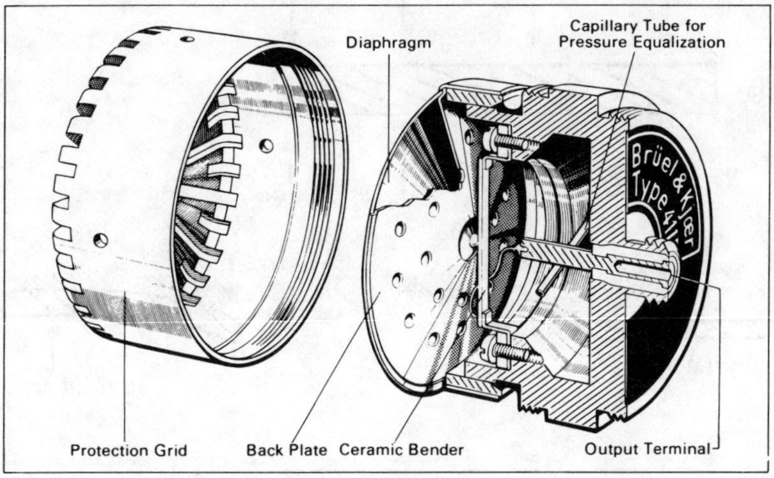

FIGURE 3.27 Cross-sectional view of piezoelectric microphone. (Courtesy of Brüel & Kjaer Co.)

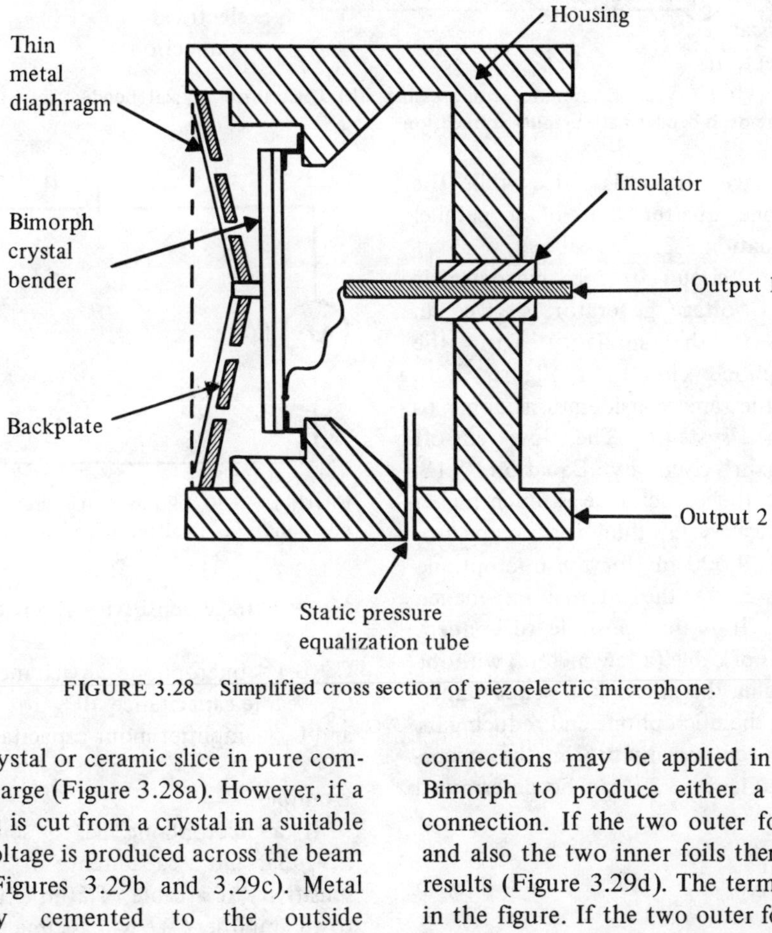

FIGURE 3.28 Simplified cross section of piezoelectric microphone.

voltage from a crystal or ceramic slice in pure compression is quite large (Figure 3.28a). However, if a thin bar or beam is cut from a crystal in a suitable orientation,* a voltage is produced across the beam as it is bent (Figures 3.29b and 3.29c). Metal foil is usually cemented to the outside surfaces of the crystal, and the two foils with the crystal in between form a capacitor of the solid dielectric type. It is usual to use two bars in conjunction to produce a *Bimorph*. Electrical connections may be applied in two ways to the Bimorph to produce either a parallel or series connection. If the two outer foils are connected and also the two inner foils then a parallel system results (Figure 3.29d). The terminals are as shown in the figure. If the two outer foils are used as the two terminals, and the center foils are connected together, a series arrangement results. Different orientations of the crystals will be needed for the parallel and series arrangements. With the series

*For a detailed discussion see References 35 and 36.

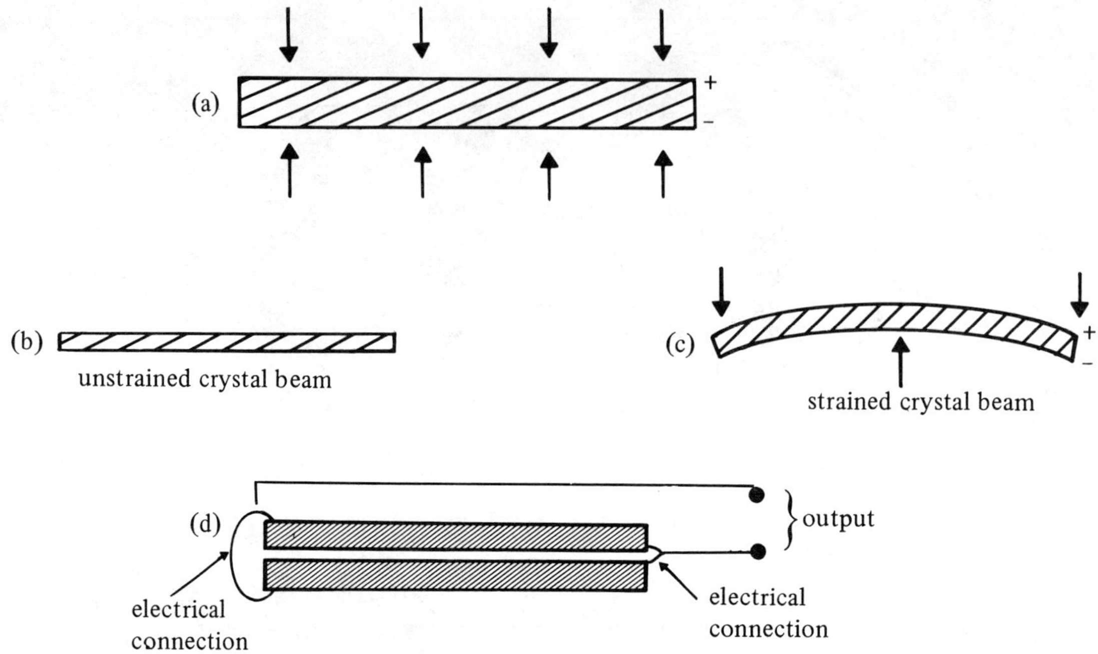

FIGURE 3.29 (a) Crystal slice under compression; (b) Unstrained crystal bender; (c) Strained crystal bender; (d) Bimorph bender with parallel connection.

Bimorph, the voltage output is twice while the capacitance is one quarter that of a parallel Bimorph, (see Equation 3.21).

The equivalent circuit for a piezoelectric microphone as a voltage generator is given in Figure 3.30. Note the similarity with the capacitor microphone circuit in Figure 3.26. Hence, many of the same considerations apply to both microphone systems. The low cut-off frequency is again given by Equation 3.18. However, because the capacitance is much higher for a piezoelectric crystal than for a capacitor microphone (it is 4,000 pF for the microphone shown in Figure 3.27), the internal impedance ($1/i\omega C$) is lower. It is thus possible to connect reasonable lengths of cable (a few meters) without appreciably changing the microphone capacitance and hence loading the microphone and reducing its output. The reduced sensitivity for the microphone shown in Figure 3.27 can be determined from the formula:[39]

$$M_{v(c)} = \frac{M_v(C + 0.1)}{C + C_c + C_i}, \qquad (3.28)$$

where

$M_{v(c)}$ is the reduced voltage sensitivity with the cable

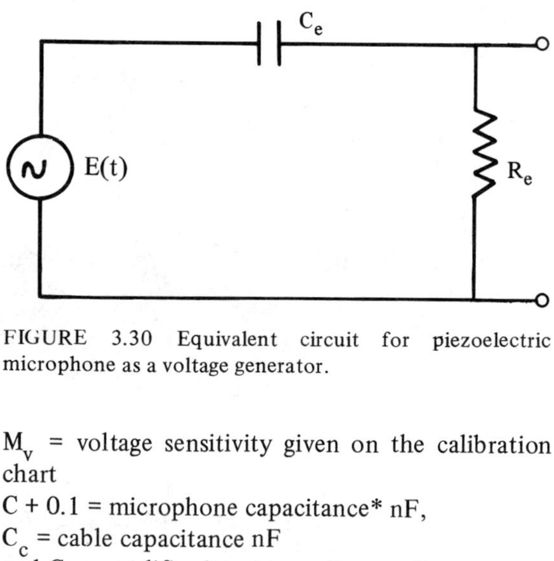

FIGURE 3.30 Equivalent circuit for piezoelectric microphone as a voltage generator.

M_v = voltage sensitivity given on the calibration chart

$C + 0.1$ = microphone capacitance* nF,

C_c = cable capacitance nF

and C_i = amplifier input capacitance nF.

Example

If a microphone has a sensitivity of 0.400 mV/μbar and a capacitance of 3.9 nF, what is the sensitivity if a cable of 50m is used to connect it to an amplifier? We will assume the amplifier input capacitance is negligible and the cable capacitance is 82 pF/m. Thus, cable capacitance = 50 × 82 pF = 4.1 nF, and

*0.1 μF was used during the sensitivity calibration. *Note*: 1 μF = 10^{-6} F; 1 nF = 10^{-9} F; 1 pF = 10^{-12} F.

$$M_{v(c)} = \frac{0.400\ (3.9 + 0.1)}{3.9 + 4.1} = \frac{0.400 \times 4.0}{8.0} = 0.200\ \text{mV}/\mu\text{bar}$$

The microphone sensitivity is halved by the addition of the cable.

The frequency response of the microphone shown in Figure 3.27 is shown in Figure 3.31. A typical calibration chart supplied by the manufacturer with each microphone is shown in the figure. Note that the upper curve shows the free-field response, while the bottom curve shows the pressure response. Because the free-field response is flat in the range of 3 to 10,000 Hz (±3dB), we would describe this microphone as a *free-field* microphone. However, Figure 3.32 shows the frequency response of a piezoelectric microphone designed to have a "flat" random incidence or grazing incidence response. We might then call this microphone a *random incidence* microphone, or since its pressure response is also probably almost flat, a *pressure* microphone. Obviously, for most noise measurements, the *free-field* microphone should be pointed at the source and the *random incidence* or *pressure* microphone should be placed at grazing incidence. Manufacturers' instructions should always be studied and followed with each different type of microphone used, or else incorrect readings will be obtained, particularly at high frequency.

In summary, the advantages of a piezoelectric microphone (compared with a capacitor microphone) are its ruggedness, its lower internal impedance (so that loading effects by cables are tolerable), less susceptibility to humidity and condensation, and low initial cost. However, it has the disadvantages of a more restricted frequency response and production of a spurious response at high acceleration levels.

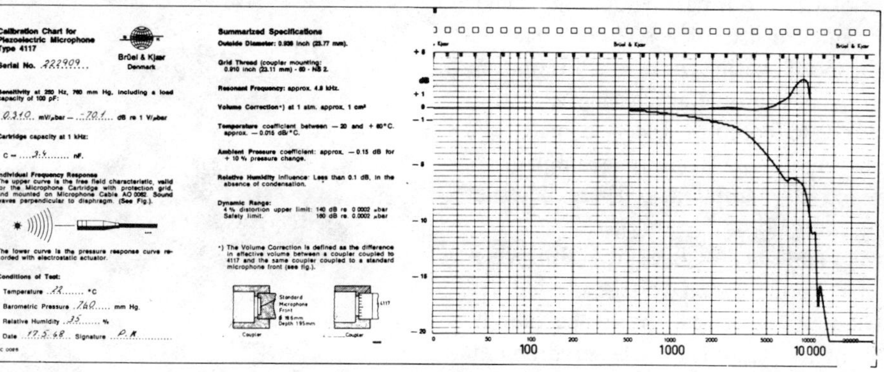

FIGURE 3.31 Typical calibration chart for piezoelectric microphone. Microphone shown in Figure 3.27. (Courtesy of Brüel & Kjaer Co.)

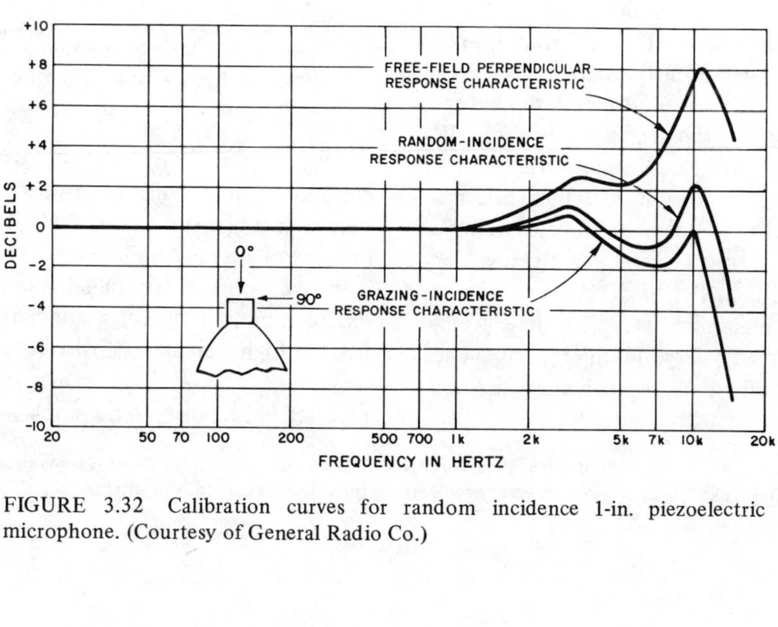

FIGURE 3.32 Calibration curves for random incidence 1-in. piezoelectric microphone. (Courtesy of General Radio Co.)

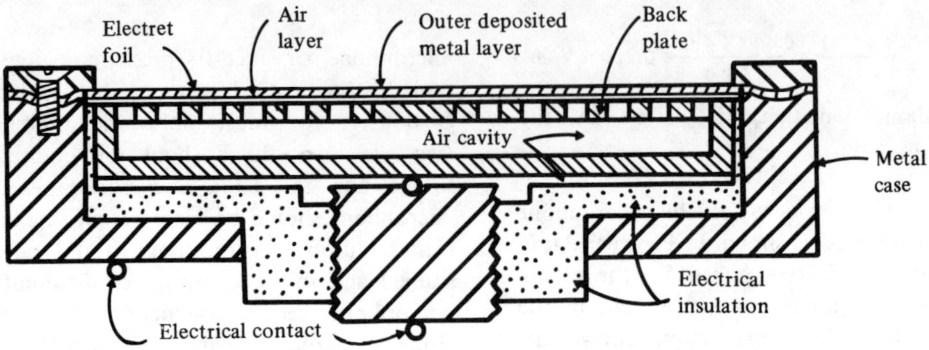

FIGURE 3.33 Cross-section of electret microphone. (From Sessler, G. M., and West, J. F., Foil electret microphones, *J. Acoust. Soc. Am.*, 40, 1433, 1966. With permission.)

3.4.3 Electret and Other Types

In the 1960s, a new type of precision microphone was developed by Sessler and West,[40,41] and has since become commercially available.* This microphone is basically a capacitor microphone, however, no DC polarization voltage is needed since the electret foil diaphragm (Figure 3.33) is permanently polarized during manufacture. The electret diaphragm consists of a stretched plastic foil with a thin metal coating (usually gold) on the outer surface. The thin metal coating and the perforated backplate comprise the two plates of the capacitor. Figure 3.34 shows examples of commercial electret microphones. These became available in the early 1970s.

The electret appears to have the following advantages: a) no polarization voltage needed, b) rugged construction, c) large capacitance (about 500 pF), and hence, loading is a lesser problem than with the capacitor microphone), and d) low cost. However, it does have the following disadvantages: a) it has a less smooth frequency response at high frequencies than the capacitor microphone (Sessler and West[41] reported a 4-cm diameter microphone with a resonance frequency of 14,000 Hz), b) it has an upper temperature limit of about 50°C, c) the sensitivity is about 10 dB lower than for comparable capacitor microphones[41] and is more temperature dependent, and d) the polarization may not be sufficiently stable over long periods of time (several years). However, it is likely that with sufficient development these drawbacks will be gradually overcome; the electret microphone is already a good microphone and holds promise of further improvements in the future.

FIGURE 3.34 One in., ½ in., and ¼ in. examples of electret microphones which are commercially available. (Courtesy of General Radio Co.)

Other microphones are used for many communication purposes but not normally for precision noise measurements. The moving coil microphone (Figure 3.35) has been in use for many years but is less valuable in noise work since its frequency response is not very smooth and is poor at high frequencies. Also its sensitivity is low. However, it does have some advantages which make it ideal for some applications. In 1931,

*Electrets actually have a history dating back over forty years. However, they have many disadvantages and have only recently become widely used. For a good review of electrets including recent developments see *J. Acoust. Soc. Am.*, 53(6), 1578, 1973.

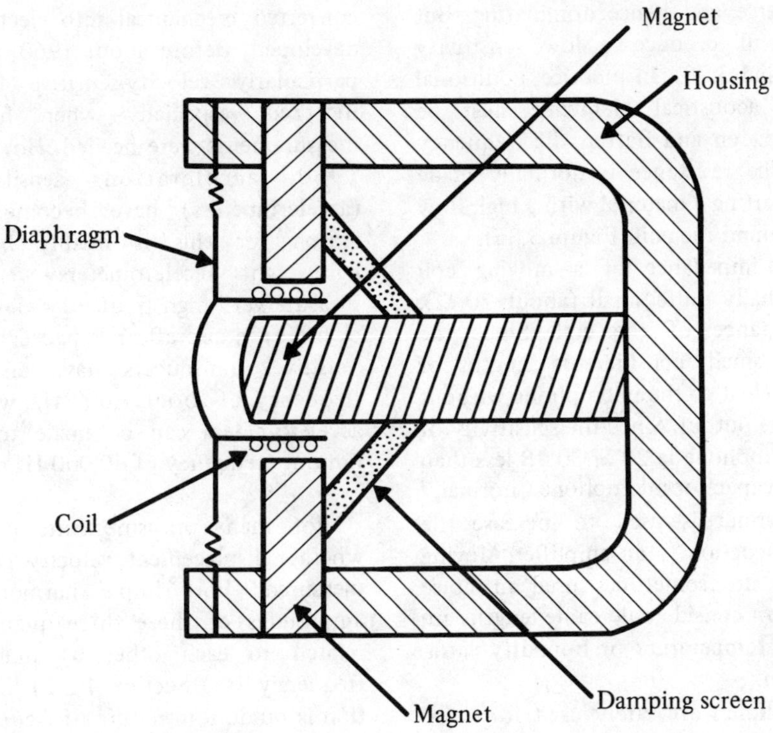

FIGURE 3.35 Cross section of a typical moving coil microphone.

Wente and Thuras[42] published a paper in which they successfully designed a moving coil microphone using electrical analogues of its acoustical/mechanical components. The moving coil microphone suffers from several resonances and by proper design, these resonances may be damped so that the pressure frequency response between about 50 and 8,000 Hz is relatively flat.[43]

The moving coil microphone works on the reverse principle of the common loudspeaker. A light diaphragm whose center portion is stiffened by giving it a slight spherical deformation is attached to a coil. The coil is situated in the annular gap of a permanent magnet. The terminal connections are made to each end of the coil. When sound pressure deflects the diaphragm and consequently displaces the coil, a voltage **E** is generated across the coil.

$$\mathbf{E} = B\ell\mathbf{u} \tag{3.29}$$

where

B is the magnetic field flux density, weber/m^2
ℓ is the length of the coil m,
u is the coil velocity, m/sec.

Ignoring electrical and acoustical impedances, assume the microphone system is modeled as a single degree of freedom system (see Section 1.2.3C). Then using Equation 1.16b and assuming simple harmonic motion, the velocity **u** is

$$\mathbf{u} = \frac{F}{\mathbf{Z}_m} = \frac{P \cdot Se^{(i\omega t + \phi)}}{\mathbf{Z}_m}, \tag{3.30}$$

where

$\mathbf{Z}_m$ is the mechanical impedance
F is the total force on the diaphragm
P is the pressure amplitude at any angular frequency ω
S is the diaphragm area.

Examining Equation 1.17 and noting $\mathbf{u} = i\omega y = i\omega A_e{}^{i\omega t}$ for simple harmonic motion, we find that the impedance $\mathbf{Z}_m$ is

$$\mathbf{Z}_m = R + i(\omega M - K/\omega). \tag{3.31}$$

For a constant voltage amplitude to be produced as the frequency ω is changed, Equations 3.29, 3.30, and 3.31 show that the impedance must be constant and completely resistive. With this simple model, this can only be achieved by making the resistance very large and hence dominating. But this obviously will produce a low sensitivity (Equations 3.29 and 3.30). In practice, additional mechanical and acoustical elements must be introduced to broaden and flatten the frequency response.[43–45] The resistance is normally made quite large by inserting a material with a high flow resistance close behind the coil (Figure 3.35).

The electrical impedance of a moving coil microphone is usually quite small (about 20 Ω), and is the resistance of the coil. Since the impedance is so small, considerable lengths of cable may be attached to the microphone without reducing its voltage output. Since the sensitivity of a moving coil microphone is 20 or 30 dB less than a piezoelectric or capacitor microphone, normally a step-up transformer is used to increase the voltage before connection to an amplifier. Moving coil microphones are sometimes used in noise measurements when considerable cable lengths are required or when temperature or humidity variations are a problem.

Carbon microphones are rarely used for noise work because of their poor frequency response characteristics. However, because of their low cost, high sensitivity, and other advantages they are almost exclusively used in telephone systems. Ribbon microphones being very directional,[46,47] are also rarely used in noise measurements, although such directionality can sometimes be an advantage. The Ribbon microphone was first developed by Gerlach in 1924. It suffers from the disadvantage of low sensitivity, although the frequency response is fairly smooth.[48]

3.5 VIBRATION MEASUREMENTS

There are three quantities which are of most interest in vibration studies. These are *displacement, velocity,* and *acceleration.* A fourth quantity *strain* is now also frequently measured.* In the early 1900s, most vibration measurements were made using mechanical or optical devices. Such devices are still used satisfactorily for low frequency measurements (a few Hz). With the advent of electronics in the 1920s, transducers which converted mechanical into electrical signals were developed. Before about 1960, displacement and particularly velocity-sensitive transducers were utilized, especially when higher frequency measurements were needed. However, since about 1960, acceleration sensitive transducers (accelerometers) have become preferred. The reason for this is mainly because excellent lightweight accelerometers were developed to measure very high frequency vibrations (5,000 Hz or more) in aircraft and spacecraft. Most velocity-sensitive transducers have an upper limiting frequency of about 1,000 Hz, while piezoelectric accelerometers can be made to have an upper limiting frequency of 40,000 Hz or more.

For many measurements it is unimportant whether displacement, velocity, or acceleration are measured. For simple harmonic motion, the amplitudes of these three quantities are simply related to each other by multiplying by the frequency (see Section 1.2.2). Even if the vibration is random in nature, if frequency filtration is used, this principle can still be applied. Alternatively, the conversion from acceleration to velocity and displacement may be made by using commercially available electronic integrators.

3.5.1 Principle of Seismic Mass Transducers

Modern day piezoelectric accelerometers and some of the earlier displacement transducers[49] work on the same principle: a seismic mass M is supported on a spring K and the whole is enclosed in a case (Figure 3.36). The damping constant R in most applications is small** and it is neglected for simplicity in the following analysis (although it could easily be included).

If the vibrating member (test piece) is undergoing a time-dependent displacement $y_2(t)$, and the mass experiences a time-dependent displacement $y_1(t)$, then considering the forces acting on the mass M,

$$K(y_2 - y_1) = M\ddot{y}_1 . \tag{3.32}$$

The situation is identical to that studied in Section 1.2.3C, except that there is no external force

*Strain gauges are becoming even more useful now that frictional unbonded gauges are available. Such gauges are bonded to one face of a rubber sheet. This gauge is reusable and may either be pressed by hand against the test piece or held in place with a special holder which applies 6 to 10 lb of force.

**This is not always true. Some accelerometers now employ damping ratios as high as 0.7 in order to extend their useful frequency range.

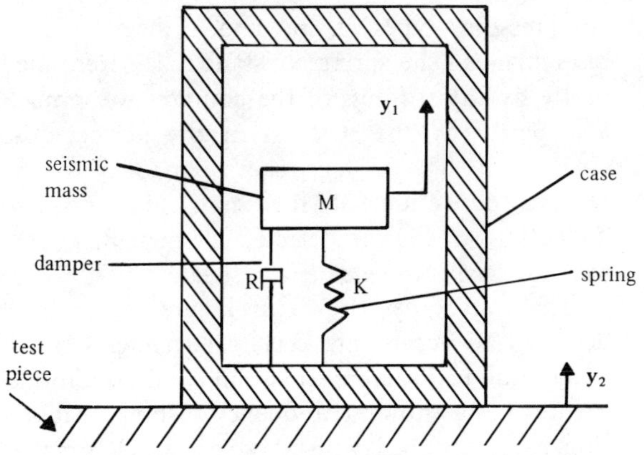

FIGURE 3.36 Idealized diagram of vibration transducer.

applied to the mass, and instead the test piece receives a displacement input.

The easiest variable to measure is the relative displacement $y_2 - y_1$. We will assume the test piece experiences simple harmonic motion

$$y_2 = B\, e^{i\omega t}; \tag{3.33}$$

and, consequently in steady state conditions, the mass experiences simple harmonic motion at the same angular frequency ω

$$y_1 = A\, e^{i\omega t}. \tag{3.34}$$

Note that **A** and **B** are written as complex quantities because y_1 and y_2 will not generally be in phase (see Section 1.2.3.C). From Equations 3.32, 3.33, and 3.34 we obtain (noting that $i^2 = -1$):

$$K(B - A)e^{i\omega t} = -M\omega^2 A\, e^{i\omega t} \tag{3.35a}$$

$$B - A = -(M\omega^2/K)A \tag{3.35b}$$

rearranging Equation 3.35b gives

$$A = -(K/(M\omega^2 - K))B, \tag{3.36}$$

and substituting Equation 3.36 into 3.35b gives

$$B - A = \frac{M\omega^2}{M\omega^2 - K}\, B. \tag{3.37}$$

It is easy to see how this result can be used to design displacement, velocity, or acceleration transducers.

A. Displacement Transducers

If we can make the output signal proportional to the relative displacement amplitude $|B - A|$, then the instrument will theoretically have a flat frequency response, providing $M\omega^2 >> K$, or $\omega^2 >> K/M$. Noting Equation 1.10, this means $\omega^2 >> \omega_n^2$, and the instrument is operated well above its resonance frequency. Thus, such instruments must be made with a large mass and a soft spring to give a low resonance frequency. Note that in this situation $|B - A| = |B|$ and obviously $|A| \approx 0$ . . . . the seismic mass remains almost stationary in space.

Instruments known as *vibrometers* where the relative displacement $y_2 - y_1$ was measured by a dial indicator, or as *vibrographs* where $y_2 - y_1$ was measured with a mechanical linkage, a stylus and recording paper were widely used until the 1950s. They were heavy and only suitable for use on heavy machinery.[49] The soft spring gave a low resonance frequency between about 0.5 and 3 Hz, but the upper frequency limit was normally about 100 Hz due to limitations in the mechanical linkages. The relative displacement with such a system can also be measured electrically in a variety of ways.

B. Velocity Transducers

Equation 3.37 shows that the seismic mass system may also be made to measure the velocity amplitude $\omega|B|$. In such a system described as a *velometer*, a coil was attached to the case and the seismic mass was made into a magnet surrounding the coil. The voltage generated by such a system is proportional to the relative velocity of the magnet and the coil (see Equation 3.29). Again examining Equation 3.37 we see that if $M\omega^2 >> K$ and the test piece velocity amplitude $\omega|B|$ is kept constant, then the relative velocity amplitude $\omega|A - B|$ will remain constant and the instrument should have a flat frequency response.

C. Acceleration Transducers

To measure acceleration conveniently with the seismic mass, the system shown in Figure 3.36 must be used in a different way. The ratio of relative *displacement* amplitude $|B - A|$ to test piece acceleration amplitude $\omega^2|B|$ is from Equation 3.37

$$\frac{|B - A|}{\omega^2\, |B|} = \frac{1}{|\omega^2 - \omega_n^2|} = \frac{1/\omega_n^2}{|1 - (\omega/\omega_n^2)|}. \tag{3.38}$$

Except for the constant $(1/\omega_n^2)$ in the numerator,

the right hand side of Equation 3.38 is identical to that of Equation 1.19 with $\delta = 0$. If we had included damping in our model, the result would have again been identical to Equation 1.19, except for the constant. Thus, we can make use of the results of Figure 1.9. Figure 1.9 shows that we should like to operate the instrument in the stiffness controlled region, where $\omega/\omega_n << 1$. Again this is obvious from Equation 3.38, since if $\omega/\omega_n << 1$, the right hand side becomes a constant (equal to $1/\omega_n^2$). To obtain a flat frequency response over a wide range, ω_n should be made large. But note that in this case the sensitivity decreases since it is approximately proportional to $1/\omega_n^2$. A large value of the frequency ω_n necessitates a very stiff spring and a small mass, the opposite to the seismic displacement and velocity transducers discussed above.

It is possible to measure the relative displacement† by attaching a strain gauge to the spring and calibrating the instrument by the change in resistance produced. This is the principle of the *strain gauge** accelerometer which is normally used with a Wheatstone bridge. Such accelerometers will usually be found to have an upper frequency limit of about 1000 or 2000 Hz. Provided the accelerometer is DC coupled, it is easy to calibrate the accelerometer by inverting it. If the strain gauge accelerometer is placed horizontally, there is no signal; placed vertically in one direction it gives a constant resistance change equivalent to +1 g** and placed in the opposite vertical position the change is equivalent to −1 g. If the accelerometer is AC coupled, such calibration is not possible. However, the most commonly used type of accelerometer is now the *piezoelectric accelerometer.*

3.5.2 Piezoelectric Accelerometers

The piezoelectric accelerometer is a seismic mass type of accelerometer which works on the principle described in Section 3.5.1C. Figure 3.37 shows cross-sectional views of four commonly used commercial types. Comparing these with the idealized design in Figure 3.36 shows that the compression types a, b, and c bear a close resemblance. The spring constant K is determined partly by the stiffness of the compressive spring S and partly by the stiffness of the piezoelectric element P. The voltage is proportional to the relative deflection of the mass M for small deflections of the piezoelectric element P. Type d uses a piezoelectric element in shear.

The four types shown in Figure 3.37 are all designed to accentuate certain advantages which are useful in different environmental conditions. Figure 3.38a shows a frequency response calibration chart supplied with a typical accelerometer. Figure 3.38b shows the frequency responses of several piezoelectric accelerometers. As already discussed in Section 3.5.1C, the heavier accelerometers have the greatest sensitivity, but the lowest resonance frequency and vice versa.

There are several characteristics of the piezoelectric accelerometer. As is shown in Figures 3.39 and 3.40, the piezoelectric accelerometer can be considered either as a charge or a voltage source. Hence, the *sensitivity* is either given in decibels re 1 mV/g or re 1pC/g*** Besides being sensitive to acceleration in the longitudinal axis due to irregularities in construction and alignment of the piezoelectric discs, the accelerometer is also slightly sensitive to vibration in the transverse axis (see Figure 3.41). Good accelerometers should have a *lateral sensitivity* less than 5% of the longitudinal (or main axis) sensitivity. The transverse sensitivity will vary in the base plane having maximum and minimum values in certain directions. The sensitivity of an accelerometer will be somewhat temperature dependent (one should always consult the manufacturer's instructions), however, provided the maximum (Curie point) temperature is not exceeded, the piezoelectric material will not be damaged and will retain its properties.

Piezoelectric accelerometers produce some output signal when subjected to acoustic signals or base strains. Normally the *acoustic sensitivity* is low (less than 1 g for a sound pressure level of 160 dB). As the test object vibrates, it will induce strain in the accelerometer base with a consequent

†Another type of accelerometer works differently. It is known as a *servo*-accelerometer and still employs the seismic mass but uses a servo-system to produce a force on the mass to duplicate the motion of the case. The current to a forcing coil is measured and when passed through a resistor, produces a voltage proportional to the acceleration.

*Piezoresistive accelerometers are now available in which the strain gauge elements are constructed from piezoresistive materials instead of thin metal foils.

**g is the acceleration of gravity, g = 32.2 ft/sec² or 9.81 m/sec².

***pC = picoColoumb = $\mu\mu$C = 10^{-12} Coloumb.

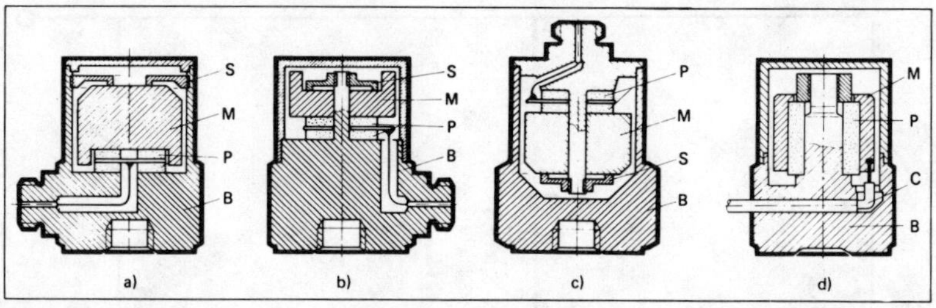

FIGURE 3.37 Schematic cross sections of four common piezoelectric accelerometer designs (a) Peripheral mounted compression design; (b) Center mounted compression design; (c) Inverted center mounted compression design; (d) Shear design. (Courtesy of Brüel & Kjaer Co.)

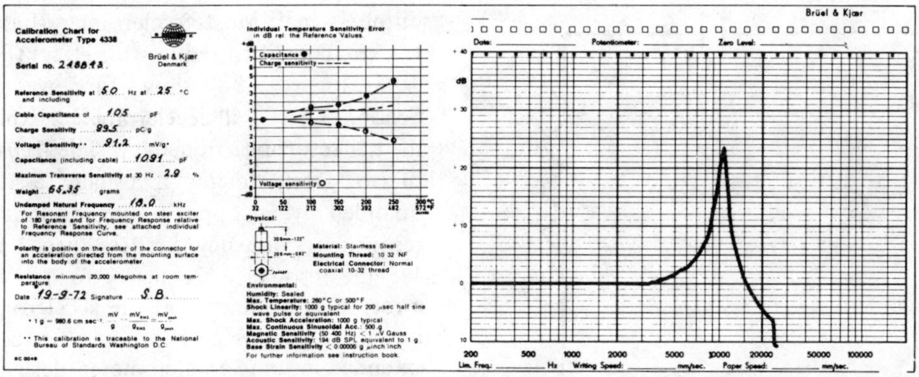

FIGURE 3.38a Calibration chart of a typical accelerometer. (Courtesy of Brüel & Kjaer Co.)

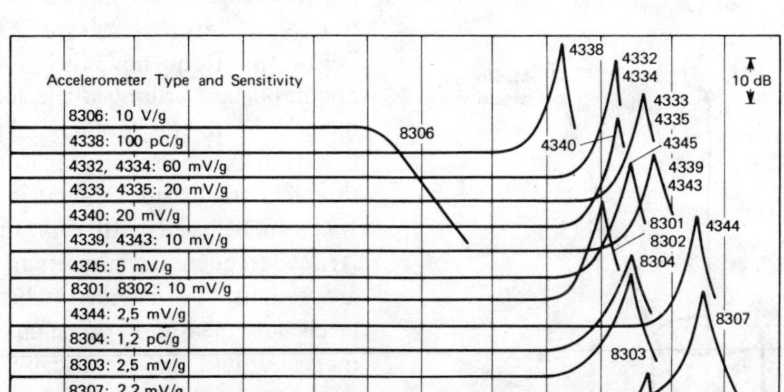

FIGURE 3.38b Frequency response characteristics of several different weight accelerometers. The model numbers and weights in grams follow: 8306, 600g; 4338, 60g; 4332 and 4334, 30g; 4333 and 4335, 13g; 4340 triaxial, 35g; 4339 and 4343, 16g; 4345, 27g; 8301 and 8302, 21g; 4344, 2g; 8304, 30g; 8303, 3.5g; 8307, 0.4g; 8305, 40g. (Courtesy of Brüel & Kjaer Co.)

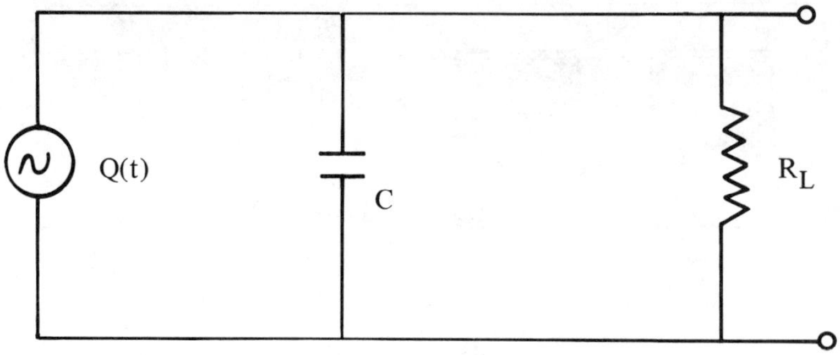

FIGURE 3.39 Equivalent circuit of piezoelectric accelerometer as a charge source.

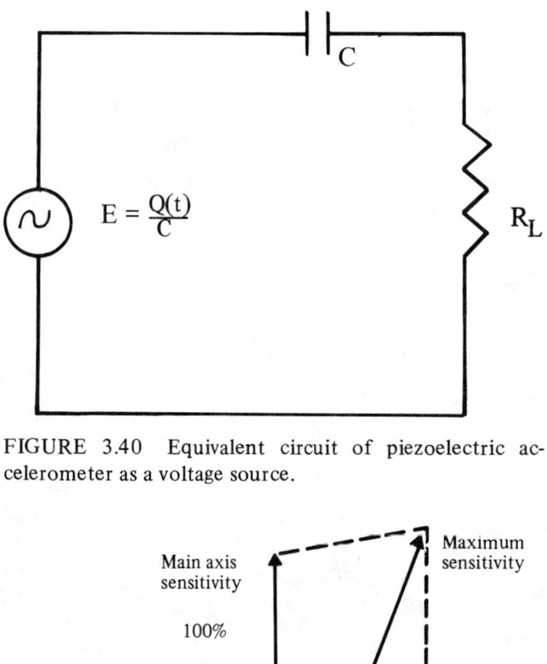

FIGURE 3.40 Equivalent circuit of piezoelectric accelerometer as a voltage source.

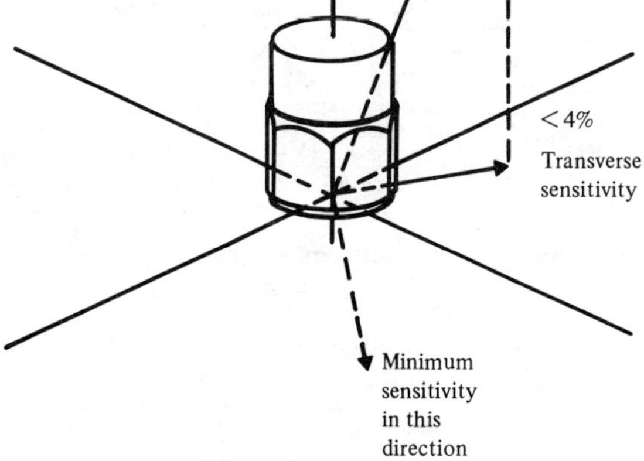

FIGURE 3.41 Transverse and longitudinal sensitivity of accelerometer.

output signal. Most accelerometer bases are made thick and rigid to reduce this effect (Figure 3.37).

As Figure 3.38b shows (and was discussed in Section 3.5.1C) all accelerometers exhibit a fundamental resonance frequency. The *frequency range* of an accelerometer is usually assumed to be bounded by an upper frequency limit of 1/3 the resonance frequency for less than 1 dB error or 1/5 the resonance frequency for less than 0.5 dB error. This assumes that the accelerometer design has a low damping. The upper frequency limit is extended in some accelerometer designs by using a high damping.

However, it should be noted that poor *mounting techniques* can have a marked effect on the frequency response above 2,000 or 3,000 Hz. Below this frequency, the mounting technique is not important as long as the accelerometer is fixed directly on to the specimen. The best technique is to attach the accelerometer to the specimen with a threaded steel stud. Cement is also acceptable and wax can be used with very small accelerometers. At low frequencies magnets may be used. *Ground loops* must be avoided when low acceleration levels are measured. In such cases the accelerometer should be electrically isolated from the specimen and attached with a nonconducting stud or cemented to a mica washer which is in turn cemented to the specimen. Cable "whip" can produce electrical noise and cables should be attached to the specimen as shown in Figure 3.42.

The *lower limiting frequency* depends on the type of preamplifier used to follow the accelerometer. Two types may be used and are commercially available. If the preamplifier is designed so that the output voltage is proportional to the input voltage it is called a *voltage amplifier*. If the output voltage is proportional to the input charge,

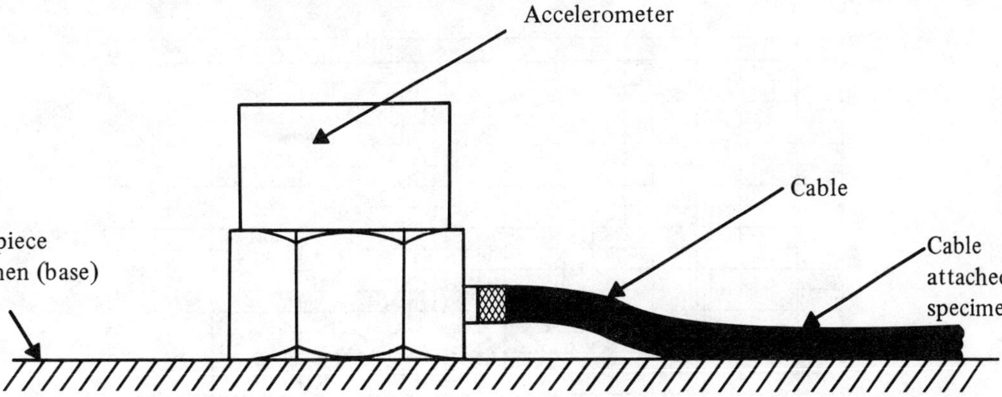

FIGURE 3.42 The accelerometer cable should be attached to the specimen to reduce cable whip noise.

it is called a *charge amplifier.* When a voltage amplifier is used, the accelerometer output voltage is very sensitive to cable capacitance. This is because typically the capacitance of a piezoelectric accelerometer is several hundred picofarad (usually between 100 and 1,000 pF). This is somewhere between the very low capacitance of a capacitor microphone and the higher capacitance of a piezoelectric microphone (see Sections 3.4.1 and 3.4.2). When a charge amplifier is used, the output voltage is not sensitive to changes in cable length. Broch[50] describes the reasons for this in some detail. If a voltage amplifier is used, the lower limiting frequency (3-dB down point) is given by

$$f_\ell = \frac{1}{2\pi RC} \tag{3.39}$$

where

R is the input resistance of the voltage preamplifier and C is the effective circuit capacitance

$$C = \frac{(C_A + C_L)\, C_C}{C_A + C_L + C_C} \tag{3.40}$$

where

C_A, C_L and C_C are the accelerometer, cable and preamplifier capacitances, respectively.

The *dynamic range* of the piezoelectric accelerometer is determined by the lower limiting frequency set by electrical noise in the system and the upper limiting frequency governed by the preloading of the accelerometer and the mechanical strength of the piezoelectric element. Smaller accelerometers usually have a higher upper limit.

We can return now to Figure 3.37 to see the advantages and disadvantages of the four designs. Type a. (peripheral mounted compression design) has the advantage of simple rugged construction with the best weight/sensitivity ratio and a high resonance frequency. However, it is sensitive to acoustic signals, temperature transients, and base strain. Type b. (center mounted compression design) has the additional advantages of higher resonance frequency, and lower sensitivity to acoustic signals, temperature transients, and base strain. Type c. (center mounted compression design) has even lower sensitivity to base strain since the center post is not connected to the base. However, it has a lower resonance frequency. The shear design, typed, has a high resonance frequency and can be easily miniaturized. The piezoelectric element is also well protected from environmental changes (acoustic pressure and temperature).

3.5.3 Integrators, and Velocity and Displacement Transducers

A. Integrators

Integrators built from resistance and capacitive components are available. The low frequency cut-off (3-dB down point) is influenced by the capacitance of the accelerometer connecting cable and preamplifier (see Equations 3.39 and 3.40). The high frequency limits are set by the capacitive coupling between the input and output of the integrator. These limits are 10,000 Hz for velocity and 4,000 Hz for displacement with the set-up in Figure 3.43. In the case when the high frequency vibration of thin metal plates is being measured, considerable care should be taken when using an accelerometer because of the *mass loading* effect. Well below its resonance frequency, the accelerometer can be assumed to act as a pure mass. The velocity of the test piece $\mathbf{u}_0$ can be assumed

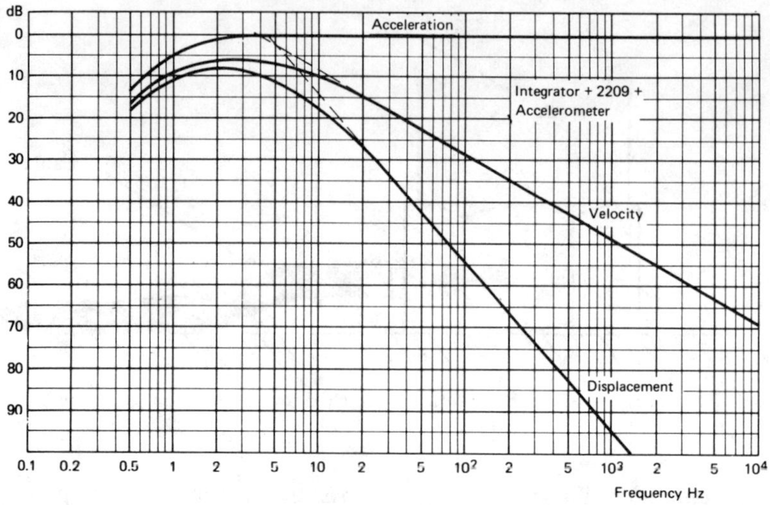

FIGURE 3.43 Frequency response curves for an accelerometer and integrator connected to a sound level meter. (Courtesy of Brüel & Kjaer Co.)

reduced to $\mathbf{u}_1$ by the addition of the accelerometer[51]

$$\frac{\mathbf{u}_1}{\mathbf{u}_0} = \frac{\mathbf{Z}}{\mathbf{Z} + i\omega M}, \tag{3.41}$$

where

$\mathbf{Z}$ = mechanical impedance of test piece at attachment point
ω = angular frequency
M is accelerometer mass.

The input point impedance for a metal plate can be assumed to be

$$\mathbf{Z} = \frac{4}{\sqrt{3}} \rho \, c_L \, h^2, \tag{3.42}$$

where

ρ = plate density
c_L = longitudinal wave speed in plate
h = plate thickness.

Substituting 3.42 in Equation 3.41 and assuming $|\mathbf{Z}| = |i\omega M|$ which gives a 3 dB reduction in velocity amplitude (Equation 3.41) gives the frequency $f_{3\ dB}$ at which the 3 dB reduction occurs as

$$f_{3\ dB} = \frac{2 \rho \, c_L \, h^2}{\sqrt{3}\pi M} \tag{3.43}$$

Equation 3.43 is plotted in Figure 3.44. This figure suggests that even lightweight accelerometers will "load" thin metal plates (e.g., a 10-g accelerometer will mass "load" an 1/8-in. aluminum plate at frequencies above about 4,000 Hz).

In order to overcome such mass loading problems noncontacting gauges have been and still are used. A particular example of where a noncontacting gauge would be invaluable would be in the study of the vibration of a loudspeaker cone.

B. Velocity Gauges

One type of noncontacting gauge available is the magnetic transducer (i.e., variable reluctance* device). Such transducers normally consist of a cylindrical permanent magnet on which is wound an insulated low resistance wire. When placed a few millimeters from a metallic test piece or a high permeability** disc cemented onto a nonmetallic test piece, such a device will produce a signal which is proportional to the separation of the device and the specimen. Figure 3.45 shows a cross-sectional view of a typical transducer and Figure 3.46 shows its sensitivity as a function of the separation distance. A static force is created by this transducer. For a 0.5 mm gap and a large iron plate test piece, the force is about 225 μN for the transducer shown in Figure 3.45.

Note that such transducers can also be used reciprocally as vibration exciters.

*Reluctance is the ratio: magnetomotive force/magnetic flux.
**Permeability is the ratio: flux density/magnetizing force.

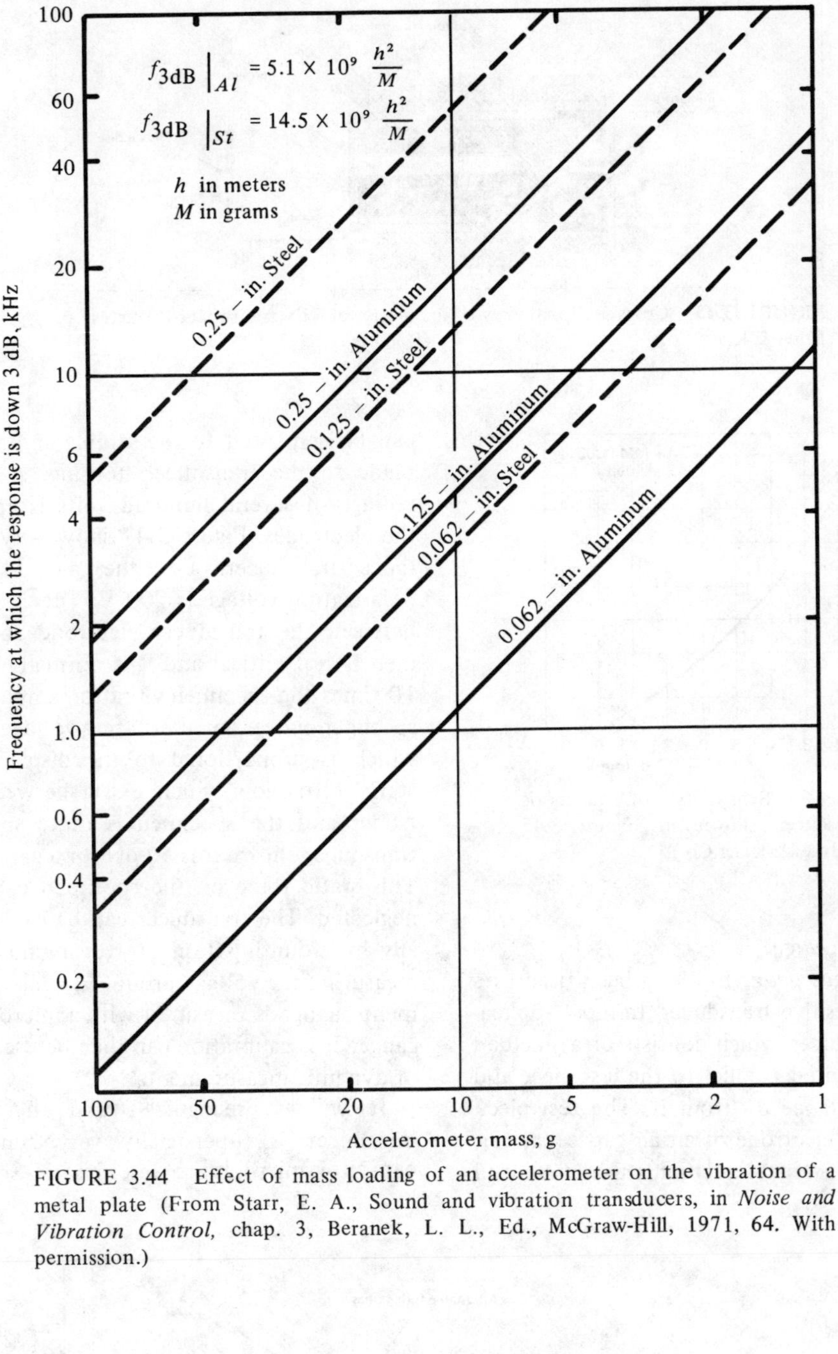

FIGURE 3.44 Effect of mass loading of an accelerometer on the vibration of a metal plate (From Starr, E. A., Sound and vibration transducers, in *Noise and Vibration Control,* chap. 3, Beranek, L. L., Ed., McGraw-Hill, 1971, 64. With permission.)

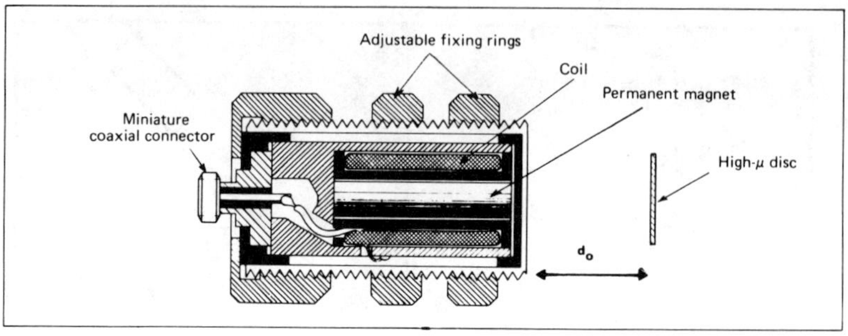

FIGURE 3.45 Cross-sectional view of magnetic velocity transducer (Courtesy of Brüel & Kjaer Co.)

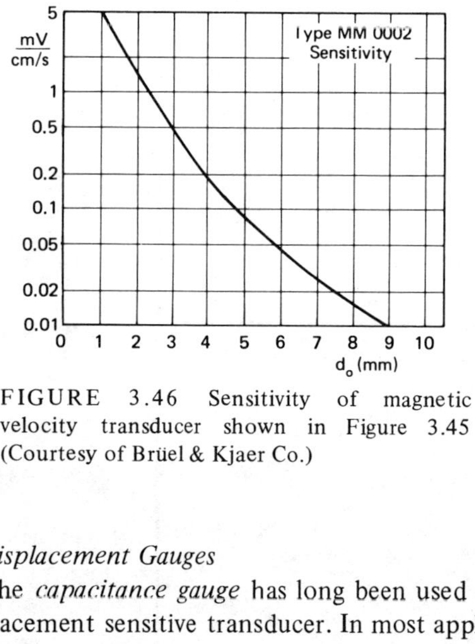

FIGURE 3.46 Sensitivity of magnetic velocity transducer shown in Figure 3.45 (Courtesy of Brüel & Kjaer Co.)

C. Displacement Gauges

The *capacitance gauge* has long been used as a displacement sensitive transducer. In most applications, the transducer which consists of a shielded electrode is mounted parallel to the test piece and at a suitable distance d_0 from it. The test piece forms the other electrode of an air-gap capacitor. If the test piece is nonconducting, a thin metal foil can be cemented to the sample and a connection made to the transducer housing. A polarization voltage of several hundred volts is applied across the electrodes. Figure 3.47 shows a typical capacitance transducer. For the model shown, the polarization voltage is 200 V. The air gap distance between the transducer electrode and the specimen is not critical and it is normally made about 10 times the specimen vibration amplitude. As the specimen moves, an alternating voltage is produced which is proportional to the displacement. The static attraction which exists between the electrodes and the specimen is quite small (for the transducer shown it is 56 μN for a gap of 0.5 mm). This static force on the specimen can usually be neglected. The transducer can be calibrated statically by mounting it on a traversing mechanism and measuring the voltage produced while the displacement change is measured with a micrometer screw gauge. This calibration can then be used with static or dynamic measurements.

It will be recognized that the capacitance transducer is superficially very similar to the capacitor microphone described in Section 3.4.1.

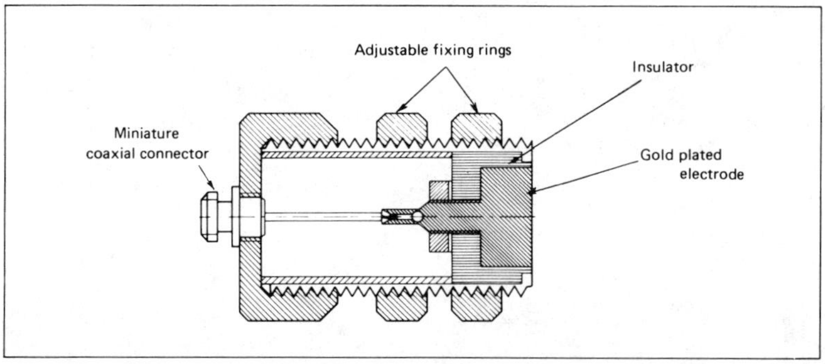

FIGURE 3.47 Cross section of a capacitance transducer (Courtesy of Brüel & Kjaer Co.)

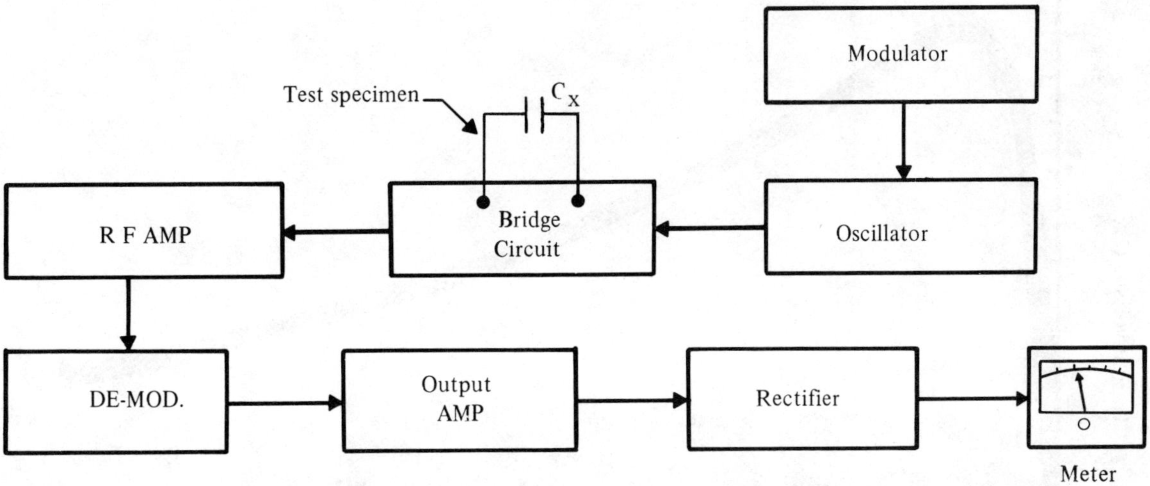

FIGURE 3.48 Bridge circuit used with a capacitance probe.

The transducer capacitance is very low* and special precautions must be taken against transducer loading. Cable capacitance problems are overcome in the design shown in Figure 3.47 by using a double screened cable. The central conductor and outer screen of the cable carry the signal while the inner screen is connected to the internal guard ring of the preamplifier. In this way the cable capacitance does not effect the output of the transducer which is only loaded by the high input impedance of the preamplifier (usually several hundred or thousand MΩ).

Other types of capacitance displacement transducers are also available. Since the change in displacement of the test piece produces a change in capacitance, this capacitance change can also be detected by the resonance circuit method or the capacitance bridge circuit. The resonance circuit method depends on the fact that as the capacitance in the circuit increases, the current of an AC supply will increase or decrease almost linearly over a small frequency range above and below the resonance frequency where it is a maxium (see Section 3.2.1). However, the following must be ensured:[52]

1. The frequency of the supply to the tuned circuit must be constant.
2. The part of the curve over which the operation takes place must be linear.
3. The amplifier must be stable with a constant gain and linear amplification.

The capacitance bridge circuit method[52] is shown in Figure 3.48. Typical frequencies of operation for the oscillator are 450 kHz with a modulation at 2.4 kHz.

Still other displacement transducers are available. One commercial transducer[53] works on inductive principles. It again is a noncontacting gauge and its head contains a pancake wound coil. Any conductive surface placed near the coil will load it. The output voltage is proportional to displacement over as much as 0.1 in. Optical proximity detectors have also come into use since about 1965. Typical displacement sensors can measure displacements from DC to 40,000 Hz. They produce a voltage proportional to the separation distance over distances up to about 0.2 in. Such a probe[54] consists of a steel jacket containing several hundred strands of glass fiber optics. Half of the strands are used to transmit light which, after being reflected by the test object, are "captured" by the remaining half of the strands. The instrument is calibrated on the job. Figure 3.49 shows a typical calibration curve. The instrument is normally operated in the shaded area (where it is linear and very sensitive). However, it can also be used in the less sensitive linear region from about 65 to 155 mil.**

*The capacitance is inversely proportional to the static separation d_0 (see Equation 3.19). For the design shown in Figure 3.47 it is 1.5 pF for d_0 = 0.5 cm.
**1 mil = 0.001 in.

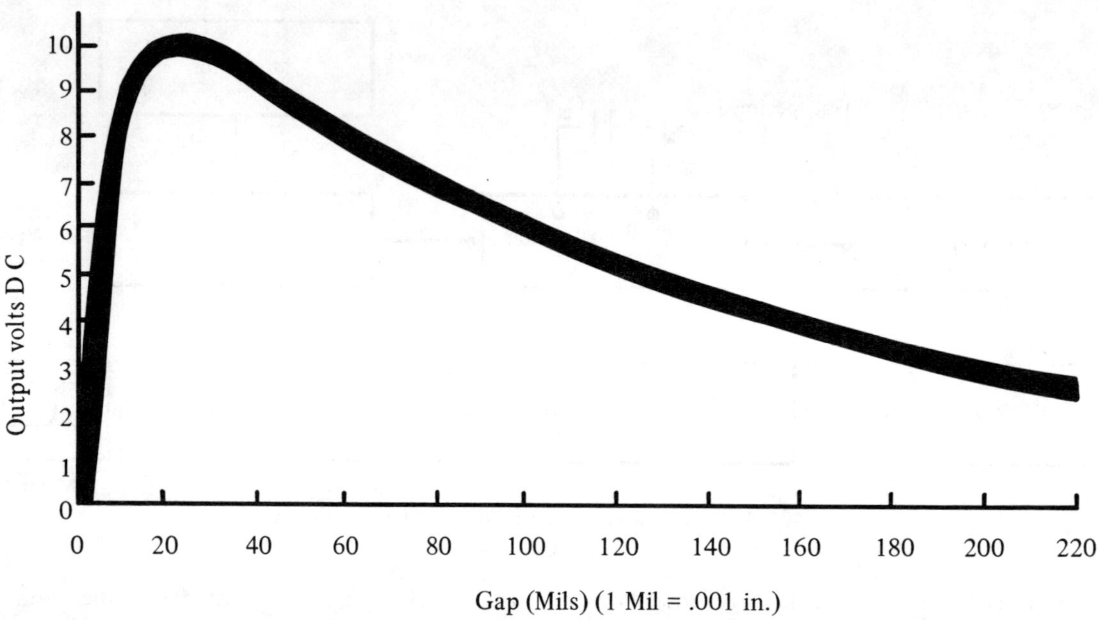

FIGURE 3.49 Calibration curve for a fiber-optics displacement sensor. (Courtesy of M.T.I. Instruments Division.)

3.6 SPECTRUM ANALYSIS

Very frequently in noise and vibration measurements we are concerned with noise or vibration signals at some particular frequency. The *frequency analysis* of time domain signals was discussed in Section 1.8. In order to determine experimentally the contribution of the overall signal in some particular frequency band we *filter* the signal. To understand what this involves we will briefly look at the theoretical behavior of some very simple idealized filters in Section 3.6.1.

3.6.1 Filter Theory

In Section 3.2.1, we briefly examined the electrical behavior of a series R-L-C circuit. Circuits made from these simple components can be used to make electrical filters. We will examine briefly some simple *passive* R-C filters in this section. By *passive* we mean that no electrical power is supplied. Practical filters are usually made from more complicated R-L-C passive filters or *active* filters[55] (involving R-C components, operational amplifiers, and a power supply). However, we can observe most of the important points concerning filters from the simple R-C passive filters examined in this section.

A. High-Pass Filters

By omitting the inductance in Figure 3.1, the circuit in Figure 3.50 is obtained. Using the theory of Section 3.2.1 and the results of Equation 3.4, we can relate the input voltage E_i and output voltage $\mathbf{E}_o$.

$$\mathbf{E} = \mathbf{Z}\,\mathbf{I}$$

$$\therefore\ \mathbf{E}_o = \mathbf{R}\mathbf{I}, \tag{3.44}$$

and

$$\mathbf{E}_i - \mathbf{E}_o = \left(\frac{1}{i\omega C}\right)\mathbf{I}, \tag{3.45}$$

thus eliminating **I**

$$H_1(\omega) = \frac{\mathbf{E}_o}{\mathbf{E}_i} = \frac{i\omega RC}{1 + i\omega RC}. \tag{3.46}$$

The magnitude of the frequency response function, $\mathbf{H}_1(\omega)$ (ignoring the phasing between $\mathbf{E}_o$ and $\mathbf{E}_i$), is

$$|\mathbf{H}_1(\omega)| = |\mathbf{E}_o/\mathbf{E}_i| = \frac{|i\omega RC|}{|1 + i\omega RC|}, = \frac{\omega RC}{[1 + \omega^2 R^2 C^2]^{1/2}}. \tag{3.47}$$

$20 \log_{10} |\mathbf{H}_1(\omega)|$ is plotted in Figure 3.51. It is observed that for high frequencies ($\omega \to \infty$), the frequency response function is $|\mathbf{H}_1(\omega)| \to 1$. It was also seen that at low frequencies, ($\omega \to 0$), $|\mathbf{H}_1(\omega)| \to \omega RC$. In this low frequency region $20 \log_{10} |\mathbf{H}_1(\omega)|$ decreases by 6 dB/octave or 20

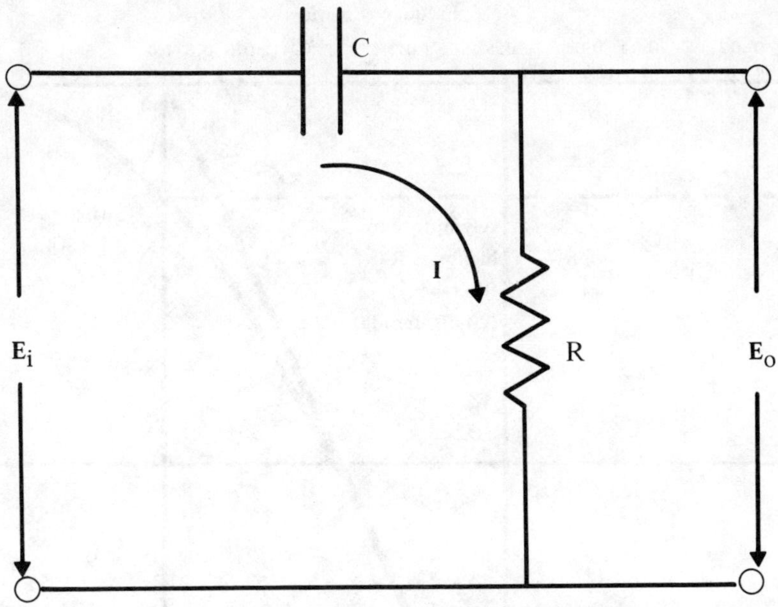

FIGURE 3.50 R-C circuit for high-pass filter.

dB/decade* decrease in frequency. The *break* frequency (or *cut-off* frequency), ω_b, is normally defined as the 3-dB down point. This occurs when $\omega RC = 1$ (see the denominator of Equation 3.47)

$$\therefore \omega_b = 1/RC. \tag{3.48}$$

Such a filter is termed a *high-pass* filter because only high frequency signals are passed without attenuation. In practice, although such a filter could be used in noise work (its performance is rather similar to the A-weighting filter described in Section 2.3.4), the roll-off at low frequency is not sufficiently rapid. An ideal high-pass filter would pass no signal below the break or cut-off frequency, f_b (see Figure 3.51), and would have a vertical "skirt". Practical filters have more complicated circuits than Figure 3.50 and normally incorporate inductances and/or active components. Such filters have been described elsewhere.[55-57]

In practice a high-pass filter might be used when it is desired to study the high frequency sound from a source and to filter out low frequency noise such as 60- or 120-Hz line frequency or low frequency ambient noise in an anechoic or reverberant room.**

B. Low-Pass Filters

Figure 3.52 shows a simple R-C circuit for a low-pass filter. By using Equation 3.4 we obtain

$$\mathbf{E}_i - \mathbf{E}_o = R\,\mathbf{I}, \tag{3.49}$$

and

$$\mathbf{E}_o = \left(\frac{1}{i\omega C}\right)\mathbf{I} \tag{3.50}$$

thus,

$$\mathbf{H}_2(\omega) = \frac{\mathbf{E}_o}{\mathbf{E}_i} = \frac{1}{1 + i\omega RC}, \tag{3.51}$$

and the magnitude of the frequency response, $\mathbf{H}_2(\omega)$, is

$$|\mathbf{H}_2(\omega)| = |\mathbf{E}_o/\mathbf{E}_i| = \frac{1}{[1 + \omega^2 R^2 C^2]^{1/2}}. \tag{3.52}$$

$20 \log_{10} |\mathbf{H}_2(\omega)|$ is plotted in Figure 3.53. With this circuit at low frequencies ($\omega \to 0$), $|\mathbf{H}_2(\omega)| \to 1$, and at high frequencies ($\omega \to \infty$), $|\mathbf{H}_2(\omega)| \to 1/(\omega RC)$. This time in the high frequency region, $20 \log_{10} |\mathbf{H}_2(\omega)|$ decreases by 6 dB/octave or 20 dB/decade increase in frequency. The break frequency (or cut-off frequency), ω_b, is given by

*A decade represents a frequency ratio of 10 to 1.

**With some sound level meters, manufacturers insert a high-pass filter with a cut-off frequency of about 20 or 25 Hz for use in the *linear* mode. Also, high-pass filters are often used when the displacement signal from a transducer is analyzed. This is because frequently the high frequency displacement is of interest and large amplitude low frequency signals must be filtered out to prevent the dynamic range of the instrumentation from being exceeded.

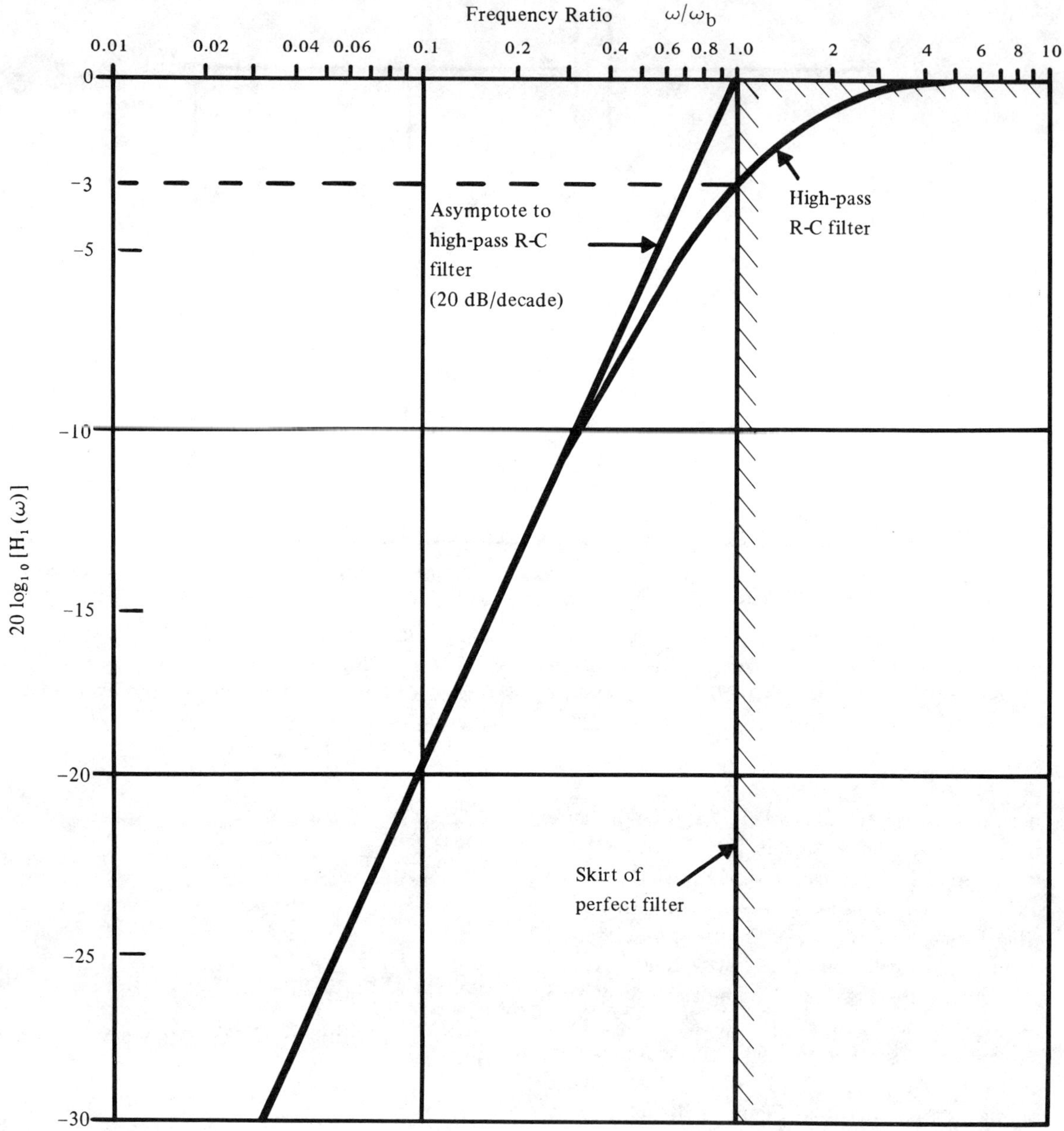

FIGURE 3.51 Logarithmic plot of the frequency response of a simple R-C high-pass filter compared with an ideal high-pass filter.

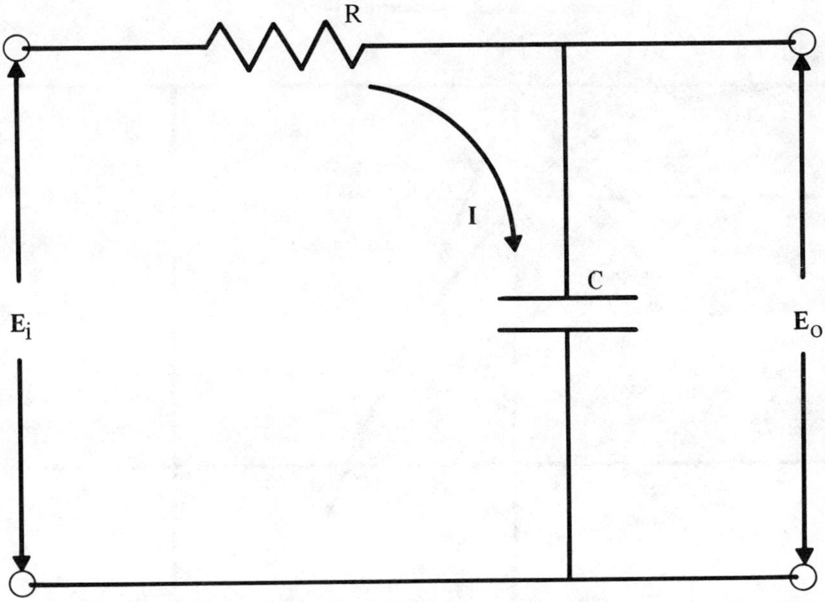

FIGURE 3.52 R-C circuit for a low-pass filter.

$\omega RC = 1$ again (see the denominator of Equation 3.52). Hence at this frequency, ω_b, the filter output is 3 dB down. Again

$$\omega_b = 1/RC. \tag{3.53}$$

This type of filter is called a *low-pass* filter because only low frequency signals are passed without attention.* In practice, the skirt of this simple RC filter is usually not steep enough and a more complicated circuit is needed.

C. Band-Pass Filters

By combining the simple high-pass and low-pass filters just discussed, a simple band-pass filter is produced. Figure 3.54 shows the circuit. By using Equation 3.4 and assuming the currents $\mathbf{I}_1$ and $\mathbf{I}_2$ as shown, since the voltage across AD is $\mathbf{E}_i$,

$$\left(\frac{1}{i\omega C_1}\right) \mathbf{I}_1 + R_1 (\mathbf{I}_1 - \mathbf{I}_2) = \mathbf{E}_i. \tag{3.54}$$

There is zero voltage between D and F, so

$$R_1 (\mathbf{I}_2 - \mathbf{I}_1) + R_2 \mathbf{I}_2 + \left(\frac{1}{i\omega C_2}\right) \mathbf{I}_2 = 0. \tag{3.55}$$

Finally,

$$\mathbf{E}_o = \left(\frac{1}{i\omega C_2}\right) \mathbf{I}_2. \tag{3.56}$$

Eliminating $\mathbf{I}_1$ in Equations 3.51 and 3.52 to obtain $\mathbf{I}_2$ which is then substituted into Equation 3.53 gives

$$\frac{\mathbf{E}_o}{\mathbf{E}_i} = \frac{i\omega\tau_1}{1 - \omega^2 \tau_1 \tau_2 + i\omega(\tau_1 + \tau_2 + R_i C_z)} \tag{3.57}$$

where $\tau_1 = R_1 C_1$ and $\tau_2 = R_2 C_2$, respectively. Note $i^2 = (\sqrt{-1})^2 = -1$.

If $\tau_1 >> \tau_2$ .... i.e. $\omega_{b2} >> \omega_{b1}$ and $R_2 >> R_1$, then $R_1 C_2$ may be neglected in Equation 3.57 which now becomes

$$\mathbf{H}_3(\omega) = \frac{\mathbf{E}_o}{\mathbf{E}_i} = \frac{i\omega R_1 C_1}{(1 + i\omega R_1 C_1)} \cdot \frac{1}{(1 + i\omega R_2 C_2)}. \tag{3.58}$$

The frequency response $\mathbf{H}_3(\omega)$ for this filter shown in Figure 3.54 is thus the product of the frequency responses of the high- and low-pass filters already discussed.

$$\therefore \mathbf{H}_3(\omega) = \mathbf{H}_1(\omega)\, \mathbf{H}_2(\omega). \tag{3.59}$$

The magnitude is given by

$$|\mathbf{H}_3(\omega)| = |\mathbf{H}_1(\omega)\, \mathbf{H}_2(\omega)| = |\mathbf{H}_1(\omega)|\, |\mathbf{H}_2(\omega)|. \tag{3.60}$$

A plot of 20 log $|\mathbf{H}_3(\omega)|$ given by Equations 3.60, 3.52, and 3.47 is shown in Figure 3.55 for the case where $\omega_{b2} = 10\, \omega_{b1}$. This type of filter is known as a *band-pass* filter since all simple harmonic

*With sound level meters, manufacturers sometimes insert a low-pass filter with a cut-off frequency of 40,000 or 50,000 Hz for use in the *linear* mode, to suppress unwanted high frequency signals.

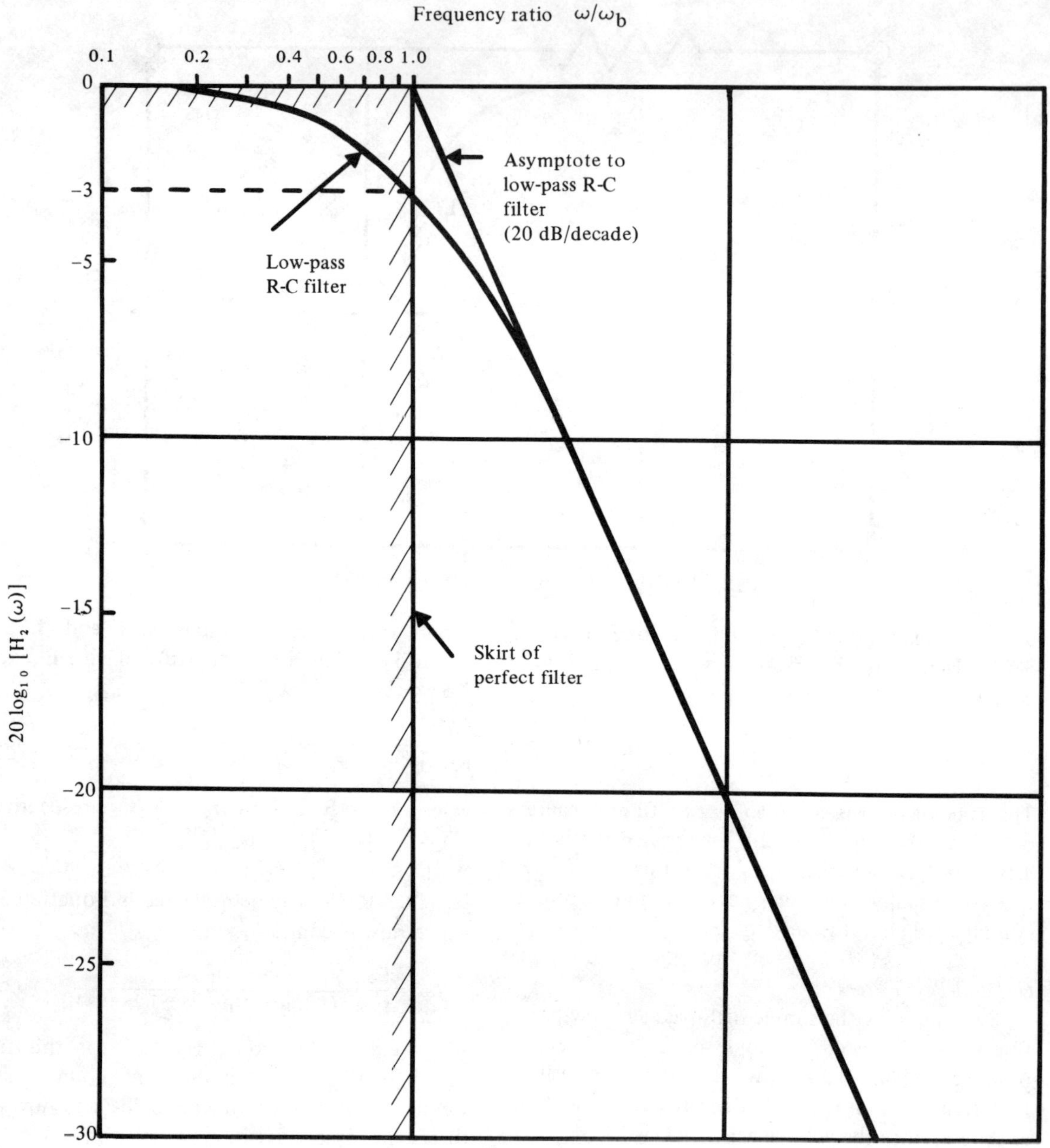

FIGURE 3.53 Logarithmic plot of the frequency response of a simple R-C low-pass filter compared with an ideal low-pass filter.

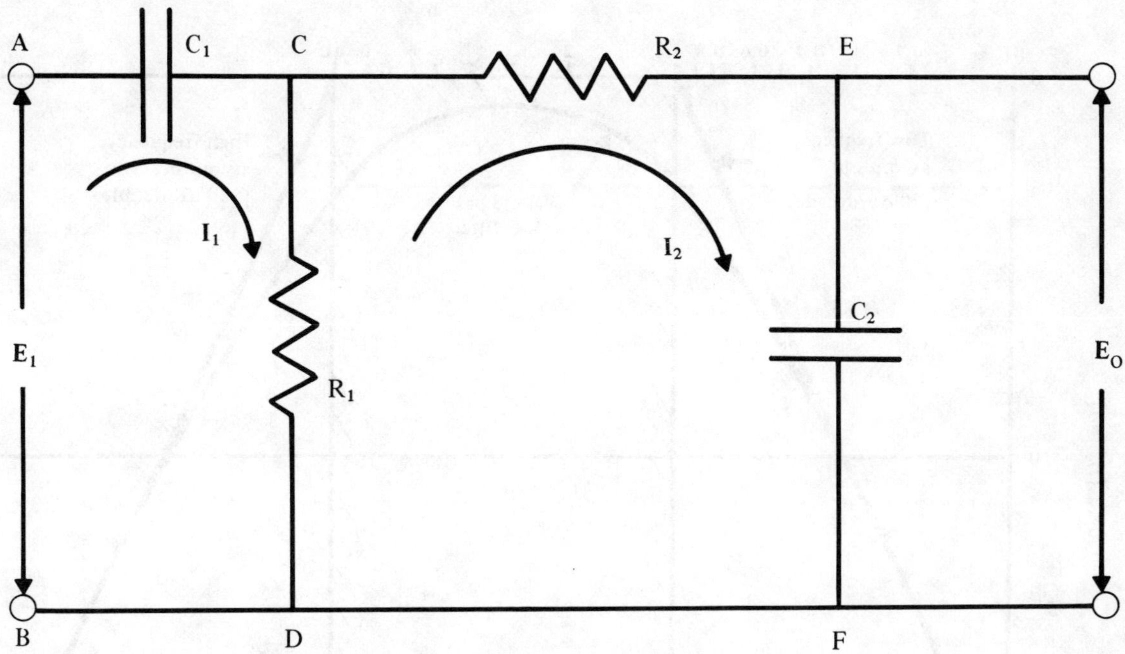

FIGURE 3.54 R-C circuit for band-pass filter.

frequency signals within the band $\Delta\omega = \omega_{b2} - \omega_{b1}$ are passed with little, if any, attenuation. All signals outside of this band are attenuated. In practice it is usually desirable to make the skirts of such a filter steeper. This requires the use of more electrical circuit elements and/or the use of active filters.

Band-pass filters are frequently used in acoustics. There are two main types: constant percentage and constant bandwidth filters. These are discussed in Sections 3.6.2 and 3.6.3.

3.6.2 Constant Percentage Filters

A. Octaves and Fractional Octaves

An *octave* is the frequency interval between two frequencies (upper and lower), f_u and f_ℓ, which have a ratio of two,

$$f_u = 2f_\ell. \tag{3.61}$$

Fractional octaves are defined in a similar way to Equation 3.61. If n is the fraction of the octave, then

$$f_u = 2^n f_\ell. \tag{3.62}$$

Thus, 1/2 octave is given by $f_u = 2^{1/2} f_\ell$, 1/3 octave by $f_u = 2^{1/3} f_\ell$, and so on. For octave and fractional octave band-pass filters with upper and lower cut-off frequencies given by f_u and f_ℓ, the pass-band or *bandwidth* Δf is

$$\Delta f = f_u - f_\ell. \tag{3.63}$$

For the special case of the octave band filter only, the interesting result is obtained from Equations 3.61 and 3.63 that $\Delta f = f_\ell$.

The center frequency, f_c, of the bandwidth of octave or fractional octave filters is defined geometrically as

$$f_u/f_c = f_c/f_\ell, \tag{3.64}$$

or

$$f_c = (f_u f_\ell)^{1/2}. \tag{3.65}$$

Using Equations 3.61 and 3.65 for octave filters gives

$$f_u = 2^{1/2} f_c, \tag{3.66a}$$

and

$$f_\ell = f_c/2^{1/2}. \tag{3.66b}$$

For fractional octave filters (where n is the fraction of the octave) the upper and lower break frequencies from Equations 3.62 and 3.65 are

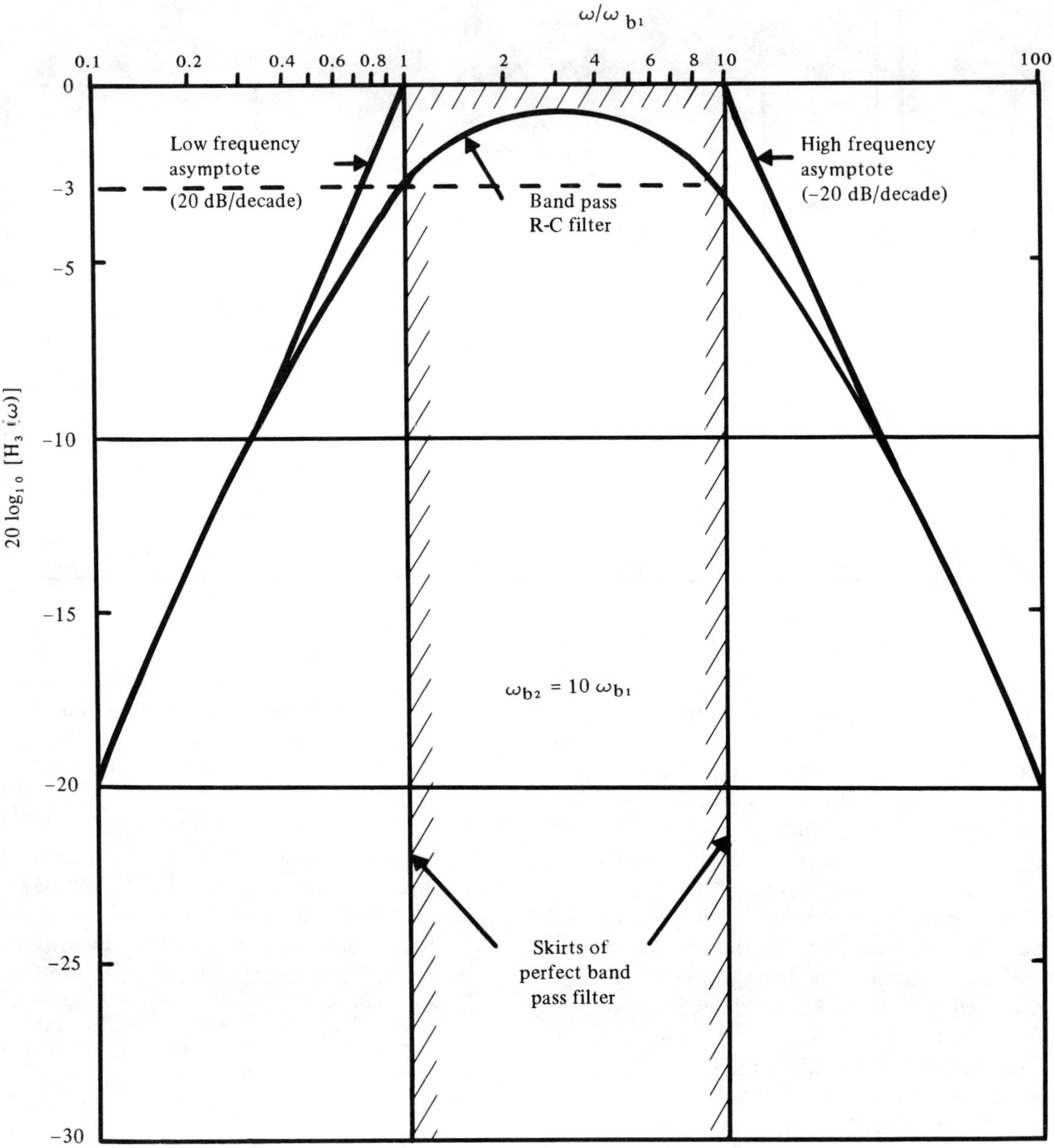

FIGURE 3.55 Logarithmic plot of the frequency response of a simple band-pass filter (see Figure 3.54) with $R_1C_1 = 10R_2C_2$.

$$f_u = 2^{n/2} f_c, \quad (3.67a)$$

and

$$f_\ell = f_c / 2^{n/2}. \quad (3.67b)$$

The bandwidth Δf is

$$\Delta f = f_u - f_\ell = (2^{n/2} - 1/2^{n/2}) f_c$$

$$\Delta f = \frac{1}{2} 2^{n/2} f_c = 2^{n/2 - 1} f_c. \quad (3.68)$$

We thus see clearly why the filter bands we have just discussed are called *constant percentage*. This is because the bandwidth is a constant percentage ($100 \times 2^{n/2-1}$) of the filter center frequency f_c. Table 3.1 shows the upper and lower break frequencies and the bandwidths for different fractional octave type filters as percentages of the center frequency. Of course the filter bandwidth does not have to be defined in terms of a fraction of an octave but can be defined simply in terms of the percentage of the center frequency. Since it is impracticable to make measurements at a large number of fixed frequencies, measurements are often made at a selected number of frequencies. The *preferred center frequencies* for such noise measurements are now based* on the frequency of 1,000 Hz (because this is such an important frequency in acoustics). The preferred center frequencies are given by

$$f_c = 1{,}000 \cdot 10^{3m/10} \text{ (for octave bands)} \quad (3.69a)$$

$$f_c = 1{,}000 \cdot 10^{3m/20} \text{ (for } 1/2\text{-octave bands),} \quad (3.69b)$$

$$f_c = 1{,}000 \cdot 10^{3m/30} \text{ (for } 1/3\text{-octave bands), etc,} \quad (3.69c)$$

where $m = 0, \pm 1, \pm 2, \pm 3$, etc. Note that $10^{3/10} \approx 2$, thus $10^{3/20} \approx 2^{1/2}$ and $10^{3/30} \approx 2^{1/3}$, etc. The ISO Recommendation** R.266 gives the following table for preferred center frequencies (Table 3.2). Note that this system has the advantage that the human hearing range of about 20 to 20,000 Hz can be covered with 10 octave band filters (with center frequencies from 31.5 to 16,000 Hz). One-third octave band frequencies have the advantage that a decade of frequency can be covered with 10 filters.

TABLE 3.1

Upper and Lower Cut-Off (Break) Frequencies and Bandwidths of Octave and Fractional Octave Filters

	Percentage of f_c		
Filter type	f_u	f_ℓ	Δf
1 octave	141	71	71
1/2 octave	119	84	35
1/3 octave	112	89	23
1/5 octave	107	93.5	13.5
1/10 octave	103.5	96.7	6.8

Octave and one-third octave band filters are used very often in noise measurements. It seems that the ear has a similar frequency resolution (see Section 2.3.2), with the bandwidth increasing as frequency increases. If fine frequency resolution is not needed or if the noise is broadband in nature, then octave band readings are sufficient. One-third octave band readings (or narrower) should be used if the noise spectrum is not smooth or if pure tones are present. For diagnostic work on machinery it may be necessary to use constant bandwidth filters (e.g., if the fan blade-passing frequency and its higher harmonics must be separated). Constant bandwidth filters are described in Section 3.6.3.

We should note that until about 1960 many octave and one-third octave readings were not taken using the preferred center frequencies given in Table 3.2, but instead using the old system given in Table 3.3. Since much useful data is available in the literature in the old system, we should be aware of its existence. We see that instead of using convenient center frequencies, the old system employed convenient upper and lower cut-off (break) frequencies. The same percentages shown in Table 3.1 still apply to the upper and lower band frequencies and to the bandwidths.

B. Practical Constant Percentage Filters

There are two main types of constant percentage filters in common use: (1) those with a fixed center frequency and bandwidth which is a certain

*See International Electrotechnical Commission (IEC) Publication 225.

**See International Organization for Standardization Recommendation ISO R.266, "Preferred Frequencies for Acoustical Measurements."

TABLE 3.2

Preferred Center Frequencies for Noise Measurements (see ISO R.266)

Preferred frequencies	One oct.	Half oct.	Third oct.	Preferred frequencies	One oct.	Half oct.	Third oct.	Preferred frequencies	One oct.	Half oct.	Third oct.
16	X	X	X	160			X	1,600			X
18				180		X		1,800			
20			X	200			X	2,000	X	X	X
22.4		X		224				2,240			
25			X	250	X	X	X	2,500			X
28				280				2,800		X	
31.5	X	X	X	315			X	3,150			X
35.5				355		X		3,550			
40			X	400			X	4,000	X	X	X
45		X		450				4,500			
50			X	500	X	X	X	5,000			X
56				560				5,600		X	
63	X	X	X	630			X	6,300			X
71				710		X		7,100			
80			X	800			X	8,000	X	X	X
90		X		900				9,000			
100			X	1,000	X	X	X	10,000			X
112				1,120				11,200		X	
125	X	X	X	1,250			X	12,500			X
140				1,400		X		14,000			
160			X	1,600			X	16,000	X	X	X

TABLE 3.3

Center, Upper and Lower Break Frequencies Used in the Old System of Octave Bands

Center frequency, Hz	Upper band frequency, Hz	Lower band frequency, Hz
53	37.5	75
106	75	150
212	150	300
425	300	600
850	600	1,200
1,700	1,200	2,400
3,400	2,400	4,800
6,800	4,800	9,600

percentage of the center frequency, and (2) those with a variable (or tunable) center frequency and a bandwidth which can be set to certain selected percentages of the center frequency. The first type of filter is perhaps in most common use. In practical instruments, many different parallel filters each with a different center frequency are assembled in one unit. The instrument is provided with a root mean square detector and a meter. The signal is fed to each filter and its output examined in turn with the detector and meter. An example of a commercial instrument working on these principles is shown in Figure 3.56a. The instrument shown is sometimes called a *spectrometer.* It can supply a polarizing voltage for a capacitor microphone, and it also has a microphone amplifier and attenuators with adjustable steps of 10 dB, a set of fixed octave and one-third octave filters, an rms or peak detector,[58] and indicating meter. Figure 3.56b shows the characteristics of the top of the one-third octave filters. As seen, IEC, ANSI, and DIN (in Germany) all have requirements for filters which are met. In this commercial instrument, the low frequency filters from 2 Hz to 200 Hz are active consisting of 3

FIGURE 3.56a Example of commercial spectrometer with filters with fixed center frequencies. (Courtesy of Brüel & Kjaer Co.)

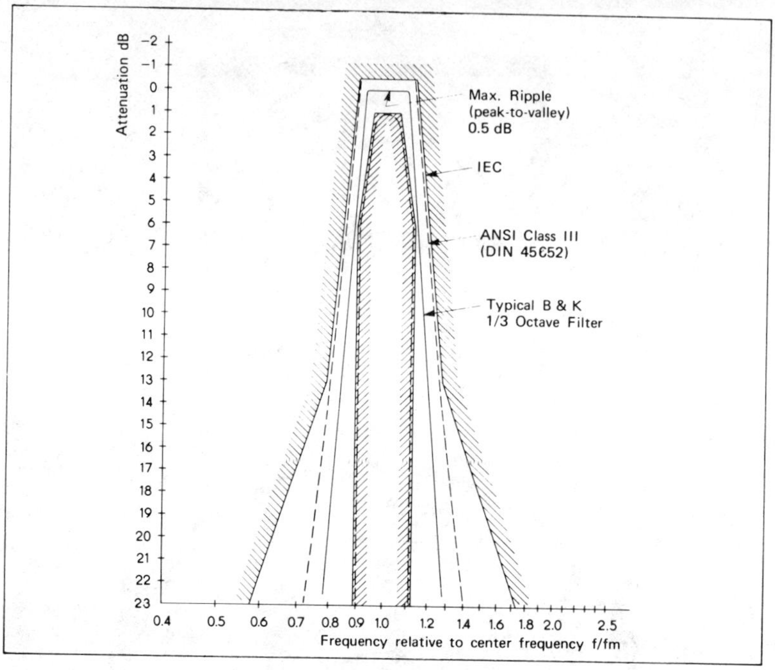

FIGURE 3.56b One-third octave filters used in filter set shown in Figure 3.56a compared with three different standard requirements. (Courtesy of Brüel & Kjaer Co.)

stagger-tuned Wien Bridge amplifiers with an emitter follower.* The other one-third octave filters are of the passive type. A typical permanent record from such a spectrometer recorded on a level recorder is shown in Figure 3.91.

Figure 3.57 shows a commercial example of the second type of filter. This instrument again has an input amplifier with a high input impedance of 2.2 MΩ and a low output impedance of less than 10Ω. However, here the filter consists of a selective amplifier system consisting of two series-connected selective amplifiers. The filtering characterisitic of the system is obtained by a frequency rejective negative feedback. The selective feedback circuit consists of a bridge tube, a resistance capacitance network of the "Wien-Bridge" type, and a cathode follower.* The attenuation of this filtering system in its different settings is shown in Figure 3.58. Note that the curves are labeled by the number of dB attenuation obtained at one octave from the center frequency. Table 3.4 gives these attenuations and the "3 dB down" bandwidth as a percentage of center frequency. The remainder of this instrument consists of an output amplifier (with an input impedance of 1.46 MΩ and an output impedance of less than 50Ω), a true rms detector and a meter. The selective amplifier filter can be tracked throughout the frequency range of 20 to 20,000 Hz.

3.6.3 Constant Bandwidth Filters

With this type of filter the bandwidth, $\Delta f = f_u - f_\ell$ is kept constant regardless of the setting of the filter center frequency. Instruments are normally constructed so that the center frequency of a single filter can be tracked effectively throughout the frequency range of interest. Often different bandwidth settings are available on the same instrument (e.g., 5, 10, 20 Hz).

The narrower the bandwidth chosen, the slower the tracking rate should be to obtain reliable

*A cathode follower is an amplifier which has little gain but converts a high input impedance to a low output impedance. It is always used to follow closely to a capacitor microphone. An emitter follower is similar but uses transistors instead of vacuum tubes used in the cathode follower. However, the emitter follower cannot offer as high an input impedance as a cathode follower. Cathode and emitter followers are used whenever low impedance devices must be connected across the output without degradation of the waveform.[1,2]

FIGURE 3.57 Example of a constant percentage narrow band filter. (Courtesy of Brüel & Kjaer Co.)

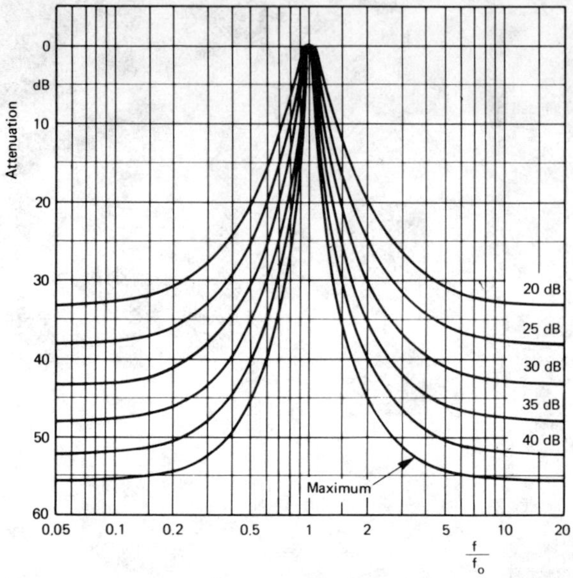

FIGURE 3.58 Attenuation of tracking filter used in the instrument shown in Figure 3.57 for different bandwidth settings. (Courtesy of Brüel & Kjaer Co.)

TABLE 3.4

Bandwidths (3 dB down) for Filters Shown in Figure 3.58

Octave selectivity, dB	3 dB Bandwidth, percentage
20	29
25	21
30	16
35	12
40	8.5
45	6

results. A rule[59] which should be used in spectrum analysis is that the duration, T, of the noise sample length (or of the analysis time) must be at least as long as the reciprocal of the bandwidth Δf,

$$T \geq 1/\Delta f. \tag{3.70}$$

Constant bandwidth filters are normally of the heterodyne type and require an oscillator to generate the frequency which is the center frequency of the tracking filter. The heterodyne principle is actually very interesting and has been discussed in detail elsewhere.[60,61] Briefly, the principle involves supplying a signal from a sine wave generator and multiplying it together with the measured signal to be analyzed. This produces an amplitude-modulated signal at the frequency produced by the generator. This is then passed through a fixed frequency band filter (see Figure 3.59). The apparent tracking of this filter is produced by altering the frequency of the sine wave generator. Figure 3.60 shows a commercial example of a heterodyne frequency analyzer. Figure 3.92 shows a permanent record of the output from a heterodyne analyzer recorded on a level recorder.

At high frequencies the constant bandwidth heterodyne analyzer will obviously have a narrower bandwidth than the constant percentage filter. Figure 3.61 shows a comparison of the bandwidths of constant percentage and constant band filters at the same frequencies.

3.6.4 Real-Time Analysis

Frequency analysis of noise signals using the analogue equipment just described can be very time-consuming. In order to speed up this process and/or to provide a continuous record of the frequency content of a signal, real-time analyzers became available in the late 1960s. There are really

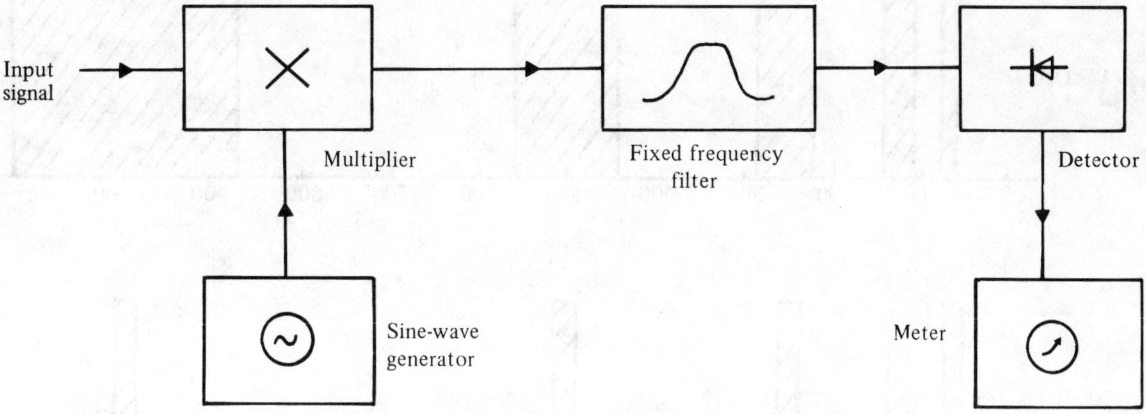

FIGURE 3.59 Simplified block diagram of a heterodyne analyzer.

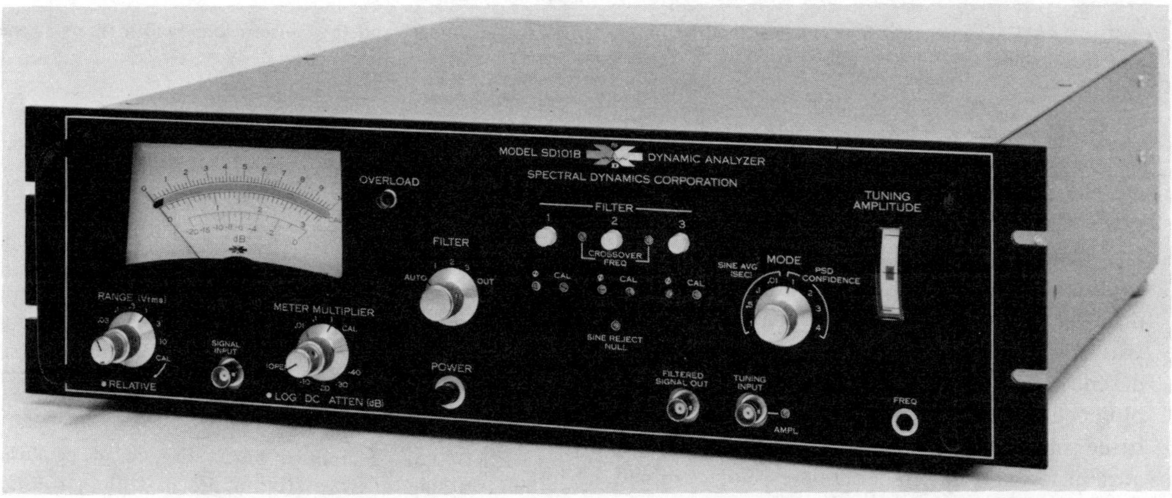

FIGURE 3.60 Example of a heterodyne analyzer. (Courtesy of Spectral Dynamics Corp.)

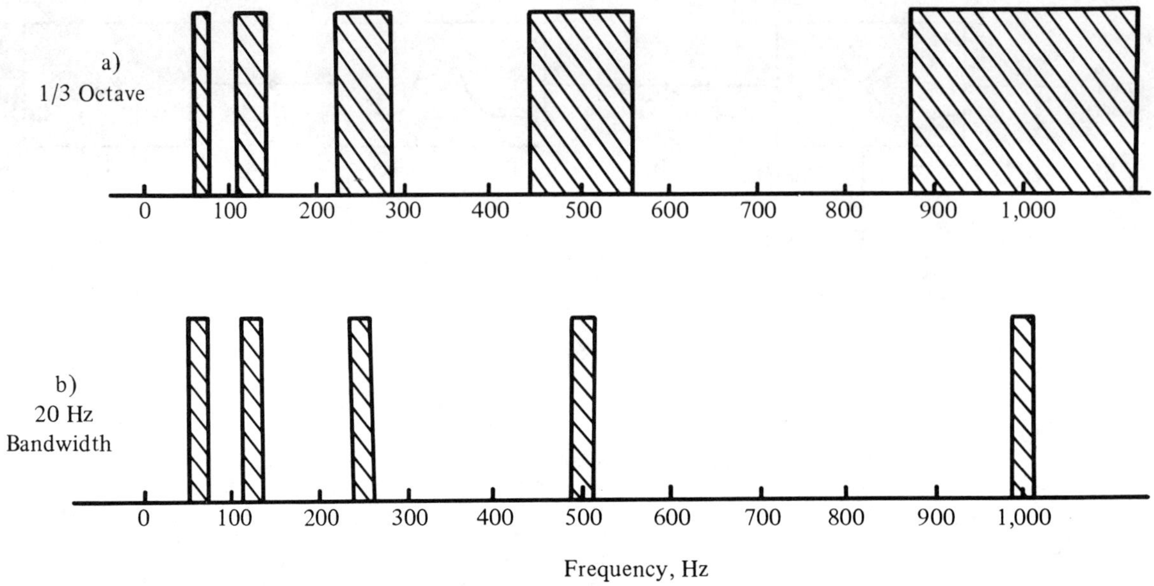

FIGURE 3.61 Comparison between bandwidths of (a) constant percentage, and (b) constant bandwidth filters at the same frequencies.

two types of real-time analyzers available. These are the *parallel-filter* and the *time compression* types which correspond to the constant percentage and constant bandwidth analyzers discussed in Section 3.6; however, they work on different principles. The cost of real-time analyzers is about two or three times that of the analog equipment described in Section 3.6.3. However, data can be obtained about 100 times as fast! If useful data are being obtained, the increase in speed is very advantageous. If incorrect data are being obtained, "mountains" of incorrect data are produced very rapidly also! The cost of the real-time analyzer, the need for rapid analysis of data, and the skill of the experimenter are all factors to be weighed before purchase of such equipment.

A. Parallel-Filter Real-Time Analyzers

Figure 3.62 shows a simplified block diagram of a *parallel-filter* real-time analyzer. The input signal is fed into a preamplifier and attenuator. (The attenuator can usually reduce the input signal in 10 dB steps up to 100 dB). The preamplifier is then connected to a large number of parallel band-pass filter channels (usually between 30 and 40 for a standard one-third octave model). Each filter channel has an rms detector, an integrator, and a storage circuit.

The electronic switch and logic circuit usually consists of a field effect transistor in each channel

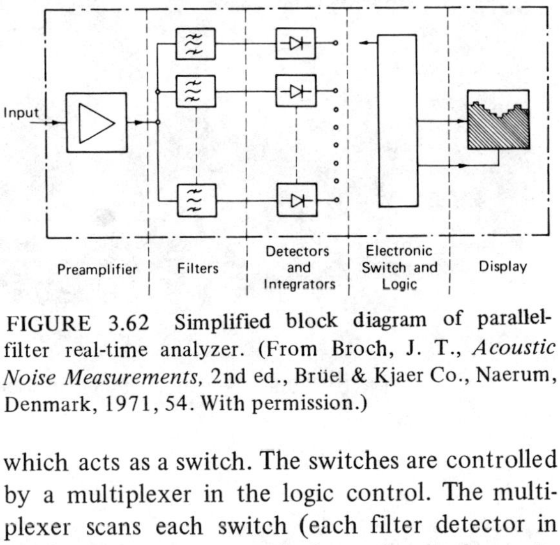

FIGURE 3.62 Simplified block diagram of parallel-filter real-time analyzer. (From Broch, J. T., *Acoustic Noise Measurements*, 2nd ed., Brüel & Kjaer Co., Naerum, Denmark, 1971, 54. With permission.)

which acts as a switch. The switches are controlled by a multiplexer in the logic control. The multiplexer scans each switch (each filter detector in turn) according to signals from the display unit. Thus each detector is examined for a fraction of a second in turn, all channels normally being scanned in a total of about 20 msec. The display is thus updated in the same time – every 20 msec.

Normally this frequency information can also be plotted out on a chart or level recorder (see Section 3.8.2). The information is retrieved from the storage units. The information can also be fed to an on-line computer. Read out of the complete spectrum by the computer is normally accomplished in a time of about 2 msec, much less than the scanning time for all the channels. More complete

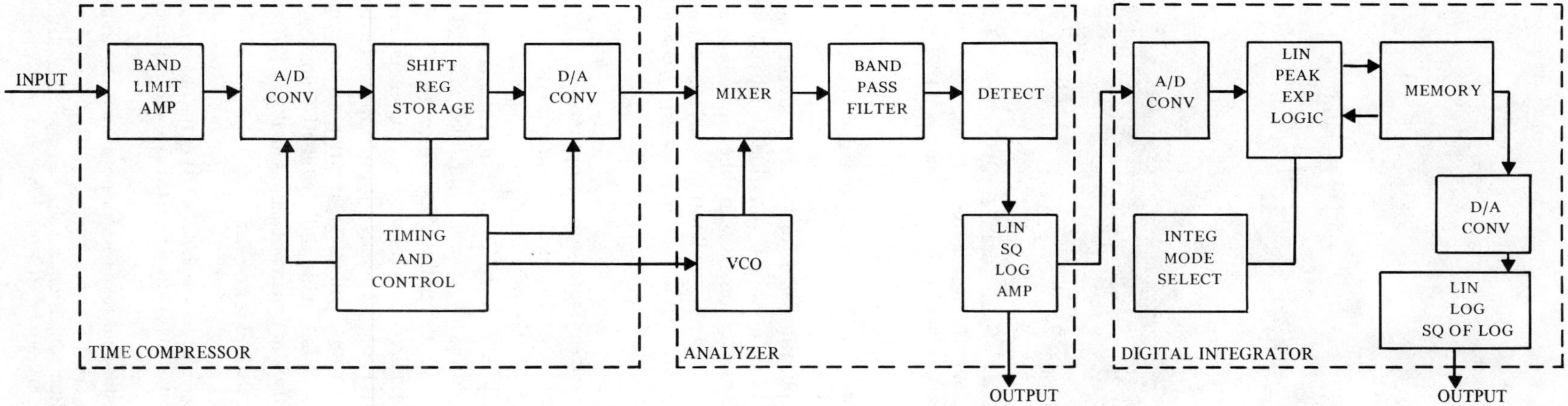

Block diagram of a Honeywell/Saicor Real-time Analyzer.

FIGURE 3.63 Block diagram of a time compression constant bandwidth real-time analyzer. (Courtesy of Honeywell Co.)

descriptions of the theory and construction of such real-time analyzers are available in the literature[62] and in the instrument manuals of several manufacturers.[13]

B. Time Compression Real-Time Analyzers

Constant bandwidth real-time analysis is normally achieved with instruments using the *time compression* principle. A block diagram of a typical time compression analyzer is shown in Figure 3.63. There are three ways of analyzing in real time. The first method, *analogue*, was discussed in Section 3.6.4A. The second method is *digital* through the use of digital computers. The third method is *hybrid*, i.e., the use of both analog and digital techniques and is the method used in time compression. The method by which this is done is described in some detail in manufacturers' literature.[13] Briefly the operation is as follows: The signal enters through an attenuator and anti-aliasing* filters. The signal is then fed to the analog to digital (A/D) converter where it is sampled at a rate several times (usually about three), the selected upper limiting frequency** of the instrument. For example, if the frequency range of 0 Hz to 1,000 Hz were chosen, the sampling rate could be 3,000 samples/sec. The signal is then converted to binary "words" which are applied to a recirculating memory. The memory has a limited storage. The "words" are read off the storage at a much higher than acquisition rate (about 100 times) and reconverted to analog form. The ratio of read-out to acquisition rate represents the time compression of the signal.

The analyzer section of the real-time analyzer (Figure 3.63) works in a similar way to the heterodyne analyzer (Section 3.6.3) and will not be discussed in detail. The frequency analysis is normally made at a large number (perhaps 200 or 400) frequencies and the data is displayed on a cathode ray tube as a 200- or 400-line spectrum. Alternatively, the data may be recorded on a level

FIGURE 3.64 Photograph of a commercial time compression real-time analyzer. (Courtesy of Spectral Dynamics Corp.)

recorder (Section 3.8.2) or fed to a digital computer (Figure 3.65).

The applications of real-time analyzers are many, including noise and vibration studies where the signal is nonstationary with time. Such applications include machinery vibration analysis, bearing noise, transient analysis, acoustic emission, speech analysis, music, and others.

3.6.5 Fast Fourier Transform Computers

In recent years the Fast Fourier Transform (FFT) Digital Fourier Analyzer has become commercially available. These Fourier analyzers are manufactured by several companies and make use of the Fast Fourier Transform (FFT) algorithm which was published by Cooley and Tukey[63] in 1965.† This algorithm is much faster than conventional digital Fourier transform algorithms and has made the digital FFT analyzer a useful, efficient means of signal analysis. So far

*Anti-Aliasing filters are low-pass filters which remove signals above the selected upper frequency range of the instrument. Aliasing errors occur when signals at high frequency (above the upper frequency limit) are interpreted as lower frequency signals below the upper frequency limit.

**It is normal with time compression analyzers to be able to select several different frequency ranges from 0 Hz to the selected upper frequency.

†Apparently Cooley and Tukey were not the first to publish such an algorithm, other similar published algorithms dating back about 50 years have also been found. However, Cooley and Tukey "rediscovered" this algorithm at a time when it could be exploited by electronic apparatus which was becoming available.

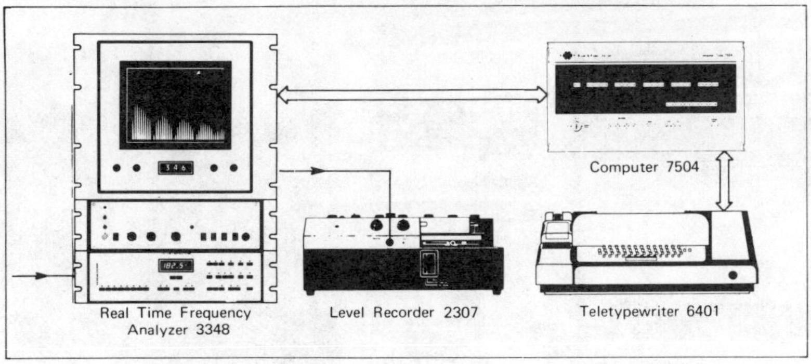

FIGURE 3.65 Read-out from a real-time analyzer may be made on a cathode ray tube, a level recorder, or computer. (Courtesy of Brüel & Kjaer Co.)

in Sections 3.6.2, 3.6.3, and 3.6.4, equipment which works on purely analog or hybrid analog/digital principles has been described. In contrast, the digital FFT analyzer makes all the analysis digitally.

Since all the analysis results are in digital form, numerous calculations can be performed by such analyzers. The Fourier transform **P**(f) of a time signal p(t) may be calculated (see Equation 1.83). In addition the auto power spectral density $G_p(f)$ (Equation 1.92) may be calculated.* If two signals $p_1(t)$, $p_2(t)$ are fed into the computer at once, then the cross-spectral density $\mathbf{G}_{12}(f)$ can be calculated,

$$\mathbf{G}_{12}(f) = \frac{1}{T}\,[\mathbf{P}_1(f)\mathbf{P}_2^*(f)]. \qquad (3.71)^{**}$$

In general $\mathbf{G}_{12}(f)$ is a complex quantity having both an amplitude and a phase. If $p_1(t)$ were an input (for example, a force) and $p_2(t)$ an output (for example, a displacement), then the *transfer function**** $H_{12}(f)$ can be computed,

$$\mathbf{H}_{12}(f) = \mathbf{P}_1(f)/\mathbf{P}_2(f). \qquad (3.72)$$

The transfer function $\mathbf{H}_{12}(f)$ is a complex quantity because it will have amplitude and phase***

In addition, the *coherence function* between the input and output may be calculated, defined by

$$\gamma_{12}^2(f) = \frac{|G_{12}(f)|^2}{G_{11}(f)\cdot G_{22}(f)}. \qquad (3.73)$$

In cases where there are multiple inputs and a single output (e.g., several cylinders (inputs) on a diesel engine), but one microphone position (output), the situation becomes more complicated.[64,66] However, in such cases the coherence function may be used to estimate the contribution to the output from each input. Measurement of the *coherence function, transfer function, cross-* and *auto-power spectral densities* are beginning to be used successfully to identify sources and predict noise levels in machinery such as diesel engines,[64] punch presses,[65] and other noise problems. The necessary theory is dealt with in detail in References 66 and 67 and in the application manuals supplied by the instrument manufacturers.

The FFT analyzer can also be used as a real-time analyzer up to a certain frequency limit (which depends on the accuracy desired). However, usually the real-time analyzers discussed in Section 3.6.4 can accomplish the same task at lower cost. A photograph of a commercial FFT analyzer is shown in Figure 3.66. Minicomputers are now becoming available which can accomplish many of the tasks of a FFT analyzer at somewhat lower cost. Computation speed reduction is not normally a problem because normally, calculations are made faster than they can be printed out.

3.7 PORTABLE INSTRUMENTATION

In many noise problems it is not possible to bring the noise source into the laboratory, and it is

*The auto power spectral densities $G_{11}(f) = \frac{1}{T}[\mathbf{P}_1(f)\,\mathbf{P}_1{}^*(f)]$ and $G_{22}(f) = \frac{1}{T}[\mathbf{P}_2(f)\,\mathbf{P}_2{}^*(f)]$ are real quantities.

**The asterisk denotes the complex conjugate.

***Note that we have already discussed the transfer function (or equivalently the frequency response function) in Section 3.6.1.

FIGURE 3.66 Photograph of a Fast Fourier Transform (FFT) Analyzer. (Courtesy of the Hewlett-Packard Co.)

necessary to make measurements on the site. One approach is to tape-record the noise and analyze it in the laboratory later (see Section 3.8.1). However, in many instances it is desirable to measure the noise at the site. If this is possible, decisions based on the noise measurements can be made at the site itself. For example, changes can be made to the noise source at the site such as the use of temporary enclosures or baffles placed near a machine and these quickly evaluated as solutions to noise problems. Noise source identification can also often be more easily accomplished near large machines by using portable equipment. For this reason there has been a considerable effort in the last several decades to produce portable noise measuring instruments. The most common and useful such instrument is the *sound level meter.* Most commonly, the sound level meter is used to make noise measurements with a microphone as its transducer. However, if the microphone is replaced with a piezoelectric accelerometer and an integrator, the instrument becomes a vibration meter capable of measuring directly acceleration, velocity, and displacement. The conventional sound level meter is discussed first.

3.7.1 Sound Level Meters

In the 1930s, as miniature vacuum tubes and Rochelle salt piezoelectric microphones became available, portable sound level meters were developed. Prior to that time, in order to conduct noise surveys (e.g., the 1929 noise survey in New York) the ear-balance method was used in which an audiometer tone was adjusted until it was judged by the listener to be equally as loud as the noise being measured.

At first it was hoped that it would be possible to produce a portable sound level meter which could be used to measure the *loudness level* of a sound. However, we now know that the ear does not add up the loudness levels of complex sounds in the manner of a rms meter. Some attempts have

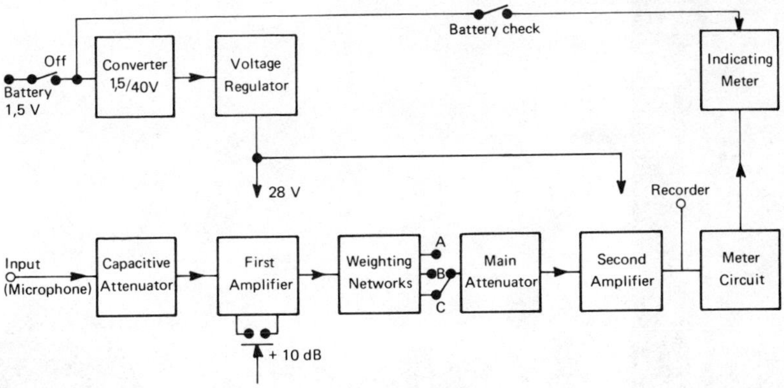

FIGURE 3.67 Block diagram of simple sound level meter. (Courtesy of Brüel & Kjaer Co.)

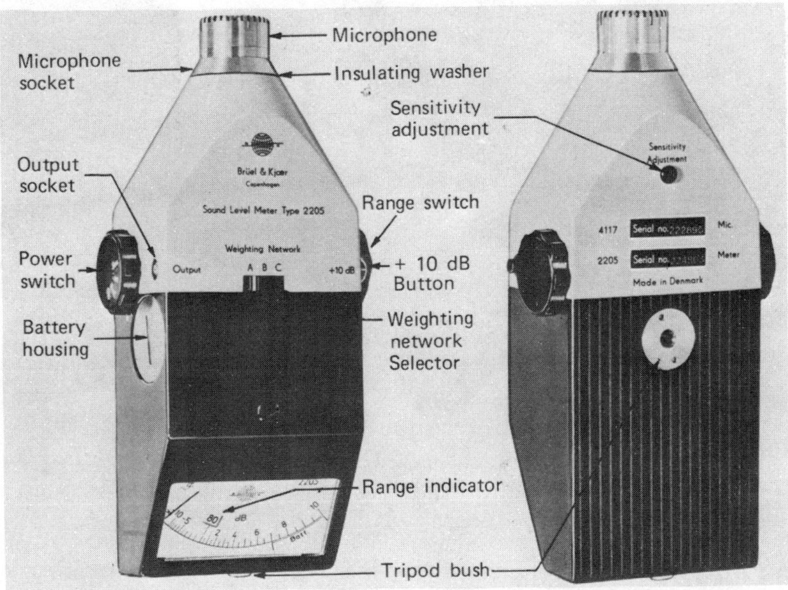

FIGURE 3.68 Photograph of simple sound level meter (Block diagram is shown in Figure 3.67.) (Courtesy of Brüel & Kjaer Co.)

been made to produce loudness level meters,[68] although they have not been completely successful. The *sound level* measured by the sound level meter does not conform, particularly, to the subjective response of the human ear. However, it is still very convenient, combining a microphone, amplifier, and meter into one portable instrument.

A. Basic Elements of Sound Level Meter

A block diagram of a typical simple modern sound level meter is shown in Figure 3.67. The example shown is of the small handheld sound level meter with a piezoelectric microphone shown in Figure 3.68. The capacitative attenuator and the main (resistive) attenuator are interconnected and can be operated by a single range switch to give 10-dB steps in attenuation. Such an arrangement is designed to give the best signal to (electrical) noise ratio. In the example shown, the first and second amplifiers have gains of 45 and 68 dB, respectively. The meter circuit is a square law rectifier; the internationally standardized *fast* and *slow* meter dampings are provided in this circuit. The indicating meter (a ribbon suspended-coil galvanometer) has a range of 20 dB, its zero position being governed by the attenuator setting. The A-, B-, and C-weighting networks (Figure 2.10) are constructed from resistance-capacitance circuits and the meter can be operated with any of these three networks. The piezoelectric micro-

a. b. c.

FIGURE 3.69 Example of ANSI types 1, 2, and 3 sound level meters manufactured by one company. (a) type 1; (b) type 2; (c) type 3. (Courtesy of General Radio Co.)

phone used in this particular sound level meter is shown in Figure 3.28.

B. Standards for Sound Level Meters

Sound level meters are built to meet various standards or recommendations.[69–71] American National Standard S 1.4-1971[69] specifies four types of sound level meters: types 1, 2, 3, and S corresponding to *Precision, General Purpose, Survey,* and *Special Purpose* types, respectively. Essentially the requirements for precision* and accuracy* are greatest for type 1 and least for type 3. The type S meter can be qualified to the performance of any of the basic types (1, 2, and 3) but is not required to have all three weighting networks (A, B, and C). Figure 3.69 shows 3 different sound level meters (types 1, 2, and 3) manufactured by one particular company. Figure 3.70 shows a type S meter (a *noise event* meter with A-weighting only). A meter satisfying ANSI type 1 requirements will normally also satisfy IEC Publication 179 requirements which are not quite so strict. Likewise, a meter satisfying ANSI type 2 requirements again should satisfy IEC Publication 123 requirements which similarly are not so strict.

Since sound level meters are used in many different applications, the choice of the sound level meter used will depend upon the precision and accuracy required, other capabilities of the instrument, and its purchase price. A type 1 sound level meter will be most expensive. Figure 3.71 shows the tolerances allowed under ANSI S 1.4-1971 for A-weighted response to random incidence sound. This standard also gives the

**Precision* may be defined as how well different measurements of the same thing agree with each other; *accuracy* refers to how well a measurement agrees with the "true" answer which would be obtained with a perfect instrument.

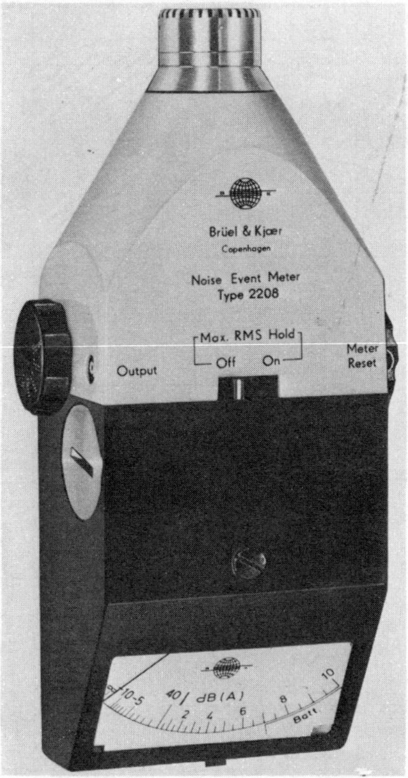

FIGURE 3.70 Type S sound level meter (noise event meter). (Courtesy of Brüel & Kjaer Co.)

tolerances allowed for B- and C-weighted responses and gives, among others, requirements for the instrument's indicating scale range, fast and slow dynamic characteristics, sensitivity to temperature, humidity, vibration, and magnetic and electrostatic fields.

C. Commercially Available Sound Level Meters

The sound level meters discussed so far in Section 3.7.1 have only been capable of measuring A-, B-, or C-weighted sound levels. Attachments can be made to a sound level meter so that octave or one-third octave band measurements can be made. Figure 3.72 shows such octave and one-third octave filter sets and Figure 3.73 shows one attached to a precision sound level meter. A sound level meter equipped with these filter sets can be very valuable in making frequency analyses during on-site measurements. The sound level meter shown in Figure 3.73 can also be used to measure impulse noise (e.g., from punch presses). It has peak hold and maximum rms hold facilities for this purpose.

In recent years sound level meters with digital displays have become available. Figure 3.74 shows an interesting sound level meter. It is a special purpose precision meter, only reading A-weighted sound levels. It has a dynamic range of 50 dB from 70 to 130 dB (A) and gives both an analogue display on a meter and a digital display.

D. Measurement Problems with Sound Level Meters

There are several measurement problems with sound level meters including undesired sensitivity to vibration, humidity, temperature, magnetic and electrostatic fields, etc. Two further problems are reflections caused by the sound level meter body itself and by the operator, and microphone noise caused by wind. As Figure 3.75 I shows the sound level meter case itself causes only small reflections, while the operator, Figures 3.75 II and III, causes larger reflections (and hence measurement errors). If fitted with an extension chord of several meters to connect it to the sound level meter the microphone has a frequency response which is essentially flat. Young[72] has discussed the reflection problem in some detail. In some cases, measurement errors even greater than those shown in Figure 3.75 are possible. To minimize such operator reflection errors the sound level meter should be held at arm's length or else an extension chord for the microphone should be used. The use of an extension chord is recommended in cases where extreme accuracy is required. With capacitor microphones the extension chord must be fitted between the cathode follower (preamplifier) and sound level meter, otherwise the microphone sensitivity is reduced and noise may be picked up.

In high winds (above about 12 mph), sound level meters should not be used, otherwise the noise to be measured tends to be masked by wind noise (unless the noise is very intense). Wind noise can be caused by *existing eddies* in the air flowing past the microphone but is mainly caused by *self noise.**

The three windscreens shown in Figure 3.76 can suppress both types of noise. The left and middle windscreens are for 1-in. and 1/2-in. microphones and are made from porous polyurethane sponge. These sponge windscreens are useful

**Self noise* is caused by the eddies created on the microphone itself as the air flows past it.

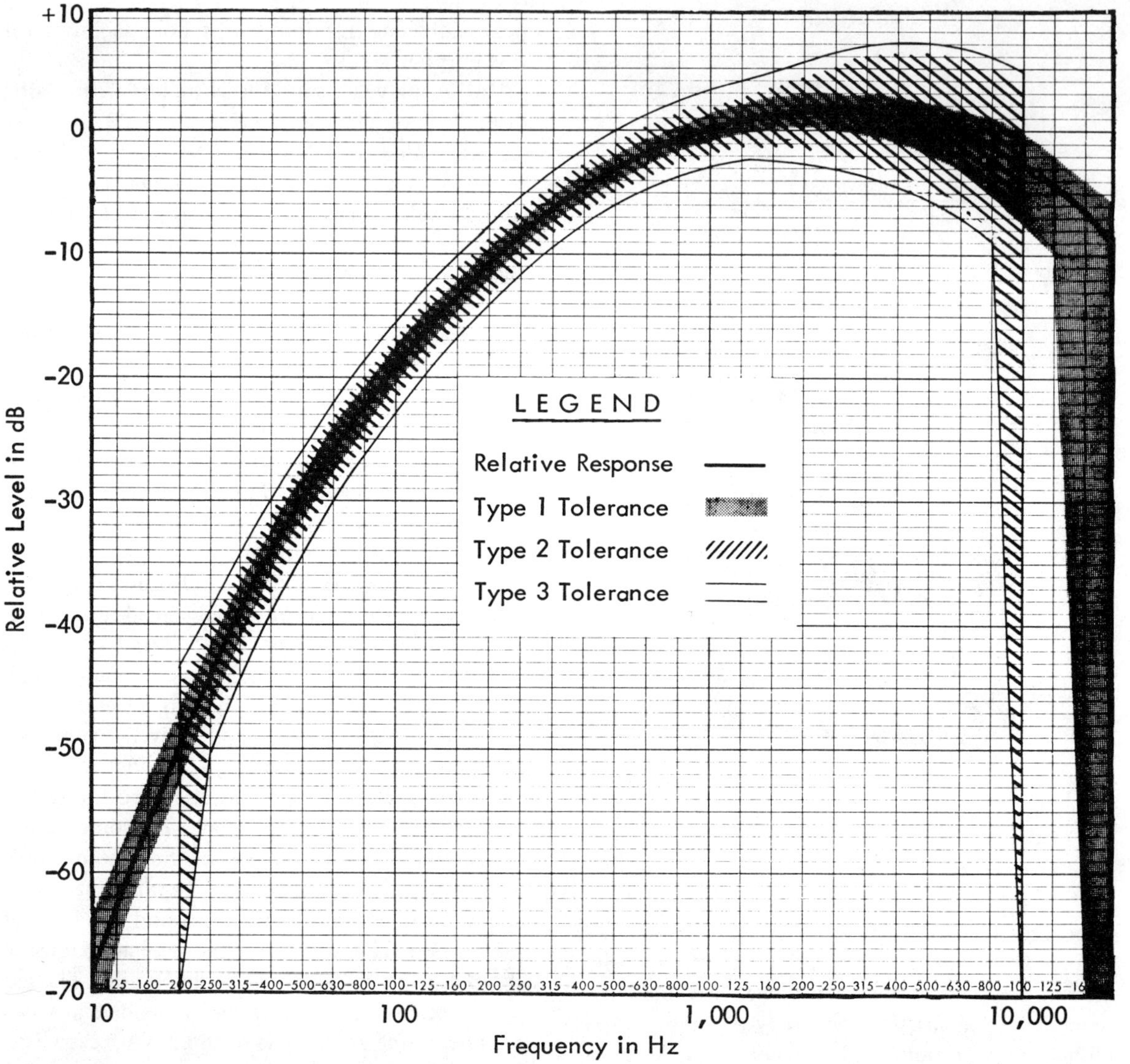

FIGURE 3.71 ANSI S1.4–1971 tolerances for A-weighted response to sound at random incidence. (This material is reproduced with permission from American National Standard Specification for Sound Level Meters, S1.4–1971, copyright 1971 by the American National Standards Institute, copies of which may be purchased from the American Standards Institute at 1430 Broadway, New York, NY, 10018.)

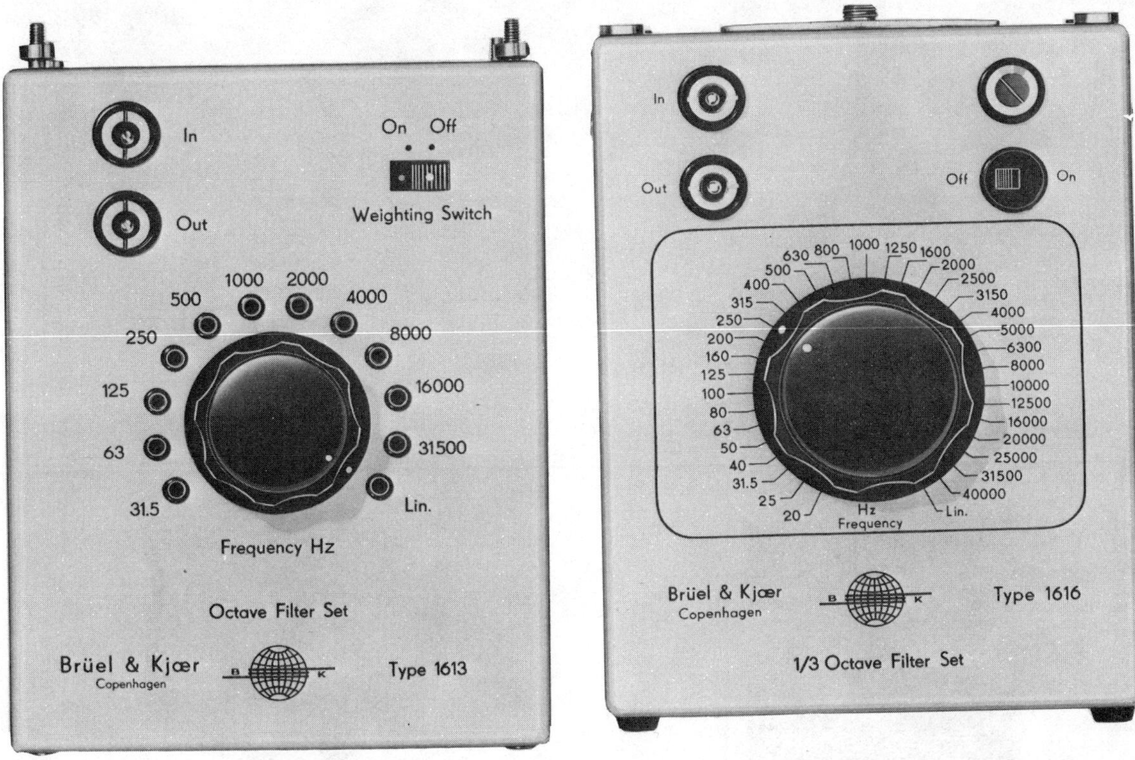

FIGURE 3.72 Octave and one-third octave filter sets for attachment to a sound level meter. (Courtesy of Brüel & Kjaer Co.)

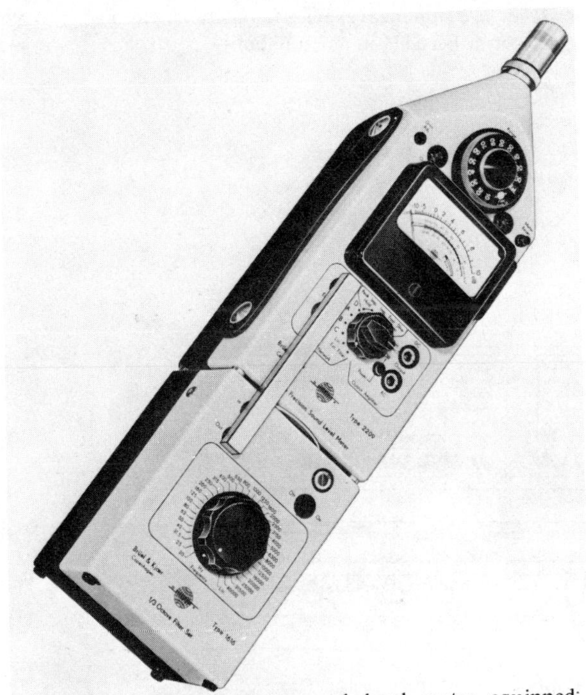

FIGURE 3.73 Precision sound level meter equipped with one-third octave filter set. (Courtesy of Brüel & Kjaer Co.)

but mainly for indoor work (e.g., with rotating microphones) or for outdoor work in the absence of rain. Rain tends to block the pores and change the frequency response of the microphone. The lowest windscreen (e) in Figure 3.76 has a wire frame covered with a nylon cloth. It has several spikes on top to prevent birds from resting on it! It is less affected by rain and can be used for permanent outdoor installations when provided with a commercial rain cover (Figure 3.77). This windscreen can be used with both 1-in. and 1/2-in. microphones. Figure 3.78 shows the wind noise measured both with and without the cloth covered windscreen. The whole question of windscreen design was discussed by Brüel[73] in 1960 in a useful review. However, since Brüel's article was written, several new windscreen designs have been produced.

In some applications the approaching air is not turbulent, e.g., in wind tunnels, and only microphone self noise is a problem. In such cases, microphone nose cones (see Figure 3.79) are substituted for the protection grid. A fine mesh grid at the base of the windscreen allows sound

FIGURE 3.74 Precision special purpose sound level meter with analogue and digital displays. (Courtesy of General Radio Co.)

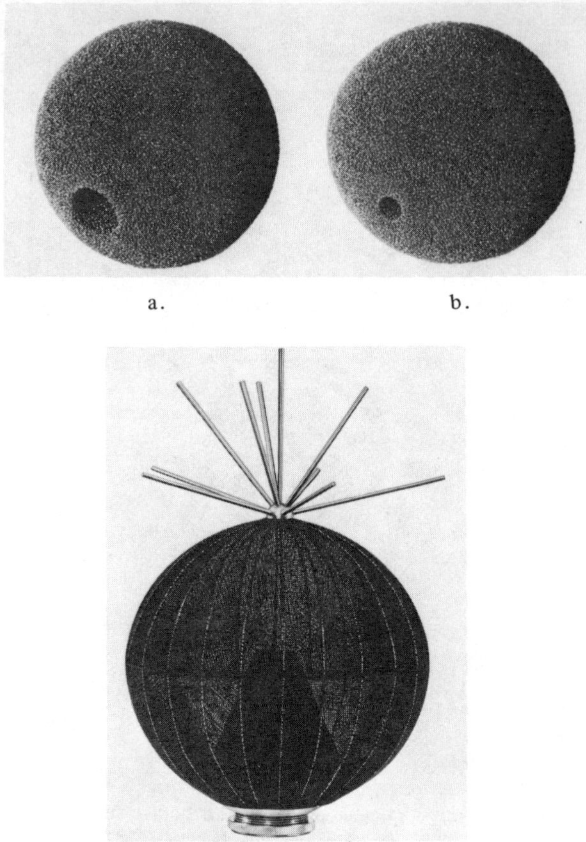

a. b.

c.

FIGURE 3.76 Three windscreens. (a) Sponge windscreen for 1-in. microphone; (b) Sponge windscreen for ½-in. microphone; (c) Wire frame-nylon cloth windscreen for both 1- and ½-in. microphones.

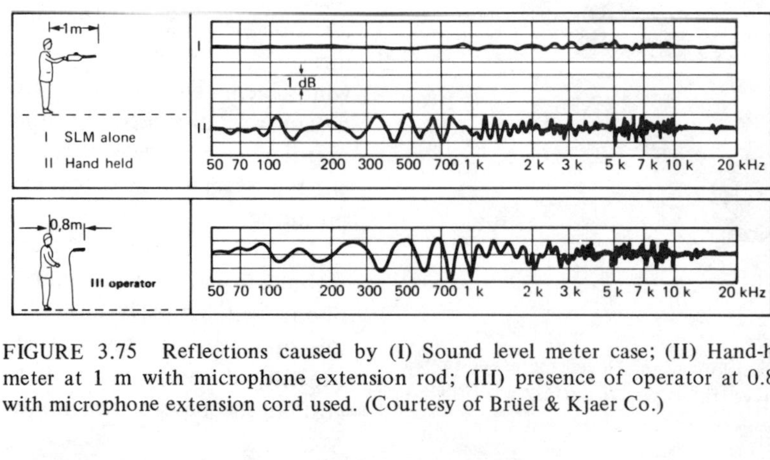

FIGURE 3.75 Reflections caused by (I) Sound level meter case; (II) Hand-held meter at 1 m with microphone extension rod; (III) presence of operator at 0.8 m with microphone extension cord used. (Courtesy of Brüel & Kjaer Co.)

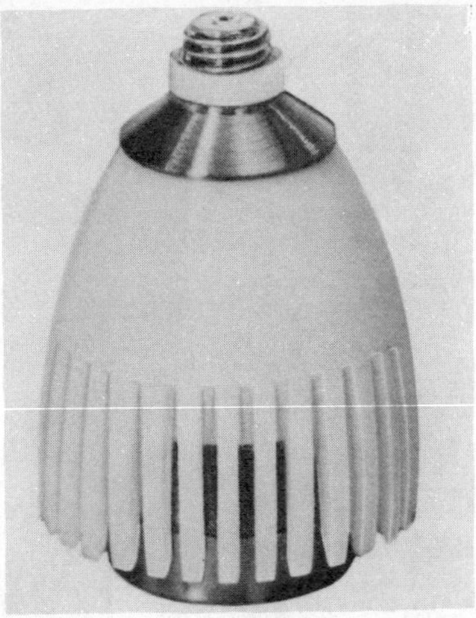

FIGURE 3.77 Rain shield for use with the wire-nylon cloth windscreen. (Courtesy of Brüel & Kjaer Co.)

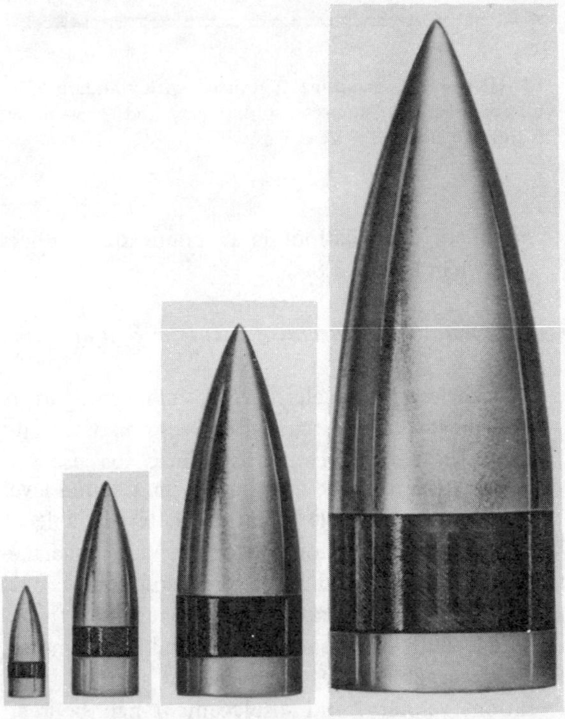

FIGURE 3.79 Microphone nose cones which can replace the protection grid and reduce self noise. (Courtesy of Brüel & Kjaer Co.)

waves to enter and reach the microphone diaphragm. The self noise is much reduced. For measurement of fan noise in a duct, nose cones are not satisfactory because a nose cone can suppress only the self noise and not the turbulence produced by the fan. Based on three years of research work by Wang and Crocker[74,75] at the Herrick Laboratories in the U.S. and Neise in Berlin, Germany, a long cylindrical nose cone has now been produced and is commercially available. This special nose cone known as a *sampling tube* can suppress both fan turbulence and self noise by about 12 to 20 dB in the frequency range 100 to 10,000 Hz. This sampling tube which has research

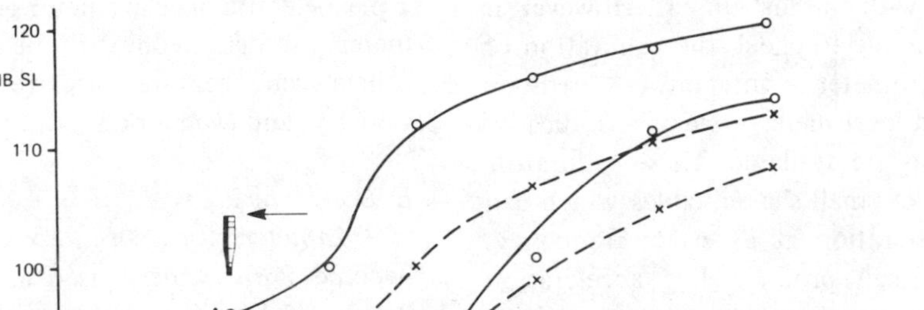

FIGURE 3.78 Wind noise (in the range 20 to 20,000 Hz) as a function of windspeed both with and without the wire-nylon cloth windscreen. (From Broch, J. T., *Acoustic Noise Measurements,* 2nd ed., Brüel & Kjaer Co., Naerum, Denmark, 1971, 62. With permission.)

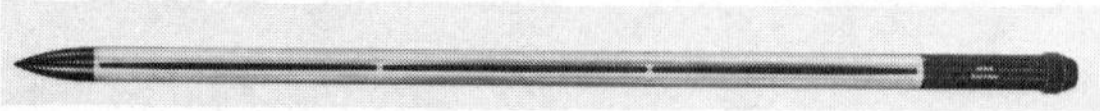

FIGURE 3.80 Sampling tube used with standard ½-in. microphone to suppress turbulence and self noise. (Courtesy of Brüel & Kjaer Co.)

and "field" applications in air conditioning ducts is shown in Figure 3.80.

E. Vibration Measurements with a Sound Level Meter

Piezoelectric accelerometers and integrators have already been discussed in Sections 3.5.2 and 3.5.3. These items can satisfactorily be used to replace a microphone and transform a sound level meter into a portable vibration meter capable of measuring acceleration, velocity, and displacement. Figure 3.81 shows a precision sound level meter and an integrator and accelerometer connected. With this particular sound level meter, interchangeable scales are provided so that acceleration, velocity, and displacement can be read directly off the meter scale with the integrator switched to give a linear response, and one or two stages of integration, respectively. An octave or one-third octave filter set can be attached to the sound level meter in order to analyze the vibration signal into frequency bands.

Normally each individual accelerometer is calibrated during manufacture and a calibration chart is supplied with the instrument. However, it is useful and desirable to check the calibration of the whole accelerometer – integrator – cathode follower – sound level meter system periodically. Various calibrators are available. These calibrators normally consist of small shaker tables which can provide 1 g acceleration at a certain frequency. Some calibrators can produce 1 g acceleration throughout the frequency range of the accelerometer by using an external oscillator system.

3.7.2 Other Portable Instruments

A. Vibration Meters

It has just been described how a sound level meter can be converted into a vibration meter. However, some portable vibration meters are available which solely measure vibration. Figure 3.82 shows such a meter. Most of these meters use

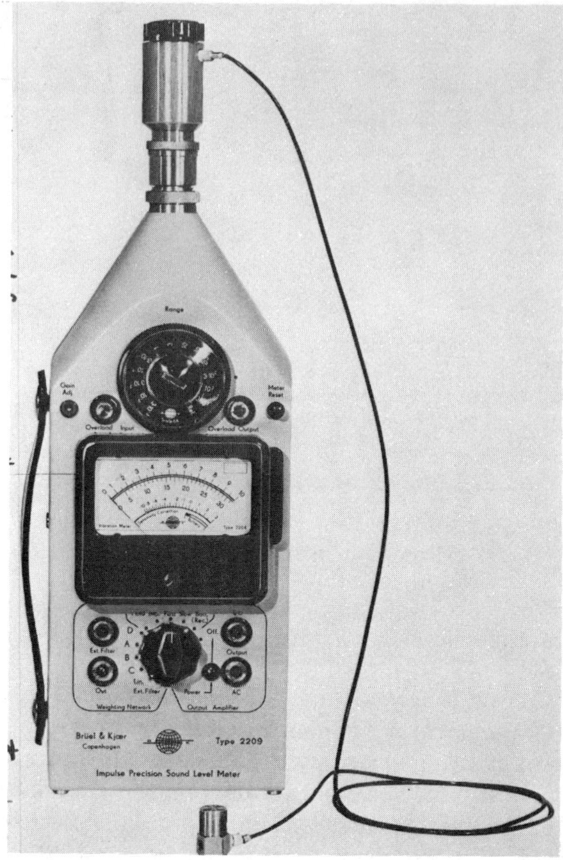

FIGURE 3.81 Precision sound level meter with integrator and accelerometer connected. (Courtesy of Brüel & Kjaer Co.)

a piezoelectric accelerometer as the pick-up. Some meters are designed just to measure velocity, while others can measure displacement, acceleration, velocity, and even jerk.*

B. Dosimeters

Dosimeters or *Noise Exposure Meters* have become fairly widely used in the U.S. since the Walsh Healey Act, 1969, and the Occupational Safety and Health Act (OSHA) Noise Regulations in 1971. They are also used in several other countries, particularly in Europe, but in these countries the noise summation process is carried out according to ISO R 1999** which states that a 3 dB (A) increase in sound level requires half the duration to give a person the same noise "dose". The OSHA regulations state that a 5 dB (A) increase in sound level needs half the duration to

**Jerk* may be defined as the rate of change of acceleration or alternatively as the third time derivative of displacement. Jerk is of interest in some shock problems.

**ISO R. 1999 Assessment of Occupational Noise Exposure for Hearing Conservation Purposes.

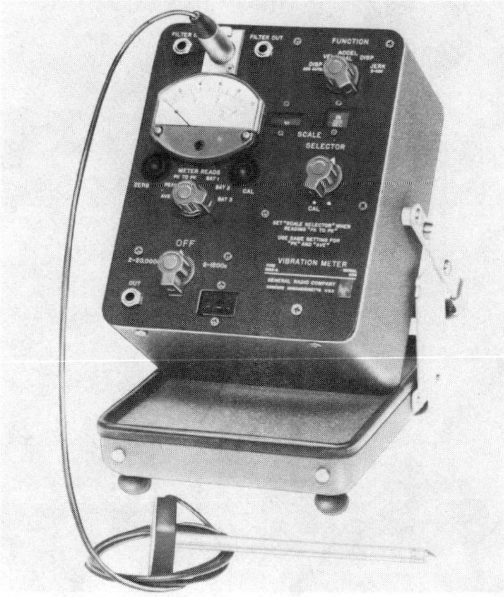

FIGURE 3.82 Vibration meter with pickup connected. (Courtesy of General Radio Co.)

give the same noise "dose" (see Section 2.6.2). Thus, dosimeters constructed in the U.S. are different from those produced in the rest of the world.

Dosimeters can be divided into two classes: *nonportable* and portable or *personal* units. The nonportable units can be located at certain selected sites in a plant and the noise dose at this location computed for each workday. Sound level meters, or microphones and spectrometers, may be used to feed the signal to the noise dose meter.

The *personal* dosimeter is really no different in concept from the above but uses miniaturized components and is carried with a worker wherever he or she goes. Figures 3.83 and 3.84 show a block diagram and a photograph of such a dosimeter and Figure 3.85 shows the unit in use. The dosimeter illustrate sums the noise exposure and displays the percentage of the daily allowable noise dose. For instance in the U.S., 100% would be displayed after 8 hr at 90 dB (A) and 200% after 8 hr at 95 dB (A). Such instruments give a much more accurate determination of the "true" noise exposure of a person who moves to several locations with different sound levels during a workday.

Dosimeters are manufactured by several different companies and vary in accuracy.[76] An assessment of the accuracy of the different types was made by Leasure[77] at the National Bureau of Standards. Some personal dosimeters must be taken and plugged into a read-out unit at the end of the workday.

3.8 DATA RECORDING

It can often save time and add convenience to record permanently noise or vibration data. This is most often done by use of a magnetic tape recorder or by a strip chart (or level) recorder. The purposes of these two types of data recording systems are really quite different.

3.8.1 Magnetic Tape Recorders

The magnetic tape recorder records the input waveform so that it can be analyzed later in the laboratory. The principle of magnetic recording is not new. According to Wood,[78] recording on magnetic wire was first introduced by Poulson about 1900 under the name "Telegraphone"; telegraph signals were recorded for subsequent transmission. In 1924, Stille, a German engineer substituted a thin metal tape for the wire and made further improvements. The British Broad-

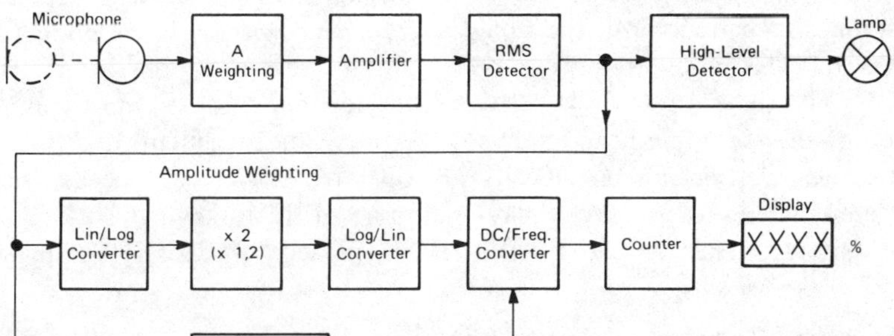

FIGURE 3.83 Block diagram of personal noise dosimeter system. (Courtesy of Brüel & Kjaer Co.)

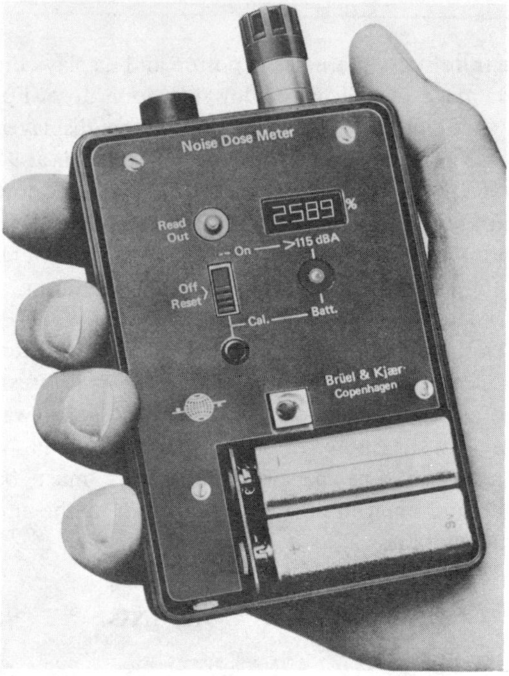

FIGURE 3.84 Photograph of personal noise dosimeter. (Courtesy of Bruel & Kjaer Co.)

FIGURE 3.85 Photograph of personal noise dosimeter in use.

casting Corporation (B.B.C.) used this system in the early 1930s under the name "Blattnerphone". More recently, thin plastic tapes on which a thin film of fine iron oxide particles have been coated on one side have been used.

The basic tape recorder system is shown in Figure 3.86. The capstan drive pulls the magnetic tape past the head assembly at a constant speed with the coated side facing the three heads.

The first head encountered by the tape is the erase head. The coil of this first head is supplied with a very high frequency alternating current (of about 50,000 Hz) which magnetically saturates the tape. As the tape passes away from the head gap the magnetic field strength decreases. The magnetic induction on the tape goes through several hysteresis loops during this passage until the final tape magnetism is zero.* This process removes any prerecorded signal.** The signal to be recorded is supplied to the coil of the recording head, and as the tape passes this head, areas of it are locally magnetized at different strengths. Finally a playback head is encountered. The passage of the magnetized tape past the playback head causes a current similar to the original signal recorded to flow. This current can be amplified to drive a loudspeaker or used for further analysis.

There are two main methods of recording on magnetic tape: *frequency modulation* (FM) and *direct recording*. These two techniques are described in several books.[79,80] The direct recording technique is used most commonly in acoustical noise measurements. Most tape recorders operate at several speeds, usually at one or more of the following: 1 7/8, 3 3/4, 7 1/2, 15, and 30 in./sec.*** Using the highest tape speed, a

*This is rather like the degaussing process carried out on ships.

**It is sometimes preferable to bulk-erase magnetic tapes. The process is similar but this more effectively removes the prerecorded signal.

***It is sometimes useful to record at one speed and play back at another. This sometimes enables the data to be analyzed by equipment with limited writing speeds or frequency ranges. In such cases corrections must be made to the frequency. The frequency must be multiplied by the tape speed ratio.

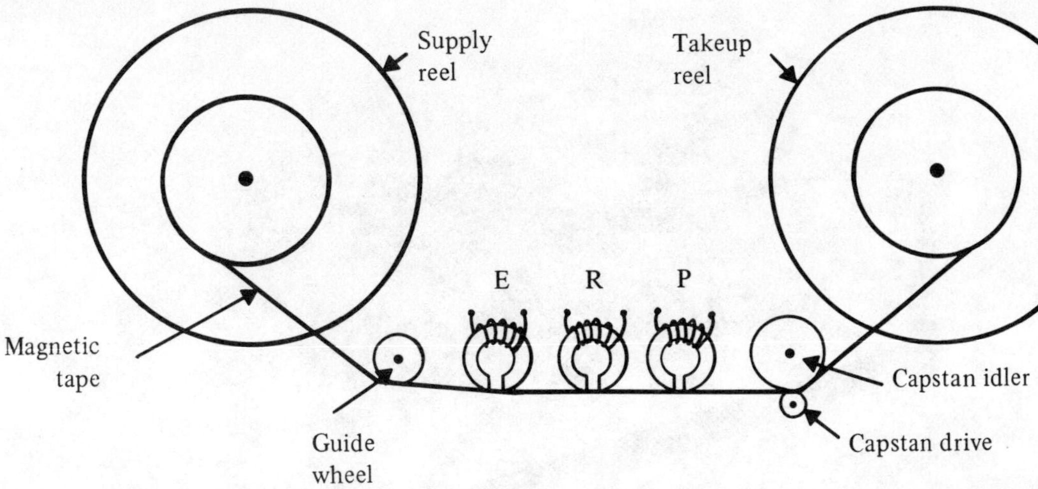

FIGURE 3.86 Basic tape recorder system; E–erase head, R–record head, P–playback head.

modern direct tape recorder can be made with a flat frequency response in the range 25 Hz to 50,000 Hz.*

The FM recording technique is used in machines where low frequency vibration signals must be recorded. A typical FM machine has a flat frequency response in the range from DC to 10,000 Hz.

Professional tape recorders are similar to the cheaper tape recorders produced for home music use, although they are more expensive because greater care has been taken to overcome the many design problems. In addition, home music recorders do not usually have a flat frequency response, since this has been altered to make it more favorable for music recording and reproduction.

The following problems are encountered with all tape recorders:

1. *Tape noise.* This comes from several sources including electronic tube noise, the type of tape used, and the hum picked up from stray magnetic fields at line frequency. This means that there will be a minimum acceptable signal level. Lower levels will be "buried" in noise.

2. *Distortion.* There is a maximum signal level governed by several factors including the amplifier design and the type of tape used, above which distortion will occur. The upper and lower levels of operation are controlled by the *distortion* and the *tape noise*, respectively. The range between these two levels is called the *dynamic range.* The dynamic range is normally between about 40 to 60 dB for good tape recorders. At considerable expense this can be extended to 70 dB.

3. *Flutter and Wow.* Flutter or wow occur since the tape cannot be driven at a completely constant speed past the head system by the capstan. Small speed changes cause fluctuations in the frequency of the signal. If these speed changes are kept less than 0.25%, this is normally satisfactory.

4. *Print Through.* Some magnetism is inevitably transferred to adjacent layers of tape causing spurious signals and "echoes". Attempts are made to keep *print through* below *tape noise.* The use of thin recording tape increases this problem.

Most tape recorders have provision so that two, four, or even more signal channels can be recorded simultaneously side by side on the 1/4-in. wide tape. In order to frequency-analyze one particular segment of a tape-recording, a section may be cut from a tape and a loop made by splicing it together carefully. This practice was more frequently used before the advent of real-time analyzers (Section 3.6.4) since frequency analyses of small sections of the tape can be made by this method. Special tape loop cassettes can be bought to simplify this procedure.

Figure 3.87 shows a photograph of a typical professional tape recorder. Figure 3.88 shows a

*Using higher tape speeds and special magnetic tapes, the upper frequency cut-off can now be extended to several hundred kHz. However, this is not normally needed with acoustical noise measurements.

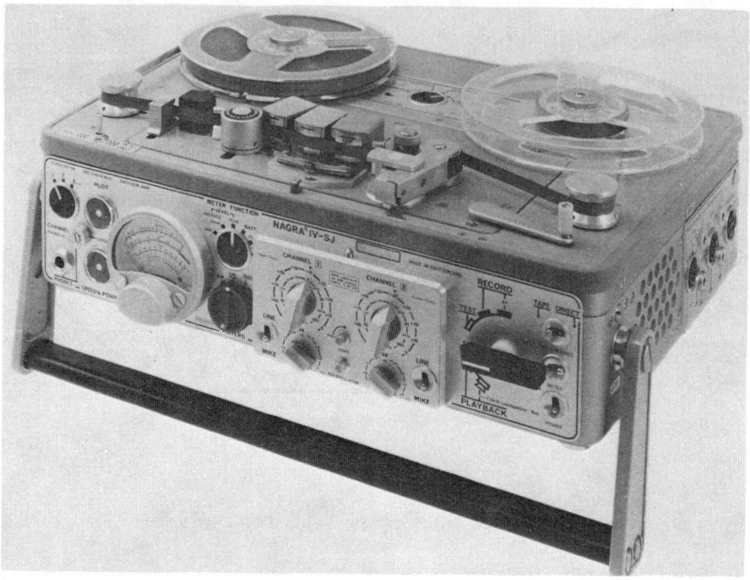

FIGURE 3.87 Professional tape recorder. (Courtesy of Nagra Co.)

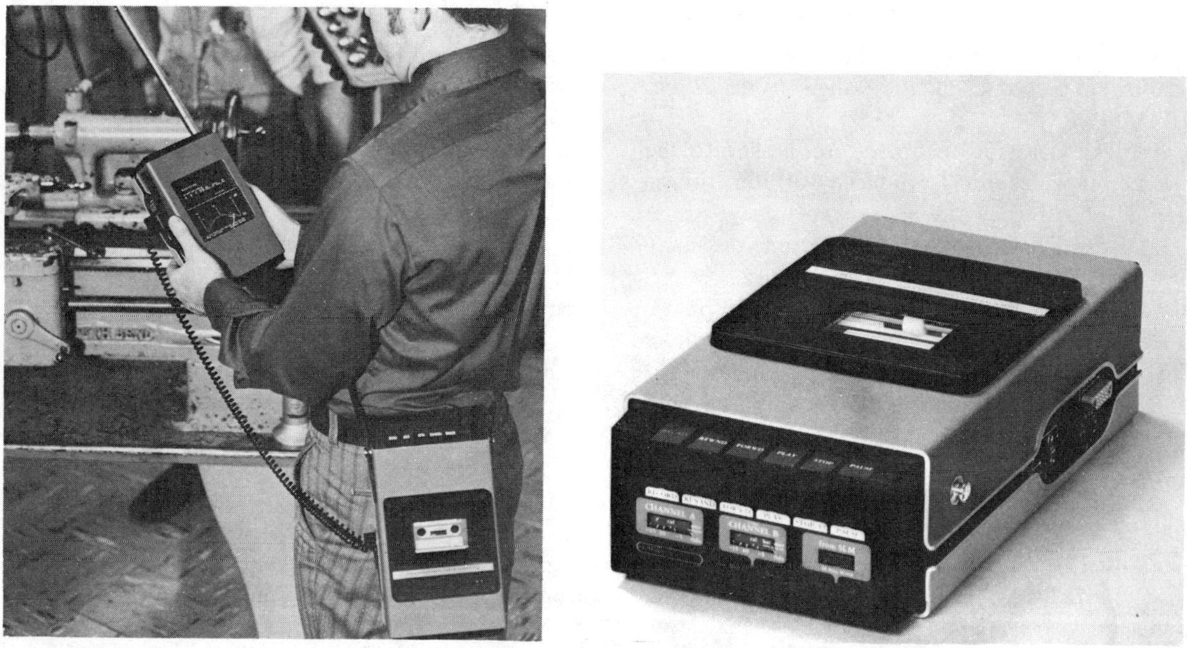

a. b.

FIGURE 3.88 Cassette tape recorder. (a) In use; (b) Instrument close-up, (Courtesy of General Radio Co.)

cassette tape recorder intended for use with a precision sound level meter. This particular tape recorder uses 0.150-in. wide tape in a cassette and has a frequency response (of ± 2 dB) in the range 50 Hz to 12,000 Hz. Although cassette tape recorders cannot offer the performance of conventional tape recorders, they have the advantage of simplicity and convenience due to light weight.

3.8.2 Graphic Level Recorders

The graphic level recorder is a laboratory instrument which is used to produce a permanent graphical record of the level of a signal (see Section 1.8). Signals directly from a microphone, accelerometer, or recorded on a tape recorder after suitable amplification may be fed to the level recorder. The graphic level recorder is suitable for making spectrograms (level plotted against frequency) or level time histories (level plotted against time). The former type of record is useful in frequency analysis, while the latter type of record is useful in any type of situation where the sound pressure level varies with time (e.g., in reverberation time measurements or traffic or aircraft noise measurements).

Many different types of recorders have been designed and described in the literature.[5] One type which has been used widely is due to Bruel and Ingard.[81] A block diagram of such an instrument which is commercially available is shown in Figure 3.89. A detailed description of its operation is available in the literature[82] and also from the manufacturers.[83] This type of instrument is a recording voltmeter and essentially functions on the servo-principle as follows: The input voltage signal is fed through a continuously variable potentiometer and through a stepped attenuator (usually having 10-dB steps) to a second range potentiometer. The range potentiometer consists of a series of over 200 resistor strips. The strips can be arranged to vary linearly or logarithmically along the range potentiometer. The position of the slider contact on the range potentiometer determines its voltage output. The output voltage from the range potentiometer is rectified and time averaged and compared with a reference voltage. If a difference exists, it is amplified and fed to the coil of the magnetic drive system causing the shaft to move. The shaft is mechanically coupled to the range potentiometer slider and the writing stylus on the recording paper. The movement of the slider on the range potentiometer will result in a change in output voltage until eventually a stable situation is reached where the rectified and averaged voltage is equal to the reference voltage. The slider and stylus then cease to move. Simultaneously, the recording paper is driven past the recording stylus at a constant speed (usually several speeds (e.g., 12) are available. Thus, a time history of the level is plotted on the recording paper. If the frequency of the input signal is changed continuously in time (or stepped into different octave bands) then a spectrogram is plotted out. Calibrated paper is available with the X-axis calibrated in frequency. Different range potentiometers, linear or logarithmic (with ranges for example of 25 dB, 50 dB, and 100 dB), are available. A secondary winding on the magnetic drive system produces a voltage which depends on the velocity of the shaft movement. This voltage signal is introduced as a negative feedback to the amplifier system. The speed of movement of the shaft can be altered by adjusting the writing speed attenuator. This adjustment changes the damping

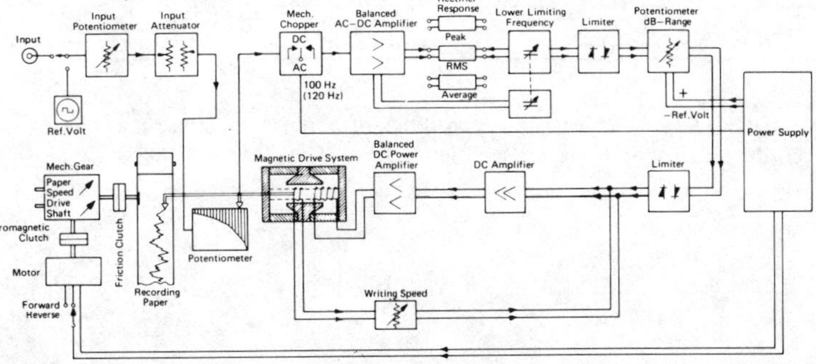

FIGURE 3.89 Block diagram of a level recorder. (Courtesy of Brüel & Kjaer Co.)

in the writing system and thus effectively it also changes the averaging or *integrating time.*

A photograph of a similar level recorder manufactured by a different company is shown in Figure 3.90.

Figures 3.91, 3.92, and 3.93 show some typical results obtained with a level recorder. Figure 3.91 shows octave and one-third octave recordings of the noise from a small compressor made with the spectrometer shown in Figure 3.56a. Figure 3.92 shows an example of a recording of the same compressor noise made from the constant percentage tracking filter shown in Figure 3.57. The 50-Hz fundamental pumping frequency and its harmonics are clearly seen in Figures 3.91b and 3.92. Lastly, Figure 3.93 shows a sound pressure level time history recording of the decay of sound in a room. From this recording the room reverberation time T_R may be calculated. This is the time for the signal to decrease by 60 dB (see Sections 1.6.2 and 4.2.2).

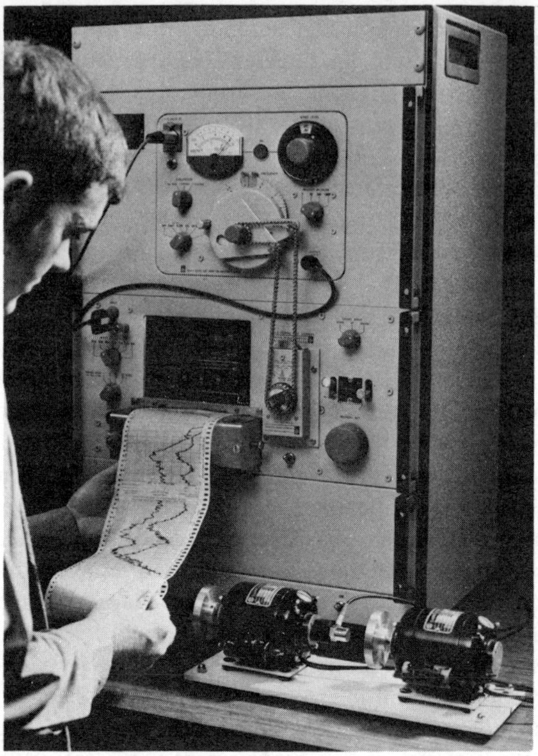

FIGURE 3.90 Photograph of a commercial level recorder. (Courtesy of General Radio Co.)

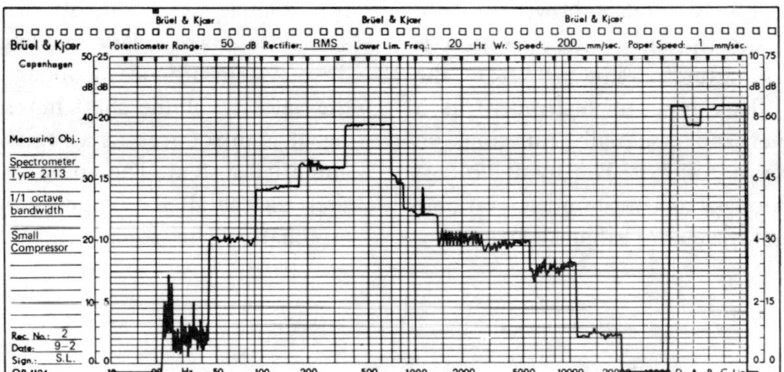

FIGURE 3.91a Octave band recordings of compressor noise made using a level recorder and a spectrometer (see Figure 3.56) (Courtesy of Brüel & Kjaer Co.)

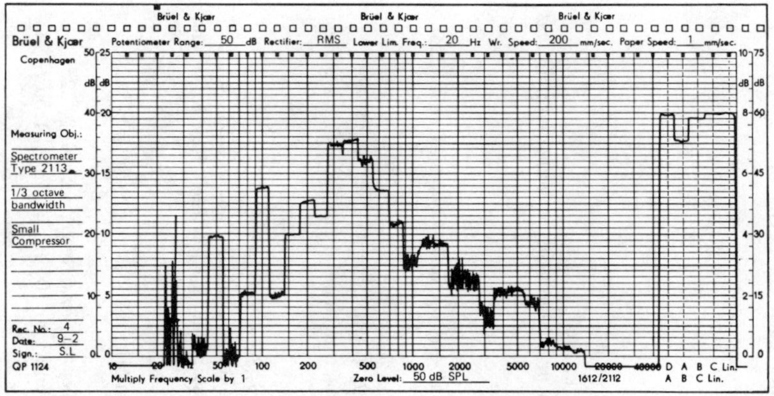

FIGURE 3.91b One-third octave band recording of compressor noise (same) compressor as Figure 3.91a) made using a level recorder and a spectrometer (see Figure 3.56) (Courtesy of Brüel & Kjaer Co.)

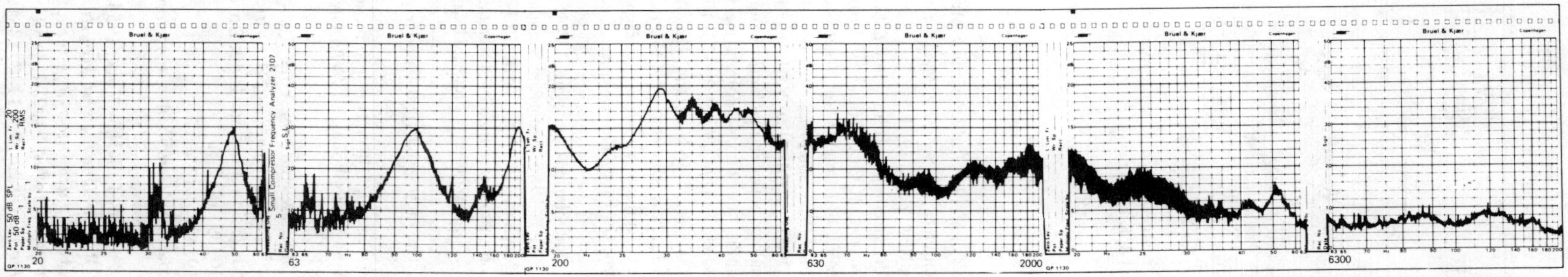

FIGURE 3.92 Recording of noise of same small compressor as in Figure 3.91a and b made using a level recorder and a constant percentage tracking filter (see Figure 3.57). (Courtesy of Brüel & Kjaer Co.)

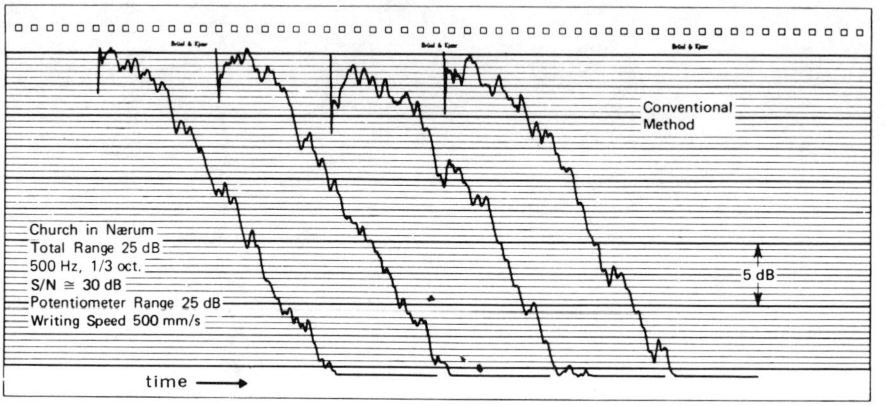

FIGURE 3.93 Four different reverberation time measurements made in a church. (Courtesy of Brüel & Kjaer Co.)

REFERENCES

1. **Keast, D. N.,** *Measurements in Mechanical Dynamics,* McGraw-Hill, New York, 1967.
2. **Magrab, E. B. and Blomquist, D. S.,** *The Measurement of Time-Varying Phenomena – Fundamentals and Applications,* Interscience, New York, 1971.
3. **Beranek, L. L.,** *Acoustic Measurements,* John Wiley & Sons, New York, 1949, chap. 5.
4. **Beranek, L. L.,** *Noise Reduction,* McGraw-Hill, New York, 1960, chap. 4–6.
5. **Peterson, A. and Brüel, P. V.,** Instruments for Noise Measurements, in *Handbook of Noise Control,* Harris, C. M., Ed., McGraw-Hill, New York, 1957, chap. 16.
6. **Beranek, L. L.,** *Acoustics,* McGraw-Hill, New York, 1954.
7. **Firestone, F. A.,** A new analogy between mechanical and electrical systems, *J. Acoust. Soc. Am.,* 4, 249, 1933.
8. **Firestone, F. A.,** The mobility method of computing the vibrations of linear mechanical and acoustical systems: Mechanical-electrical analogies, *J. Appl. Phys.,* 9, 373, 1938.
9. **Olson, H. F.,** *Dynamical Analogies,* Van Nostrand, New York, 1943.
10. **Kinsler, L. E. and Frey, A. R.,** *Fundamentals of Acoustics,* John Wiley & Sons, New York, 1950.
11. **Anon.,** *Acoustics Handbook,* Application Note 100, Hewlett-Packard Co., Palo Alto, Calif., 1968, sect. V.
12. **Brüel, P. V. and Rasmussen, G.,** Free field response of condenser microphones, Brüel & Kjaer Tech. Rev. No. 1, and Rev. No. 2, 1959.
13. **Brüel & Kjaer** Instruments, Inc. Master Catalogue, 1974, Cleveland, Ohio.
14. American Standard ANSI S1.10–1966, American Standard for Calibration of Microphones.
15. **Beranek, L. L.,** *Acoustics,* McGraw-Hill, New York, 1954, 377.
16. **Kinsler, L. E. and Frey, A. R.,** *Fundamentals of Acoustics,* 2nd ed., John Wiley & Sons, New York, 1962, 325.
17. **Rudnick, J. and Stein, M. N.,** Reciprocity free-field calibration of microphones to 100 Kc in air, *J. Acoust. Soc. Am.,* 20(6), 818, 1948.
18. **Sivian, B. and O'Neil, R.,** Diffraction by rigid circular plate, square plate and semi-infinite screen, *J. Acoust. Soc. Am.,* 3, 483, 1932.
19. **Müller, Z., Black, U. and Davis, S.,** Diffraction produced by obstacles and plates, *J. Acoust. Soc. Am.,* 10, 6, 1938.
20. **Rasmussen, G.,** Measurement Microphones, in *INTER-NOISE 72 Proceedings,* Crocker, M. J., Ed., Washington, D. C., October 4–6, 1972, 387.
21. **Beranek, L. L.,** *Acoustics,* McGraw-Hill, New York, 1954, 144.
22. **Kinsler, L. E. and Frey, A. R.,** *Fundamentals of Acoustics,* 2nd ed., John Wiley & Sons, New York, 1962, 294.
23. **Wood, A.,** *Acoustics,* Blackie & Son, London, 1940, Dover Reprint, New York, 1960, 513.
24. **Wente, E. C.,** *Phys. Rev.,* 10, 39, 1917.
25. **Wente, E. C.,** *Phys. Rev.,* 19, 498, 1922.
26. **Gayford, M. L.,** *Electroacoustics, Microphones, Earphones and Loudspeakers,* American Elsevier, New York, 1971, 162.
27. **Beranek, L. L.,** *Acoustics,* McGraw-Hill, New York, 1954, 157.
28. **Kinsler, L. E. and Frey, A. R.,** *Fundamentals of Acoustics,* 2nd ed., John Wiley & Sons, New York, 1962, 298.
29. **Kinsler, L. E. and Frey, A. R.,** *Fundamentals of Acoustics,* 2nd ed., John Wiley & Sons, New York, 1962, 89.
30. **Curie, J. and Curie, P.,** *Comptes Rendus,* 91, 294, 1880.
31. **Curie, J. and Curie, P.,** *Comptes Rendus,* 93, 1137, 1881.

32. **Mason, W. P.,** *Piezoelectric Crystals and Their Application to Ultrasonics,* Van Nostrand, New York, 1950.
33. **Cady, W. G.,** *Piezoelectricity,* McGraw-Hill, New York, 1946.
34. **Wood, A.,** *Acoustics,* Blackie & Son, London, 1940, Dover Reprint, New York, 1960, 519.
35. **Beranek, L. L.,** *Acoustics,* McGraw-Hill, New York, 1954, 165.
36. **Hueter, T. F. and Bolt, R. H.,** *Sonics,* John Wiley & Sons, New York, 1955.
37. **Kinsler, L. E. and Frey, A. R.,** *Fundamentals of Acoustics,* 2nd ed., John Wiley & Sons, New York, 1962, 304.
38. **Hansen, K. S.,** Details in the Construction of a Piezoelectric Microphone, Brüel & Kjaer Tech. Rev. No. 1, 1969.
39. **Brüel & Kjaer,** Piezoelectric Microphone Type 4117 Manual, Naerum, Denmark, 1968.
40. **Sessler, G. M.,** Electrostatic microphones with electret foil, *J. Acoust. Soc. Am.,* 35, 1354, 1963.
41. **Sessler, G. M. and West, J. F.,** Foil electret microphones, *J. Acoust. Soc. Am.,* 40, 1433, 1966.
42. **Wente, E. C. and Thuras, U.,** *J. Acoust. Soc. Am.,* 3, 34, 1931.
43. **Beranek, L. L.,** *Acoustics,* McGraw-Hill, New York, 1954, 155.
44. **Kinsler, L. E. and Frey, A. R.,** *Fundamentals of Acoustics,* 2nd. ed., John Wiley & Sons, New York, 1962, 307.
45. **Gayford, M. L.,** *Electroacoustics, Microphones, Earphones and Loudspeakers,* American Elsevier, New York, 1971, 143.
46. **Jones, G. R., Hempstock, T. I., Millholland, K. A., and Stott, M. A.,** *Teach Yourself Acoustics,* The English Universities Press Ltd., London, 1967, 10.
47. **Beranek, L. L.,** *Acoustics,* McGraw-Hill, New York, 1954, 178.
48. **Meyer, E. and Neumann, E. G.,** *Physical and Applied Acoustics,* Academic Press, New York, 1972, 289.
49. **Myklestad, N. O.,** *Fundamentals of Vibration Analysis,* McGraw-Hill, New York, 1956, 101.
50. **Broch, J. T.,** *Mechanical Vibration and Shock Measurements,* Brüel & Kjaer Co., Naerum, Denmark, undated, 66–68.
51. **Starr, E. A.,** Sound and vibration transducers, in *Noise & Vibration Control,* Beranek, L. L., Ed., McGraw-Hill, 1971, chap. 3.
52. Robert Shaw-Fulton Controls Co., Instructions for Proximity Meter-Capacitance Gauge Number 9501, Issue 4, Anaheim, California.
53. Bently Nevada Corp., Proximity Measuring System 300 Series Instruction Manual, Minden, Nevada, 1962.
54. M.T.I. Instruments Division, Bull. #02-200(11–66), Latham, N.Y., 1966.
55. **Magrab, E. B. and Blomquist, D. S.,** *The Measurement of Time-Varying Phenomena-Fundamentals and Applications,* Interscience, New York, 1971, chap. 2.
56. **Skilling, H. H.,** *Electrical Engineering Circuits,* John Wiley & Sons, New York, 1957, chap. 19.
57. **Studer, J. J.,** *Electronic Circuits and Instrumentation Systems,* John Wiley & Sons, New York, 1963, chap. 5.
58. **Magrab, E. B. and Blomquist, D. S.,** *The Measurement of Time-Varying Phenomena-Fundamentals and Applications,* Interscience, New York, 1971, chap. 4.
59. **Keast, D. N.,** *Measurements in Mechanical Dynamics,* McGraw-Hill, New York, 1967, 182.
60. **Keast, D. N.,** *Measurements in Mechanical Dynamics,* McGraw-Hill, New York, 1967, 146.
61. **Magrab, E. B. and Blomquist, D. S.,** *The Measurement of Time-Varying Phenomena – Fundamentals and Applications,* Interscience, New York, 1971, 73.
62. **Soeburg, J.,** Real Time Analysis, Brüel & Kjaer Tech. Rev., No. 4, 1969, 3.
63. **Cooley, J. W. and Tukey, J. W.,** An Algorithm for the Machine Calculation of Complex Fourier Series, in *Mathematics of Computation,* 19, 90, April 1965.
64. **Chung, G., Crocker, M. J., and Hamilton, J. F.,** Measurement of the coherence function and the prediction of diesel engine noise, in press, *J. Acoust. Soc. Am.,* 1975.
65. **Koss, L. K. and Alfredson, R. J.,** Identification of transient sound sources on a punch press, *J. Sound Vib.,* 34(1), 11, 1974.
66. **Bendat, J. S. and Riensol, A. G.,** *Random Data: Analysis and Measurement Procedures,* Interscience, New York, 1971.
67. **Hodgson, T. H.,** Diagnostic test for locating noise sources: Part II, Signal processing techniques, *Noise Contr. Eng.,* 3(1), 36, 1974.
68. **Bauer, B. B., Torick, E. L., and Allen, R. G.,** The measurement of loudness level, in *Noise Control Engineering,* Crocker, M. J., Ed., Purdue University, Lafayette, Ind., 1972, 150.
69. American National Standard, Specification for Sound Level Meters, ANSI S1.4–1971.
70. International Electrotechnical Commission, Precision Sound Level Meters, IEC Publ. 179, 1965.
71. International Electrotechnical Commission, Recommendations for Sound Level Meters, IEC Publ. 123, 1961.
72. **Young, R. W.,** *Sound,* 1, 17, 1962.
73. **Brüel, P. V.,** Aerodynamically Induced Noise of Microphones and Windscreens, Brüel & Kjaer Tech. Rev., No. 2, 1960, 3.
74. **Wang, J. S. P. and Crocker, M. J.,** Tubular windscreen design for microphones for in-duct fan sound power measurements, *J. Acoust. Soc. Am.,* 55, 568, 1974.
75. **Crocker, M. J. and Wang, J. S. P.,** In-duct fan sound power measurement systems, ASHRAE Trans., 1974
76. **Christiensen, L. S. and Hemingway, J. R.,** Sources of Error in Noise Dose Measurements, Brüel & Kjaer Tech. Rev., No. 3, 1973, 3.

77. **Leasure, W., Fisher, R. L., and Cadoff, M. A.,** Evaluation of commercial integrating-type noise exposure meters, National Bureau of Standards, Rep. NBSIR 73–417, EPA–550/9–73–007, Washington, D.C., 1973.
78. **Wood, A.,** *Acoustics,* Blackie & Son, London, 1940; Dover Reprint, New York, 1960, 509.
79. **Magrab, E. B. and Blomquist, D. S.,** *The Measurement of Time-Varying Phenomena – Fundamentals and Applications,* Interscience, New York, 1971, 183.
80. **Meyer, E. and Neumann, E. G.,** *Physical and Applied Acoustics,* Academic Press, New York, 1972, 352.
81. **Brüel, P. V. and Ingard, K. U.,** *J. Acoust. Soc. Am.,* 21, 91, 1949.
82. **Magrab, E. B. and Blomquist, D. S.,** *The Measurement of Time-Varying Phenomena – Fundamentals and Applications,* Interscience, New York, 1971, 181.
83. Brüel and Kjaer Co., Instructions and Applications Manual for Level Recorder Type 2305, Naerum, Denmark, 1970.

Chapter 4

ACOUSTICS OF ENCLOSURES

TABLE OF CONTENTS

4.1 INTRODUCTION

Understanding the behavior of sound in large enclosures is fundamental to the acoustical design of such spaces. This chapter describes the mechanisms for the growth and decay of sound in rooms, and examines the nature of the steady-state sound field. The concept of reverberation is discussed and equations for the reverberation time of a room are developed.

It has long been known that rooms exhibit acoustic resonances that can cause them to be boomy, for example, at certain low frequencies. These resonances are normal room modes of vibration and can be easily calculated. Their frequency distribution and the effect of room symmetry on this distribution are examined.

Finally, this chapter deals with the design of reverberation and anechoic testing rooms.

4.2 GROWTH AND DECAY OF SOUND IN ENCLOSURES

4.2.1 Growth of Sound

If a sound source is suddenly turned on in a live room (i.e., one with little absorption), sound waves will travel away from the source and eventually strike one of the boundaries whereupon some of the energy will be absorbed and the remainder will be reflected back into the room. This process will continue until a state of equilibrium is established in which the rate at which acoustic energy is fed into the room equals the rate at which it is dissipated by the effect of sound absorption at the boundaries and in the air. After a number of successive reflections the average energy density, ϵ (Section 1.6.3), becomes constant throughout the room and the sound field produced by these successive reflections is said to be reverberant and diffuse. In this condition there should be equal probability of sound rays arriving from all directions at each point in the room at any instant.

It is apparent that if it were not for absorption in the room, the sound field would increase to an infinite intensity. The situation is analogous to filling a bucket with water supplied at a constant rate. If none of the water were absorbed or lost at the boundaries of the bucket (i.e., through a hole in its bottom or side) then the water level would increase and the water would soon overflow. Only in a state of equilibrium, when the rate of water supplied equals the rate lost, would there be a constant water level established. Further, the larger the rate of loss becomes (due to absorption), the lower the final water level will be. This is, of course, exactly analogous to the reverberant sound field within a large enclosure. The reverberant sound level in the room is inversely proportional to the total absorption within the air space and at the boundaries. Similarly if this sound absorption is large, then the final equilibrium level is quickly attained, while if there is only a small amount of absorption, the growth is comparatively slow. Conversely, the decay of the sound level is also slow for rooms with little absorption. Such rooms are often said to be live. It is found that it is in rooms of this type, the application of acoustic ray theory gives good results.

In general, when dealing with relatively small enclosures of simple shape, one can usually obtain an exact solution through wave analysis by solving the appropriate wave equation and using the known boundary conditions. However, when dealing with large irregular enclosures, such an exact analysis is not normally possible. In this case, very useful information can, however, be obtained from the application of statistical techniques. Although such methods do not give us exact solutions, they do give fairly accurate average values (of, for example, sound level) and very often this is all the acoustical engineer requires.

As described in Chapter 1 (also see Reference 2), if a diffuse reverberant sound field is set up in an enclosure with energy density ϵ then the rate at which energy falls on any unit area of the walls of the enclosure from all directions is given by the intensity I (see Equation 1.69).

$$I = \epsilon c/4. \tag{4.1}$$

The intensity of such diffuse sound at the walls is therefore one quarter of that for a plane wave of equal energy density incident upon a surface at normal incidence.

We have already briefly discussed the absorption of sound in enclosures in Section 1.6. We will now consider this is more detail. Again suppose $\alpha_1, \alpha_2, \alpha_3$ . . . are the respective statistical random incidence energy absorption coefficients (see Equation 1.61) (i.e., giving the fraction of randomly incident sound energy absorbed by the material) for different materials having areas S_1, S_2, S_3 . . . which form the interior boundaries of the enclosure, then the rate at which energy can be absorbed, W_D by these materials is

$$W_D = \frac{\epsilon c}{4}(\alpha_1 S_1 + \alpha_2 S_2 + \alpha_3 S_3 + \ldots \text{etc.}) \tag{4.2}$$

Writing the total absorption, A, of such a room as (Equation 1.63):

$$A = \alpha_1 S_1 + \alpha_2 S_2 + \alpha_3 S_3 + \ldots \text{etc.} \tag{4.3}$$

then the above rate of energy absorption, W_D, is given by

$$W_D = \epsilon c A/4. \tag{4.4}$$

This is the rate at which energy is absorbed at the interior surface of the room. The rate at which energy can increase in the reverberant sound field, W_R, is therefore given by

$$W_R = W - W_D \tag{4.5}$$

where W is the rate at which energy is being supplied to the room from the sound source. Since the total reverberant energy is ϵV, where V is the volume, then $W_R = \frac{d(\epsilon V)}{dt} = V\frac{d\epsilon}{dt}$ and we obtain from Equation 4.5:

$$V\frac{d\epsilon}{dt} = W - \epsilon\frac{cA}{4}. \tag{4.6}$$

This equation can be easily solved by separating the variables and integrating $\int dt = \int \frac{V d\epsilon}{W - \frac{\epsilon c A}{4}}$ and applying the boundary conditions $\epsilon = 0$ when $t = 0$.

$$\epsilon = \frac{4W}{Ac}\left(1 - e^{-\frac{Ac}{4V}t}\right). \tag{4.7}$$

Using $\epsilon = \langle p^2\rangle_t/\rho c^2$ (see Equation 1.65), the mean square acoustic pressure $\langle p^2\rangle_t$, at time t after the sound source is switched on, is given by

$$\langle p^2\rangle_t = \frac{4W\rho c}{A}\left(1 - e^{-\frac{Ac}{4V}t}\right). \tag{4.8}$$

After a sufficiently long time has elapsed, the steady state reverberant mean square sound pressure and energy density will be attained, i.e., as $t \to \infty$, Equation 4.8 gives:

$$\langle p^2\rangle_t = \frac{4W\rho c}{A} \tag{4.9}$$

and so

$$\epsilon = \frac{4W}{Ac}. \tag{4.10}$$

These two equations indicate that the final sound pressure level throughout the room is independent of the room volume and hence the shape of the room. This would not be necessarily true for irregularly curved shaped rooms having special focusing effects or for rooms with patches of highly absorbing material distributed unevenly throughout the surface. In such a case the energy density and sound pressure would clearly vary as these surfaces were approached. It should also be noted that Equation 4.8 is only valid if there are sufficient reflections occurring during the sound build-up, to keep a constant energy density throughout the room. The room must therefore be fairly live for Equations 4.8 through 4.10 to apply.

4.2.2 Decay of Sound

If we again consider a live room with a uniform energy density, ϵ, and we suppose that the sound source is cut off at time t when $\epsilon = \epsilon_o$ then

$$-\frac{V d\epsilon}{dt} = \frac{Ac\epsilon}{4},$$

consequently by integrating,

$$\epsilon = \epsilon_o e^{-\frac{Ac}{4V}t}. \tag{4.11}$$

According to the definition of reverberation time, as given by Sabine, this is the time, T_R (sec), taken for the sound pressure level to drop by 60 dB. This can be obtained by putting $\epsilon/\epsilon_o = 10^{-6}$ and then $t = T_R$ so that

$$10^{-6} = e^{-\frac{Ac}{4V}\cdot T_R}$$

from which

$$T_R = \frac{24\,V}{Ac \log e}. \tag{4.12}$$

In metric units, c = 341 m/sec, and so

$$T_R = \frac{0.161\,V}{A}, \tag{4.13a}$$

and in English units, c = 1,120 ft/sec and

$$T_R = \frac{0.049\,V}{A} \tag{4.13b}$$

Here V is the room volume in cubic feet and A is the total absorption area in sabins (ft^2). This defines the average Sabine absorption coefficient $\bar{\alpha}_{SAB}$ in terms of the total surface area S:

$$A = S\,\bar{\alpha}_{SAB}. \tag{4.14}$$

This is identical to the equation first proposed by Sabine in 1895[1] (see Equation 1.64). It may be seen, however, that even if all the surfaces of the room had perfect absorption (α of unity), Equations 4.13a and b would yield a finite value for the reverberation time. Yet using the previous analysis we can see that since there would then be no reflected sound, the reverberation time should be zero. In order to overcome this problem a new approach must be sought for rooms in which the average absorptive coefficient is no longer small. This usually occurs when $\bar{\alpha}$ exceeds about 0.2.

Let us consider a traveling packet of energy with sound rays of intensity I_0 striking a wall surface for the first time. The intensity of the reflected ray will be $I_0\ (1 - \bar{\alpha})$. After a further encounter with another wall surface, its intensity will be $I_0\ (1 - \bar{\alpha})^2$, and so on, until after n reflections and time t, its intensity will be

$$I(t) = I_0\ (1 - \bar{\alpha})^n. \tag{4.15}$$

It is useful at this point to define the average distance which a sound ray will travel in the enclosure between successive reflections, as the mean free path, d. The total energy removed from the enclosure per second then will be $\epsilon V\ \bar{\alpha}\ c/d$. But the total energy absorbed per second will be $\epsilon c\ S\bar{\alpha}/4$. By equating these two results, the mean free path is found to be

$$d = 4V/S, \tag{4.16}$$

where S is the total surface area of the room. The average number of reflections expected after a time t seconds is therefore ct/d and Equation 4.15 may be written as

$$I(t) = I_0\ (1 - \bar{\alpha})^{ct/d} = I_0\ (1 - \bar{\alpha})^{\frac{cS}{4V}t}. \tag{4.17}$$

Using the exponential identity $(1 - \bar{\alpha}) = e^{\ln(1 - \bar{\alpha})}$, we obtain

$$I(t) = I_0\ e^{\ln(1 - \bar{\alpha}) \cdot \frac{cS}{4V}t}. \tag{4.18}$$

This represents the exponential decay of the traveling wave packet being considered. If we further assume that all such packets or rays have similar time histories of encounters with the enclosure boundaries, then the mean square pressures can be written as (see Equation 1.33):

$$\langle p^2(t)\rangle_t = \langle p_0^2\rangle_t \cdot e^{\ln(1 - \bar{\alpha})\frac{cS}{4V}t}. \tag{4.19}$$

Using the same definition as before for the reverberation time, i.e., $\langle p^2(t)_t\rangle / \langle p_o{}^2\rangle_t = 10^{-6}$ for $t = T_R$ gives

$$T_R = \frac{0.049\ V}{-S\ \ln(1 - \bar{\alpha})} \tag{4.20}$$

in English units. This equation was independently obtained by Eyring[3] and by Norris[4] and is generally known as the Norris-Eyring Reverberation Time Equation. It should be noted that if $\bar{\alpha} = 1$, then $T_R = 0$, as it should be (unlike the Sabine Equation 4.13), and that for small values of $\bar{\alpha}$, the term $[-S\ \ln(1 - \bar{\alpha})] \to S\bar{\alpha}$. It must also be remembered that the above derivation is based on the assumption of a perfectly diffuse field. This assumption is not generally true in real rooms, especially if they contain a large amount of absorption. If a number of different absorbing materials are present in the room, then, according to the analysis suggested by Millington[5] and Sette[6], the above Norris-Eyring equation may be written in terms of the energy absorption coefficient α_i of each surface as

$$T_R = \frac{0.049\ V}{\sum_i - S_i\ \ln(1 - \alpha_i)} \tag{4.21}$$

where

α_i = energy absorption coefficient for i^{th} surface
S_i = area of ith surface (ft^2)
V = room volume (ft^3).

It should be noted that the $\bar{\alpha}$ in Equations 4.20 and 4.21 is the statistical energy absorption coefficient (see Equation 1.61) (i.e., the ratio of absorbed to incident sound energy falling on an infinite sample of the material). However, the absorption coefficients of most materials are usually measured and then published from reverberation-time measurements using Equation 4.13 and allowing $A = S\bar{\alpha}_{SAB}$. The values obtained by this method usually exceed the true statistical (energy) absorption coefficients by as much as 50%. Thus it is entirely reasonable[7] to replace $-S\ \ln(1 - \bar{\alpha})$ in Equation 4.21 by $S\bar{\alpha}_{SAB}$ where $\bar{\alpha}_{SAB}$ is the average sabine absorption coefficient for the whole room. Strictly speaking, one should calculate the $\bar{\alpha}$ in the Norris-Eyring equation from the area-weighted geometric mean $\bar{r}_g$* of the energy reflection coefficients r_i for the whole enclosure

*We can define a *statistical reflection coefficient* r_i for an ith surface as the ratio of sound energy reflected from that surface of area S_i to the total sound energy incident upon it, when the incident sound field is diffuse. There is therefore an exact relationship between the statistical absorption coefficient α_i and reflection coefficient r_i such that $\alpha_i + r_i = 1$. It can be shown (see Reference 7) that $\bar{\alpha} = 1 - \bar{r}_g$, where $\bar{r}_g$ is the area-weighted geometric mean of the energy reflection coefficients r_i for the whole room, defined as $\bar{r}_g = \prod_{i=1}^{i=n} r_i^{(S_i/S)} = [\prod_{i=1}^{i=n} (1 - \alpha_i)^{S_i}]^{1/S}$. On the other hand, the area-weighted arithmetic mean reflection coefficient $\hat{r}_a$ can be defined as $\hat{r}_a = 1 - \bar{\alpha}_a = \Sigma\ S_i(1 - \alpha_i)/S$.

and then use $\bar{\alpha} = 1 - \bar{r}_g$. Hence $\bar{\alpha}$ is only the arithmetic mean if all the α_i are equal.[3]

A further problem in using the Millington-Sette summation method is that if any surface, regardless of its size, has an $\alpha_i \approx 1$, then $T_R \to 0$, which of course does not occur in practice.

It is therefore recommended that the Sabine equation be used for most design problems, except those in which there is a completely nonuniform distribution of absorption throughout the enclosure.

4.2.3 Effect of Air Absorption on Reverberation Time

So far, it has been assumed that energy is only lost by reflections at the boundaries of the room. If the room is small, then the number of reflections per second will be large and therefore the mean free path will be also small. In this case, loss of energy by the sound waves during their propagation through the air will generally be very small compared to the boundary absorption. However, in a large enclosure the mean free path will be large and the energy loss due to atmospheric absorption cannot be neglected, especially at high frequencies.

The intensity of a plane wave traveling through a fluid decreases according to

$$I(t) = I_0 \, e^{-mct} \qquad (4.22)$$

where m is the attenuation constant. If this effect is introduced into Equation 4.17 the result is

$$I(t) = I_0 \, e^{-(\frac{A}{4V} + m)ct} \qquad (4.23)$$

and the resulting equation for the reverberation time of a live room therefore becomes

$$T_R = \frac{0.049\ V}{A + 4\ mV} \qquad (4.24)$$

where m is the attenuation constant, ft^{-1}. Similarly for larger values of boundary absorption:

$$T_R = \frac{0.049\ V}{[\sum_i - S_i \ln(1 - \alpha_i)] + 4mV}\ . \qquad (4.25)$$

Some recent measured values of the energy attenuation constant, m, for air are shown in Figure 4.1, as a function of relative humidity, for different frequencies at a temperature of 20°C (68°F). The strong dependence on relative humidity and frequency is very evident. In many large spaces, and especially those with 10 to 30% relative humidity (i.e., swimming pools and ice skating rinks), it is essential to include the air absorption effect into the calculation of the room reverberation time.

Example

Consider a swimming pool of volume = 600,000 ft^3 with a total boundary absorption 9,000 sabins at 4,000 Hz and a relative humidity of 30% at 20°C. From Figure 4.1 we have m = 0.0035 at 4,000 Hz. Neglecting air absorption gives

$$T_R = \frac{0.049 \times 600{,}000}{9{,}000} = 3.27\ \text{sec},$$

while including atmospheric absorption yields

$$T_R = \frac{0.049 \times 600{,}000}{9{,}000 + (0.0035 \times 4 \times 600{,}000)} = 1.69\ \text{sec}.$$

An error of this magnitude could have a large effect on the prediction of speech intelligibility and the design of a public address system for the pool area.

4.2.4 Effects of Area and Location of Absorbing Material on Reverberation Time

The original experiments conducted by Sabine[1,8] showed that the effective absorption coefficient of open windows increased as the total window area was decreased. Since then, various experimenters[9,10] have shown quite clearly that the effective absorption per unit area changes with the area of material present in the enclosure during the test measurements. For example, in one set of measurements made by Kuttruff,[11] an absorption coefficient of 0.18 was obtained when the test material covered one entire wall of the reverberant room used for the experiment, yet when the same material was put on adjacent perpendicular walls the effective absorption coefficient increased to 0.59! Undoubtedly, some of this "area effect" can be attributed to sound diffraction at the edges of the material. However, this appears to be very difficult to predict. In any case this assumption does not account for many of the observed large differences in the reverberation time of a room with a given amount of surface absorption material placed in a number of different locations within the room. This fact is not accounted for in the previous equations for the reverberation time of a room (Equations 1.64, 4.12, 4.13, 4.20, 4.21, 4.24, and 4.25).

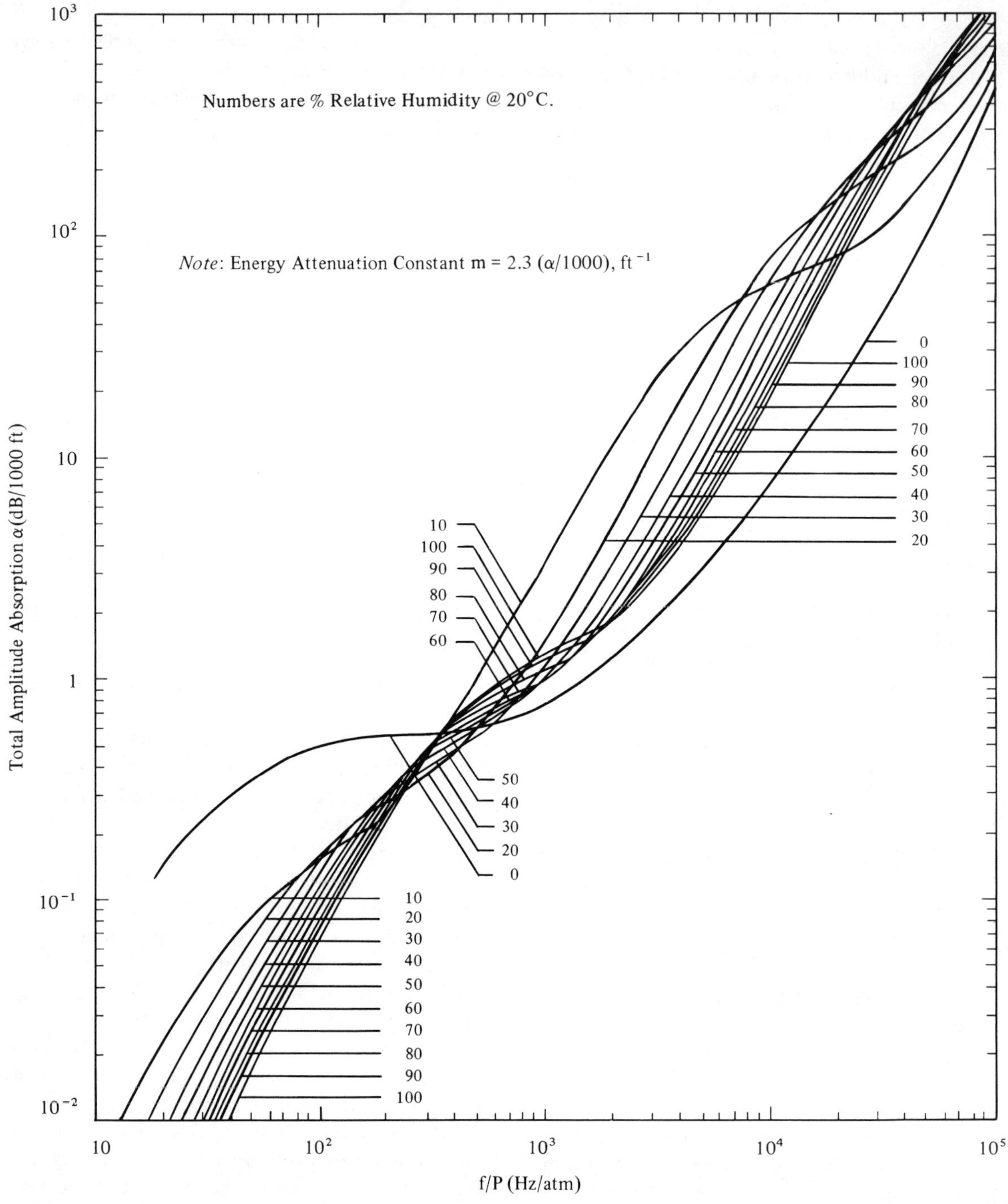

FIGURE 4.1. Total absorption of sound in air as a function of frequency, f, and atmospheric pressure, P. All the curves are for 20°C (68°F) and cover the range 0 to 100% relative humidity. (From Evans, L. B., Bass, H. E., and Sutherland, L. C., Atmospheric absorption of sound: Theoretical predictions, *J. Acoust. Soc. Am.*, 51, 1565, 1972. With permission.)

One of the most important problems in the preceding analysis lies in the method used to average the absorption coefficients of different materials in the room. Equation 4.3 assumes that this averaging process may be done simply by calculating the arithmetic mean. However, as Andrea[12] concluded, this assumption implies that for sound which is reflected from a material of a given absorption coefficient, the probability of encountering a material of identical absorption coefficient on the next reflection is equal to the ratio of the area of that material to the total surface area present. There is nothing to justify this assumption for hard-walled reverberant rooms. Andrea suggests that in the case of parallel walls, the average absorption coefficient for each individual wall may be obtained from the application of the arithmetic rule, but that the average coefficient for two walls is the geometric mean of the coefficients of the individual walls.

In an attempt to overcome the above problems of being able to account for the location of absorbing materials in a room in the calculation of its reverberation time, Fitzroy proposed[13] that the Eyring equation should be modified to consider the possibility that the sound field decays by a process of three simultaneous Eyring-type decays between the pairs of parallel boundaries. He suggested that

$$T_R = \left(\frac{S_x}{S}\right)\left[\frac{0.049\ V}{-S\ \ln(1-\bar{\alpha}_x)}\right] + \left(\frac{S_y}{S}\right)\left[\frac{0.049\ V}{-S\ \ln(1-\bar{\alpha}_y)}\right] + \left(\frac{S_z}{S}\right)\left[\frac{0.049\ V}{-S\ \ln(1-\bar{\alpha}_z)}\right] \qquad (4.26)$$

where

S_x	=	total side wall area (ft^2)
S_y	=	total floor-ceiling area
S_z	=	total end wall area
$\bar{\alpha}_{x,y,z}$	=	average absorption coefficient for S_x, S_y, S_z, areas, respectively
S	=	$S_x + S_y + S_z$ (total surface area).

The application of Equation 4.26 is found to give better agreement between measured and predicted reverberation time values, especially for rooms with nonuniform absorption distribution, than obtained from the Sabine, Erying, or Millington equations. Table 4.1 summarizes such a comparison.

In practice, however, diffusion and the effect of nonuniform absorption distribution will be modified somewhat by reflections from many surfaces.

TABLE 4.1

Comparison of Reverberation Times Measured in Rooms Containing Nonuniform Distribution of Absorption to the Predicted Values Using the Eyring (Equation 4.21) and the Fitzroy (Equation 4.26) Reverberation Time Equations.

					Reverberation times* T_R (Sec)		
Case	Dimensions, ft	Areas	Total Absorption Sabins ft^2	Average $\bar{\alpha}$	Eyring	Fitzroy	Measured
I	12 × 30 × 36	v. 2284	890	0.39	0.63	2.62	2.55
		t. 960	29	0.03			
		l. 720	22	0.03			
II	14 × 55 × 57	v. 6270	2,532	0.40	0.64	1.73	1.70
		t. 1936	265	0.14			
		l. 1540	46	0.03			
III	20 × 54 × 87	v. 8748	1,291	0.15	1.51	1.62	1.57
		t. 3240	657	0.20			
		l. 2160	617	0.29			

v. = Vertical, floor/ceiling areas
t. = Transverse, side-wall areas
l. = Longitudinal, end wall areas

*Taken from Reference 13.

Care must therefore be taken in applying the Eyring equation (4.26) to ensure the above assumptions are met for a given room; otherwise, considerable discrepancies (as shown in Table 4.1) will often arise. Even the Fitzroy equation will give erroneous results if any pair of walls has a high value of $\bar{\alpha}$ (>0.6) and in such cases a better value is obtained by replacing $-\ln(1-\bar{\alpha})$ with $\bar{\alpha}$ in Equation 4.26 as follows:

$$T_R = \frac{0.049\ V}{S^2}\left[\frac{1}{\bar{\alpha}_x} + \frac{1}{\bar{\alpha}_y} + \frac{1}{\bar{\alpha}_z}\right] \tag{4.27}$$

Finally a remark must be made concerning room shape. It is found that the mean free path, d, varies from one room shape to another.[14,15] If d = FV/S then, as previously stated in Equation 4.16, the statistical value, F = 4, is the one generally adopted in general calculations of reverberation time. This factor F changes with room shape; for a cube $F = 2\sqrt{3} = 3.46$, while for a cylinder of height equal to its diameter, $F = 3\sqrt{2} = 4.24$, and for a sphere F = 6 (for a hemisphere F – 4.5). Many other measurements have concluded that for a variety of rectangular-shaped rooms, F varies from 3.8 to 4.2. Consequently this gives a corresponding range of uncertainty in the values obtained from the Eyring-type equations, i.e., Equations 4.20, 4.21, and 4.25.

4.3 TOTAL SOUND FIELD IN AN ENCLOSURE

Any continuous sound source emitting acoustic power into an enclosure will produce two types of acoustic fields at any given position within the enclosure. In the first instance there is the direct field produced by direct sound radiation from the source to the receiver. As stated earlier in Section 1.5, the mean square pressure $\langle p_D^2\rangle_t$ resulting from such radiation will depend upon the type of source (monopole, dipole, etc.), the distance from the source, r, and the directivity factor DF at the particular angles θ and ϕ (see Section 1.5.5), in the following manner:

$$I = \frac{\langle p_D^2\rangle_t}{\rho c} = \frac{W\cdot DF}{4\pi r^2}$$

so that

$$\langle p_D^2\rangle_t = \frac{W\rho c}{4\pi r^2}\cdot DF\,. \tag{4.28}$$

On the other hand, the second type of sound field is due, as we have discussed previously in this chapter, to sound reflections from the boundaries of the enclosure which create the reverberant field. The mean square reverberant sound pressure $\langle p^2{}_R\rangle_t$ was given in Equation 4.9 as

$$\langle p_R^2\rangle_t = \frac{4W\rho c}{A}\,. \tag{4.29a}$$

Using the Eyring analysis it was previously shown (Equation 4.20) that

$$A = -S\ln(1-\bar{\alpha}). \tag{4.29b}$$

The reverberant pressure might be obtained from a different approach as follows.

In the reverberant steady-state condition, the power being injected into a room must be equal to the power being absorbed by the boundaries. The energy contained in the reverberant field will be $\langle p^2{}_R\rangle_t\ V/\rho c^2$ and in one second this will undergo c/d collisions with the absorbing boundaries, a fraction $\bar{\alpha}$ of the energy being absorbed in each collision. Hence the power loss, W_R, from the reverberant field is

$$W_R = \frac{\langle p_R^2\rangle_t}{\rho c^2}\cdot\frac{c}{d}\cdot\bar{\alpha} = \frac{\langle p_R^2\rangle_t}{\rho c^2}\cdot\frac{cS\bar{\alpha}}{4V}\,. \tag{4.30}$$

This power loss must be balanced by the power supplied to the reverberant field, $W(1-\bar{\alpha})$, so that we have

$$\langle p_R^2\rangle_t = 4W\rho c\left(\frac{1-\bar{\alpha}}{S\bar{\alpha}}\right) = \frac{4W\rho c}{R}\,, \tag{4.31}$$

where R is known as the room constant defined as:

$$R = \frac{S\bar{\alpha}}{1-\bar{\alpha}} \tag{4.32}$$

ft^2 (in English units). By adding Equations 4.28 and 4.31 the total mean square sound pressure $\langle p_T^2\rangle_t$ can be found

$$\langle p_T^2\rangle_t = \rho c W\left(\frac{DF}{4\pi r^2} + \frac{4}{R}\right). \tag{4.33}$$

By comparing Equations 4.29a and 4.31 we see that these give identical results for small values of $\bar{\alpha}$ (i.e., $\bar{\alpha} \ll 1$), since for small $\bar{\alpha}$, $A \to R = S\bar{\alpha}$. However, with higher values of absorption some discrepancies become apparent.

Expanding terms we obtain:

$$S\ln(1-\bar{\alpha}) = S\left(\bar{\alpha} + \frac{\bar{\alpha}^2}{2} + \frac{\bar{\alpha}^3}{3} + \ldots \text{etc.}\right), \tag{4.34}$$

and

$$S\bar{\alpha}/(1-\bar{\alpha}) = S(\bar{\alpha} + \bar{\alpha}^2 + \bar{\alpha}^3 + \ldots \text{etc.}). \qquad (4.35)$$

So that neglecting terms smaller than $\bar{\alpha}^2$, we observe that Equation 4.29b gives an "error" of $S\bar{\alpha}^2/2$ compared to the sabine absorption $S\bar{\alpha}$, while Equation 4.32 gives a difference of $S\bar{\alpha}^2$, i.e., twice the error. Strictly speaking, only energy absorption coefficients should be used in the Room Constant, whereas the coefficients tabulated for most materials are Sabine coefficients. The use of the Room Constant also tends to overestimate the effect of large patches of sound absorbing material.[7]

In view of all these uncertainties it is recommended that for most practical engineering applications, $S\bar{\alpha}$ be substituted for R in Equation 4.33 to give

$$\langle p_T^2 \rangle_t = W\rho c \left(\frac{DF}{4\pi r^2} + \frac{4}{S\bar{\alpha}} \right). \qquad (4.36)$$

Here $S\bar{\alpha}$ is obtained from an arithmetic average of the Sabine absorption coefficients and the corresponding material areas ($S\bar{\alpha} = \Sigma S_i \alpha_i$).

In English units, the above equation may be written more conveniently as

$$L_{pT} = L_W + 10 \log \left(\frac{DF}{4\pi r^2} + \frac{4}{S\bar{\alpha}} \right) + 10.5 \qquad (4.37)$$

where

- L_{pT} = total sound pressure level, dB, 2×10^{-5} N/m²
- L_W = sound power level dB, 10^{-12} watt
- DF = directivity factor of the source, at angles θ and ϕ to axis (see Section 1.5.5)
- r = distance from source, ft
- S = total surface area of room, ft²
- $\bar{\alpha}$ = average Sabine absorption coefficient for the room including objects and people. Air absorption may be included by adding 4mV/S to $\bar{\alpha}$.

It is often useful to be able to calculate directly the sound pressure level L_{pR} due to the reverberant sound (which has suffered at least one reflection). This may be written in English units as

$$L_{pR} = L_W - 10 \log (S\bar{\alpha}) + 16.5. \qquad (4.38)$$

The sound pressure level, L_{pD}, due to the direct field is given by

$$L_{pD} = L_W - 20 \log r + DI_{\theta,\phi} - 0.5 \qquad (4.39)$$

where $DI_{\theta,\phi}$ = 10 log DF, the directivity index, dB (see Section 1.5.5) is given by:

$$DI_{\theta,\phi} = 10 \log (DF). \qquad (4.40)$$

Equations 4.37, 4.38, and 4.39 can give us direct physical insight into the particular type of sound field to which one would be exposed at a given point in a room. As long as we are situated close enough to the sound source and $DF/4\pi r^2 > 4/S\bar{\alpha}$, then we would be in the direct field. Since the direct field is independent of the sound absorption in the room it would be of no benefit to add more absorption to the room in an attempt to lower the sound level so long as the receiver remains in the direct field. Since the reverberant sound level is dependent upon the absorption $S\bar{\alpha}$, additional absorption is of benefit to people exposed to high reverberant levels. However, it must be remembered that the reverberant sound pressure level will drop by only 3 dB for every doubling of the total quantity of sound absorption, A. Hence it is essential to know how much absorption is present in the room before deciding to add more. For example, if a large room (floor area > 1,000 ft²) contained only 200 sabin absorption, at say 1,000 Hz, then it would be clearly feasible to double or perhaps quadruple this amount and realize a 3 or 6 dB decrease in the reverberant sound level. However, if the same room had initially contained 1,000 sabin then it very well may have been uneconomical or even impossible to double that amount of absorption, especially to obtain only a 3 dB reduction in the reverberant sound level! Figure 4.2 shows the relationship of direct and reverberant sound pressure levels in a room as a function of distance from the source and total absorption. We can see that the direct sound pressure level drops off according to the inverse square law of Equation 4.39 at 6 dB/doubling of distance, while the reverberant level is independent of distance from the source but strongly dependent upon the amount of absorption in the room. The total sound level experienced is the sum of these two levels.

The distance at which the direct and reverberant sound levels become equal is sometimes called

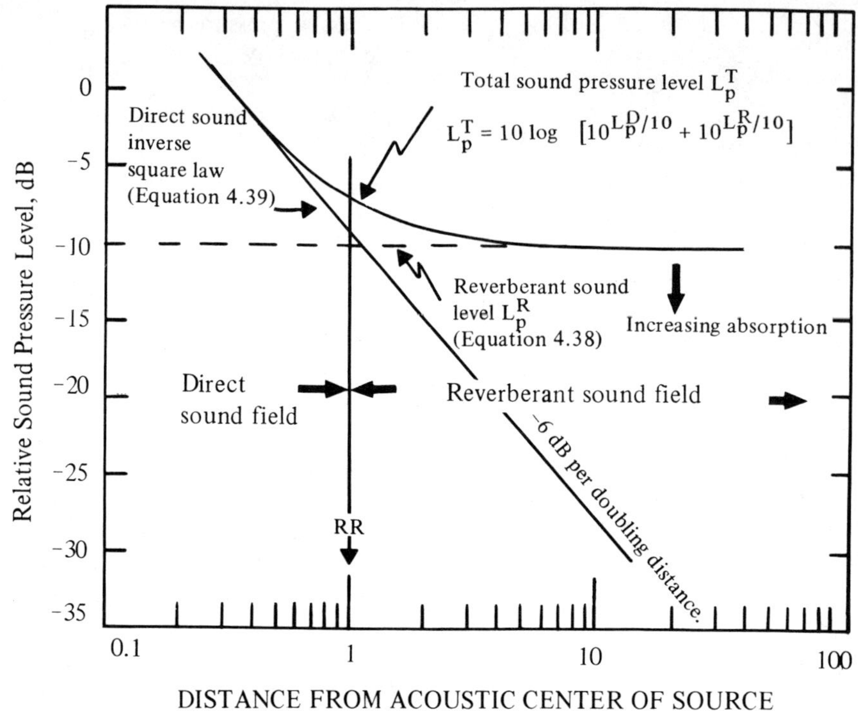

FIGURE 4.2. The total sound field in a large enclosure is composed of two secondary sound fields: the *direct* field which dominates close to the source and is independent of absorption, and the *reverberant* field which is dependent upon absorption (but not distance). The distance from the source at which these two fields produce the same sound pressure level is known as the *radius of reverberation,* RR.

the critical distance or the *Radius of Reverberation* RR and from Equation 4.37 can be seen to be given by

$$RR = [DF \cdot S\bar{\alpha}/16\pi]^{1/2} \tag{4.41}$$

At distances less than RR from the source, the receiver is in the direct field. As the amount of total absorption is increased, so is the radius of reverberation. In the limit, we can observe that the direct field represents the maximum reduction we can get by increasing the amount of absorption. This is illustrated in Figure 4.3. In many real rooms, however, the sound field may not be diffuse, as assumed so far in this chapter, or the absorption may be confined to one heavily absorbing wall. Such examples may be found in many modern offices and other open plan rooms. In such cases it is often found that the total sound pressure level does not become constant at the distance from the source that would be expected from the above discussions. Instead the level often drops a further 4 to 6 dB before leveling off.[7]

4.4 STANDING WAVES IN ENCLOSURES

All of the analysis presented so far in this chapter has been based on the concept that sound waves can exist as energy packets that travel through the room and contribute to the reverberant sound field in a particular statistical manner. This approach is generally known as geometric or ray acoustics. Although this approach is often able to provide useful information, it cannot provide a complete explanation of the behavior of sound in an enclosure. An alternative approach, wave acoustics, is based on a study of the natural modes of oscillation of the room. The appropriate wave equation is written and solved for the particular room and initial time boundary conditions. As long as the enclosure is simple in shape (i.e., a rectangular box, shell, or tube, etc.) then the exact solution can be obtained and it will describe both the transient and steady-state behavior of the sound field.

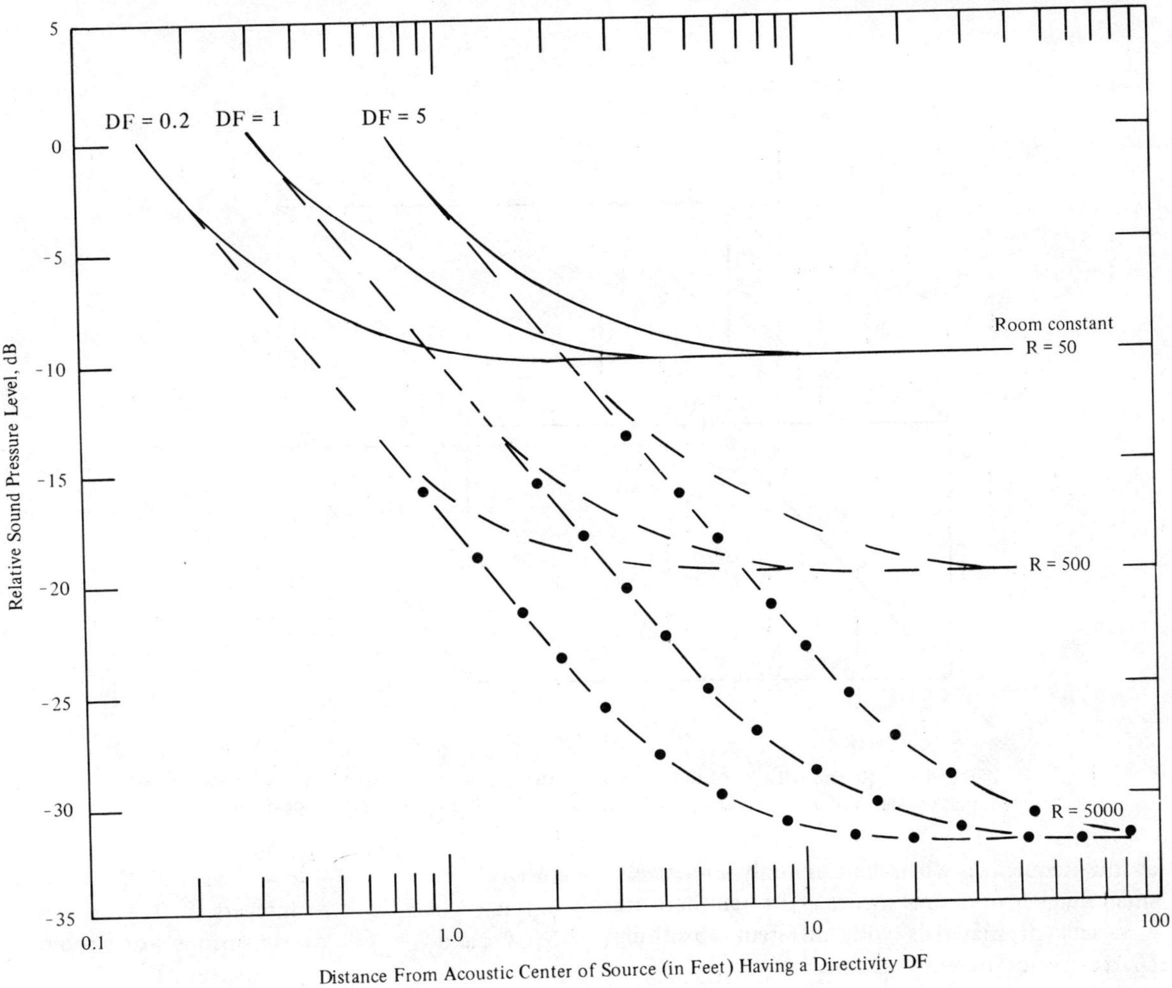

FIGURE 4.3. This shows the relative sound pressure level ($L_p - L_w$) decay as a function of the room constant, R, and source directivity, DF. Note that for a given source and room constant, the sound level eventually becomes constant; this is the reverberant sound pressure level.

4.4.1 Natural Modes of Vibration

Let us consider a rectangular enclosure with perfectly rigid walls containing a small acoustic source at the corner of the space (at $x = y = z = 0$) as shown in Figure 4.4. If we were to trace all the possible paths that sound waves radiated by the source could follow on their journey around the room, we would find that there are certain paths that arrive back at their point of origin. These paths are responsible for the setting up of room resonances or normal modes of vibration (at the normal frequencies) within the enclosure. The normal modes can be divided into three groups as follows:

a. Axial (one-dimensional) – waves which move parallel to one axis.

b. Tangential (two-dimensional) – waves which move in one of the major planes of the room (i.e., tangential to one pair of walls but oblique to the other two pairs).

c. Oblique (three-dimensional) – waves which are oblique to all pairs of walls.

Usually all the normal modes will have different rise and decay times because of the different room absorption for each mode. Thus, if the sound source is suddenly shut off, it is likely that one would see a short initial drop followed by a series of increasingly longer decays, as the less heavily damped modes successively decay. Hence, unlike geometric acoustics, it is unlikely that wave theory would give a single value for reverberation time. This type of stepwise decay is particularly evident

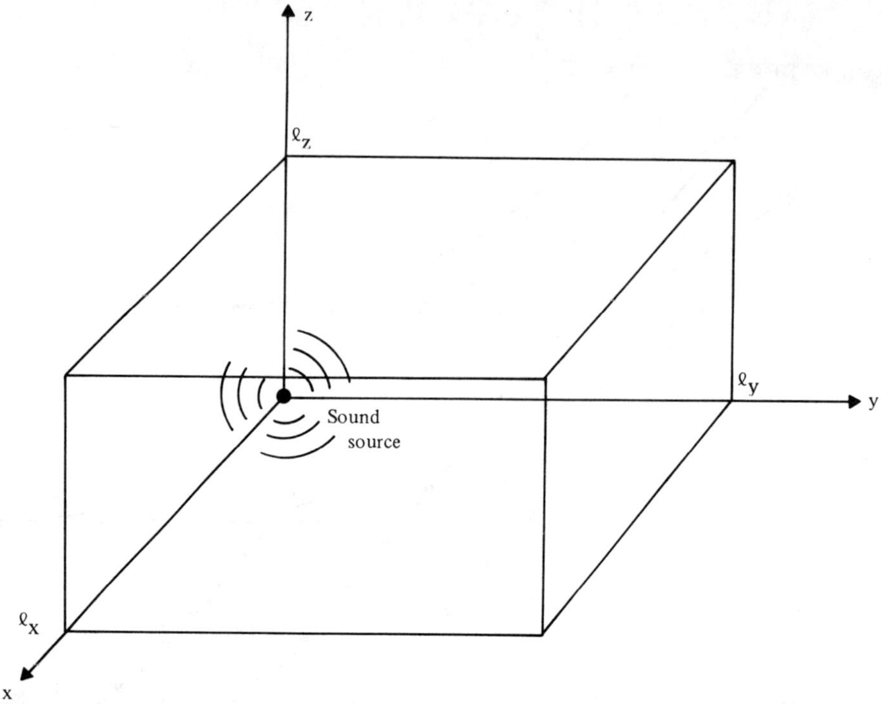

FIGURE 4.4. Room with a sound source positioned in one corner (at the origin of the coordinate system) of a rectangular room with dimensions ℓ_x, ℓ_y, and ℓ_z), respectively.

at low frequencies when there are only a relatively small number of normal modes and when there are a variety of materials with different absorbing characteristics present in the room.

If all of the walls are perfectly rigid then the particle velocity must be zero at each wall. Using this boundary condition, the general three-dimensional wave equation of the form (compare with Equation 1.22 for the case of one-dimensional waves):

$$\nabla^2 p = \frac{\partial^2 p}{\partial x^2} + \frac{\partial^2 p}{\partial y^2} + \frac{\partial^2 p}{\partial z^2} = \frac{1}{c^2}\frac{\partial^2 p}{\partial t^2}, \qquad (4.42)$$

$$p = \underset{\sim}{A}\, e^{i(\omega t - k_x x - k_y y - k_z z)} \qquad (4.43)$$

can be solved[2,16] to provide an expression for the sound pressure, p, produced by any mode in the room. It is also found that the above zero particle velocity boundary condition limits the frequencies of the normal modes so that the only modes which can exist obey the following expression:

$$f_n = \frac{c}{2}\left[\left(\frac{n_x}{\ell_x}\right)^2 + \left(\frac{n_y}{\ell_y}\right)^2 + \left(\frac{n_z}{\ell_z}\right)^2\right]^{1/2} \qquad (4.44)$$

where

n_x, n_y, and n_z = integers, 0, 1, 2 . . . ∞
ℓ_x, ℓ_y, and ℓ_z = dimensions of room along axes, ft
c = speed of sound, ft/sec.

If n_x, n_y, and n_z are all odd numbers the modes are called odd modes, while if all are even integers they are called even modes.

As an example, let us consider the normal modes of vibration within a rectangular room 10 × 15 × 20 ft in size. Table 4.2 gives all the possible natural frequencies for modes under 100 Hz using Equation 4.44 and c = 1,100 ft/sec. One can see that the frequency spacing becomes smaller with increasing frequency and that there may be *degenerate modes* present (i.e., where two or more modes have the same characteristic frequency but different values of n_x, n_y, and n_z). Modes which are close to each other in frequency can easily "beat," while degenerate modes can greatly increase the response of the room at particular frequencies where degeneracy occurs. This can give rise to the "boomy" sensation (at low frequencies) which is often found in regular shaped rooms of similar wall dimensions.

TABLE 4.2

Characteristic Frequencies (less than 100 Hz) for a 10 × 15 × 20 ft Rectangular Room, for c = 1,100 ft/sec.

n_x	n_y	n_z	f_n
0	0	1	27.5
0	1	0	36.7
0	1	1	45.8
0	0	2	55.0
1	0	0	
1	0	1	61.5
0	1	2	66.1
1	1	0	
1	1	1	71.6
0	2	0	73.3
1	0	2	77.8
0	2	1	78.3
0	0	3	82.5
1	1	2	86.0
0	1	3	90.3
0	2	2	91.7
1	2	0	
1	2	1	95.7
1	0	3	99.2

The position of the sound source within the room is also an important parameter since for many source positions certain types of modes may not be excited. For example, if the source is located in one of the corners of the room, then it is possible to excite every normal mode, while if the source is located at the center of a rectangular room then only the even modes (one eighth of the total number of possible modes) can be excited. Similarly if we keep the position of the source constant and measure the sound pressure throughout the room we see differences in level depending on where we are standing in the room relative to the normal modes. In this way the room superimposes its own acoustic response characteristics upon those of the source. Hence we cannot measure the true frequency response of a sound source (e.g., loudspeaker) in a reverberant room because of the effect of the modal response of the room. This interference can be removed by making all the wall surfaces highly sound absorbent ($\alpha >$ 0.99). Then all the modes are sufficiently damped so we are able to measure the true output of the source. Such rooms are called *anechoic* (no echoes). These will be discussed later in Section 4.6.

4.4.2 Distribution of Normal Modes

It is very often useful to know not only the frequencies of the normal modes of a room but also how many modes exist below a given frequency or in a given frequency band. It is also useful to know how many modes of a given type (axial, tangential, or oblique) exist in a given frequency band, since while modes of different types usually have different reverberation times, those of the same type have essentially the same reverberation times.[16]

We may rewrite Equation 4.44 as

$$f_n^2 = \left(\frac{cn_x}{2\ell_x}\right)^2 + \left(\frac{cn_y}{2\ell_y}\right)^2 + \left(\frac{cn_z}{2\ell_z}\right)^2 \qquad (4.45)$$

This equation suggests that the frequency of a normal mode of a rectangular room may be considered to be a vector (in frequency space) having x, y, and z components, $cn_x/2\ell_x$, $cn_y/2\ell_y$, and $cn_z/2\ell_z$, respectively. Therefore, any normal mode can be considered to be a point in frequency space with the magnitude of the line joining this point to the origin being the value of the frequency of that mode and the direction of the line giving the direction cosines* of the wave required to generate the mode in the room. These modes therefore form a rectangular lattice in frequency space as shown in Figure 4.5. If we wished to compute the number of room modes with natural frequencies lying below a given frequency we could compute all the modes with frequencies less than the chosen frequency limit and count them. This would be very tedious and time-consuming, especially at high frequencies. It is for such purposes that the modal lattice method of representation becomes extremely useful. It can be seen that the volume of frequency space enclosed by one rectangular lattice block in Figure 4.5 is $c^3/(8\ell_x \ell_y \ell_z)$ or $c^3/8V$, where V is the volume of the room. Therefore, the number of modes, N, enclosed by a frequency vector, f, will be approximately given by the volume of the octant enclosed

*The "direction cosines" of a vector define its direction in space such that $\cos\theta_x = x/r$, $\cos\theta_y = y/r$, and $\cos\theta_z = z/r$, where θ_x, θ_y, and θ_z are the angles between the vector and the axes and where x, y, and z are the vector components such that the magnitude of the vector $r = (x^2 + y^2 + z^2)^{1/2}$.

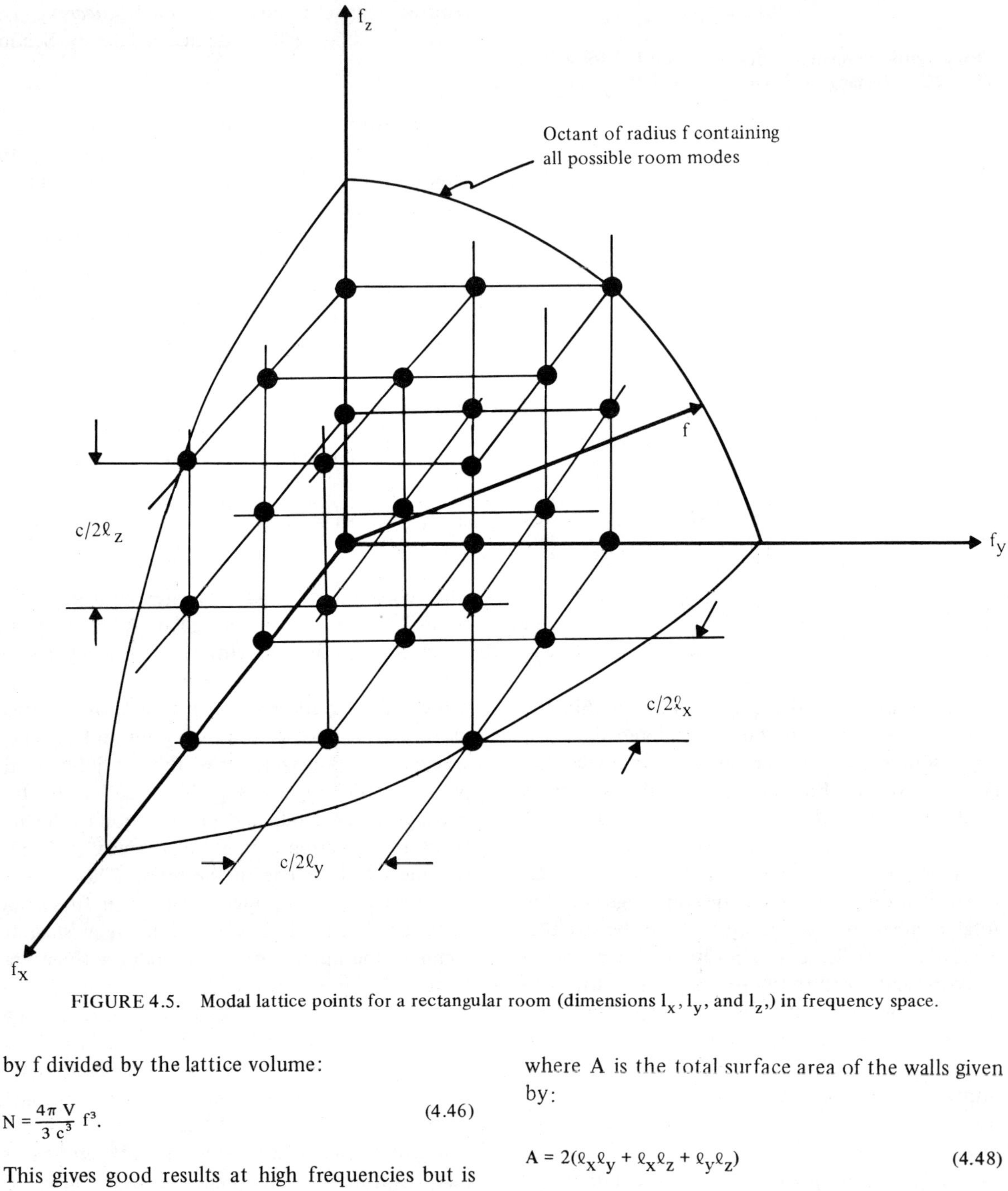

FIGURE 4.5. Modal lattice points for a rectangular room (dimensions l_x, l_y, and l_z,) in frequency space.

by f divided by the lattice volume:

$$N = \frac{4\pi V}{3 c^3} f^3. \tag{4.46}$$

This gives good results at high frequencies but is not very accurate at low frequencies since some error is associated with neglecting the frequency space outside the octant $\pi f^3/6$ which corresponds to modes both along the axes and in the planes formed by the axes. If these corrections are included, the following more accurate expression is obtained:[16]

$$N = \frac{4\pi V}{3 c^3} \cdot f^3 + \frac{\pi A}{4 c^2} \cdot f^2 + \frac{L}{8 c} \cdot f, \tag{4.47}$$

where A is the total surface area of the walls given by:

$$A = 2(\ell_x \ell_y + \ell_x \ell_z + \ell_y \ell_z) \tag{4.48}$$

and L is the sum of all the edges of the room, i.e.,

$$L = 4(\ell_x + \ell_y + \ell_z) \tag{4.49}$$

As long as the room is not too symmetrical, the above equation yields values of N within one or two modes.

Very often it is even more useful to know the number of modes ΔN in a given frequency band

Δf. An approximate value for ΔN is obtained by differentiating the above equation, with respect to f, to give:

$$\Delta N = \left(\frac{4\pi V}{c^3} f^2 + \frac{\pi A}{2 c^2} f + \frac{L}{8 c} \right) \Delta f. \tag{4.50}$$

The strong dependence of the number of modes per unit bandwidth $\Delta N/\Delta f$ on frequency, f, and room volume, V, should be noted. Also as the frequency and volume are increased, the number of modes in a frequency band rapidly increases and therefore the variation in the direction cosines in a frequency band increases. This results in a more random sound field. Hence, one would expect those reverberation time equations which were derived on the assumption of a uniform random sound field (e.g., Equation 4.20) would give better results at high frequencies and for large rooms. This appears to be the case.

4.4.3 The Effect of Room Symmetry

As previously stated, the more symmetrical a room becomes, the more irregular its modal response becomes. This is because of the increase in the number of degenerate modes which occur as the room becomes increasingly symmetrical. This effect can be demonstrated by considering two rooms with the same volume but different dimensions. For example, let us choose a room with the following dimensions:

$$\ell_y = p\ell_x,\ \ell_z = \ell_x/p$$

so that the volume, $V = (\ell_x)^3$. The natural frequencies are then given by

$$f_n = \frac{c}{2\,\ell_x} \sqrt{n_x^2 + \frac{n_y^2}{p^2} + p^2\, n_z^2} \tag{4.51}$$

The spacing of the modal lattice will therefore change with p (as the relative room dimensions change). For the most symmetrical case (a cube) $p^2 = 1$, while when $p^2 \neq 1$, the room is not symmetric. Table 4.3 shows the possible modal frequencies, and their corresponding integers for n_x, n_y, and n_z for a cubic room and for two rooms for which $p^2 = 2$ and $p^2 = 3$. Therefore, these irregular rooms have dimensions, $\ell_x:\ell_x\sqrt{2}:\ell_x/\sqrt{2}$ and $\ell_x:\ell_x\sqrt{3}:\ell_x/\sqrt{3}$, respectively. In the case of the cubic room it can be seen that threefold and sixfold degeneracies are quite common at low frequencies (independent of room volume). This results in the room response being quite irregular at low frequencies with large gaps in some frequency bands in which there are no modes at all. On the other hand, for the cases of $p = \sqrt{2}$ and $p = \sqrt{3}$, there are never more than threefold degeneracies and the possible modes are therefore more evenly spaced. If we chose a value of $p^2 = 7^{1/3}$, for example, we would have no degeneracies at all. However, it must be remembered that we can never get an absolutely uniform modal separation in a rectangular room no matter what value we choose for p^2, since the modal lattice will always remain rectangular.[16] Although we could achieve a more random spacing of these lattice points in frequency space by having irregular shaped walls, we cannot necessarily conclude that this would lead to a more uniform response. However, irregular walls do tend to distribute the sound energy more uniformly thoughout a room.

4.5 DESIGN OF REVERBERATION ROOMS

In the previous sections in Chapter 4 we have discussed the ways in which the sound field in a room may be modified by changing both room dimensions and shape, and by altering the nature of the wall surface treatments. If one wishes to perform acoustical testing in such a room to, for example, measure the sound power output of a given source such as a machine then it is desirable that both of the following conditions be met:

(a) The room should not affect the sound output characteristics of the source. Since the power radiated by a simple source is equal to the product of the rms velocity and the radiation resistance, this means that the room should be able to provide a radiation resistance as close as possible to the free-field value for all frequencies of interest and throughout the room. If a sound source is placed in a room and positioned close to the boundaries, then the reflected sound will interfere with the direct sound and will either reduce or increase the effective radiation resistance (depending on phase). However, as long as the source is placed well away (greater than a wavelength) from the walls and the room is relatively large, then the space-frequency averaged radiation resistance will remain unchanged and equal to the free-field value. The effect of sound reflections on the radiation resistance can be reduced by

TABLE 4.3

Modal Frequencies and Corresponding Mode Numbers for Cubic and Noncubic Rooms All Calculated for ℓ_x = 15 ft, c = 1,100 ft/sec.

	Cubic room p^2 = 1				p^2 = 2 Noncubic rooms				p^2 = 3			
	f, Hz	n_x	n_y	n_z	f, Hz	n_x	n_y	n_z	f, Hz	n_x	n_y	n_z
20												
									21.2	0	1	0
					25.9	0	1	0				
30												
		0	0	1								
	36.7	0	1	0	36.7	1	0	0	36.7	1	0	0
		1	0	0								
40												
										0	2	0
									42.3			
										1	1	0
					44.9	1	1	0				
50												
		0	1	1		0	0	1				
	51.8	1	0	1	51.8							
		1	1	0		0	2	0				
									56.0	1	2	0
					57.9	0	1	1				
60												
						1	0	1		0	0	1
	63.5	1	1	1	63.5				63.5			
						1	2	0		0	3	0
									66.9	0	1	1
					68.6	1	1	1				
70												
		0	0	2		0	2	1		1	0	1
	73.3	0	2	0	73.3				73.3	1	3	0
		2	0	0		2	0	0		2	0	0
										0	2	1
									76.3	1	1	1
						0	3	0		2	1	0
					77.8	2	1	0				
80												
		0	1	2								
		0	2	1								
		1	0	2								
	81.9	1	2	0	81.9	1	2	1				
		2	0	1								
		2	1	0						0	4	0
									84.7	1	2	1
										2	2	0
					86.0	1	3	0				
		1	1	2		2	0	1				
	89.8	1	2	1	89.8				89.8	0	3	1
		2	1	1		2	2	0				
90												
						0	3	1	92.3	1	4	0
					93.5							
						2	1	1		1	3	1
									97.0	2	0	1
										2	3	0
									99.3	2	1	1
100												
					100.4	1	3	1				
		0	2	2		0	0	2				
	103.7	1	0	1	103.7	0	4	0				
		2	2	0		2	2	1		0	4	1
									105.8	0	5	0
										2	2	1
						0	1	2				
					106.9	2	3	0				
110												
		0	0	3								
		0	3	0								
		1	2	2		1	0	2				
	110.0	2	1	2	110.0	1	4	0	110.0	3	0	0
		2	2	1		3	0	0				
		3	0	0								

positioning near-field reflectors in the room. This will be discussed in more detail later in the chapter.

(b)The sound field should be diffuse so that spatial variations in the sound pressure level observed throughout the room are reduced to an acceptable minimum (i.e., ±1 dB). This is primarily governed by the modal structure within the room. The modes should be evenly spaced in frequency (clumps of degenerate modes are undesirable) and there should be a sufficient number in a frequency band so that there will be a statistically reliable reverberant sound field at the lowest frequency of interest.

4.5.1 Low Frequency Cut-off

In order to estimate the lowest usable frequency in a well-designed reverberation room, a measure of the modal spacing and overlap can be found from the *Modal Overlap Index,* M, defined as

$$M = B_r \cdot n(f) \tag{4.52}$$

where:

B_r = bandwidth of a typical mode, (i.e., the width (in Hz) of a mode at those points on its resonance curve which are 3 dB below the peak value of modal response).

n (f) = modal density, modes/Hz.

Using $n(f) = \omega^2 V/\pi c^3$ (the first term of Equation 4.50 with $n(f) = \Delta N/\Delta f$ and $\omega = 2\pi f$) and $B_r = 2.2/T_R$, we get

$$M = \frac{2.2\,\omega^2 V}{\pi c^3 T_R}\,. \tag{4.53}$$

From which the room "cut-off frequency," f_c, can be obtained as

$$f_c = \left[\frac{Mc^3 T_R}{8.8\,\pi V}\right]^{1/2}. \tag{4.54}$$

This cut-off frequency is the lowest frequency at which the reverberant sound field can be considered to be statistically reliable. However, there is some difference of opinion regarding the appropriate value of M which should be used in Equation 4.54. Schroeder[17] has suggested the value of M = 3, which means that the spacing between modes is less than 1/3 bandwidth of a typical mode. This could also be regarded as having a 3:1 modal overlap. Using this value of M = 3, one obtains the now well-used and accepted *Schroeder Cut-off Frequency,* $f_{c,S}$ as:

$$f_{c,S} = 11{,}885\,(T_R/V)^{1/2}. \tag{4.55}$$

However, other researchers have shown that lower values of M are still often acceptable. Scharton et al.[18] found that using M = 1/3 provided good modal overlap and M = 1 was even conservative. In this case, the desired value for M can be substituted into Equation 4.54 rewritten as:

$$f_c = 7{,}000\,(MT_R/V)^{1/2}. \tag{4.56}$$

It can be seen that in order to achieve a lower cut-off frequency, either the volume of the room has to be increased, or the reverberation time T_R decreased. However, if the room has a short reverberation time then it might prove difficult to produce the desired sound pressure level in the room and, as will be shown in the next section, the spatial variations may become too large to be acceptable. The design of a particular reverberation room is therefore a process in which the optimum solution must be found which is compatible with several conflicting requirements.

4.5.2 Spatial Variance in the Reverberant Field

In addition to the above cut-off frequency another important characteristic of the room is the spatial variance, v_n^2, in the reverberant sound field in the room. As the variance in the sound pressure level increases, more measurements have to be made at different locations in order to give adequate confidence in the values obtained.

Lubman[19] has given the following expression for the variance* v_n^2, in dB:

$$v_n^2 = (2/z)\tan^{-1}(z) - (1/z^2)\log(1+z^2),\ \text{dB}, \tag{4.57}$$

where

*The spatial variation in the sound pressure level observed throughout a reverberant room can be described using statistics through the *variance* v_n^2 and the *standard deviation* v_n of the set of measurements. If L_i is the ith sound pressure level observed, then the variance is given by $v_n^2 = [\Sigma(L_i - \bar{L})^2]/n(n-1)$ where $\bar{L} = (\Sigma L_i)/n$, the sample mean of n observations. Lubman also gives[19] the spatial variance observed in a reverberant room containing a multitone sound source as $v_n^2 = \Sigma A_n/(\Sigma A_n)^2$, where A_n is the amplitude of the sound pressure squared, p^2, of the nth component of the sound field.

$z = B/B_r$,
B = bandwidth of signal,
B_r = bandwidth of typical room mode.

The ratio B/B_r can also be written in terms of the room reverberation time, T_R, as $B/B_r = 0.455\ BT_R$. It can be shown from Equation 4.57 that if $BT_R << 1$, then $v_n^2 \rightarrow 1$. In other words, as the signal bandwidth becomes very small (i.e., a pure tone) the variance tends to unity. This is exactly what one would expect[19] for excitation by a pure tone.

Figure 4.6 shows the monotonic variation of v_n^2 with BT_R from Equation 4.57. As $BT_R >> 1$, the variance is lowered. A good approximation to the result in Equation 4.57 for this case is

$$v_n^2 = [1 + BT_R/6.9]^{-1}. \qquad (4.58)$$

This gives values for v_n^2 within ± 10% if $BT_R > 10$ and to within ± 2½% if $BT_R > 20$. Clearly as $BT_R >> 1$ we must be exciting a large number of room resonances and so we would expect the variance to decrease.

Once we can estimate the variance, we can then decide on the number of spatial samples of the sound pressure level which should be made in the room in order to be able to estimate the space-averaged sound pressure level with a given degree of confidence. Again, Lubman has given the number of samples, N, required for the precise estimate of the space-averaged mean-square pressure as:

$$N \geqslant \left(\frac{2.5\ v_n}{e}\right)^2 ; \qquad (4.59)$$

where

v_n^2 = spatial variance
e = allowable sampling error
2.5 = a constant chosen for 98.75% certainty that the sample mean will differ from the true mean by less than e.

If we wished to be sure that the estimated space average should differ from the true average by less than ± 1 dB, then since 1 dB = 25.9% change in power (see Sections 1.4.1 and 1.4.2), we would substitute $e = 0.259$ in the above equation together with the estimated value of v_n^2 (from Equation 4.58). We should than obtain the number of samples required to have 98.75% confidence in the result that our estimate of the space-average mean square pressure was within ± 1 dB of the true space average.

Equation 4.59 shows that the number of

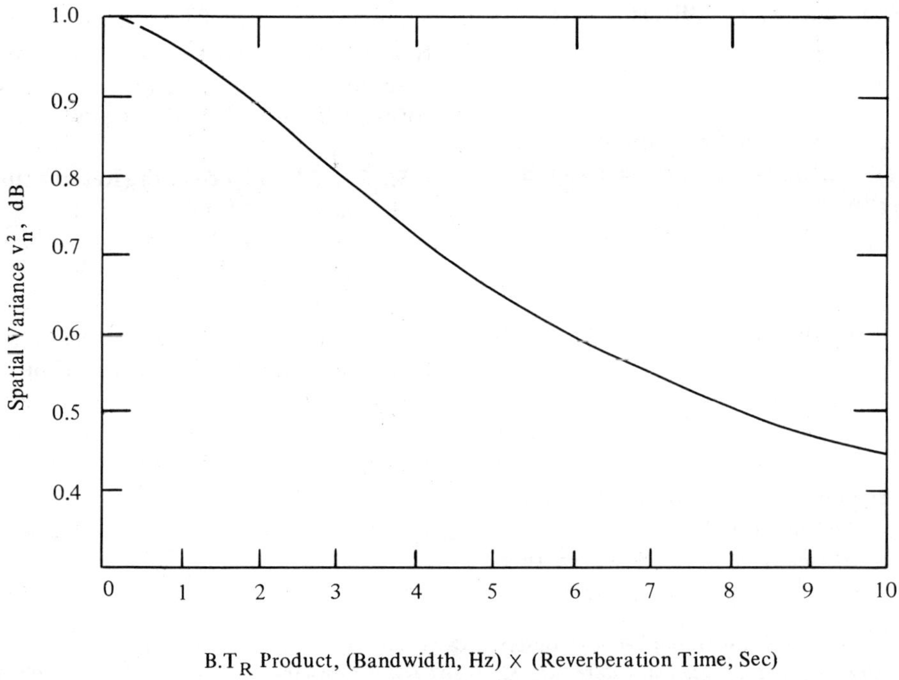

FIGURE 4.6. Spatial variance for a flat spectrum noise signal of bandwidth B in a room of reverberation time T_R sec. (See Reference 19.)

samples required is proportional to the variance v_n^2. Thus if we double either the signal bandwidth or reverberation time (if $BT_R >> 1$) then from Equation 4.58 we see that we can expect about one half the variance and hence only one half as many measurements would be required to find the space-averaged sound pressure level to a given degree of precision. Figure 4.7 may be used to

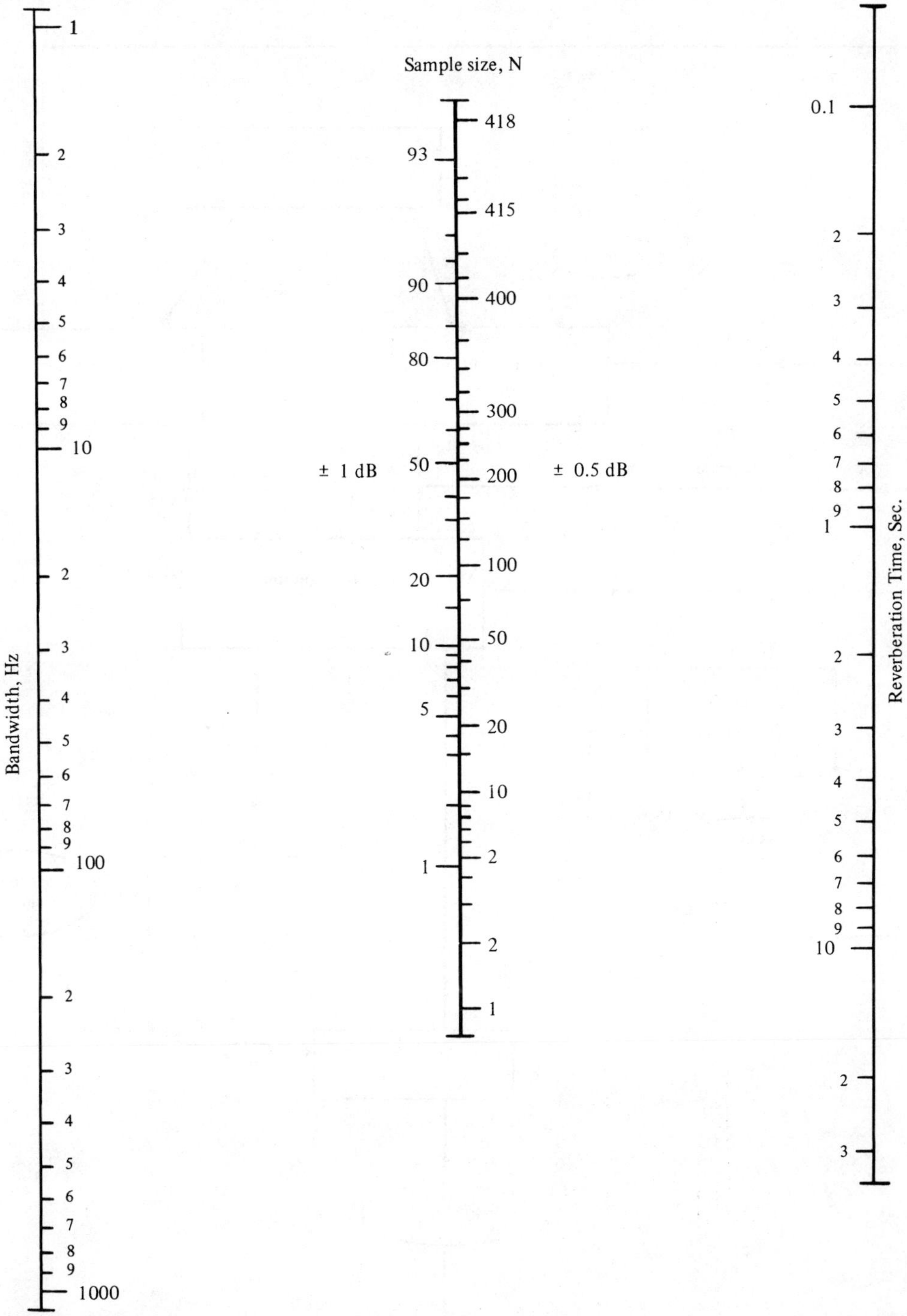

FIGURE 4.7. Nomogram (Reference 19) for calculating the number of samples required, N, in order to achieve 98.75% confidence in the measured space average for allowable sampling errors of 1 dB or 0.5 dB, respectively.

estimate the sample size required in order to determine the space-averaged sound pressure level to within either ±0.5 dB or ±1 dB with 98.75% confidence. In such a case only 1.25% of the measurements will differ from the true mean by more than 0.5 or 1 dB, respectively.

Figure 4.8 shows the complex iteration procedure that the designer of reverberation chambers

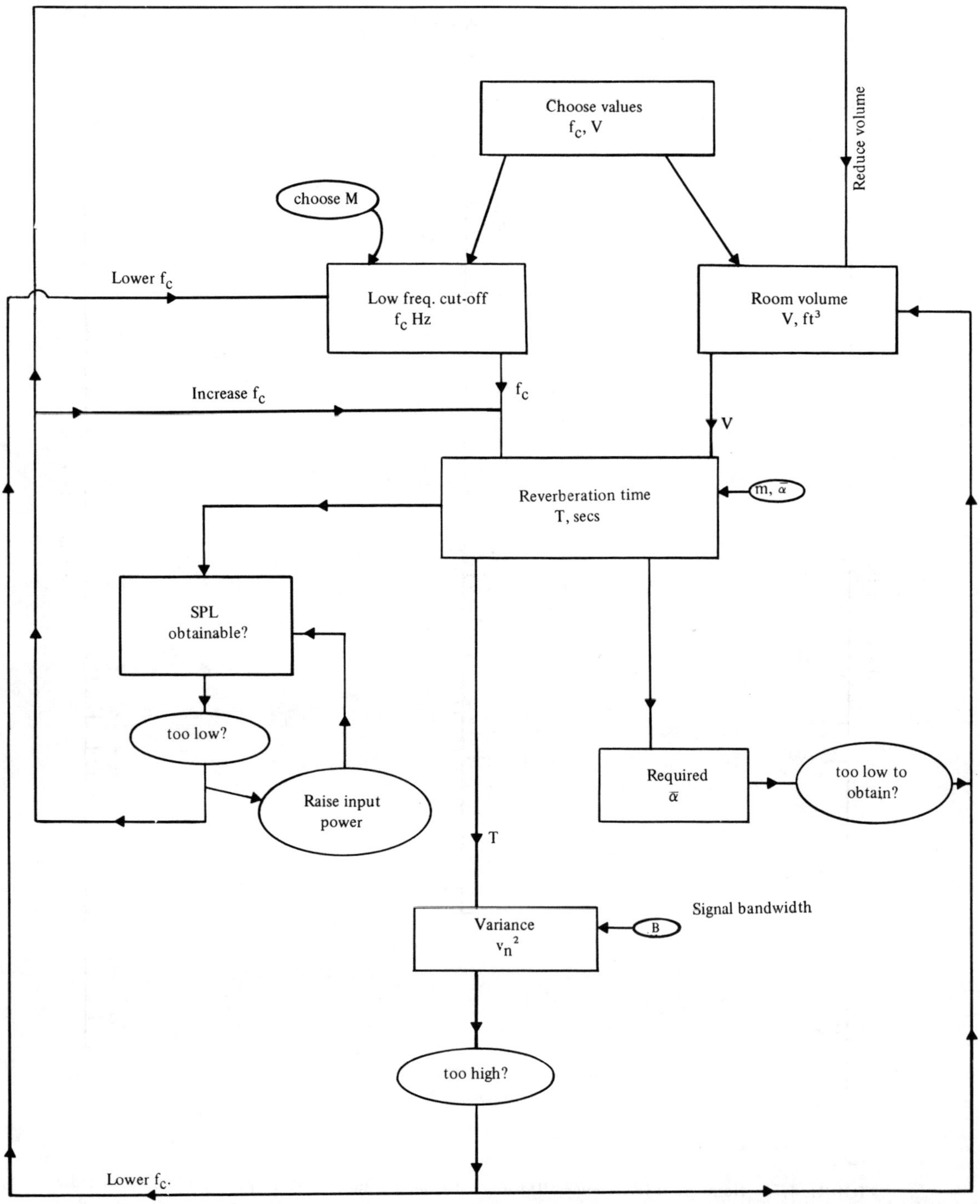

FIGURE 4.8. Iteration procedure for reverberation room design.

should follow in order to get the maximum acoustical performance from such a room. Towson[31] has also produced a good set of design rules and the reader is directed to them.

4.5.3 Techniques Used to Improve Diffusion

As previously stated, diffusion is probably the most desirable acoustic characteristic in almost any reverberation room. Generally, as long as we are above the Schroeder cut-off frequency (M = 3) and have obtained an acceptable value of variance through the reverberation time, we should then have satisfactory diffusion. However, attempts are constantly being made to increase diffusion. These are generally included in one or more of the following:

a. Use of increased absorption.
b. Use of stationary or rotating diffusers.
c. Use of modal enrichment panels.
d. Use of irregularly shaped rooms.

For the case of increased absorption, the low frequency absorption in the room is increased to broaden the modal responses of low frequency modes. In this way a pure tone can excite nearby modes just outside the frequency band being observed. Thus, since more modes are involved, this makes the room appear to have better diffusion and at the same time to smooth out the room transfer function.[18] However, there is a corresponding increase in the spatial variance throughout most of the frequency spectrum and this approach has to be carried out with care.

In many reverberation rooms stationary diffusers are often used. These are large panels of irregularly curved shapes which are suspended from the ceiling in a random orientation so as to "break up" the sound field and add to the existing diffusion. These panels must be comparable in their dimensions with the lowest wavelength of interest (i.e., $1 > c/2fd$) and be heavy enough to reflect these low frequency waves. This usually requires that the diffusers be at least 1 lb/ft^2 in weight. These stationary panels have the effect of forcing the eigenvalues (especially of relatively small chambers) downwards, depending upon the panel orientation and density. This downward shift is more evident with the low frequency modes.[20,21] Figure 4.9 shows the effect of increasing the number of stationary panel diffusers on the measured values of absorption coefficient obtained during the construction of the 200 m^3 parallel-walled Acoustics Laboratory of Rockwool AB, Skövde Sweden.[30] Each panel diffuser consisted of a laminated plastic sheet, 3 mm thick with dimensions 125 × 80 cm suspended from the ceiling. It can be seen that the absorption coefficient measured varied with the number of diffusers installed. Clearly the effect of increasing the number of diffusers from 20 to 30 was very small, indicating that the diffusion was already good with 20 diffusers.

The use of rotating diffusers has also been fairly widespread in reverberation test rooms. Indeed in 1915 Wallace Clement Sabine[1] used and described the use of a rotating vane diffuser in his experiments to measure the transmission loss of building partitions using the arrangement shown in Figure 4.10. Modern rotating diffusers are often cylindrical or semiconical in shape, again having dimensions which are large compared to the wavelength of the lowest frequency measured. Such rotating vanes are particularly useful for pure tone and narrow band measurements. In cases where multitone or relatively wide band excitation is employed, then rotating vanes frequently become unnecessary because such excitation gives rise to small spatial variance anyway and their effect becomes too small to be of importance. Although it is not completely certain just how a rotating diffuser works, it can be generally said that it essentially "warps" the geometry of the room so that the measuring devices see several rooms during their integration time. If a noise source is located near rotating vanes, there are strong reflections set up which give strong fluctuating image sources close to the real source.[22] This helps to reduce the effect of the room upon the sound source, probably by reducing the uncertainty in the average radiation impedance presented to the source. The extreme fluctuation in the radiation resistance is therefore avoided as either the source location or frequency is changed.[23] Lubman[19,25] found that a rotating diffuser acted as a modulator, being able effectively to transform a pure tone into a multitone, thus increasing the effective bandwidth of the signal and lowering the spatial variance. The speed of rotation is also found to be fairly important and most diffusers are rotated at about 30 rpm. In addition, care must be taken to ensure that they do not produce an excessive background noise. For a more detailed study of the effect of rotating diffusers on

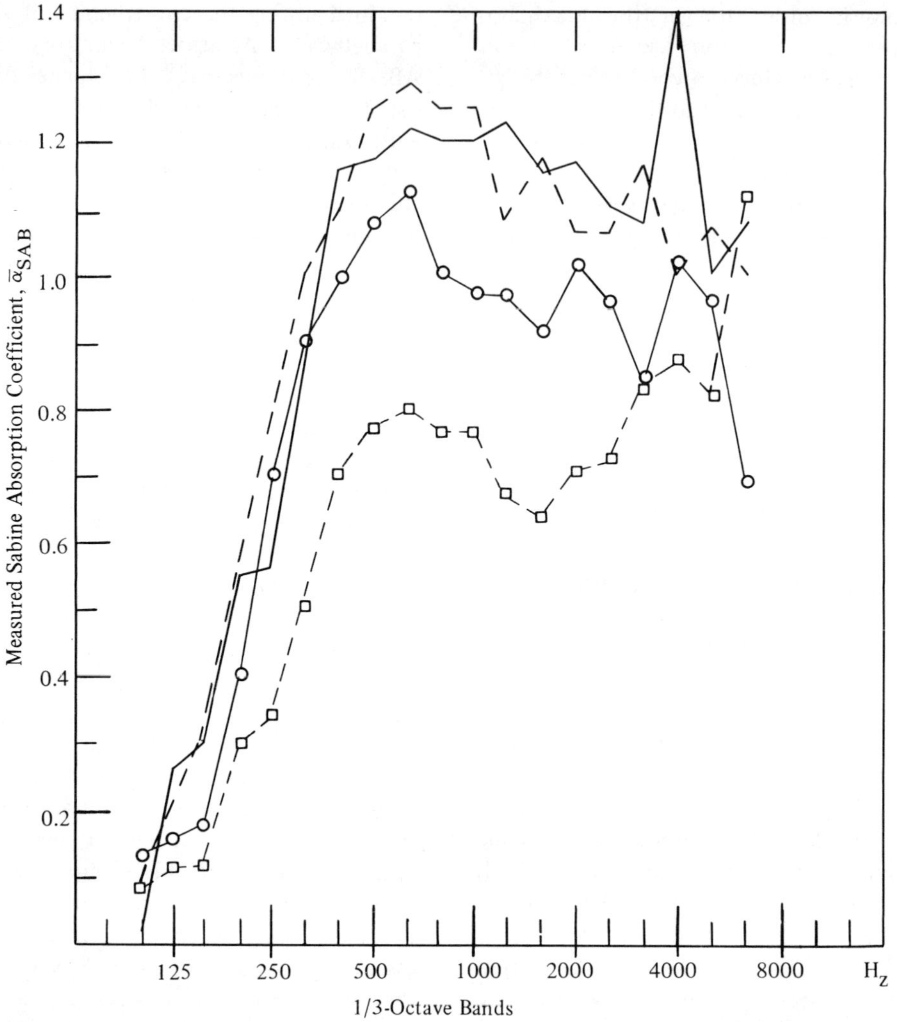

FIGURE 4.9. Effect of increasing area of suspended diffusers in a 200 m^3 (7,062 ft^3) reverberation chamber upon the measured absorption coefficient of an identical test specimen. (See Reference 30.) Diffuser area: —— 30 m^2; - - - 20 m^2; ○—○ 10 m^2; □- - -□ none.

the sound field, the reader is referred to the work of Ebbing and Ingalls,[20,23,24] and Lubman[19,25] and Tichy[26-28].

Some success has been reported in the use of modal enrichment panels[18] to improve the low frequency performance characteristics of small reverberation chambers. In this work, honeycomb panels were placed inside a small reverberation chamber to enrich the apparent acoustic modal density. In order for such panels to be effective, it is necessary that they should exhibit a high modal density and be strongly coupled to the acoustic field – especially at low frequencies. To obtain strong coupling at low frequencies the panel must have:

a. Low surface density.
b. Low coincidence frequency.
c. Small internal damping.

The cross section of the panels used by Scharton is shown in Figure 4.11. The total weight of the surface aluminum, core, and adhesive was 0.43 lb/ft^2. Four panels were suspended 6 in. from the 4 walls of a small 135-ft^3-reverberation chamber and 1 panel was supported on rubber foam 6 in. off the floor. These panels had a total surface area of 90 ft^2. The modal density of the panels was calculated to be equal to the acoustic modal density of 250 Hz and to be greater below this frequency. Below 250 Hz the overall modal

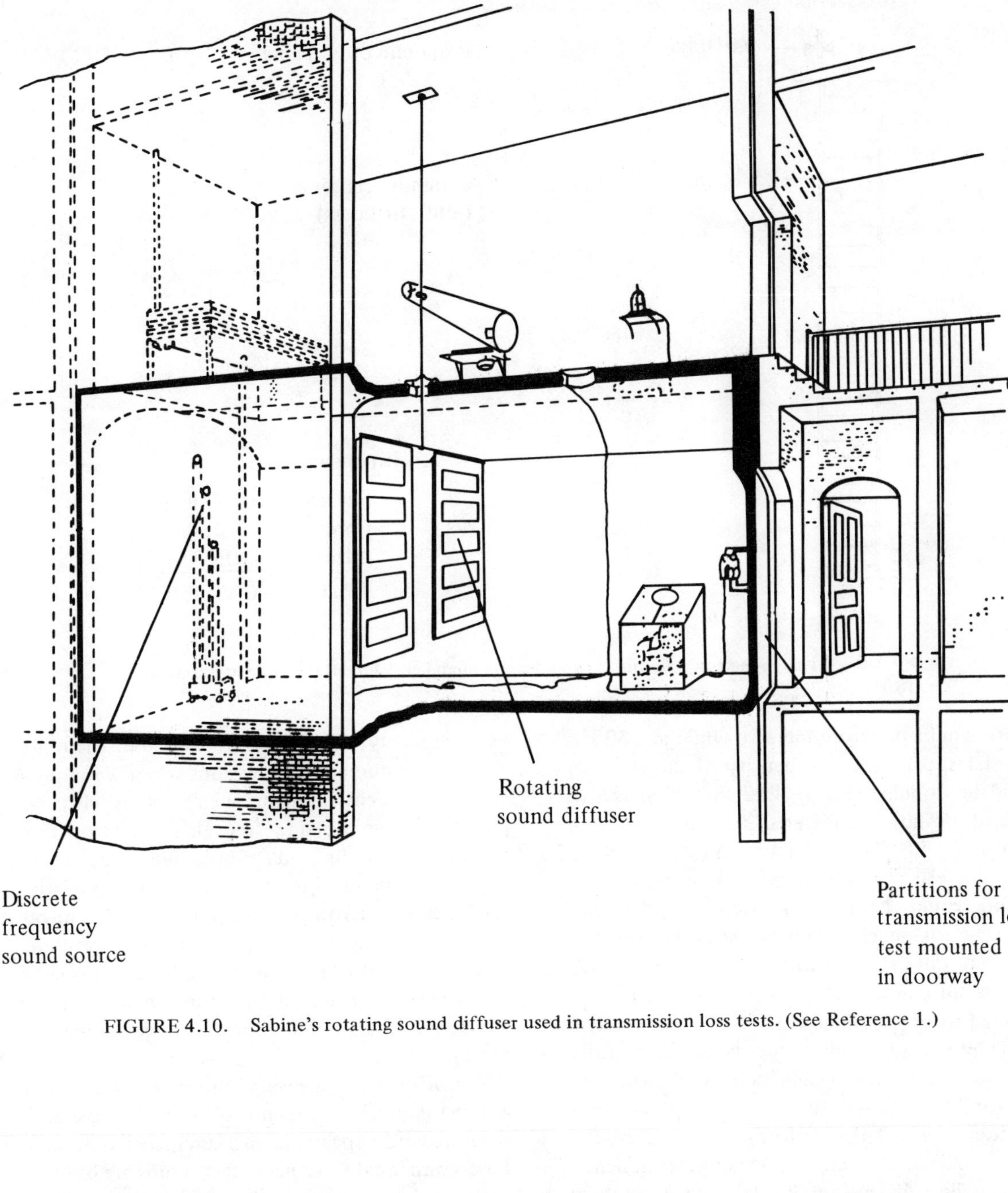

FIGURE 4.10. Sabine's rotating sound diffuser used in transmission loss tests. (See Reference 1.)

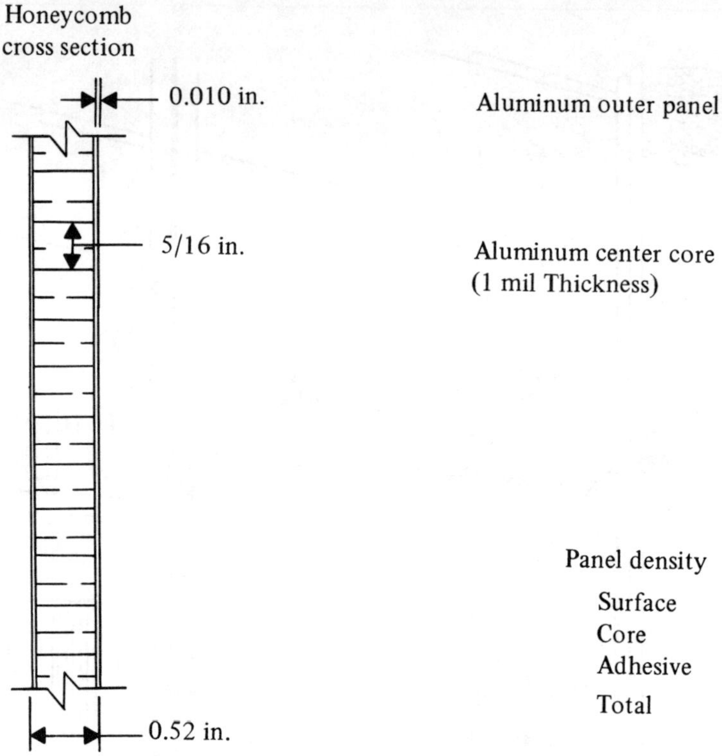

Panel density

Surface	0.24	lb/ft²
Core	0.16	lb/ft²
Adhesive	0.03	lb/ft²
Total	0.43	lb/ft²

FIGURE 4.11. Scharton's aluminum honeycomb modal enrichment panel. (See Reference 18.)

density would therefore increase while at 250 Hz, the total apparent modal density of the chamber should be doubled. Figure 4.12 shows the spatial standard deviation measured for the untreated room and with the modal enrichment panels in place. The effect is well marked, showing that this method could have many useful applications, especially for small chambers. Although the low frequency cut-off frequency, f_c, is increased by the use of these panels, this can usually be expected to be only minimal and of little concern.

The effect of room shape is also not fully understood; however, some recent work, again by Scharton,[18] indicated that room shape did not significantly affect the performance of reverberation chambers. Rectangular chambers designed with baseline dimensions ℓ_x, ℓ_y, and ℓ_z such as those suggested by Sepmeyer[29] in the ratio $(1{:}3\sqrt{4}{:}3\sqrt{2})$ should provide equal acoustic performance to any other irregularly (nonrectangular) shaped room.

4.6 DESIGN CONSIDERATIONS FOR ANECHOIC ROOMS

Although we often need acoustic test rooms to be completely reverberant and diffuse, there are also a large number of occasions when we require an acoustic environment which is as close as possible to an outside free field situation (see Section 1.5.4). In other words, we would often like to measure certain acoustic characteristics (such as directivity patterns) of sound sources such as machines, loudspeakers, etc. outdoors where there would be no reflections whatsoever and where there would therefore be no extreme interference on the radiation from the source (see Section 1.5.5).

Unfortunately, the variability and influence of weather conditions does not allow us to make such accurate and repeatable measurements outdoors. To accommodate this need, test rooms are built in which there are no acoustic reflections, or at least where they are so small in amplitude as to be insignificant, and these are known as anechoic rooms, i.e., rooms without echoes. Sound sources placed in such a room would therefore be expected to behave exactly as they would if allowed to radiate into a free field condition. In a truly anechoic room we could expect to measure the true radiation directivity characteristics of the source.

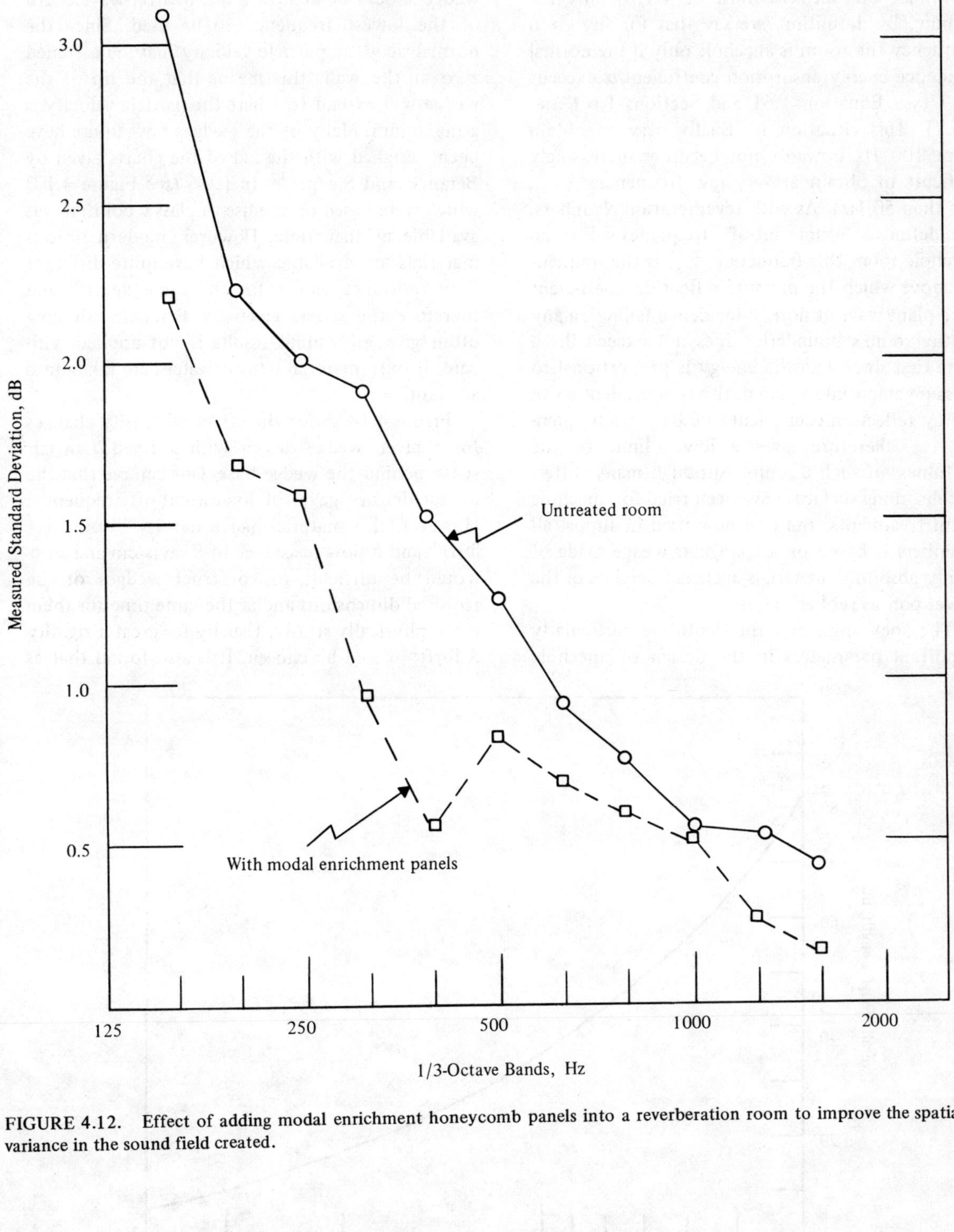

FIGURE 4.12. **Effect of adding modal enrichment honeycomb panels into a reverberation room to improve the spatial variance in the sound field created.**

In order to achieve anechoic conditions in a room, all the surfaces must be very highly absorbing. By definition, we say that for any given frequency the room is anechoic only if the normal incidence energy absorption coefficient α exceeds 0.99 (see Equation 1.61 and Sections 1.6.1 and 1.6.2). This situation is usually easy to obtain from 100 Hz upwards but becomes increasingly difficult to obtain at very low frequencies (e.g., less than 50 Hz). As with reverberation chambers, we define a room cut-off frequency. For an anechoic room this frequency, f_{ca}, is the frequency above which the pressure reflection coefficient for a plane wave at normal incidence falling on any of the room's boundaries does not exceed 0.10. Note that since acoustic energy is proportional to pressure amplitude squared, this is equivalent to an energy reflection coefficient of 0.01. This frequency, f_{ca}, therefore gives a lower limit to the usefulness of such a room. Although many different absorbing surfaces have been tried for anechoic room treatments, the one now used in almost all chambers is based on a triangular wedge made of highly absorbing materials such as fiberglass or the newer porous rubber foams.

The apex angle and the depth are particularly important parameters in the design of anechoic wedges. Generally speaking, the depth of the wedge should be at least a one quarter wavelength of the lowest frequency to be used. Since the normal acoustic particle velocity may be assumed zero at the wall, this means that the tip of the wedge will extend to where the particle velocity is a maximum. Many of the wedges now in use have been designed with the aid of the charts given by Beranek and Sleeper[32] in 1946 (see Figure 4.13) which were based on the use of glass wool that was available at that time. However, modern fibrous materials are available which have quite different flow resistance values for the same density and therefore the charts given by Beranek can now often give misleading results if not applied with care, if only material density values are taken into account.

Figure 4.14 shows the effect of density changes for a given wedge design with a fixed 7 in. air space behind the wedge base. One can see that the lowest density gave the lowest cut-off frequency. However, this material had a density of only 1.1 lb/ft^3 and a flow resistance of 8 rayls/cm and so it would be difficult to construct wedges of the required dimensions and at the same time for them to be physically sturdy. Usually for greater rigidity 3 lb/ft^3 would be chosen. It is also found that as

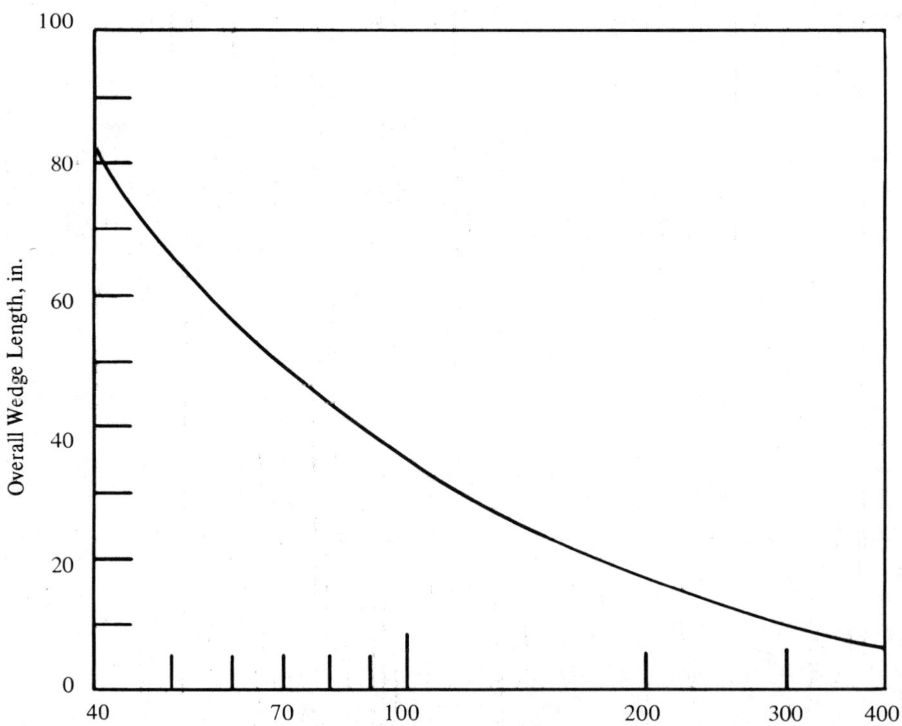

FIGURE 4.13. Simple design chart for anechoic chamber wedge selection.

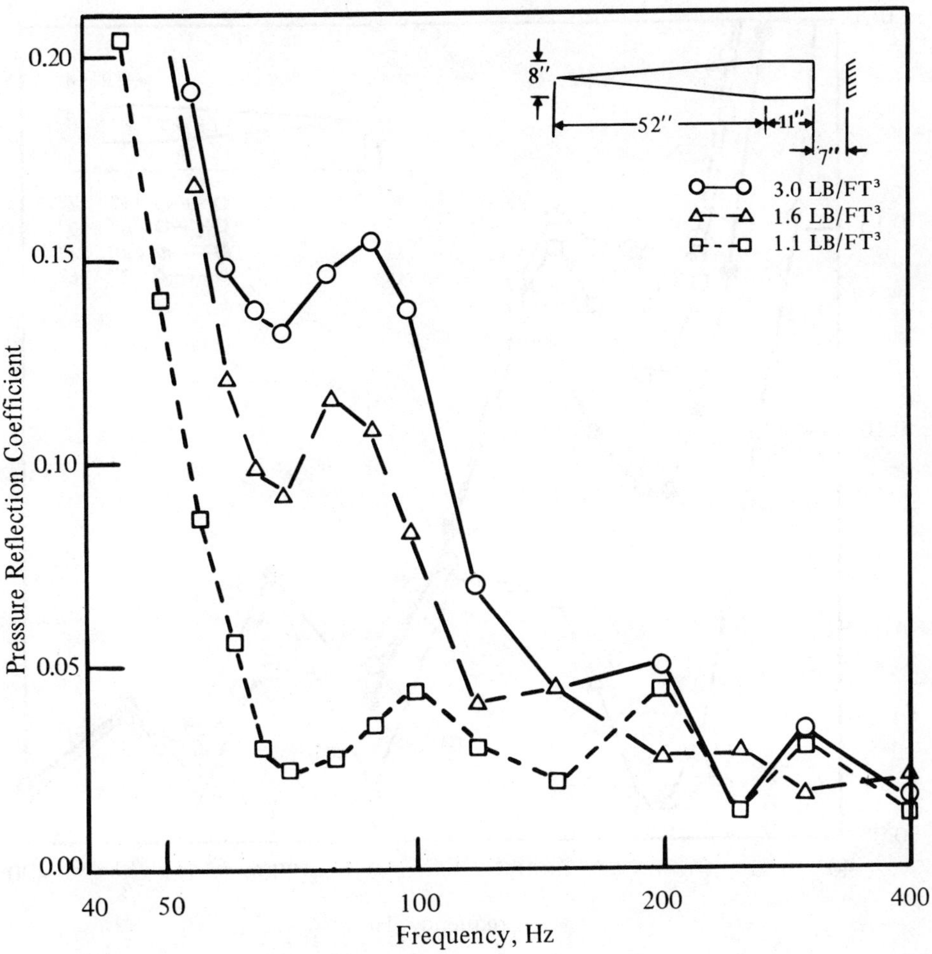

FIGURE 4.14. Pressure reflection coefficient vs. frequency for three sample wedges of glass-wool: (1) Density 3.0 lb/ft^3, flow resistance 24 cgs rayls/cm; (2) Density 1.6 lb/ft^3, flow resistance 10 cgs rayls/cm; (3) Density 1.1 lb/ft^3, flow resistance 8 cgs rayls/cm. (See Reference 33.)

the base of the wedge is decreased in depth and/or the backing air space is increased, then the cut-off frequency is lowered.[33] This is shown in Figure 4.15 from which one would clearly choose a 16 in. air space and 2 in. base to achieve the lowest cut-off frequency. However, since the wedges have to be anchored to a steel wire cage frame by their bases, and since their bases therefore rest against each other, there must usually be more than a 2 in. base to prevent the wedges from sagging. This problem can be overcome by substituting a relatively long base of lower density material for the short heavy base[33] as shown in Figure 4.16. For a much more extensive discussion of anechoic wedge design, the reader is advised to consult References 33 and 34.

Once the anechoic room has been constructed, it becomes necessary to determine the cut-off frequency and to verify the predicted value. This may be done by placing a point source of sound at one end of the room and measuring the sound pressure level as a function of increasing distance from the source to determine any substantial deviation (usually taken to be more than ± 1 dB) from the expected free-field inverse square law decay (see Figure 1.21). This is usually conducted with pure tone test frequencies from 31.5 Hz to 20,000 Hz.

Even below the cut-off frequency it is possible to obtain inverse square law characteristics between the source and some locations sufficiently removed from the walls. This distance from the

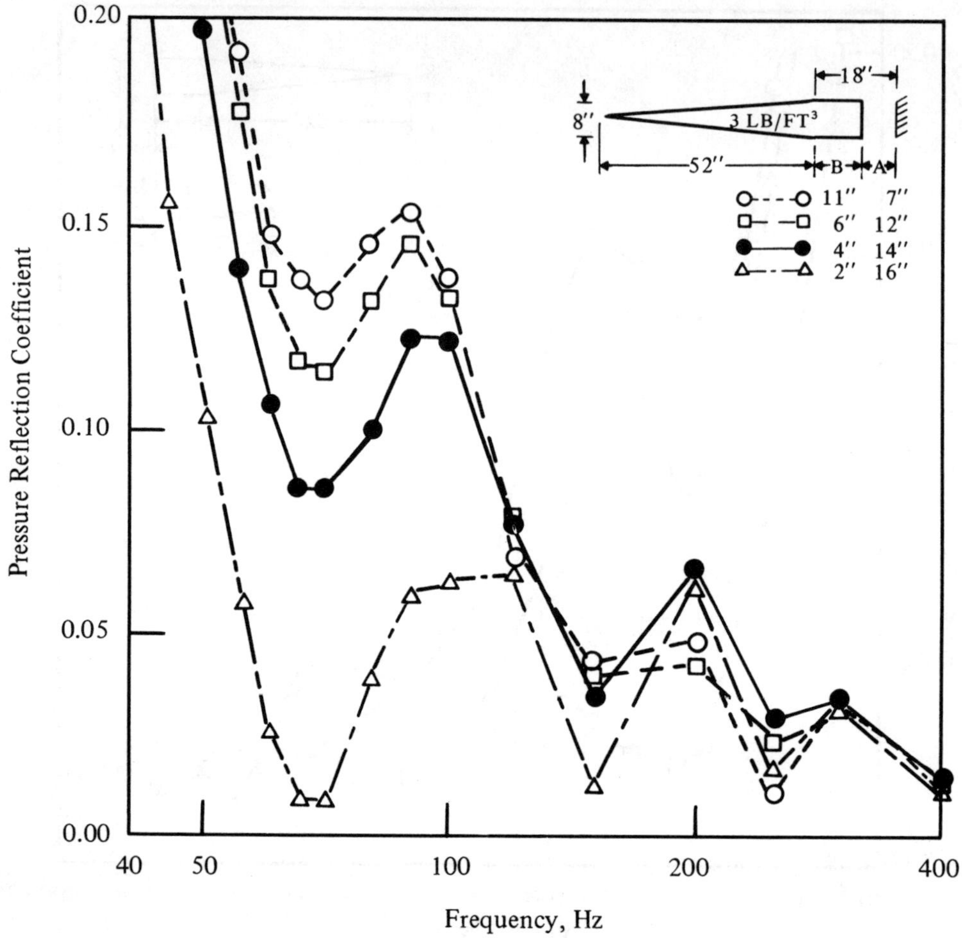

FIGURE 4.15. Pressure reflection coefficient vs. frequency for glass-wool wedge with base length and air-space length (A + B) constant. (See Reference 33.)

walls will naturally become larger as the frequency of the source becomes lower. Since inverse square law conditions can be obtained over certain distances, even below the cut-off frequency, this can often result in one's being able to use a smaller chamber at a lower construction cost. Figure 4.17 shows a typical set of such measurements for an anechoic room[35] of dimensions 23 × 24 × 20 ft with wedges having a calculated cut-off frequency of f_{ca} = 90 Hz. It can be seen that fairly good free-field conditions prevail up to 13 ft from the source down to a frequency of 80 Hz (to within ± 2 dB) while at distances of less than 6.5 ft from the source, excellent low frequency performance, well below the calculated cut-off, is apparent.

The maximum deviation, ΔdB, from the inverse square law behavior can be estimated from the following equation given by Diestel:[36]

$$\Delta dB = 20 \log (1 + 7rRS^{-1/2}), \qquad (4.60)$$

where

r = distance between source and receiver
R = reflection coefficient of wall surface
S = interior room surface area.

In addition to the above criteria, anechoic rooms must also be able to provide a very low background noise level within the room when the door is closed. The exterior walls, therefore, have to be able to provide sufficient sound transmission loss to give background noise levels below NC (Noise Criterion) 15 to 20 in the chamber. Depending upon location and external noise levels this requires up to a double cavity wall having a STC (Sound Trans-

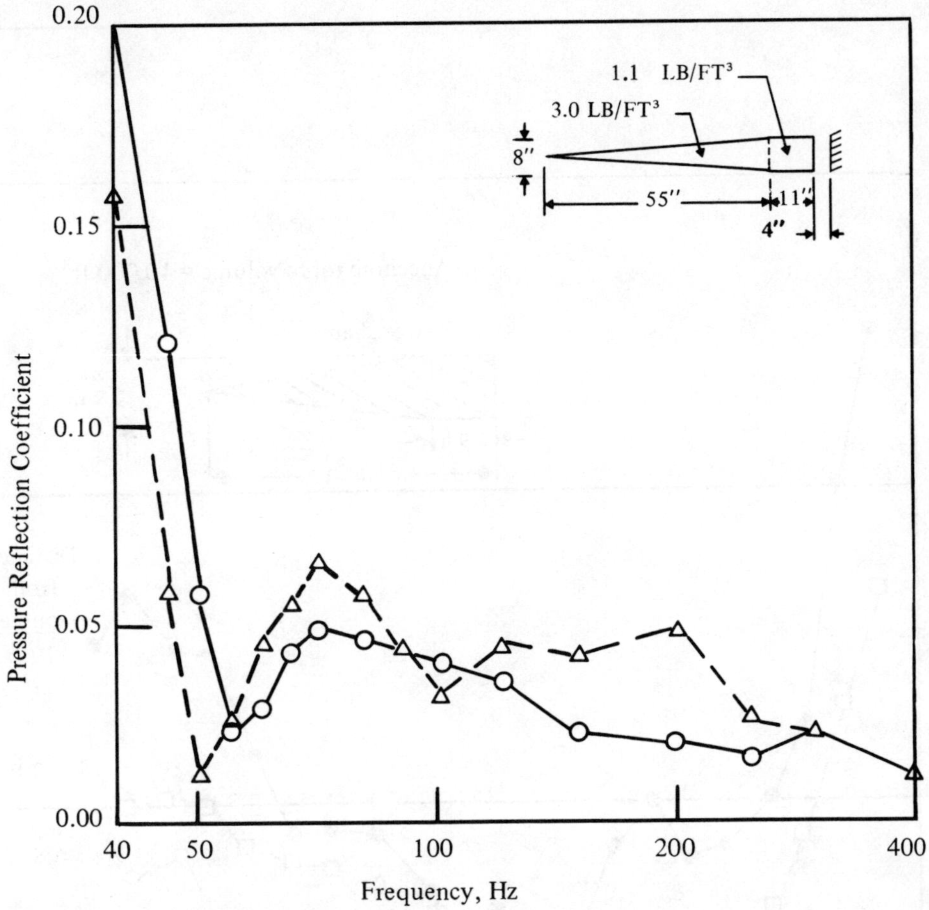

FIGURE 4.16. Pressure reflection coefficient vs. frequency for two samples of the "hybrid" glass-wool wedge. (See Reference 33.)

mission Class) rating not less than STC 50. Similarly it is usual to vibration isolate the whole chamber from the surrounding building structure using isolators such as metal springs fixed between metal plates mounted on a cork base[30] or special neoprene isolators. The isolators are normally chosen to have a low natural frequency such as 5 to 7 Hz. Particular care must be taken to provide vibration-isolators and high sound transmission loss walls where anechoic rooms are built in environments of high vibration and noise, e.g., in factories or plants. In such cases ground vibration may be produced by vehicles, loading trucks, railroads or punch presses.

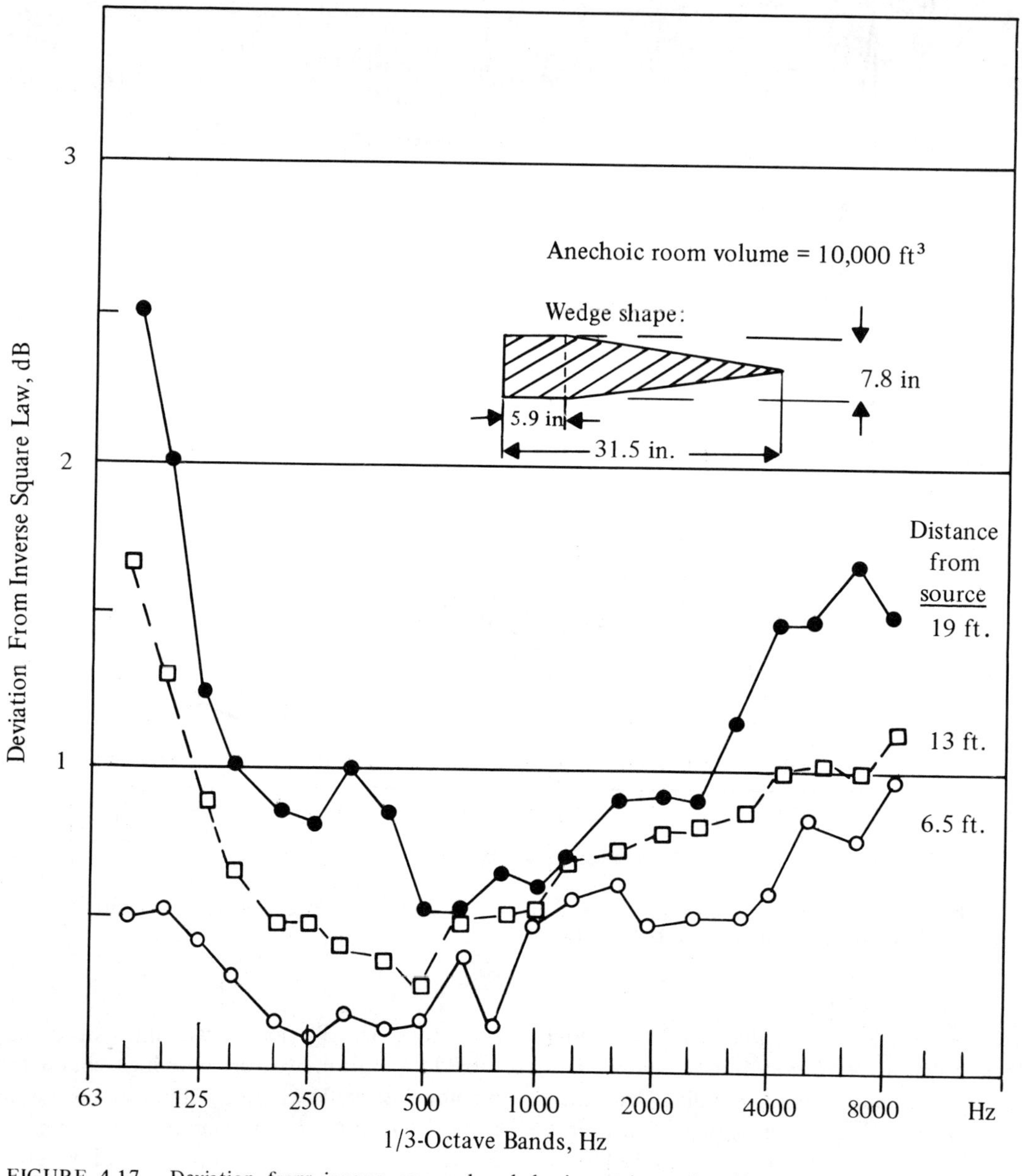

FIGURE 4.17. Deviation from inverse square law behavior at increasing distance from a source in an anechoic room having a cut-off frequency of 90 Hz. (See Reference 35.)

REFERENCES

1. **Sabine, W. C.,** *Collected Papers on Acoustics,* Harvard University Press, 1922, Dover Reprint, 1964, Chap. 1.
2. **Kinsler, L. E. and Frey, A. R.,** *Fundamentals of Acoustics,* John Wiley & Sons, 1966, Chap. 14.
3. **Eyring, C. F.,** Reverberation time in dead rooms, *J. Acoust. Soc. Am.,* 1, 217, 1930.
4. **Norris, R. F.,** *J. Acoust. Soc. Am.,* 3, 361, 1932. Also see Knudsen, V. O., *Architectural Acoustics,* John Wiley & Sons, New York, 1932, 603.
5. **Millington, G.,** A modified formula for reverberation, *J. Acoust. Soc. Am.* 4, 69, 1932.
6. **Sette, W. J.,** A new reverberation time formula, *J. Acoust, Soc. Am.* 4, 193, 1933.

7. **Beranek, L. L.,** *Noise and Vibration Control,* McGraw-Hill, New York, 1972.
8. **Sabine, P. E.,** *Acoustics and Architecture,* McGraw-Hill, New York, 1932, 58, 309.
9. **Northwood, T. D., Grisaru, M. T., and Medcof, M. A.,** Absorption of sound by a strip of absorptive material in a diffuse sound field, *J. Acoust, Soc. Am.,* 31, 595, 1959.
10. **Morris, R. M., Nixon, G. M., and Parkinson, J. S.,** Variations in sound absorption coefficients as obtained by the reverberation chamber method, *J. Acoust. Soc. Am.,* 9, 234, 1937–1938.
11. **Kuttruff, H.,** Eigenschaften und auswertung von nachhaukurven, *Acustica,* 8, 273, 1958.
12. **Andrea, C. A.,** The effect of position on the absorption of materials for the case of a cubical room, *J. Acoust. Soc. Am.,* 3, 535, 1932.
13. **Fitzroy, D.,** Reverberation formula which seems to be more accurate with non-uniform distribution of absorption, *J. Acoust. Soc. Am.,* 31, 893, 1959.
14. **Fokker, A. D.,** *Physica,* 4, 262, 1924.
15. **Knudsen, V. O.,** *Architectural Acoustics,* John Wiley & Sons, New York, 1948, 141.
16. **Morse, P. M., and Ingard, K. U.,** *Theoretical Acoustics,* McGraw-Hill, New York, 1968.
17. **Schroeder, M. K.,** Effect of frequency and space averaging on the transmission response of multimode media, *J. Acoust. Soc. Am.,* 46, 2, 1969.
18. **Scharton, T. D., Rentz, P. E., Lubman, D., and White, P. H.,** Techniques for Improving the Low-Frequency Performance of Small Reverberation Chambers, Bolt Beranek & Newman, Rep. 1867, 1970.
19. **Lubman, D.,** Spatial Averaging in Sound Power Measurements, presented at the 76th meeting of the Acoustical Society of America, Cleveland, Ohio, Nov. 1968.
20. **Ingalls, D. J.,** The use of stationary near field diffusers for improving low frequency measurements in a reverberant room, *J. Acoust. Soc. Am.,* 49, 1971.
21. **Ingalls, D. J.,** The Effects of Large Panels on the Eigenvalues of Rectangular Rooms, paper presented to the Acoustical Society of America, San Diego, Calif., 1969.
22. **Schultz, T. J.,** Sound Power Measurements in a Reverberant Room, paper presented to the Acoustical Society of America, Houston, Texas, 1970.
23. **Ebbing, C. E.,** Experimental evaluation of moving sound diffusers for reverberant rooms, *J. Sound Vib.,* 16, 99, 1971.
24. **Ebbing, C. E. and Ingalls, D. T.,** Experimental evaluation of microphone traverses in a reverberant room, *J. Acoust. Soc. Am.,* 47, 116(A), 1970.
25. **Lubman, D.,** Spatial averaging in sound power measurements, *J. Sound Vib.,* 16, 43, 1971.
26. **Tichy, J.,** The effect of rotating vanes on the sound field in reverberation chambers, *J. Acoust. Soc. Am.,* 49, 89(A), 1971.
27. **Tichy, J.,** The effect of boundary conditions on the statistical properties of the sound field in enclosures, *J. Acoust. Soc. Am.,* 50, 98(A), 1971.
28. **Tichy, J. and Baade, P. K.,** Effect of rotating diffusers and sampling techniques on sound-pressure averaging in reverberant rooms, *J. Acoust. Soc. Am.,* 56, 137, 1974.
29. **Sepmeyer, L. W.,** Computed frequency and angular distribution of normal modes of vibration in rectangular rooms, *J. Acoust. Soc. Am.,* 37, 3, 1965.
30. **Friberg, R.,** "The Acoustic Laboratory of Rockwool AB, Skovde, Sweden", Paper E-5-12, Proc. 6th Int. Congr. Acoust., Tokyo, Japan, 1968.
31. **Towson, P. G.,** A study of the criteria for the design of reverberation chambers, *Arch. Sci. Rev.,* December, 1966.
32. **Beranek, L. L. and Sleeper, H. P., Jr.,** The design and construction of anechoic sound chamber , *J. Acoust. Soc. Am.,* 18, 140, 1946.
33. **Koidan, W., Hruska, G. R., and Pickett, M. A.,** Wedge design for national bureau of standards anechoic chamber, *J. Acoust. Soc. Am.,* 52, 1071, 1972.
34. **Velizhanina, K. A. and Rzhevkin, S. H.,** The investigation of sound-absorbent structures for the anechoic chamber of the physics department of the Moscow State University, *Sov. Phys. Acoust.,* 3, 21, 1957. (English translation, *Akust, Zh.,* 3, 28, 1957.)
35. **Myneke, H.,** The New Acoustics Laboratory of the Catholic University at Leuven, Paper E-5-11, Proc. 6th Int. Congr. Acoust., Tokyo, Japan, 1968.
36. **Diestel, H. G.,** Zur Schallausbreitung in reflexionsarmen Räumen, *Acustica,* 12, 113, 1962.

Chapter 5

ARCHITECTURAL ACOUSTICS

TABLE OF CONTENTS

5.1 INTRODUCTION

In Chapter 4 we saw that the behavior of the sound field within an enclosure can be analyzed both by using three-dimensional wave theory and in some cases by geometric ray tracing techniques. In many applications of acoustics in architecture, geometric techniques yield valuable results and can provide an excellent method of acoustical analysis of the proposed architectural form. There are, however, limitations to this method as we have already seen in Chapter 4. However, for a large room, ray tracing can be used with confidence in the mid to high frequency range where the laws of geometric acoustics are valid, while at low frequencies there is need for concern as to the validity of the technique and where close examination is required.

5.2 SOUND REFLECTION

Geometric acoustics assumes that we can represent outgoing spherical waves from a sound source by "rays" travelling perpendicular to the advancing wave front. This is similar to the optical ray concept used to describe, for example, the behavior of optical lenses and instruments. Indeed, just as in optics, sound rays are found to obey the following laws of reflection when they strike a rigid smooth surface:

1. Both the incident and reflected rays lie in the same plane.
2. The angle θ_1, which the incident ray makes with the normal to the reflecting surface is equal to the angle of reflection θ_2 (see Figure 5.1).

In addition, if the reflecting surface is a plane, then the acoustical "image" of the source can be located from the fact that: The image will lie on a line perpendicular to the plane through the source and will be *equidistant* from it. This is illustrated in Figure 5.2. It is this law which is particularly useful in architectural ray tracing since it enables us to construct ray paths between the sound source and any given receiving point without having to construct and measure angles of reflection.

All of the above laws of reflection are only valid, however, if the reflecting surface is very much larger than the wavelength of the sound reflected from it. This is very important to remember in the architectural design of small

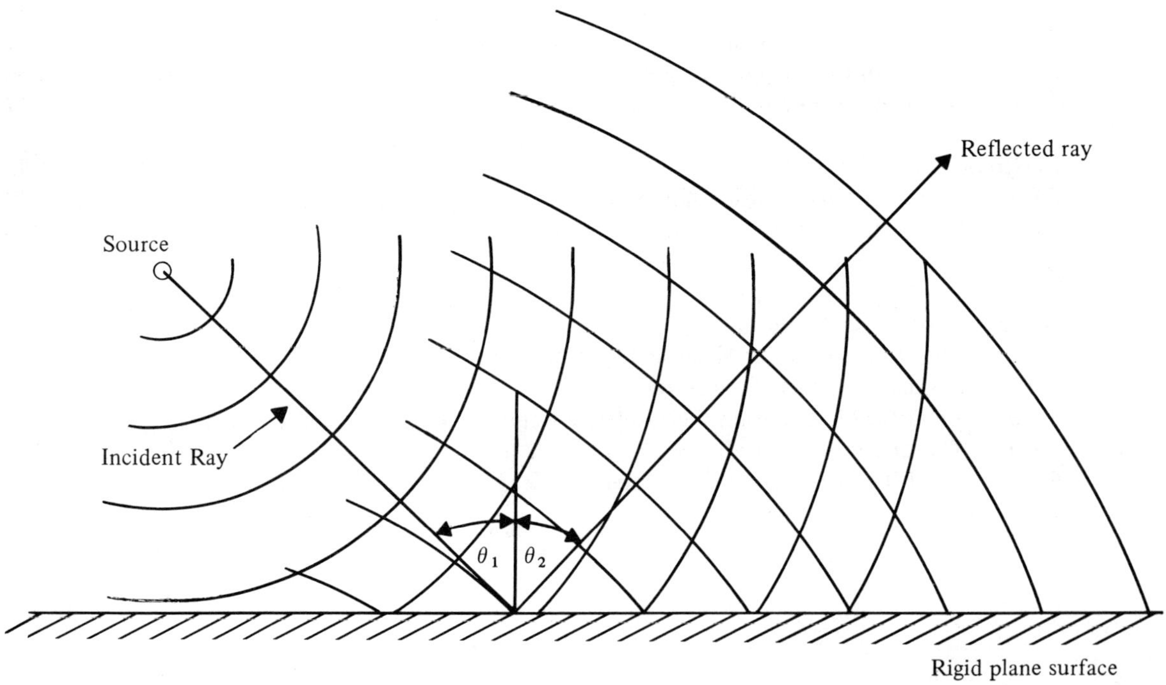

FIGURE 5.1 Sound reflection from a rigid plane surface. The angle made by the incident ray and the normal equals that made by the reflected ray.

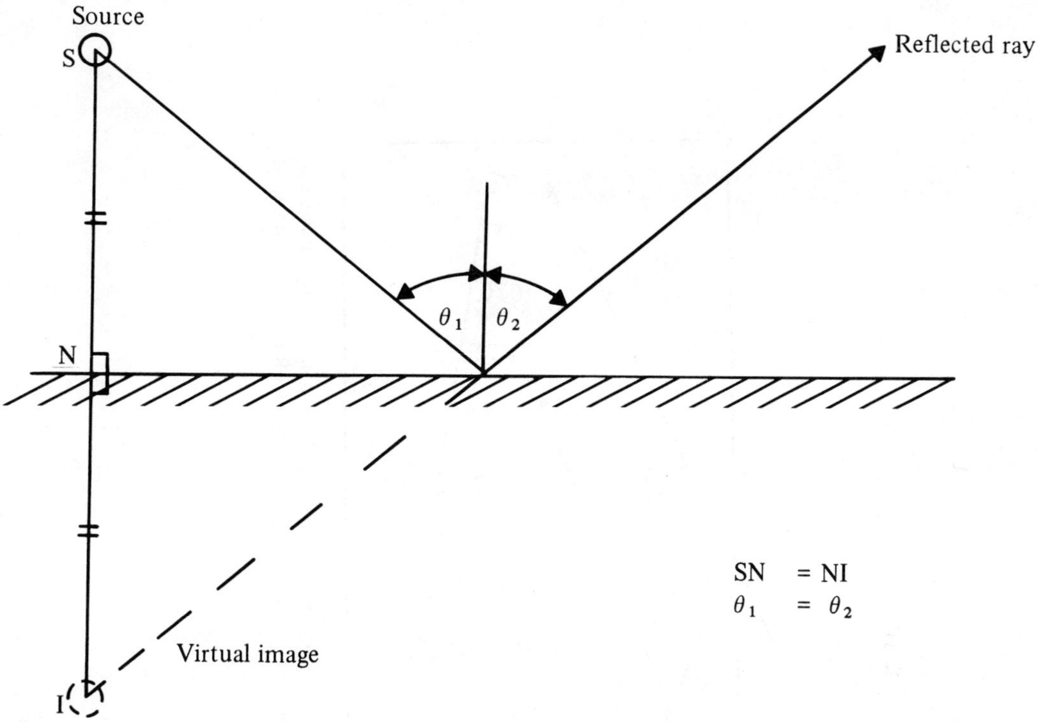

FIGURE 5.2 The path of a sound ray may be traced by locating the position of the virtual image of the source such that SN = NI where the line SI is normal to the reflecting plane.

rooms or in auditoria, where special ceiling panel reflectors are often used to reflect sound from the sound source (e.g., orchestra) to the rear of the room. Since the physical size of reflecting panels is often limited by a variety of parameters, this then also limits their low frequency sound reflection characteristics. Indeed, if the size of the reflector is smaller than a wavelength, then the sound field will merely bypass the obstacle and no reflection will occur. When a ceiling or side wall is composed of several inclined planes, the ray tracing technique can be used to determine the angles of inclination required for these planes in order to "focus" sound to the rear part of the room where some reinforcement may be required. Figure 5.3 shows this technique applied to a ceiling composed of 3 planes. In practice, however, many more planes may be used although the technique is unchanged.

If the reflecting surface is not a plane, but curved, then the rays will be focused in a similar manner to that found in optics for reflection by curved mirrors. In the case of a concave reflecting surface, sound rays will be focused to a point forming a real acoustic image. This can be most undesirable acoustically, since it will result in a "hot spot" within the room which will be very loud while it would be considerably quieter at other points away from the focus in the room. Figure 5.4 shows that the image produced by a concave reflector can be located by using the fact that rays traveling parallel to the axis will be reflected through the focus (equal to one half of the radius of curvature R) and conversely, rays travelling through the focus will be reflected parallel to the axis. Again, in order for this focusing effect to take place the dimensions of the concave surface must be larger than a wavelength (see Equation 1.27) and the surface must be smooth and reflective. This effect must be avoided in architectural acoustics since it leads to hot spots and echoes. Many auditoria have concave rear walls or balconies with concave front walls which would give rise to echoes at the front of the audience or even on the stage if not properly treated to be made highly absorbent. Domes and shell ceilings can cause similar effects and almost always result in an acoustic problem.

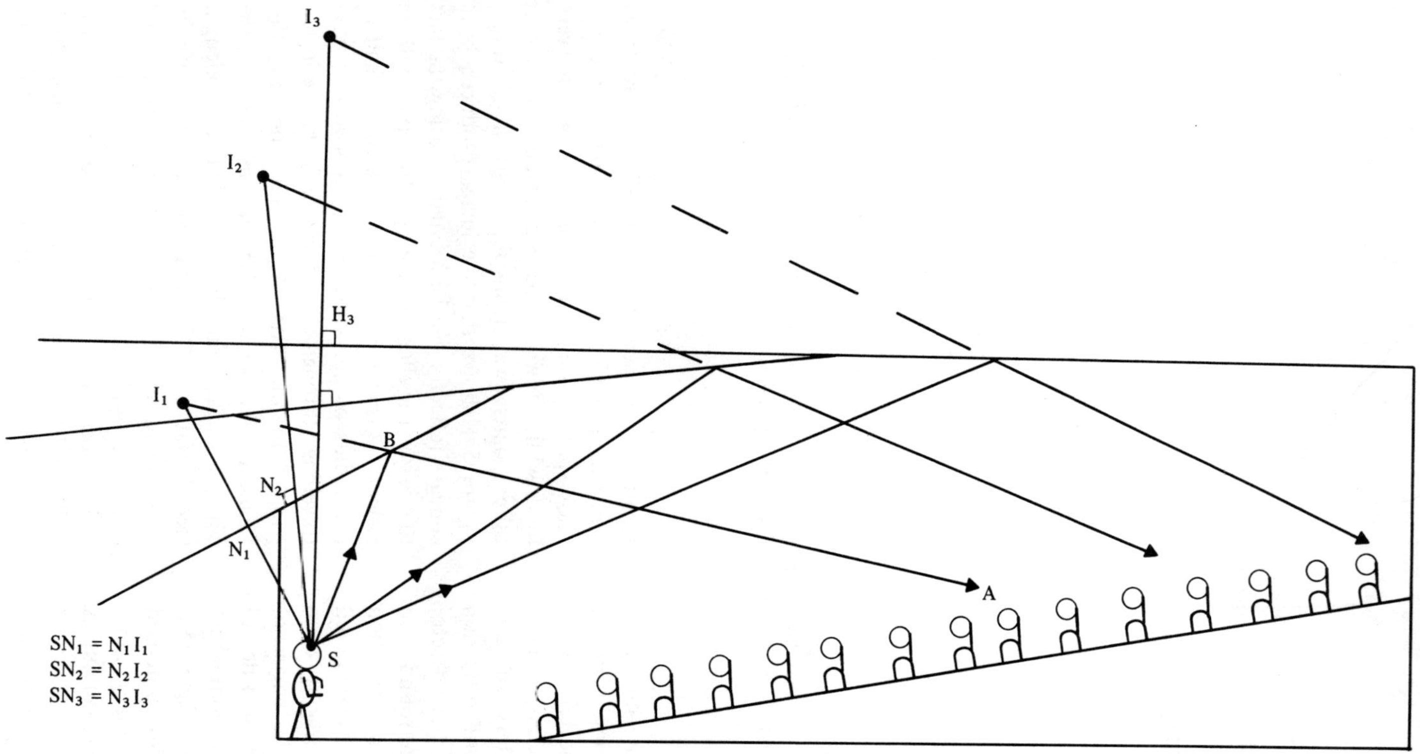

FIGURE 5.3 The construction of reflected ceiling rays can be accomplished by locating each of the images I_1, I_2, and I_3 and then drawing in possible paths to the audience from the sound source. For example, first locate I_1 then join A and I_1 to give B, then join B to S to give the total path from the first inclined plane at SBA.

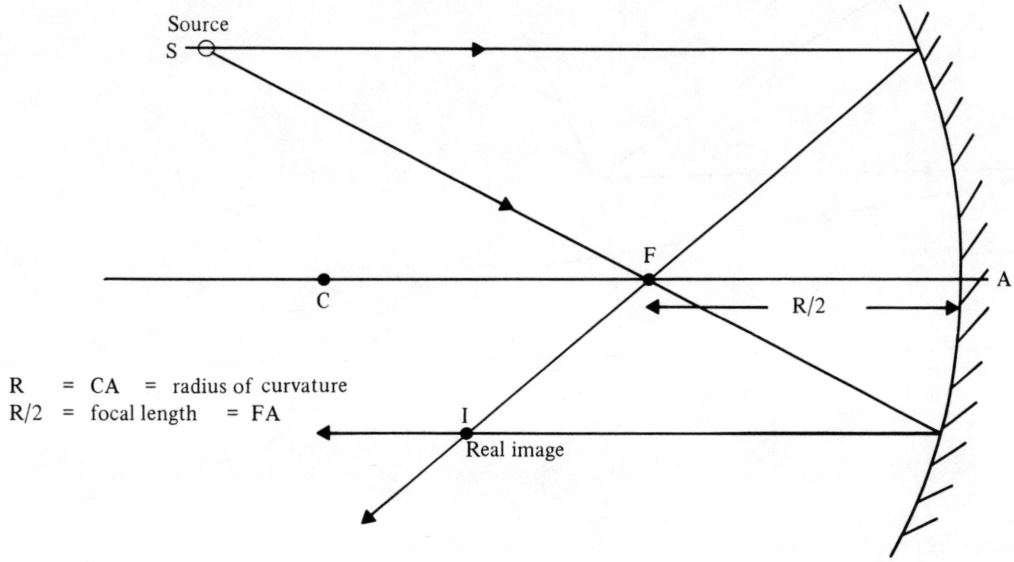

FIGURE 5.4 Rays reflected from a concave surface can be traced by drawing one ray parallel to the axis and reflected through a point F on the axis equal to one half the radius of curvature and one ray through F and reflected parallel to the axis. The intersection of these rays gives the image. Note that the size of the reflecting surface must be large compared to a wavelength of the incident sound.

Convex reflecting surfaces behave as diffusers since they disperse the reflected rays (to form a virtual image behind the reflector). This effect is shown in Figure 5.5. Once again, the size of the reflector must be large compared to a wavelength. Hence, if convex surfaces are to be used to give good diffusion they must be of adequate size to be effective. This means that they should have a chord length of some 10 ft with a radius of curvature of at least 25 to 30 ft in order to be effective above approximately 100 Hz.

5.3 SOUND DIFFRACTION

One may recall from optics that *diffraction,* is the "bending" of light waves at the edges of objects due to interference of secondary wave sources which gives rise to diffraction fringes. Note the difference between diffraction and *refraction* discussed in Sections 1.7.1 and 1.7.2. This phenomenon of diffraction also occurs in acoustics when obstacles are small compared to the wavelength of the sound incident upon them (see Figure 5.6). Diffraction is the change of direction of propagation of sound waves due to their passage around an obstacle or through an opening. This bending (or scattering) phenomenon is therefore more pronounced for low frequency sounds than for those of higher frequency. For example, if we consider the effect of a 3-ft-diameter pillar in a large room, we find that at low frequencies an acoustic shadow would not be formed behind the pillar since almost all the sound would diffract around it. On the other hand, high frequency sound waves would not be able to diffract significantly and an acoustic shadow would be formed, causing a substantial drop in the intensity of the high frequency sound received. This effect can often be observed in deep galleries beneath balconies where the balcony edge acts as a diffracting element bending low frequency sounds to the gallery audience, while high frequency sounds are lost due to the acoustic shadow thrown by the balcony. Sound diffraction is of major importance in the design and performance of indoor and outdoor acoustical barriers.

In the process of sound passing through an opening there is little diffraction if the opening is large compared to a wavelength (i.e., a stage opening). However, if the opening is small (i.e., a 3-ft door), then at low frequencies of, for example, 50 Hz, the 22-ft wavelength waves would be very much diffracted, spreading in all directions spherically from the other side of the door opening. High frequencies, on the other hand, would be able to pass directly through the door opening unaffected by diffraction.

Diffraction and scattering can also be caused by

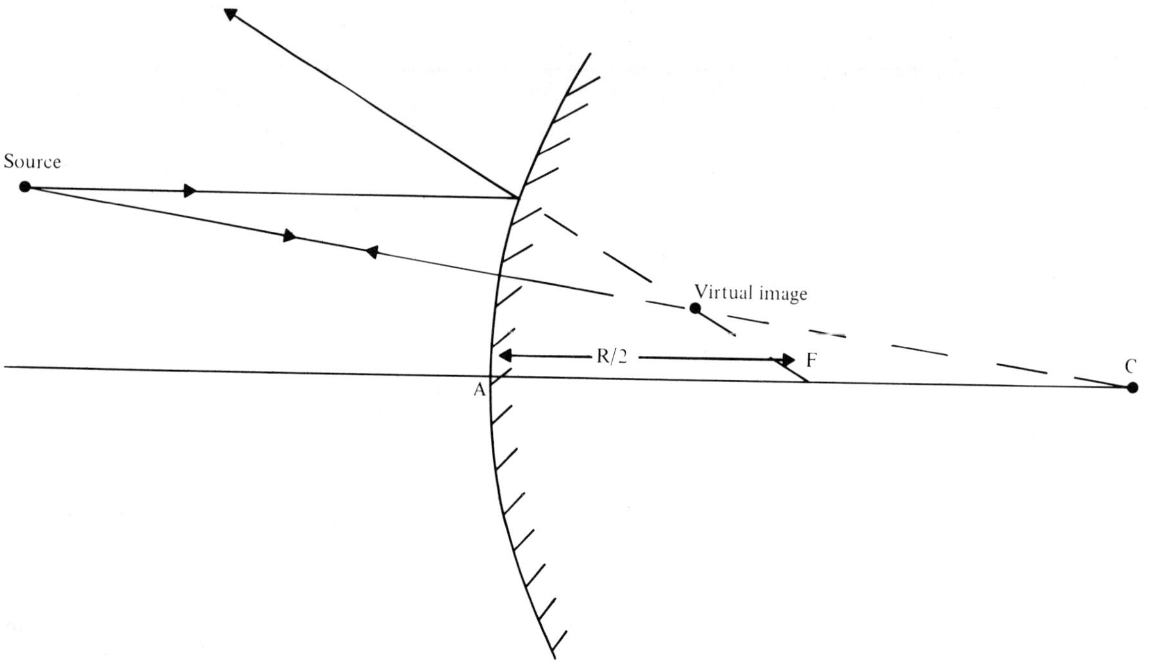

FIGURE 5.5 Reflection from a convex surface (of a size greater than a wavelength) gives a virtual image and the reflected rays are therefore dispersed.

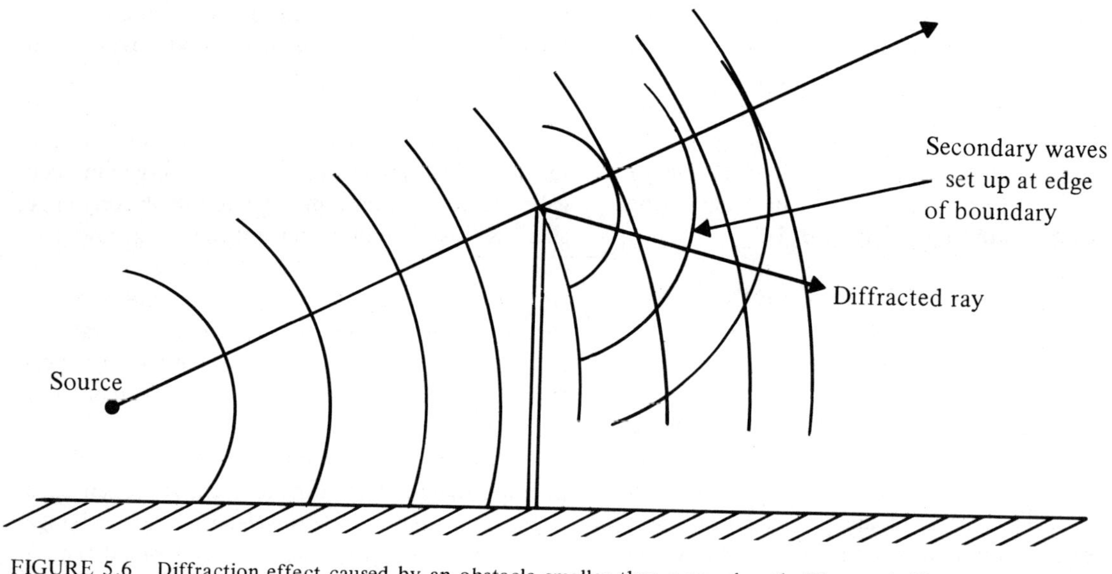

FIGURE 5.6 Diffraction effect caused by an obstacle smaller than a wavelength. The ray is "bent at the edge of the object" by the formation of secondary waves along the incident wavefront. This causes an interference effect – diffraction.

patches of absorbing material on a surface in a room and by irregular shapes and projections on wall surfaces. This aids in increasing the amount of diffusion in the room which, as we shall see later, is very much sought after in listening rooms. Great care must therefore be exercised if the geometric laws of reflection are to be applied to, for example, a coffered ceiling, since a very complex reflection and diffraction pattern will result that cannot be explained by these simple laws. However, as long as the reflecting surfaces are large and flat, these laws do provide a very useful tool to the architect and acoustician.

5.4 SOUND-ABSORBING MATERIALS

When sound waves strike a boundary separating two media, some of the incident energy is reflected from the surface and the remaining energy is transmitted into the second medium. Some of this energy is eventually converted by various processes into heat energy and is said to have been absorbed by that medium. The fraction of the incident energy absorbed is the absorption coefficient α (see Sections 1.6.1 and 1.6.2) which theoretically ranges from zero to unity. In practice, values of $\alpha > 1.0$ are sometimes measured. This anomoly is due to the theoretical derivation of the equations describing the absorption process. Although all materials do absorb some incident sound, the term "acoustical material" has been primarily applied to those materials which have been produced for the specific purpose of providing high values ($\alpha > 0.2$) of absorption. The major uses of acoustical materials are almost invariably found to include the reduction of reverberant sound pressure levels and the reduction of reverberation time in enclosures (rooms).

Since about 1965, the use and variety of available specialized acoustical materials has greatly increased. This has been due mainly to both increased technology and public awareness and concern about noise in everyday life. In turn, this has led many public bodies and commercial public service operations to realize the benefits of providing good acoustical conditions for their clients. The architect and acoustical engineer now have a wide choice of sound-absorbing materials which not only provide the desired acoustical properties, but also offer an extremely wide variety of colors, shapes, sizes, light reflectivities fire ratings, and methods of attachment – to say nothing of the costs of purchase, installation, and upkeep!

Another parameter often of interest in assessing the usefulness of an acoustic absorber is the *Noise Reduction Coefficient* (NRC). As discussed in Section 1.6.1, this is simply the average of the measured absorption coefficients for the 250-, 500-, 1,000-, and 2,000-Hz octave bands expressed to the nearest 0.05. This NRC value is often useful in the determination of the applicability of a material to a particular situation.

Although a wide range of sound-absorbing materials exist which provide absorption properties dependent upon frequency, composition, thickness, surface finish, and method of mounting, they can be divided into several major classifications. These will now be discussed separately.

5.4.1 Porous Sound Absorbers

Porous materials are characterized by the fact that the nature of their surfaces is such that sound energy is able to enter the materials by a multitude of small holes or openings. They consist of a series of tunnel-like pores and openings which are formed by interstices in material fibers, or by foamed products. (Usually the more open and connecting these passages are, the larger the values of the sound-absorbing efficiency of the material.) If, on the other hand, the pores and penetrations are small and not joined together, then the material becomes substantially less efficient as a sound absorber. Included in this broad category of porous absorbers are fibrous blankets, hair felt, wood-wool, acoustic plaster, a variety of spray-on products, and certain types of acoustical tiles.

When a porous material is exposed to incident sound waves, the air molecules at the surface of the material and within the pores of the material are forced to vibrate and in doing so lose some of their original energy. This is because part of the energy is converted into heat due to frictional and viscous losses of air molecules at the walls of the interior pores and tunnels within the material. At low frequencies these changes are isothermal, while at high frequencies they are adiabatic. In fibrous materials, much of the energy can also be absorbed by scattering from the fibers and by the vibration caused in the individual fibers. In this way, high values of absorption coefficient in excess of 0.95 can be observed. Depending upon how α is measured, values in excess of unity can also be found; just how this happens will be discussed later. The values of α observed are usually strongly dependent upon (a) frequency, (b) thickness, and (c) method of mounting. These should always be considered in the choice of a particular material.

Figure 5.7 shows typical sound absorption characteristics for a blanket type fibrous porous material placed (A) against a hard wall, and (B) with a 4-in. air space between the material and the wall. In both cases the absorption properties are substantially better at high frequencies than low. When the same material is backed by an air space, the low frequency absorption is improved without significantly changing the high frequency characteristics. Figure 5.8 shows the effect of increasing the thickness of the material on a solid backing. Again, increased low frequency absorption is observed for increased thickness.

An additional effect to consider is that of surface treatment, since most materials will become discolored or dirty after prolonged exposure for a number of years and may require cleaning or refinishing. Because the surface pores must be open to incident sound in order for the porous material to function, it is essential that they should not be blocked with paint or any other surface coating treatment. The effects of brush painting porous materials are usually more severe than spray painting; however, the general effect is to lower the absorption coefficient to about 50% of its unpainted value particularly at high frequencies. In addition, the absorption peak is shifted downwards in frequency as observed by Price and Mulholland.[11] As more coats of paint are applied, the paint membrane becomes more dense and more pores are sealed, the result of which is to shift the absorption peak even lower in frequency and magnitude.

It is useful now to describe briefly the physical parameters used to account for the sound absorbing and attenuating properties of porous materials. These include specific flow resistance, porosity, volume coefficients of elasticity of both air and the skeleton, structure factor, and specific acoustic impedance. These will now be described separately.

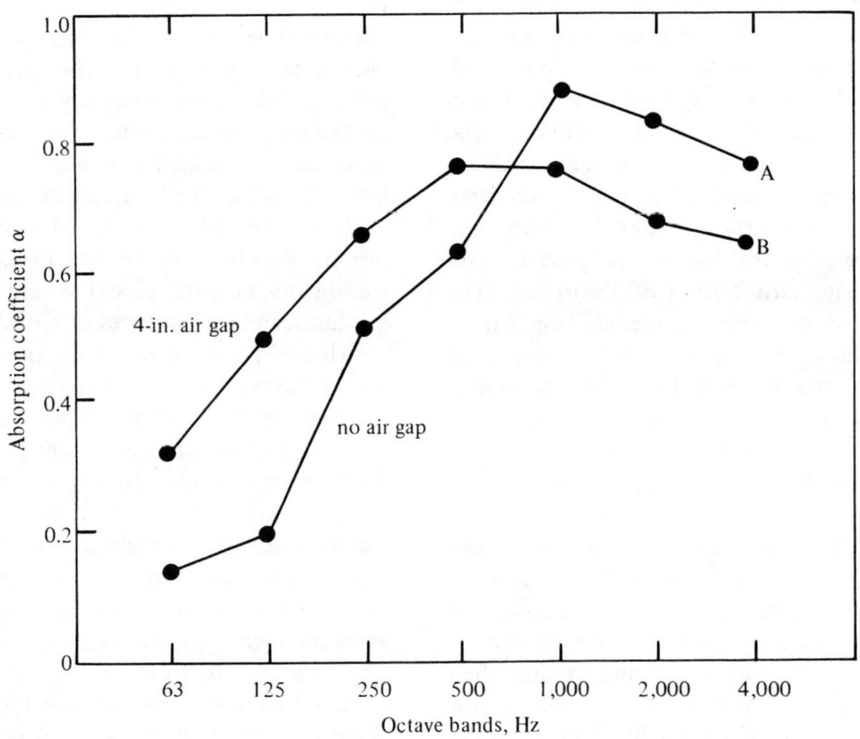

FIGURE 5.7 Typical coefficient vs. octave band frequency characteristics for a 1-in. thick fibrous absorbing material. Curve A is for the material laid directly on a rigid backing, while curve B shows the effect of introducing a 4-in. air gap.

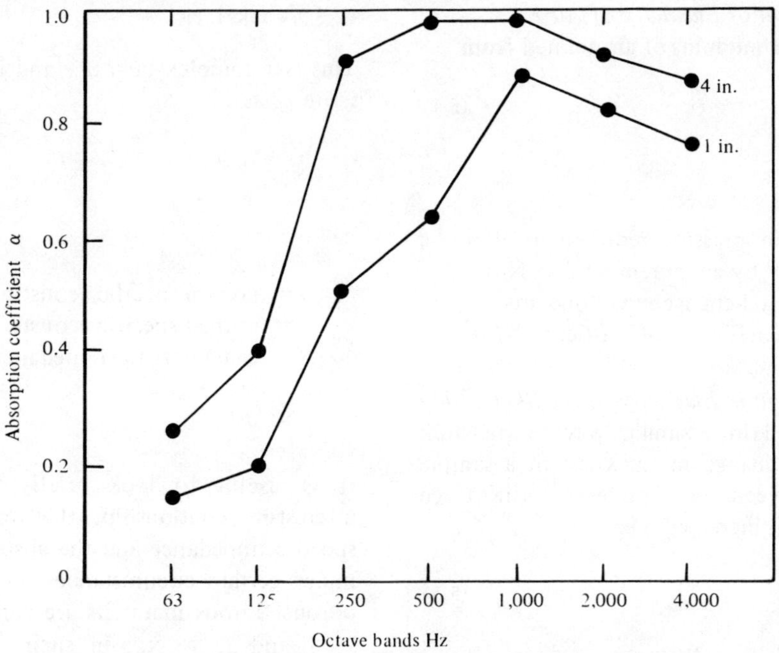

FIGURE 5.8 Effect of increased thickness upon the absorption coefficient of a typical fibrous material (solid backing, 3 lb/ft³ density).

Specific Flow Resistance R

This accounts for the resistance offered to air flow through the media. It is defined in the mks system as

$$R = \frac{\Delta p}{\Delta T} \cdot \frac{1}{v} \text{ mks rayls/m}, \tag{5.1}$$

where

R = specific flow resistance, mks rayls/m
Δp = differential sound pressure created across a sample of thickness ΔT, measured in the direction of air flow, N/m²
v = particle velocity in sample, m/sec
ΔT = incremental thickness, m.

To convert R from mks rayls/m to cgs rayls/in. divide the former value by 400. For example, 4×10^3 mks rayls/m = 10 cgs rayls/in.

Typical values for porous fibrous materials vary from 4×10^3 to 4×10^4 mks rayls/m for a density range of 1 to 10 lb/ft³. Generally speaking, if the specific flow resistance R becomes very large, then most of the incident sound falling on the material will be reflected, while if R is too small, then the material will offer only very slight viscous losses to sound passing through it and so it will provide only little sound absorption or attenuation. Although the absorption is proportional to thickness, it is generally found that for a given flow resistance value R, the optimum thickness, T, of material is approximately given by $T = 100/\sqrt{R}$.

Porosity Y

The porostity of a porous material is defined as the ratio of the void space within the material to its total displacement volume as

$$Y = \frac{V_a}{V_m}, \tag{5.2}$$

where

V_a = volume of air in void space in sample
V_m = total volume of sample.

For a material composed of solid fibers the porosity can be estimated[1] from

$$Y = 1 - \frac{M_s}{V_s \rho_F}, \tag{5.3}$$

where

M_s = total mass of sample
V_s = total volume of sample
ρ_F = density of fibers.

Volume Coefficient of Elasticity of Air K

This is the bulk modulus of air defined from

$$\Delta P = -K\frac{\Delta V}{V} \tag{5.4}$$

where

ΔP = change in pressure required to alter the volume V by an increment ΔV, N/m^2

ΔV = incremental change in volume, m^3

K = volume coefficient of elasticity, N/m^2.

Volume Coefficient in Elasticity of the Skeleton Q

This is defined in a similar way to the bulk modulus as the change in thickness in a sample sandwiched between two plates as the force applied on them is increased, i.e.,

$$dF = -Q\frac{dT}{T}S, \tag{5.5}$$

where

dF = incremental force applied to the sample (Newtons)

dT = incremental change in thickness of sample, m

T = original thickness, m

Q = volume coefficient of elasticity of skeleton, N/m^2

S = sample area, m^2.

Structure Factor k

It is found that in addition to the flow resistance, the composition of the inner structure also affects the acoustical behavior of porous materials. This has been dealt with by Zwikker and Kosten[2] and treated as an effective increase in the density of the air in the void space of the material. Beranek[1] reports that flexible blankets have structure factors between 1 and 1.2 while rigid tiles have values between 1 and 3. He also shows that for homogeneous materials made of fibers with interconnecting pores there is an approximate relationship between the porosity Y and structure factor k as

$$k = 5.5 - 4.5Y. \tag{5.6}$$

Specific Acoustic Impedance z_0

This is defined as the ratio of the sound pressure to particle velocity at the surface of the material, for a sample of infinite depth, when plane sound waves strike the surface normally.

$$z_0 = p/v \text{ mks rayls.} \tag{5.7}$$

This is a complex quantity and it is customary to write z_0 as

$$z_0 = (r_n + ix_n)\rho_n c, \tag{5.8}$$

where

r_n = normal specific acoustic resistance

x_n = normal specific acoustic reactance

ρc = characteristic impedance of air (415 mks rayls)

i = $\sqrt{-1}$.

It is useful to look briefly at some of the interesting relationships that exist between the specific impedance and the absorption coefficient under certain circumstances. For example, many porous fibrous materials are composed so that $r_n \gg 1$ and $r_n \gg x_n$; in such circumstances the absorption coefficient α_θ for a sound wave striking a surface at angle θ to the normal is given by

$$\alpha_\theta = \frac{4r_n \cos\theta}{(r_n \cos\theta + 1)^2}. \tag{5.9}$$

Furthermore, for a diffuse sound field, the random incidence absorption coefficient α is given by[3]

$$\alpha = \frac{8}{r_n}\left[1 + \frac{1}{1+r_n} - \frac{2}{r_n}\log_e(1+r_n)\right]. \tag{5.10}$$

If $r_n > 100$ this equation can be substantially simplified to

$$\alpha = 8/r_n. \tag{5.11}$$

This equation can therefore be used only for materials with small absorption coefficients.

The use of the complex acoustic impedance, rather than absorption coefficient, allows a much more rigorous treatment of room reverberation time analysis. Although often considerably more complex than the theory described in Chapter 4, this technique does provide more accurate values of reverberation times in rooms containing uneven distribution of absorption material, even if either pair of opposite walls are highly absorbing or if opposite walls are composed of one very soft and one very hard wall.[4]

Beranek shows[1,5] that the specific acoustic impedance of a rigid tile can be written in terms of the previously defined fundamental parameters as

$$z_0 = \left(\rho \frac{kK}{Y}\left[1 - i\frac{R}{\rho k\omega}\right]\right)^{\frac{1}{2}}, \qquad (5.12)$$

where, as before,

K = volume coefficient of elasticity of air, N/m^2
k = structure factor
R = flow resistance, mks rayls/m
Y = porosity
ρ = density of air,
i = $\sqrt{-1}$
ω = $2\pi f$, angular frequency.

It can therefore be seen from Equations 5.8, 5.11, and 5.12 that the absorption coefficient is proportional to the porosity and flow resistance of the material (provided $r_n > 100$ and $r_n \gg x_n$), and is inversely proportional to the density and structure factor. For soft blankets the expressions for the acoustic impedance become very much more complex and the reader is directed to References 1, 2, and 5 for a much more complete analysis. In another method of analysis a fibrous medium is considered to be composed of an array of parallel elastic fibers in which a scattering of incident sound waves takes place to result in conversion to viscous and thermal waves by scattering at the boundaries. This technique was first used by Lord Rayleigh, William and Strutt[6] and has since been refined by several researchers.[7,8] More recently Attenborough and Walker have included effects of multiple scattering in the theory[9,10] which is now able to give good predictions of the impedance of porous fibrous materials.

5.4.2 Panel or Membrane Absorbers

When a sound source is turned on in a room, a complex pattern of room modes is set up, each having its own characteristic frequency as described by Equation (4.44). These room modes are able, in turn, to couple acoustically with structures in the room, or even the boundaries of the room, in such a way that acoustic power can be fed from the room modes to other structural modes in, for example, a panel hung in the room.

A simply supported plate or panel can only vibrate at certain allowed frequencies f_{mn} and these are given by

$$f_{m,n} = \frac{hc_L}{4\sqrt{3}\,\pi}\left[\left(\frac{m\pi}{l_1}\right)^2 + \left(\frac{n\pi}{l_2}\right)^2\right], H_z, \qquad (5.13)$$

where

f_{mn} = characteristic modal frequency
m,n = integers; 1,2,3 . . . etc.
c_L = longitudinal wave speed in plate material, ft/sec
h = thickness of plate, ft
l_1,l_2 = dimensions of plate, ft.

If m = n = 1, then this gives the first allowable mode called the fundamental mode $f_{1,1}$. The above equation is valid only for plates with simply supported edges. For a plate with clamped edges the fundamental mode occurs at approximately twice the frequency as calculated from Equation 5.13. Thus a room mode, at or close to the fundamental frequency of a plate hung in the room, will excite the plate fundamental mode. In this way the plate will be in a resonant condition and therefore have a relatively large vibration amplitude. In turn, this will cause the plate to dissipate some of its energy through damping and radiation. The plate therefore acts as an absorber having maximum absorption characteristics at its fundamental frequency (and higher order modes) which will depend upon the geometry of the plate and its damping characteristics. In all practical cases this effect takes place at low frequencies, usually in the range 40 to 300 Hz. Particular care has to be taken, therefore, that any panels which may be hung in a room to improve reflection or diffusion are not designed in such a manner that they act as good low frequency absorbers and have a detrimental effect on the acoustics of the space.

If a panel is hung in front of a hard wall at a small distance from it, then the air space acts as a compliant element (spring) giving rise to a resonant system comprised of the panel's lumped mass and the air compliance. This is analogous to an electrical series resonance circuit made up of an inductive element L, represented by the acoustic mass M_A of the panel and a capacitance C, represented by the acoustic compliance C_A of the air space backing the panel (also see Section 3.2). This is shown in Figure 5.9.

The resonance frequency of such a circuit is (see Equation 3.6b)

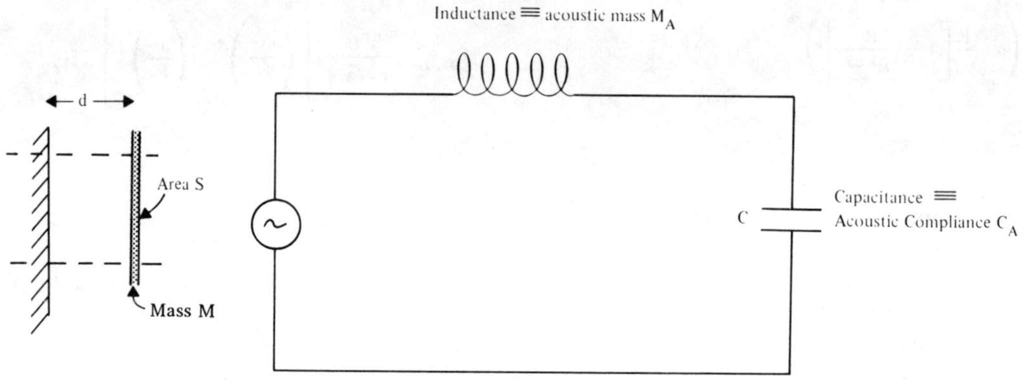

FIGURE 5.9 Equivalent electrical circuit for an acoustic panel resonator. Here $M_A = M/S^2$ and $C_A = d/\rho c^2$.

$$f_R = \frac{1}{2\pi}\sqrt{\frac{1}{LC}}\ \text{Hz.} \tag{5.14}$$

Using $L = M_A = M/S^2$ and $C = C_A = d/\rho c^2$, per unit area

where

M = mass of panel per unit area
d = depth of air space
S = area of panel
ρc = characteristic impedance of air
c = sonic velocity.

After using unit corrections we obtain

$$f_R = \frac{170}{\sqrt{Md}},\ \text{Hz} \tag{5.15}$$

where

f_R = resonance frequency, Hz
M = mass/ft^2
d = air space depth, in.

Hence, a thin panel of 1 lb/ft^2 weight placed a distance of 1 in. from a rigid wall would have a resonance frequency of 170 Hz. Figure 5.10 gives the resonance frequencies of such air-backed panels as a function of their surface weight and depth of air space. As in the case of a simply supported panel, a spaced panel absorbs energy through its internal viscous damping. Since its vibration amplitude is largest at resonance, its sound absorption is maximum at this frequency. Usually this absorption can be both further increased in magnitude and extended in its effective frequency range (i.e., giving a broader resonance peak) by including a porous sound absorbing material, such as a fiberglass blanket, in the air space contained by the panel. This effectively introduces damping into the resonant system. In some cases the absorbing material may be attached to the back of the panel itself. Figure 5.11 shows the effect of introducing a 1-in.-thick fibrous blanket into the 1¾-in. air space contained by a 1/8-in. plywood sheet. The change is quite significant at low frequencies in the region of the resonance peak where both the magnitude and width of the peak are increased. On the other hand, there is little effect at higher frequencies.

We can see from the above discussion that resonant panels can be very efficient low frequency absorbers. In addition, their nonperforated surfaces are durable and can be painted with no effect on their acoustical properties. It is important that this type of sound absorber be recognized, as such, in the design of an auditorium. Failure to do so, or to underestimate its effect, will lead to excessive low frequency absorption, and the room will have a relatively short reverberation time. The room will then be considered to be acoustically unbalanced and will lack warmth. Typical panel absorbers found in auditoria include: gypsum board partitions, wood paneling, windows, wood floors,

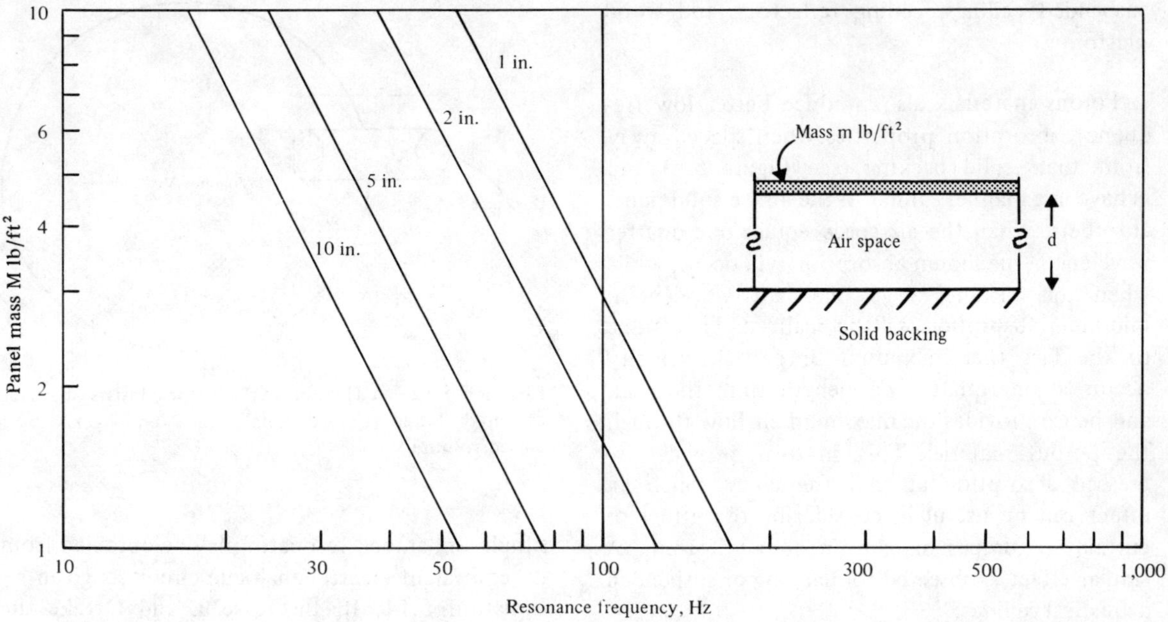

FIGURE 5.10 Resonance frequency of sound-absorbing flexible panels as a function of mass and air cavity depth, d.

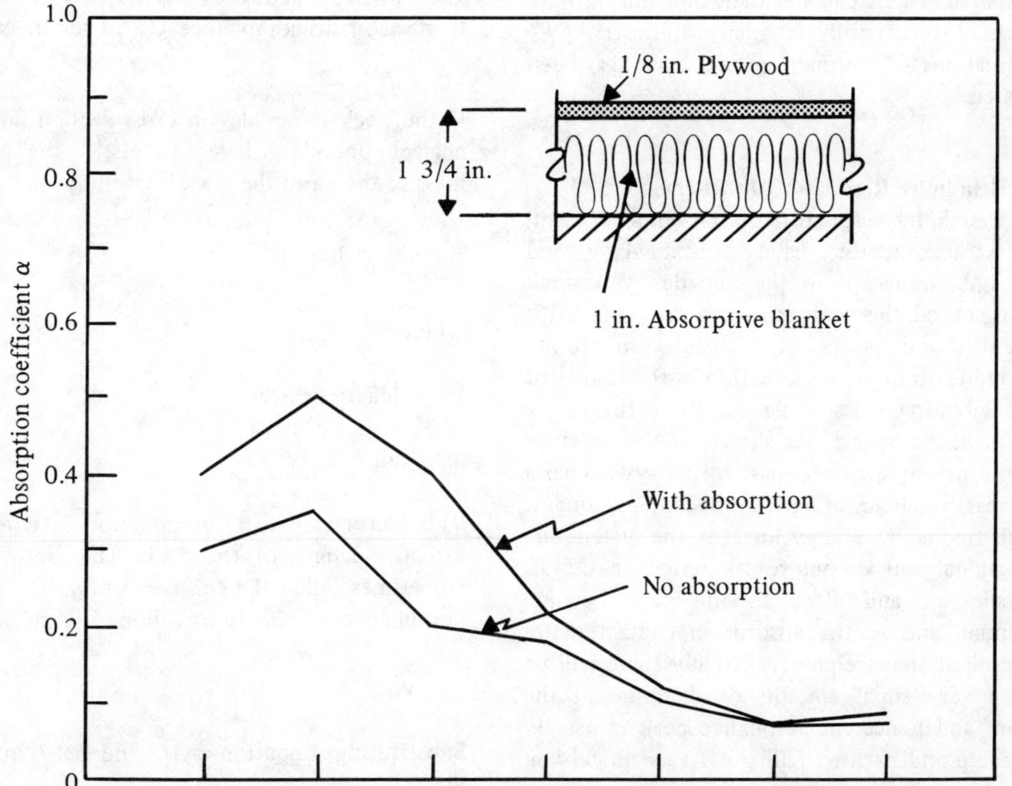

FIGURE 5.11 The effect on the absorption coefficient of placing a 1-in. thick sound-absorptive blanket in the air space (1 3/4-in. deep) behind a flexible 1/8-in. plywood panel.

suspended ceilings, ceiling reflectors, and wood platforms.

Porous materials also produce better low frequency absorption properties when spaced away from their solid backing (see Figure 5.7) and behave in a manner similar to the above solid panel absorbers. When the air space equals one quarter wavelength, maximum absorption will occur, while when this distance is a one half wavelength, minimum absorption will be realized. This is due to the fact that maximum air particle velocity occurs at one quarter wavelength from the wall, and hence provides the maximum air flow through the porous material. This, in turn, provides increased absorption at that frequency. Such an effect can be useful in considering the effect of curtains or drapes hanging in an auditorium. A similar effect is observed for hanging or suspended acoustical ceilings.[12]

In order to achieve well-balanced low frequency absorption, a selection of different size and thickness spaced panels can be used, and indeed have been used successfully in many auditoria.[13–16] Combinations of resonant panels have also been suggested.[17]

5.4.3 Helmholtz Resonator Absorbers

A Helmholtz resonator, in its simplest form, consists of an acoustic cavity contained by rigid walls and connected to the outside by a small opening called the neck as shown in Figure 5.12. Incident sound causes air molecules to vibrate back and forth in the neck section of the resonator like a vibrating mass while the air in the cavity behaves like a spring. As shown in the previous section, such an acoustic mass spring system has a particular frequency at which it becomes resonant. At this frequency, energy losses in the system due to frictional and viscous forces acting on the air molecules in and close to the neck become maximum, and so the absorption characteristics also peak at that frequency. Usually there will be only a very small amount of damping in the system, and hence the resonance peak is usually very sharp and narrow, falling off very quickly on each side of the resonance frequency. This is observed in practice for if we blow across the top of the neck of a bottle we hear a pure tone developed rather than a broad resonance.

As before, the resonance frequency of the simple undamped resonator may be obtained from its equivalent electric analogue circuit as given by Equation 5.14. In this case L and C take the following values:

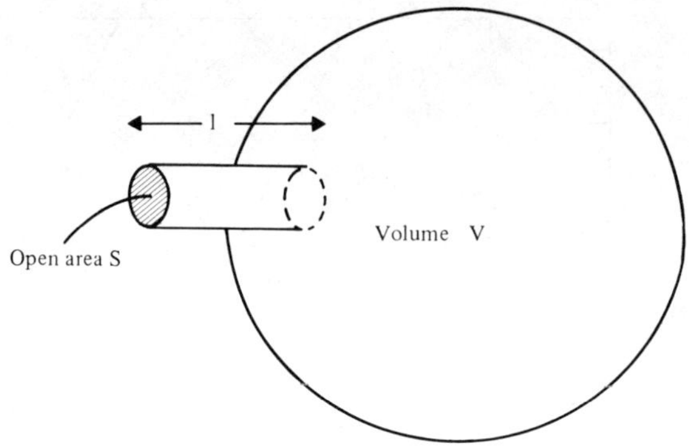

FIGURE 5.12 A Helmholtz resonator consists of a neck of length l and corss-sectional area S, and backed by a closed volume V.

L = effective acoustic mass M_A of air in neck;
C = acoustic compliance C_A of air in cavity.

If the neck is circular in cross section and if we neglect boundary layer effects,[18] the effective mass of the air in the neck is given by

$$M_A = \rho(1 + 1.7r)/\pi r^2, \tag{5.16}$$

where

l = length of neck
r = radius of neck
ρ_o = air density.

The factor (l + 1.7r) in Equation 5.16 gives the effective length of the neck. The factor 1.7r is sometimes called the *end correction.* The acoustic compliance of a cavity of volume V is given by

$$C_A = V/\rho c^2. \tag{5.17}$$

Substituting Equation 5.16 and 5.17 into 5.14 yields

$$f_R = \frac{c}{2\pi}\sqrt{\frac{S}{(1 + 1.7r)V}}, \text{ Hz}, \tag{5.18}$$

where

f_R = undamped resonance frequency, Hz
S = area of neck, ft^2
l = length of neck, ft
r = radius of neck, ft
V = cavity volume, ft^3
c = acoustic velocity, ft/sec.

Figure 5.13 gives a nomogram for the computation of f_R for a general Helmholtz resonator where the effective neck length is L in. Thus for circular holes, L = 1 + 1.7r, in.

Although Helmholtz type resonators could be built for any frequency, their size is such that they are used mainly for low frequencies in the region 50 to 400 Hz. Because of their sharp resonance peaks, undamped resonator absorbers have particularly selective absorption characteristics. Therefore, they are used primarily in situations where a particularly long reverberation time is observed at one frequency. This frequency may correspond to a well-excited low frequency room mode, and an undamped resonator absorber may be used to reduce this effect without changing the reverberation at other, even nearby, frequencies. They are also used in noise control applications where good low frequency sound absorption is required at a particular frequency. In this respect special Helmholtz resonators, constructed from hollow concrete blocks with an aperature or slit in their faces, are used in transformer rooms and electrical power stations to absorb the strong 120-Hz noise produced therein.

Figure 5.14 gives a nomogram for computing the resonance frequency for slotted 8-in. concrete blocks. Such resonators are not an entirely new idea and, in fact, were used in some churches built in Sweden almost 1,000 years ago. Several clay vases, often called *sound vases,* have been found in a number of old churches in Scandinavia, Russia, and France. These vases were located in the walls and ceilings of the churches forming resonator absorbers, the frequency depending on the size of

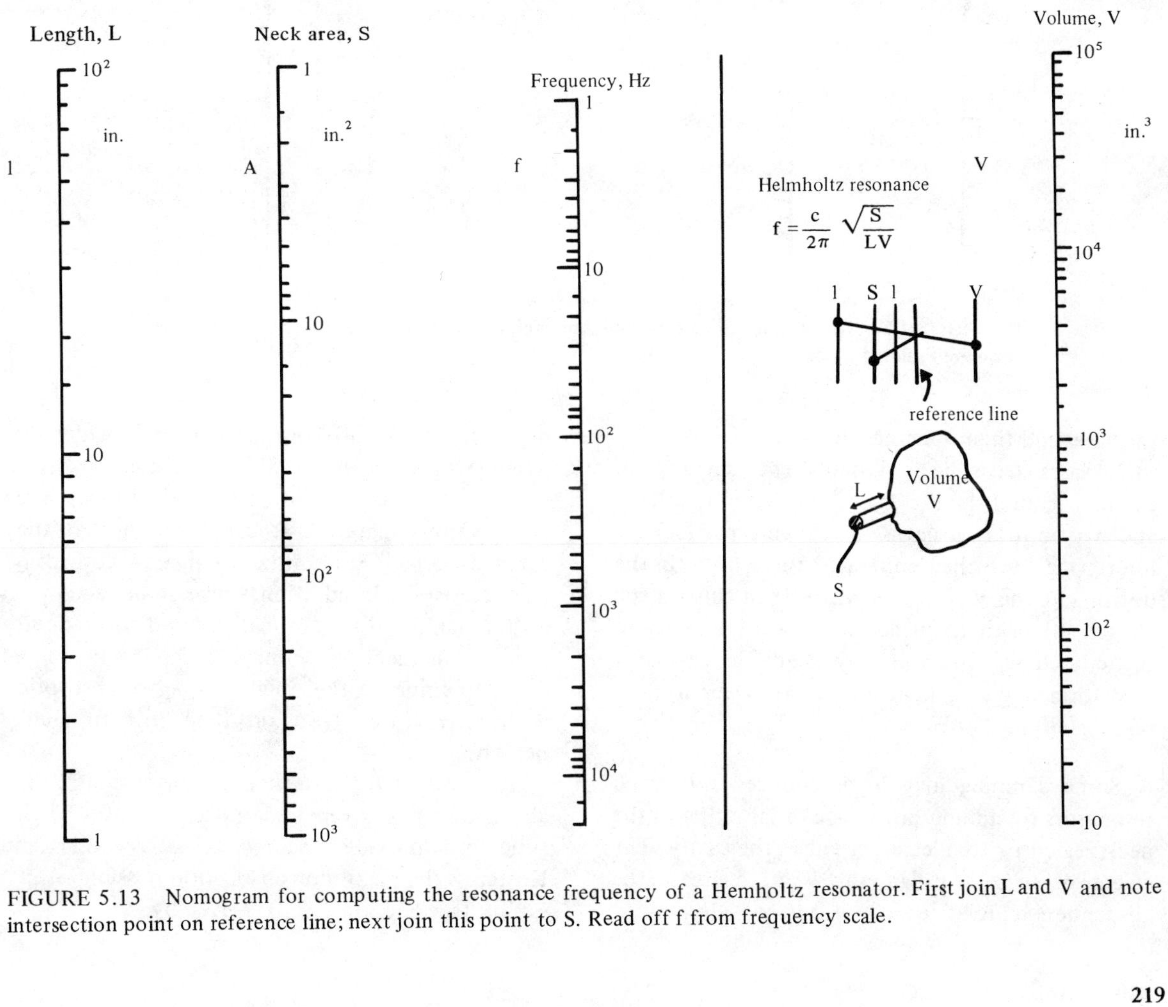

FIGURE 5.13 Nomogram for computing the resonance frequency of a Hemholtz resonator. First join L and V and note intersection point on reference line; next join this point to S. Read off f from frequency scale.

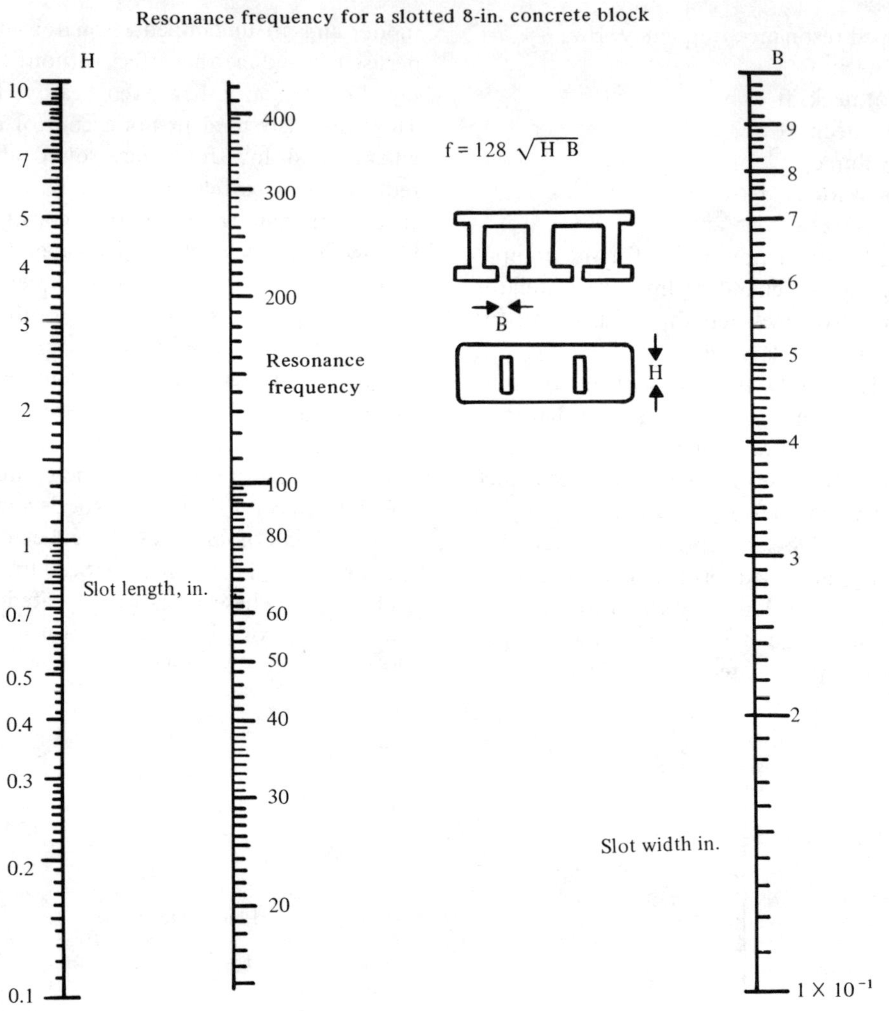

FIGURE 5.14 Nomogram for computing the Helmholtz frequency for slotted hollow concrete building block.

the vases and their openings. Recently a number of such vases have been found in a small 15th Century church at Pleterje,[19] between Ljubljana and Zagreb in Yugoslavia. These were particularly interesting since they contained three holes in the bottom of the vase which was turned toward the church, while the mouth faced toward a cavity in the church wall. In this way a double resonator was formed giving high absorption over a wide range of 80 to 250 Hz.

Some damping may be introduced into such resonators by adding porous material either in the neck region or to a lesser extent in the cavity. The effect of increased damping is to decrease the absorption value at resonance but considerably to broaden the absorption curve over a wider frequency range. Figure 5.15 shows measured absorption coefficients for slotted concrete blocks filled with porous material. It can be seen that they offer especially good low frequency absorption characteristics. In addition, they can be faced with a thin blanket of porous material and covered with perforated metal sheeting, as shown in Figure 5.16, to improve their high frequency absorption with only little effect on their low frequency performance.

It is useful to note that there is a limit to the absorption any given undamped resonator of this type can provide. According to Zwikker and Kosten,[2] the maximum absorption possible, A_{max} sabins, (see Section 1.6.2) is given by

$$A_{max} = 0.159\ (c/f_R)^2,\ \text{sabins} \quad (5.19)$$

where

c = sonic velocity, ft/sec
f_R = resonance frequency, Hz.

Therefore, if f_R = 150 Hz, for example, then we cannot expect to realize more than A_{max} = 8.5 sabins per unit. Equation 5.19 also shows that as the resonance frequency becomes lower, more absorption can be obtained from the resonator.

In certain circumstances, the performance of a Helmholtz type resonator can be drastically influenced by the effect of its surrounding space. This is particularly found in the case for such resonators mounted in a highly absorbing plane, when interference can take place between the sound radiated from the resonator and reflected from the absorbing plane.[20] In such circumstances special care has to be taken.

5.4.3.1 Perforated Panel Absorbers

As described above, single Helmholtz resonators have very selective absorption characteristics and are often expensive to construct and install. In addition, their main application lies at low frequencies. Perforated panels offer an extension to the single resonator absorber and provide a number of functional and economic advantages. When spaced away from a solid backing, a perforated panel effectively is made up of a large number of individual Helmholtz resonators, each consisting of a "neck," constituted by the perforation of the panel, and a shared air volume formed by the total volume of air enclosed by the panel and its backing. The perforations are usually holes or slots and, as with the single resonator, porous material may be included in the air space to introduce damping into the system. Perforated panels are mechanically durable and can be designed[12,21-23] to provide good broad-band

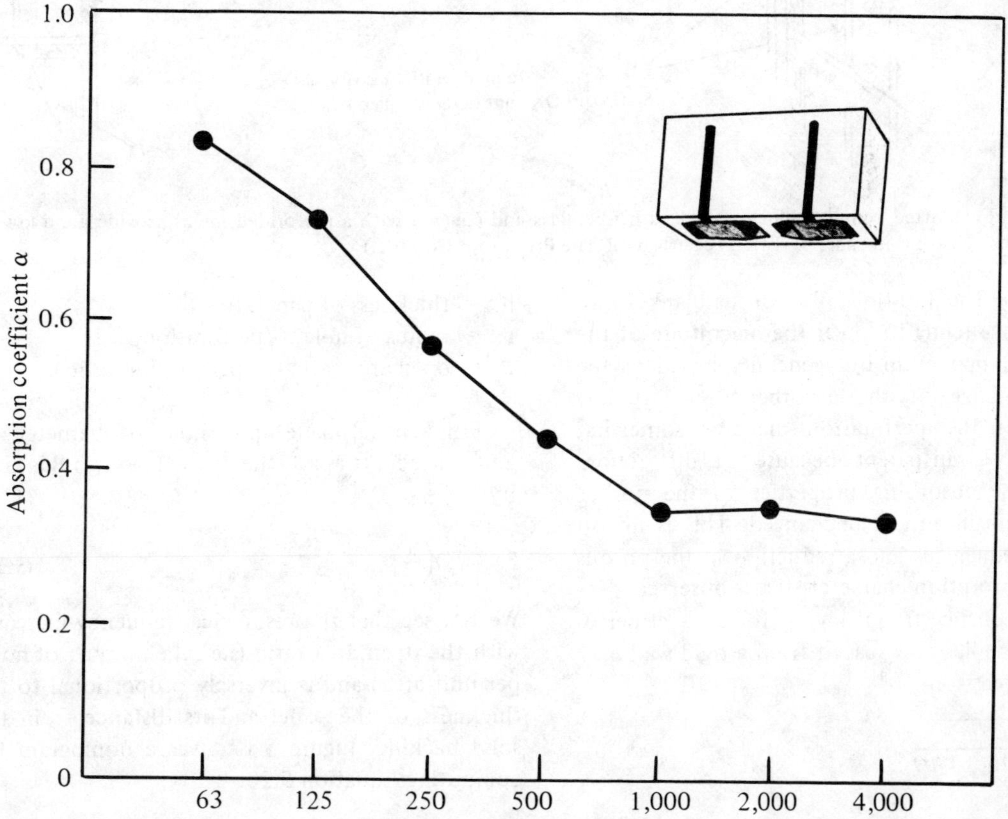

FIGURE 5.15 Absorption coefficients vs. frequency for a slotted 8-in. concrete block filled with an incombustible fibrous material. (Courtesy of The Proudfoot Co., Inc.)

Sound Absorption Coefficients

125 Hz	250 Hz	500 Hz	1,000 Hz	2,000 Hz	4,000 Hz	NCR Range
0.80	0.97	1.02	0.90	0.77	0.71	85–95

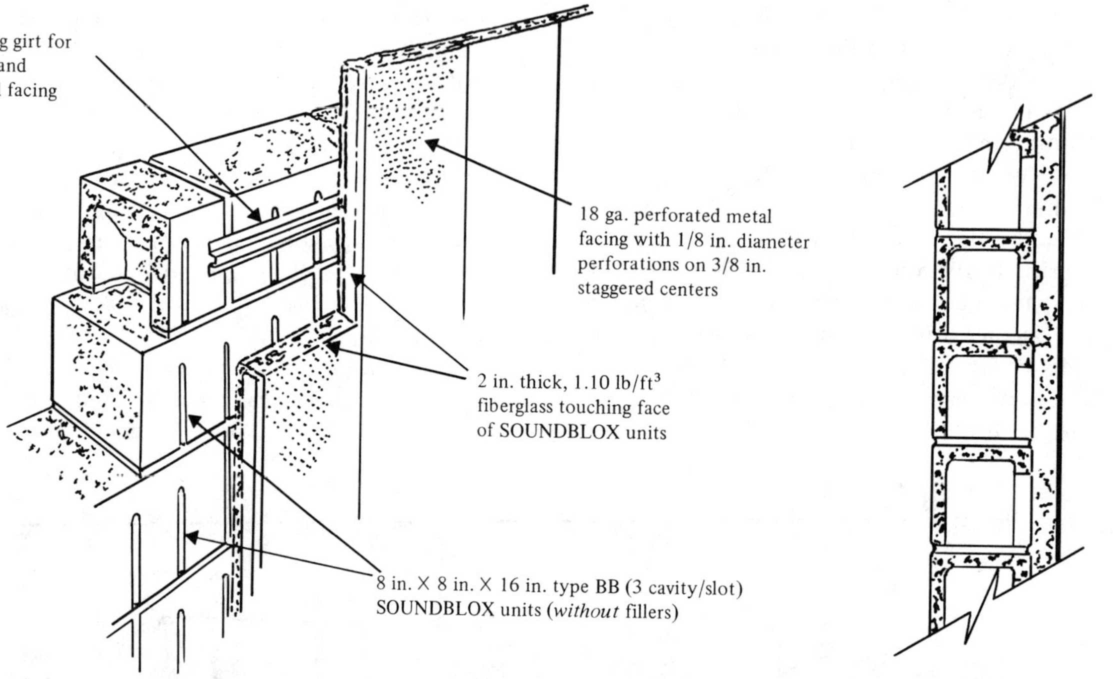

FIGURE 5.16 Slotted concrete blocks faced with fiberglass and covered with a perforated metal provide good low and high frequency absorption characteristics. (Courtesy of The Proudfoot Co., Inc.)

absorption. The addition of a porous blanket into the air space tends to lower the magnitude of the absorption maximum but generally broadens the effective range of the absorber.[24-26] At low frequencies the perforations become somewhat acoustically transparent because of diffraction, and so the absorbing properties of the porous blanket remain almost unchanged. This is not so at high frequencies and a reduction in the porous material absorption characteristics is observed.

The resonance frequency f_R for a panel perforated with holes and spaced from a rigid wall may be calculated from:

$$f_R = \frac{c}{2\pi}\sqrt{\frac{P}{D(h + 1.7r)}}, \text{ Hz}, \tag{5.20}$$

where

f_R = resonance frequency, Hz
c = sonic velocity, ft/sec
D = distance from wall, ft
h = thickness of panel, ft
r = radius of hole, – perforation, – ft
P = open area ratio (or perforation ratio).

For a panel made up of holes of diameter a ft and spaced s ft apart, the open area ratio P is given by

$$P = (\pi/4)\left(\frac{a}{s}\right)^2. \tag{5.21}$$

We can see that the resonance frequency increases with the open area ratio (i.e., the number of holes per unit area) and is inversely proportional to the thickness of the panel and its distance from the solid backing. Figure 5.17 gives a nomogram for computing Equation 5.20.

Example

A perforated panel with a 10% open area (P = 0.10) and thickness ¼-in. is installed 6 in. in front of a solid wall. If we neglect the end effect and put

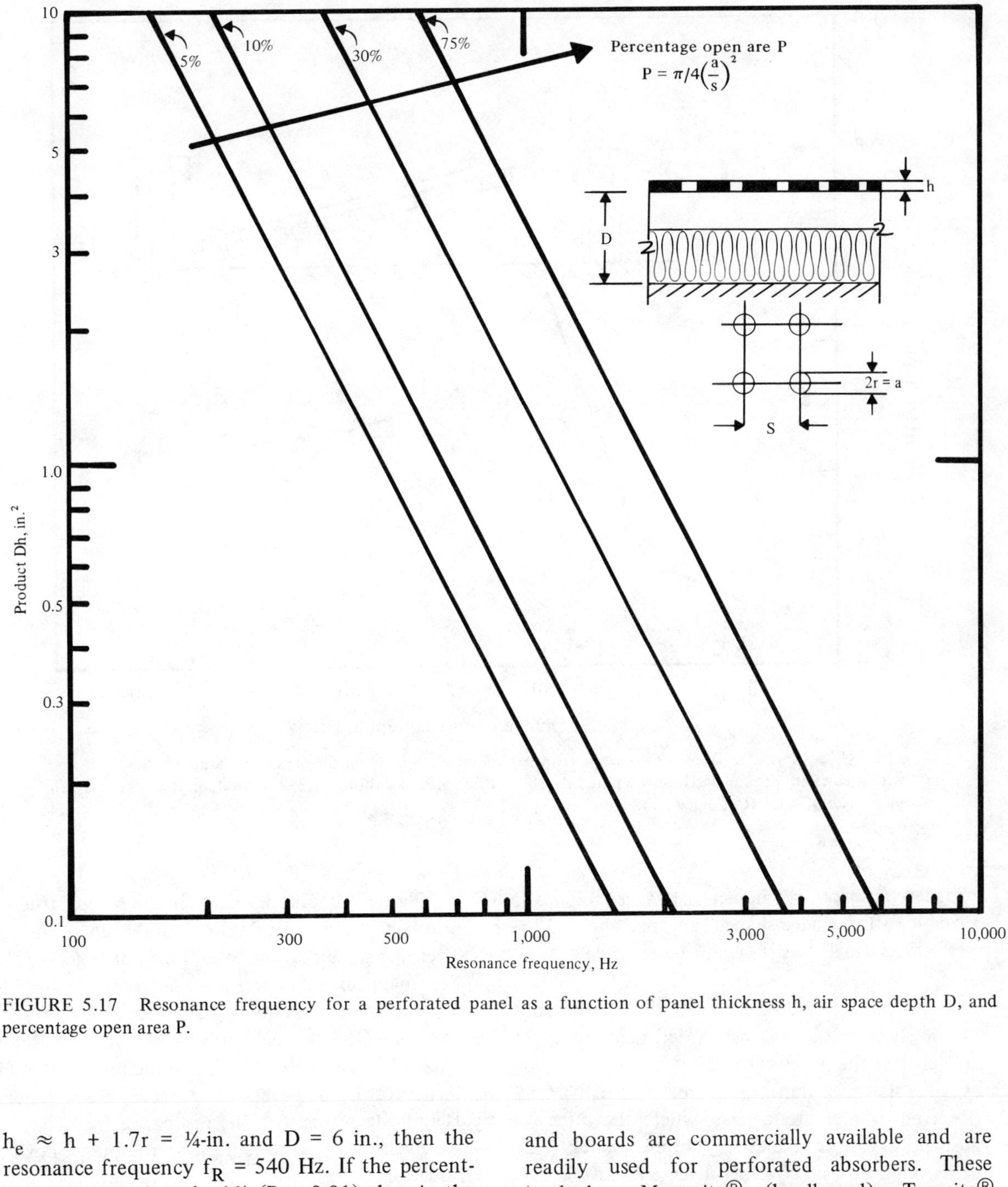

FIGURE 5.17 Resonance frequency for a perforated panel as a function of panel thickness h, air space depth D, and percentage open area P.

$h_e \approx h + 1.7r$ = ¼-in. and D = 6 in., then the resonance frequency f_R = 540 Hz. If the percentage open area is only 1% (P = 0.01) then in the same situation, f_R = 171 Hz.

In practical situations, air spaces up to 12 in. can be used with open areas ranging from 1 to 30% and thicknesses from 1/8 to 1 in. These particular restrictions would allow a resonance frequency range of 60 to 4,600 Hz. Many perforated panels and boards are commercially available and are readily used for perforated absorbers. These include: Masonite® (hardboard), Transite® (asbestos sheet), wood and plywood panels, and a variety of plane and corrugated metal facings. Some perforated sheets are available which consist of a number of different size perforations on one sheet. This can be useful to give broader absorption characteristics.

Another technique commonly used to achieve

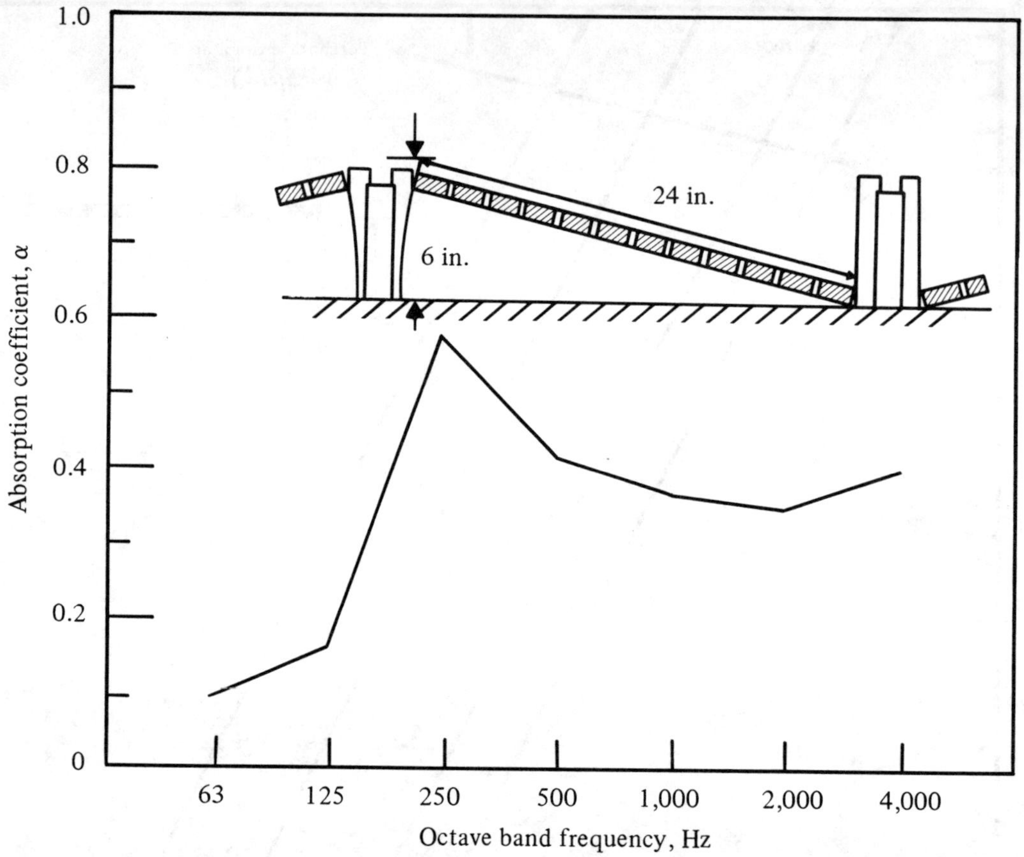

FIGURE 5.18 A variable air space perforated panel gives broader absorption characteristics than one with a constant air depth. Here the panel is 5/8 in. thick and has aholes of 3/8 in. in diameter spaced 1 3/8 in. on center. (See Reference 27).

broader absorption characteristics is to use a variable, often wedge-shaped, air space behind the perforated panel[27] as shown in Figure 5.18.

It is common practice to install a thin (2 mil) plastic sheet behind the perforated panel to cover and protect the porous absorbing material. As long as this sheet is thin, its effect is usually only observed at high frequencies where the effective absorption can be reduced by approximately 10% of its value with no covering sheet present.

5.4.3.2 Slit Absorbers

Another type of perforated resonator is the *slit* or *slat absorber.* These are made up of wooden battens fixed fairly close together and spaced at some distance from a solid backing. Porous material is usually introduced into the air cavity (see Figure 5.19). This type of resonator is fairly popular architecturally since it can be constructed in many ways, offering a wide range of design alternatives. Equation 5.20 does not hold for slat absorbers since the perforations are very long. For a detailed discussion of the properties of such absorbers the reader is advised to consult References 12, 28, and 29. The principle of the slit resonator is, however, the same as for a general Helmholtz resonator. The resonance frequency f_R is given[30] by the solution to the following equation:

$$f_R = \frac{c}{2\pi}\left[\frac{a}{S\left(d + \frac{2a}{\pi}\left[1.12 + \log_e\left\{\frac{c}{\pi a f_R}\right\}\right]\right)}\right]^{1/2} \quad (5.22)$$

where

f_R = resonance frequency, Hz
c = sonic velocity, ft/sec
a = slit width, ft
d = slit depth, ft
S = cross-sectional area of space behind slats formed by each slat (i.e., slat width W × air space D), ft^2.

A typical acoustical design procedure would be to choose a resonance frequency f_R and a slat of width W and thickness d and then determine the required value of air space from S = W × D for a chosen value of slit width a. For example, let us suppose that we want f_R = 200 Hz and that the resonator should be constructed of slats measuring 1 × 3 in. Therefore, we have W = 3 in. and d = 1 in. Since the slit width should be approximately 10 to 20% the width of the slat, let us try a slat spacing (i.e., slit width) of 0.3 in. Writing down all the known values in Equation 5.22 we have

f_R = 200 Hz
W = 3 in. = 0.25 ft
d = 1 in. = 0.083 ft
a = 0.3 in. = 0.025 ft
c = 1,100 ft/sec.

By rewriting Equation 5.22 we get

$$S = W \times D = \frac{a}{\left(\frac{2\pi f_R}{c}\right)^2 \left[d + \frac{2a}{\pi}\left(1.12 + \log_e\left[\frac{c}{\pi a f_R}\right]\right)\right]} \ ft^2, \quad (5.23)$$

so that

$$S = W \times D = \frac{0.025}{\left(\frac{2\pi \times 200}{1100}\right)^2 \left[0.083 + \left(\frac{2 \times 0.025}{\pi}\right)\left(1.12 + \log_e\left\{\frac{1100}{\pi \times 0.025 \times 200}\right\}\right)\right]},$$

W × D = 0.11 ft and since W = 0.25 ft, we require D = 5.5 in.

We see, therefore, that for the above example of a slit resonator, an air space of 5.5 in. is required to provide a resonance frequency of 200 Hz. If the calculation is repeated for a 100-Hz resonance frequency, it is found that an air space of some 20 in. is required.

5.4.4 Suspended Absorbers

This class of sound absorbers is known by several names including *suspended absorbers, functional absorbers,* and *space absorbers.* They generally refer to sound-absorbing objects and surfaces which can be easily suspended, either as single units or as a group of single units within a room. They are particularly useful in rooms in which it is difficult to find enough surface area to attach conventional acoustical-absorbing materials either through simple lack of available space or interference from other objects or mechanical services such as ducts and pipes, in the ceiling space. It is relatively easy and inexpensive to

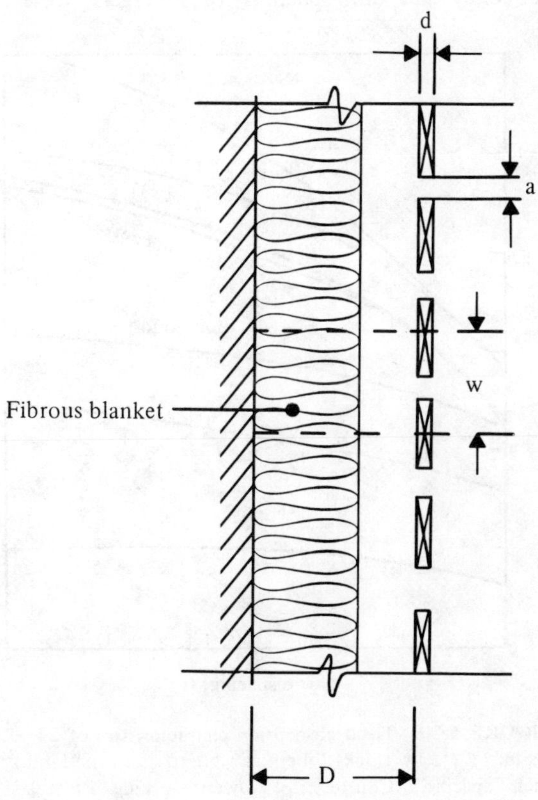

FIGURE 5.19 A slat type of resonator absorber.

install them, without interference to existing equipment. For this reason they are often used in noisy industrial installations such as assembly rooms or machine shops.

Functional absorbers are usually made from highly absorbing materials in the form of a variety of three-dimensional shapes, such as spheres, cones, double cones, cubes, and panels. These are usually filled with a porous absorbing material. Since sound waves fall on all their surfaces and because of diffraction, they are able to yield high values of effective absorption coefficients. It is, however, more usual to describe their absorption characteristics in terms of their total absorption in sabins per unit, as a function of octave band frequencies, rather than to assign an absorption coefficient to their surfaces. One also finds that when a group of functional absorbers are installed, the total absorption realized from the group depends upon the spacing between the individual units to a certain extent. Once a certain optimum spacing has been reached the effective absorption per unit does not increase. Figure 5.20 shows the variation of absorption per unit as a function of frequency and unit spacings for Fiberglas Noise-Stop Baffles®. These consist of 24 X 48 in. Fiberglass boards 1½ in. thick, and covered with 2 mil opaque plastic film, suspended end to end in rows from 2 to 7 ft apart.

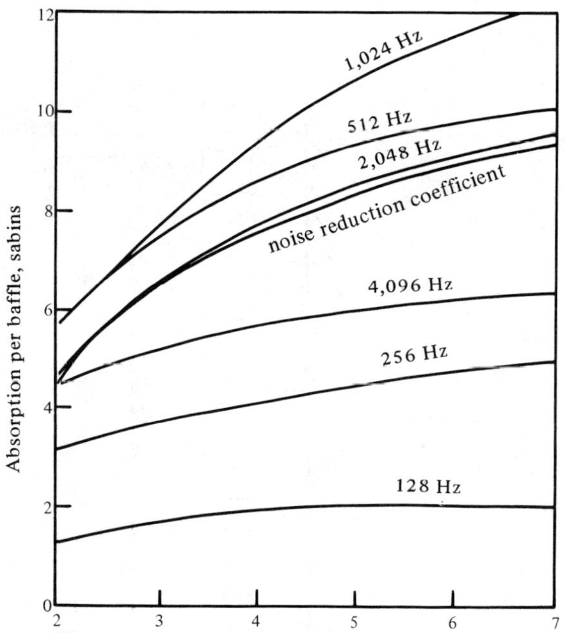

FIGURE 5.20 Total absorption characteristics of 24 X 48 in. (1 1/2 in. thick) Fiberglas® board as a function of their spacing. (Courtesy of Owens-Corning Fiberglas Corporation.)

For a more detailed discussion of functional absorbers the reader is advised to consult References 31 to 34.

5.4.5 Acoustical Spray-On Materials

These consist of a range of materials formed from mineral or synthetic organic fibers mixed with a binding agent to hold the fibrous content together in a porous manner and also to act as an adhesive. During the spraying application the fibrous material is mixed with a binding agent and water to produce a soft lightweight material of coarse surface texture with high sound-absorbing characteristics. Due to the nature of the binding agent used, the material may be easily applied directly to a wide number of surface types including wood, concrete, metal lath, steel, and galvanized metal. When sprayed onto a solid backing, this type of spray-on material usually exhibits good mid and high frequency absorption and, as shown in Figure 5.21, when applied to, for example, metal lath with an air space behind it, the material then also exhibits good low frequency absorption. As would be expected from previous discussions of porous absorbers (in Section 5.4.1), the absorption values increase with greater depth of application, especially at low frequencies. Spray-on depths of up to 2 in. are fairly common.

The absorption characteristics of such materials are very much dependent upon the amount and type of binding agent used and the way it is mixed during the spray-on process. If too much binder is used, then the material becomes too hard and therefore a poor absorber, while on the other hand, if too little binder is employed, the material will be prone to disintegration and since it will have a very low flow resistance, it will also not be a good absorber. Some products use two binders (for example, K13® by National Cellulose Corporation, Texas). One is impregnated into the fibrous material during manufacture, while the other is in liquid form and included during application. The fibrous material (containing its own adhesive) and the liquid adhesive are applied simultaneously to the surface using a special nozzle. The material is

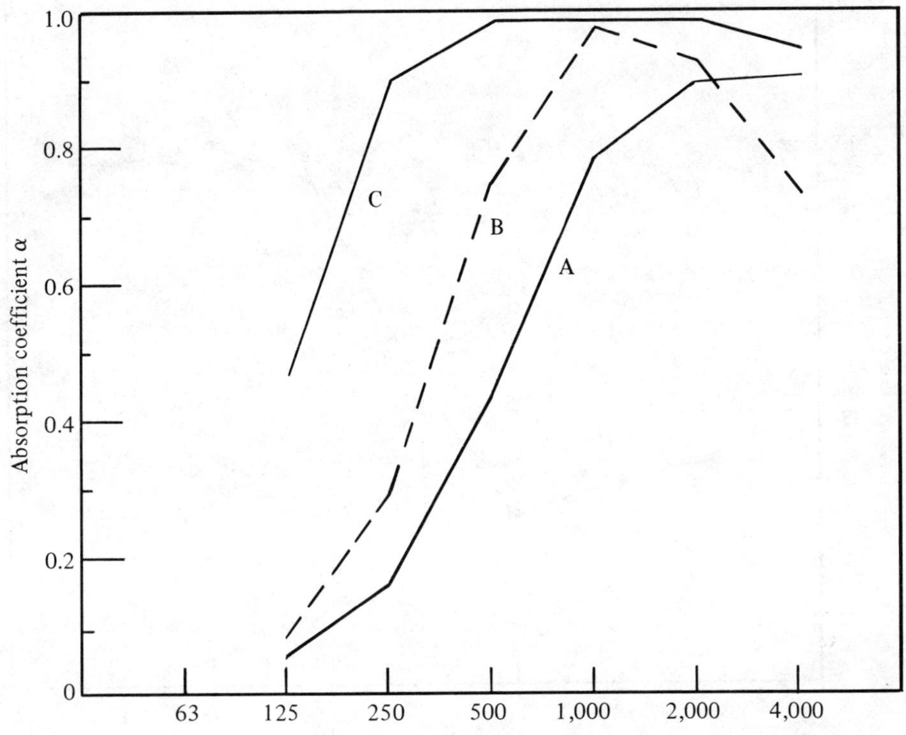

FIGURE 5.21 Absorption coefficients of a fibrous spray-on material known as K-13®. Curve A – 5/8 in. sprayed on solid backing; curve B – 1 in. on solid backing; curve C – 1 in. sprayed on metal lath. (Courtesy of the National Cellulose Corporation.)

particularly resistant to disintegration or shrinkage, and furthermore is fire resistant and posesses excellent thermal insulation properties. If visually acceptable, spray-on materials of this type can be successfully used as good broad-band absorbing materials in a variety of architectural spaces including schools, gymnasia, auditoria, shopping centers, pools, sports stadia, and in a variety of industrial applications such as machine shops and power plants. Disadvantages generally include the difficulty to clean and redecorate the material, although some manufacturers claim that their product can be spray painted without loss of acoustical performance. Such claims should be supported by data before proceeding to paint such a surface.

5.4.6 Acoustical Plaster

This term has been applied to a number of combinations of Vermiculite® and binder agents such as gypsum or lime. They are usually applied either to a plaster base or to concrete, and must have a solid backing. Because of this, acoustical plasters have poor low frequency absorption characteristics (also due to the thickness of application which rarely exceeds ½ in.). This can be slightly improved by application to metal lath. The material may be applied by a hand trowel or by machine. However, the latter tends to compact the material and give lower absorption characteristics. The surface of acoustical plasters can be sprayed with water-thinned emulsion paint without any significant loss of absorption, although brushed oil-based paints should never be used. Figure 5.22 shows typical sound absorption characteristics for a ½ in.-thick hand trowel applied acoustical plaster taken from a variety of published manufacturers data.

5.5 MEASUREMENT OF ABSORPTION COEFFICIENTS USING REVERBERANT ROOMS

As we have previously seen, in order to evaluate the acoustical performance of an enclosure, it is necessary to know the sound absorbing properties, as a function of frequency, of all the surfaces and objects within it. Although some approximate values for absorption coefficients and resonance

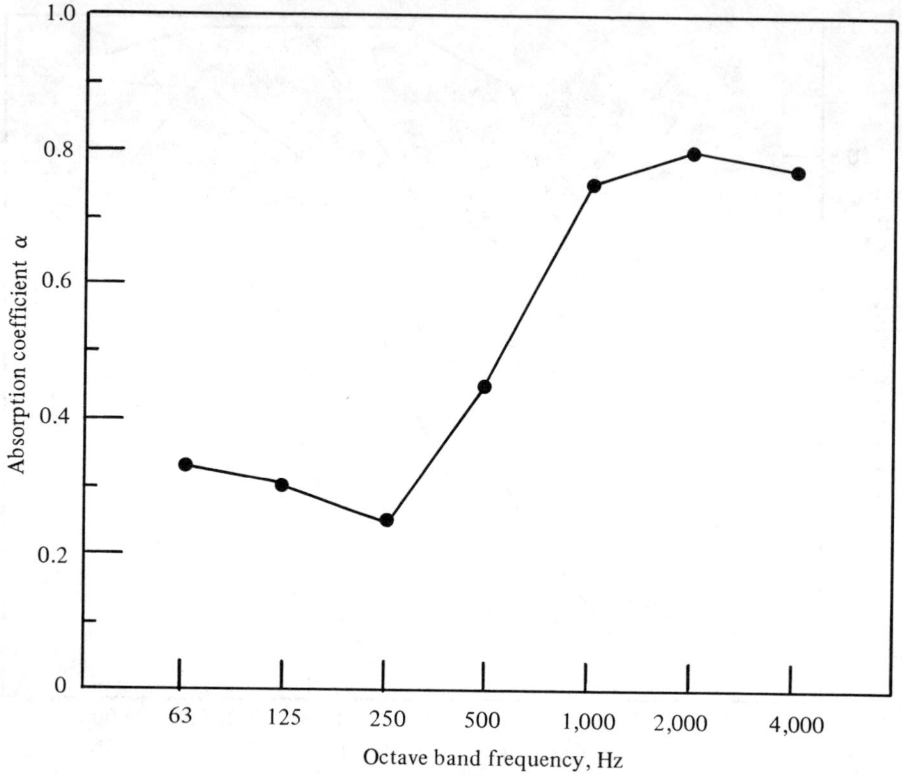

FIGURE 5.22 Absorption coefficient of a 1/2-in. thick acoustical plaster.

frequency characteristics can be estimated from the geometry, flow resistance, porosity, and other physical factors, it is clearly necessary to have actual measured values of absorption coefficients for a variety of materials and different constructions.

Since the reverberation time of a room depends upon the absorption present within it, this gives us a method to measure the absorption coefficient of a chosen material by observing the change in the reverberation time of the room caused by the introduction of the specimen. This method is particularly useful since large specimens of absorbing material can be measured which can be built and mounted in the same manner as they will be used in a real building, and therefore should be more representative than measurements made on small samples.

If the reverberation time of a large, empty, and highly reverberant room is T_1 sec and the sound field is completely diffuse, then from Equation 4.24 we can write

$$T_1 = \frac{0.049V}{A_1 + 4m_1 V}, \text{ sec,} \tag{5.24}$$

where

T_1 = reverberation time of empty room (no sample)

A_1 = total absorption area with empty walls. This is given by $A_1 = S_T\alpha_o$ where S_T is the total surface area of the room and α_o is the effective sabine absorption coefficient of the room surfaces

m_1 = effective energy attenuation constant of the air for the empty room

V = volume of room, ft^3.

Suppose that we now introduce a sample of absorbing material of surface area S, then at the same frequency the reverberation time will change to T_2 sec given by

$$T_2 = \frac{0.049V}{A_2 + 4m_2 V}, \text{ sec,} \tag{5.25}$$

where

T_2 = reverberation time with sample in place
A_2 = total absorption area of room and sample
m_2 = air attenuation constant with sample in room.

The change in the total effective absorption is $\Delta A = A_2 - A_1$ and is obtained from Equations 5.24 and 5.25 above as

$$\Delta A = 0.049V\left(\frac{1}{T_2} - \frac{1}{T_1}\right) - 4V(m_2 - m_1). \qquad (5.26)$$

If we assume that the room temperature and humidity have remained unchanged throughout the measurements, then we can put $m_1 = m_2$ so that

$$\Delta A = 0.049V\left(\frac{1}{T_2} - \frac{1}{T_1}\right). \qquad (5.27)$$

Since that change in total absorption area is due to the sample of area S ft^2 and effective absorption coefficient α_e covering a wall surface of absorption coefficient α_o, this increase is given by

$$\Delta A = S(\alpha_e - \alpha_o). \qquad (5.28)$$

Equations 5.27 and 5.28 therefore give the effective absorption coefficient of the sample as

$$\alpha_e = \alpha_o + \frac{0.049V}{S}\left(\frac{1}{T_2} - \frac{1}{T_1}\right). \qquad (5.29)$$

Again, the above equation assumes that T_1 and T_2 are measured at the same frequency. It must be realized that the values obtained for α_e are not a true measure of the energy absorption coefficient of the material for random incidence sound falling on the sample since many of the factors such as size and location of the sample influence the values obtained. We only really obtain a measure of the influence that the sample had on reverberation time of the test room. In some circumstances it could be argued, as we saw in Chapter 4, that the sabine reverberation equation does not give accurate results and that perhaps some other formula such as the Eyring equation should be used (see Equation 4.25). This question is discussed in detail by Powell.[35] If the Eyring equation (based on complete diffuseness) is used, then the effective absorption coefficient of the sample can be calculated from

$$\alpha_e = \alpha_o + \frac{S_T}{S}\left\{1 - \exp\left[\frac{-0.049V}{S_T}\left(\frac{1}{T_2} - \frac{1}{T_1}\right)\right]\right\}. \qquad (5.30)$$

However, values of absorption coefficients listed by all manufacturers and testing laboratories are sabine values calculated from Equation 5.29. Provided the test room has a large volume and long reverberation time, and the test sound field was diffuse, the values obtained are usually fairly accurate. Nevertheless, despite the fact that measurements made in different reverberation rooms on the same absorbing sample often yield different values of absorption coefficient, the results of such tests have been used successfully by architectural acousticians in many situations.

As we have seen in Chapter 4, a reverberation test room should have a large volume and long reverberation time to increase the acoustic modal density and decrease spatial variance in the sound field. To improve diffuseness, stationary or rotating vanes should be used. Since the effective absorption coefficient of a material depends to some extent upon its area,[36,37] due to diffraction at its edges, it is necessary for the test sample to be sufficiently large so that diffraction effects are minimized.

The American Society for Testing Materials (ASTM) has set a standard for such measurements under ASTM C423-66 "Sound Absorption of Acoustical Materials in Reverberation Rooms." This standard covers the acoustical performance of the testing reverberant room and the sample size and mounting, and governs the method for measuring the reverberation times. The room volume V ft^3 should be such that

$$V > 4\lambda_e^3, \qquad (5.31)$$

where λ_e is the longest wavelength used in the test. In addition, the smallest dimension of the room should exceed $2\lambda_e$ and the ratio of the largest to the smallest dimension should be less than 2:1. This therefore requires rooms to have volumes in excess of 10,000 ft^3 in order to operate at 100 Hz. The ASTM standard also specifies that the average absorption coefficient of the empty room at all test frequencies should not exceed 0.06, including the effect of air absorption. This requirement, together with the above volume restrictions, guarantees that the reverberation time of the whole usable frequency range (125 to 4,000 Hz, octave bands) should be greater than 3.6 sec – assuming a

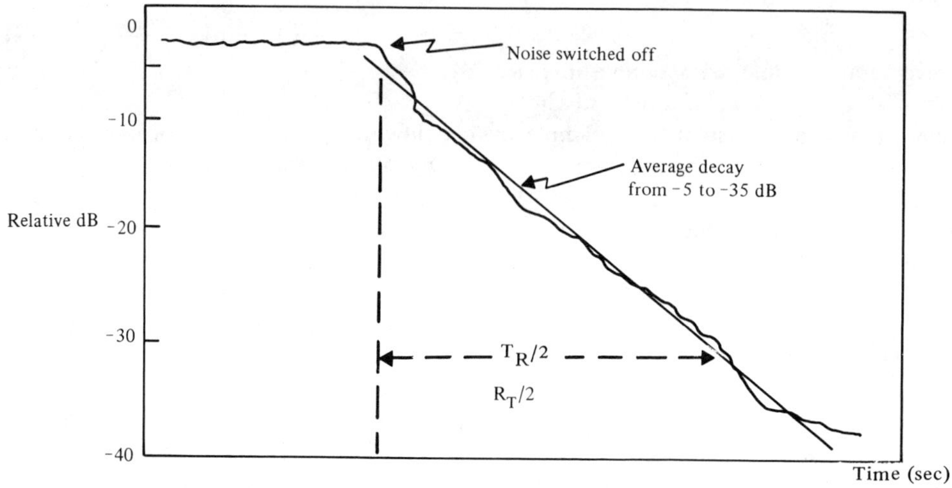

FIGURE 5.23 Logarithmic decay of sound in an enclosure after sound source is turned off. The reverberation time is taken as twice the time to decay from –5 to –35 dB.

2,400 ft^2 surface area. One third octave bands of white or pink noise are used to create the sound field in the room and the slope of the resultant sound pressure level decay curve should be measured over a decay range of not less than 30 dB, starting from at least 5 dB down from the beginning of the decay (see Figure 5.23). The test sample should be not less than 48 ft^2 although sample areas of 72 ft^2 (8 × 9 ft) are recommended. The measured absorption coefficients should be rounded off to the nearest multiple of 0.01, while the NRC (average of 250 to 2,000 Hz) should be rounded to a NRC of 0.05.

The International Standards Organization also has set a testing standard for reverberation room absorption testing under ISO Recommendation 354, "Measurement of Absorption Coefficients in a Reverberation Room." The size of the testing chamber is again governed by the lowest test frequency desired, but in order to measure down to 100 Hz the room should be in excess of 225 m^3 (7,945 ft^3) and also satisfy $V > l^3/6.8$, where l is the greatest diagonal length of the room. The sample size is restricted to rectangular plane samples of area from 10 m^2 to 12 m^2 (i.e., 108 to 130 ft^2). Strips of materials are to be avoided and the ratio of sample length to breadth must be not less than 0.7. The reverberation times of the test room itself must exceed 5 sec at 125, 250 and 500 Hz, 4.5 sec at 1,000 Hz, 3.5 sec at 2,000 Hz and 2 sec at 4,000 Hz. Either one third or one half octave bands of white noise may be used and warble tones may also be used, provided the frequency of deviation from the tone is ± 10% for $\leqslant$500 Hz and ± 50 Hz for $>$500 Hz – the modulation frequency being set at 6 Hz. The use of stationary or rotating vane diffusers is again recommended.

Tables of absorption coefficients are published by the Acoustical and Insulation Materials Association (AIMA) in the U.S. in their bulletin "Sound Absorption Coefficients of Architectural Acoustical Materials" for a variety of acoustical materials in which numerous mounting systems employed. Table 5.1 (see p. 251) lists absorption coefficients for a variety of commonly found architectural finishes. The AIMA tables are very often made up after receiving absorption results from a number of independent testing laboratories for the same sample of material. It is found that even when all of the requirements of either the ISO 354 or ASTM C423–66 standards are met, there can still be quite large differences observed between the values obtained for the absorption coefficients measured in different laboratories for the same sample. Koyasu[38] recently reported the results of such a so-called "round robin" series of tests performed by 13 research laboratories in Japan from 1965 to 1966. Figure 5.24 shows the result of this exercise from which quite large deviations in the measured values can be seen. These were found to be due to variations in the diffuseness of the testing rooms and to human error and bias in reading the decay curves. In many cases some of the error in the actual measurement of reverberation time could have been reduced[39] and a new

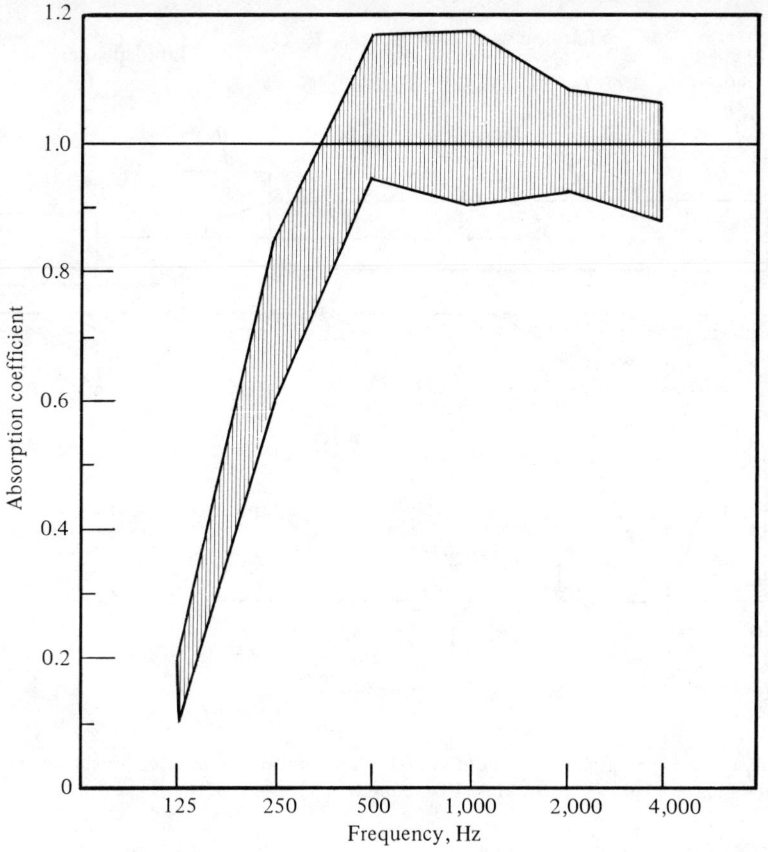

FIGURE 5.24 This shows the spread of values obtained for the absorption coefficients of a single 108-ft^2 sample as measured during a round robin series of tests. The values were measured by 13 independent laboratories. Note that experimental values of the Sabine coefficient can exceed unity. (From Koyasu, M., Investigations into the Precision Measurement of Sound Absorption Coefficients in a Reverberation Room, Proc. 6th Int. Congr. on Acoustics, Tokyo, 1968, paper E-5-8, pE 189.)

method has been proposed by Schroeder[40] using an integrated sound pulse technique which substantially reduces some of the experimental error involved in the standard method of measuring the sound pressure level vs. time decay curves.

5.6 ABSORPTION COEFFICIENT AND IMPEDANCE MEASUREMENT USING A STANDING WAVE TUBE

The standing wave or impedance tube provides a convenient laboratory method of measuring both the complex acoustic impedance and acoustic absorption coefficient of small samples for sound waves normally incident upon their surfaces. Because of the sample size restrictions, this method is only useful for porous materials and some resonant perforated absorbers. It cannot be used for resonant panels or large slit resonators since their absorption characteristics depend upon the size of the sample.

The sample material is placed in front of a heavy termination at one end of a rigid walled tube (Figure 5.25), while a loudspeaker is mounted along the axis at the other end. The loudspeaker is fed with pure tone signals (at one third octave center frequencies) and this then radiates plane waves down the tube towards the sample; as long as the diameter of the tube is small compared with the sound wavelength, transverse modes cannot be set up within the tube. The plane waves are then partially reflected by the sample and travel back along the tube towards the loudspeaker. This results in a longitudinal interference pattern consisting of standing waves set up within the tube. A microphone connected to an extension probe tube is moved along the axis to measure the variation in sound pressure within the

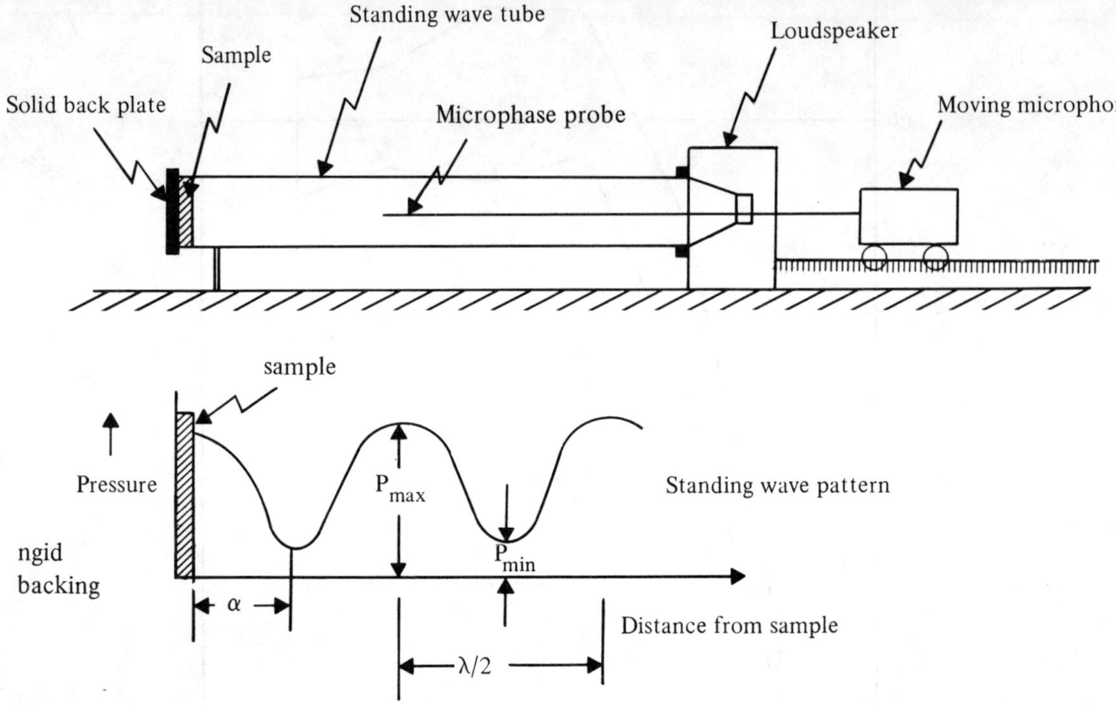

FIGURE 5.25 Standing wave apparatus used to determine both the normal incidence absorption coefficient and complex impedence of a sample of material placed at the end of the tube.

standing wave tube. From measurements of the ratio of maximum to minimum sound pressure within the tube, the absorption coefficient of the sample, at normal incidence, can be calculated. By measuring the distance between the sample and the first standing wave minumum sound pressure in addition, then the complex acoustic impedance of the specimen can also be calculated.

At any point the incident sound pressure p_i of a plane wave travelling down the tube from the loudspeaker to the specimen can be put in the form

$$p_i = A \cos \omega t, \tag{5.32}$$

where A is the amplitude and ω the angular frequency, $\omega = 2\pi f$. The reflected wave, at the same point in the tube, has a pressure p_r given in terms of its amplitude B and distance α from the sample surface as

$$p_r = B \cos \omega \left(t - \frac{2\alpha}{c}\right). \tag{5.33}$$

The total sound pressure p_T at a distance α from the sample is therefore,

$$p_T = p_i + p_r = A \cos \omega t + B \cos \omega\left(t - \frac{2\alpha}{c}\right). \tag{5.34}$$

The maximum pressure will therefore occur when $\alpha = \lambda/_2$ and will be equal to

$$P_{max} = (A + B) \cos \omega t, \tag{5.35}$$

and the minimum pressure will occur when $\alpha = \lambda/4$ to give

$$P_{min} = (A - B) \cos \omega t. \tag{5.36}$$

The Standing Wave Ratio, n, is defined as the ratio of maximum to minimum sound pressures within the tube so that

$$n = \frac{A + B}{A - B}. \tag{5.37}$$

Since, by definition, the reflection coefficient $r = B/A$ we have

$$r = \frac{n - 1}{n + 1}. \tag{5.38}$$

Using the relation $\alpha = 1 - r^2$ gives the sound absorption coefficient as

$$\alpha = \frac{4n}{(n + 1)^2}. \tag{5.39}$$

Hence, by measuring the standing wave ratio n, one can directly calculate the normal incidence absorption coefficient. We can see from Equation 5.39 above that if $n = \infty$ (i.e., A = B), then $\alpha = 0$, and if $n = 1$ (i.e., $B = 0$), then $\alpha = 1$ as we would expect.

It is found that the normal incidence absorption coefficient values measured in an impedance tube are generally lower than the random incidence values obtained from the reverberation room method. At low frequencies the difference is only slight while at high frequencies the tube values are generally 50% lower than those measured in a room.

It is important to ensure that the physical restrictions on the size of the tube are obeyed. These are namely that the length of the tube shall exceed $\lambda/4$ and that the diameter of the tube shall not exceed $0.58\,\lambda$ (to ensure plane waves, and therefore no transverse waves, in the tube). Hence a 4-in. (10 cm) diameter, 3-ft long tube would have a useful range of 90 to 1,800 Hz. In order to make measurements over the range of 90 to 6,000 Hz, 2 different size tubes are required; a smaller one for high frequencies and a larger one for low frequencies.

The standing wave tube can also be used to measure the parts of the complex acoustic impedance $z_o = r_n + jx_n$ as defined by Equations 5.7 and 5.8. It is easy to show that the complex impedance of the sample can again be written in terms of the reflection coefficient r and the position of the first minimum from the sample as

$$z_o = \rho c \left[\frac{1 + r\,e^{i\theta}}{1 - r\,e^{i\theta}}\right] . \qquad (5.40)$$

where

θ = phase angle = $2kd - \pi$
k = wave number $2\pi/\lambda$
d = distance from specimen to first minimum
r = reflection coefficient given by Equation 5.38.

Once r and d are determined, the resistive and reactive components (r_n and x_n) of the complex impedance may be determined from Equation 5.40. The most convenient method of solution is from a graphical solution using a *Smith Chart* as shown in Figure 5.26. This example illustrated is for $\theta = 30°$ and $r = 0.5$. Using the scale for r at the bottom of the chart, a length corresponding to this value of r is marked off along the radius of the major circle at $\theta = 30°$ to give the real and imaginary parts of the impedance as $r_n = 1.9\,\rho_o c$ and $x_n = 1.2\,\rho c$.

It is also interesting to note the normal incidence relationship between acoustic impedance and absorption coefficient as follows[3]

$$\alpha_o = \frac{4r_n\rho c}{(r_n + \rho c)^2 + x_n^{\,2}} . \qquad (5.41)$$

Dubout and Daven[42] discuss the relationship between the statistical absorption coefficient and complex impedance and the reader is directed to that work for more in-depth study.

Although the above methods for using a standing wave tube to measure both the absorption coefficient and complex impedance are fairly accurate, a number of refinements in the methods have been suggested to include the effect of viscous losses at the walls of the tube and in the gas within the tube. These are described in detail in References 43 to 46.

5.7 GENERAL REQUIREMENTS FOR ACOUSTICAL DESIGN OF ROOMS

The architecture of spaces used for public debate or entertainment has always concerned itself with the problem of providing "adequate acoustics." This has meant that rooms designed for speech should be able to provide the listener with a high degree of intelligibility throughout the room, while rooms designed for music should be able to provide both the player and the listener with optimum sound enrichment and blending. Furthermore, a room which exhibits good acoustical qualities for speech does not necessarily meet the requirements for satisfactory musical appreciation and vice versa.

Such architectural acoustics described above are by no means a new problem. In fact, it is fascinating to observe the remarkable acoustics of the best of the classical open air Greek theaters constructed over 2,000 years ago. Superb speech intelligibility was attainable over large distances and these open theaters could also be used for solo or unison singing, chanting, and for solo musical instruments. The theater at Epidaraus in Greece,

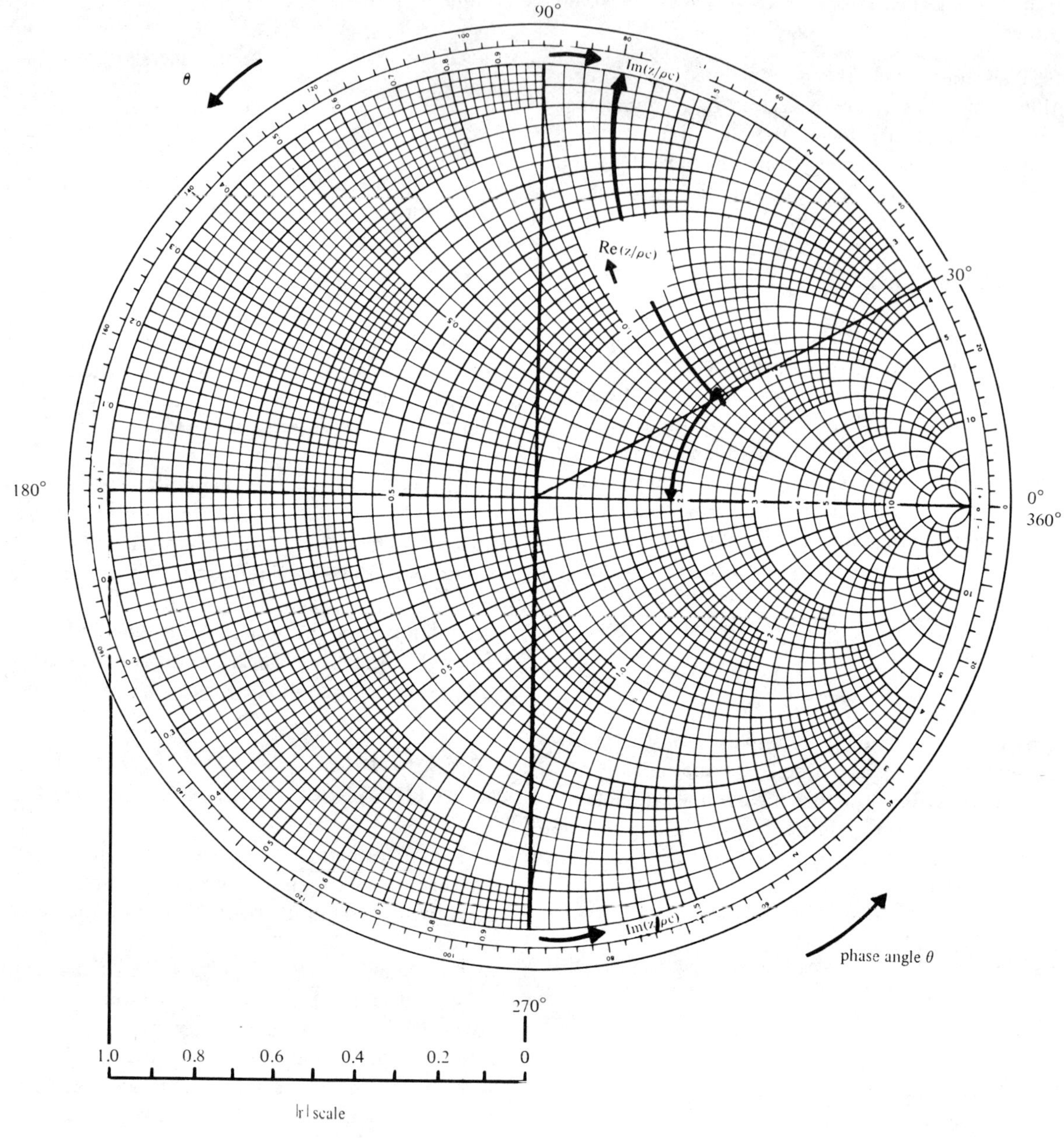

FIGURE 5.26 Use of Smith Chart to determine real and imaginary parts of complex impedence from |r| and θ.

designed in the middle of the 4th Century B.C., is one of the finest of all the Greek or Roman open air theaters[47,48] being able to accommodate an audience of 14,000. The theater is 387 ft in diameter with the orchestra section being a complete circle of 67 ft diameter. Studies of this theater have shown that the major factor influencing the acoustics is the perfection in the geometric location of the seating relative to the stage and performance area. This results in all of the audience receiving many early reflected sounds (i.e., sounds arriving within 50 msecs of the direct sound) scattered by both the stone seats and from the heads of the audience. As we shall see later, such short time delay reflections substantially aid in the speech intelligibility observed. Although it is most probable that the Greek architects designed their theaters primarily with visual sight-line requirements in mind and knew very little of the acoustic theory involved, the results were excellent outdoor theaters for speech. Indeed as we shall see, these early theaters possessed (even if through good luck rather than design) many of the qualities that the modern acoustician strives to create in theaters today.

The designer of a listening room, whether for speech or music, must attempt to synthesize many often conflicting requirements for audience comfort and safety, good vision, and good hearing, and at the same time produce a space that is aesthetically pleasing. In rooms such as concert halls, the acoustical requirements very much control the form of the space in terms of the size, volume, shape, audience layout, and surface finishes. In order to achieve excellent, or at least very good, acoustics, the designer must fully appreciate and understand the physical processes involved in the propagation and dissipation of both steady state and transient sounds within the room. In addition, it is absolutely necessary to understand the mechanisms of hearing and aural perception. This is not to say that great concert halls have not been or cannot be designed without such knowledge. Such knowledge is required, however, if one is to have confidence in one's design (acoustically), rather than to wait until the first performance and hope for the best.

Let us now examine the requirements of rooms which are to be used for either speech or music reception or both. These are:

(a) **Adequate loudness*** – The sound arriving at the listeners must be of sufficient intensity so that they can hear, understand, and appreciate the sound.

(b) **Uniform sound distribution** – Each member of the audience should be subjected to the same sound intensity independent of seating location. This requirement eliminates so-called hot and dead spots in the room.

(c) **Optimum reverberation** – In order to understand speech fully, each component of the speech must be heard by the listener. If the room has a long reverberation time, then the speech components overlap and a loss of intelligibility results. Similarly for music enjoyment and appreciation, a certain amount of reverberation is required in order to obtain the quality and blend of the music. It is, however, much more difficult to describe the acoustical qualities of a room used for music since many are subjective and therefore cannot be described in measurable physical quantities. Optimum reverberation is not only required for good subjective reception but also for an efficient performance.

(d) **Freedom from acoustical defects** – Many physical characteristics of listening rooms have a direct adverse effect on the quality of the sound received by the listener. These are generally known as *acoustical defects* and include: resonance, distortion, flutter echoes, long delayed reflections, focusing, and acoustical shadows.

(e) **Freedom from excessive background noise** – It is clearly important that the sound received by a listener should not be mixed with detectable exterior or interior unwanted noises, such as traffic, rail or aircraft noise from outside the building, or by mechanical system noise within the building. Such sounds may completely ruin what may otherwise be a room with excellent acoustics by distracting the listener's attention and by masking the wanted sound to produce lower intelligibility of speech and poor subjective reception of music.

Let us now consider the implications that the above requirements have on the overall design of a room used for speech or music reception.

5.7.1 Loudness – Its Effect on Room Shape and Volume

It is essential that every member of an audience receive adequate and uniform loudness. Since the intensity or sound pressure level of the direct sound propagating from a source decays with distance according to the inverse square law, a

*The terms *loudness* and *loudness level* used throughout Chapter 5 are used in a qualitative manner and do not have the strict scientific meanings given in Section 2.3.1. The qualitative use of these terms is common practice in architectural acoustics.

point is reached beyond which audience members cannot be located, in order that they should receive signals of sufficient loudness. By sufficient loudness, it is meant that the direct signal must be loud enough to be heard and understood above the reverberant and background noise in the room. The lowest allowable signal level that the most distant audience member should receive must probably be not less than about 45 dB (although it depends upon the background noise level). This means that if the sound source is an unamplified human voice then the maximum distance from the source should be around 90 ft. For a full orchestra, since the sound power output is increased, distances up to 160 ft can be allowed. Many people, however, feel most comfortable with the loudness in concert halls at distances of between 60 and 70 ft from the orchestra. If closer than this, the audience experiences sound which becomes too loud, and if further away, then it is not loud enough for aural comfort.

This places immediate restrictions upon the audience layout. In order to get as many people as close as possible to the stage, several floor plan types have been proposed, ranging from conventional rectangular to fan-shaped rooms. In the case of a fan-shaped floor plan, more people can be positioned within a given distance from the source than in a rectangular room of the same floor area. The angle subtended by the apex of such rooms, however, should not exceed 120°, otherwise the front seats at the sides of the room will be affected by the decrease in the voice power level because of its horizontal directivity pattern especially at high frequencies. For rectangular rooms, similar considerations for the front seats limit the length-to-width ratio to between 2:1 and $\sqrt{2}$:1.

If the complete audience cannot be seated on one floor because of the restrictions imposed by the inverse square law, then more floors (i.e., balconies) have to be added at the rear of the theater. These serve the dual purpose of bringing the audience closer to the stage and also providing better sight-lines. It is not unusual to find concert halls with two balconies – again each constructed to provide increased vision and loudness for the audience. The architect of such large acoustical spaces must remember, however, that the balcony must not degrade the acoustics of the rear section of the main floor by shielding that portion of the audience from useful reflections caused by the ceiling (see Figure 5.27). As a rule of thumb, the depth of the balcony should not exceed the open height of the balcony.

The requirements for good vision usually go hand in hand with those for good direct sound reception from the stage or speaker.* It is therefore essential to provide either a raised sound source (speaker) or raked seating in all but the smallest (less than 1,000 ft^2 floor area) rooms used for speech or music reception. Providing clear sight-lines to each audience member not only improves his visual and aural perception, but tends to reduce the effective total absorption in the room due to sound passing over the heads of the audience at grazing incidence. In order to ensure adequate visual sight-lines a minimum clearance of 4 in. is required between the sight-lines from the one row to the next. This is based on the fact that the eyes are approximately 3 in. below the top of the head. The first few rows of audience can usually be level (depending upon the relative elevation of the sound source) and still maintain a good line for both sight and sound. As shown in Figure 5.28, in order to maintain a sight-line clearance of h feet for level seating where the eye level is H_e feet above the ground, the maximum distance from the source is given by

$$d = \frac{s}{h}(H_s - H_e) \tag{5.42}$$

where

d = maximum distance from source, ft
H_s = height of source above ground, ft
H_e = eye height above ground level, ft
s = spacing of adjacent seating rows, ft
h = sight-line clearance required, ft.

By taking an average eye level of $H_e = 44$ in. $= \frac{44}{12}$ ft, and h = 4 in. = 1/3 ft we obtain:

$$d = s(3H_s - 11) \text{ ft.} \tag{5.43}$$

For example, if the speaker's lips are 5 ft above the ground and the seat spacing is 3 ft, then the maximum distance for level seating should not exceed 12 ft. However, if the speaker stands on a small platform so as to raise his lips to 7 ft, then

*Since both visual and acoustical requirements in an auditorium affect its geometry, these two requirements cannot be separated. Hence, several paragraphs following in Section 5.7.1 are now devoted to providing adequate visual sight-lines.

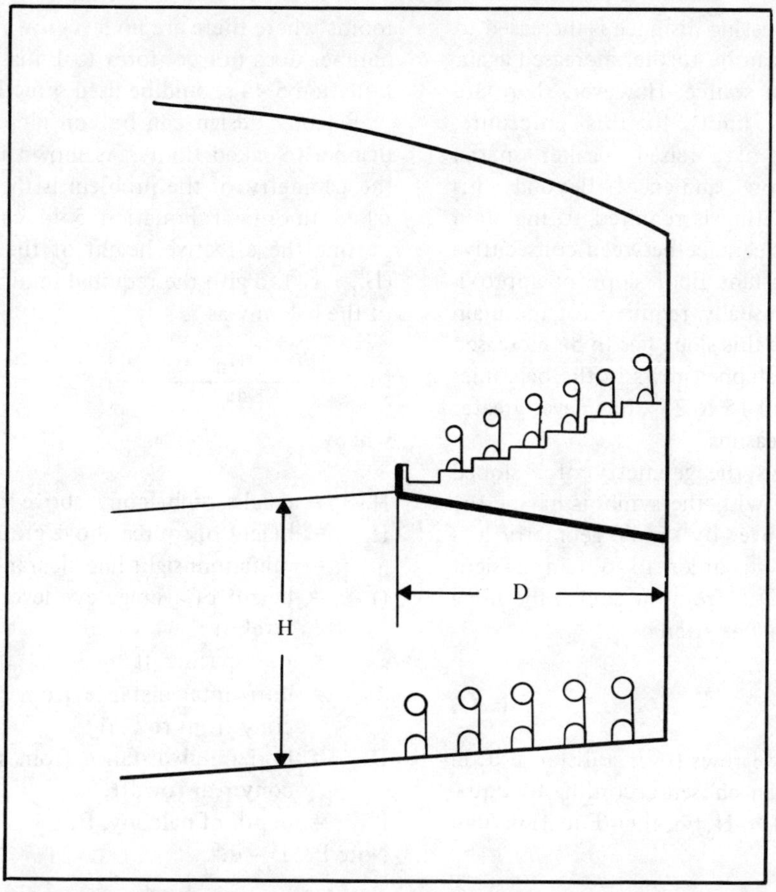

FIGURE 5.27 A balcony in an auditorium used for music should be positioned so that its depth D does not exceed its opening height H. For speech and opera, the depth may be larger, but must not exceed twice the height. (From Beranek, L. L., *Music, Acoustics and Architecture,* John Wiley & Sons, New York, 1962. With permission.)

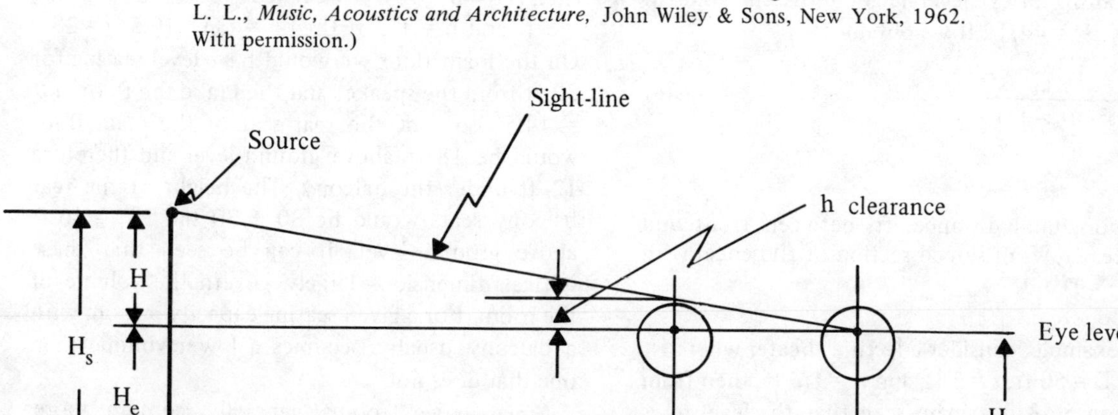

FIGURE 5.28 Sight-line requirements dictate a minimum clearance of h ft between the eye level and top of the head between members of the audience on consecutive rows. This requirement can only be fulfilled for up to a maximum distance from the speaker if the audience is seated on a horizontal plane.

the maximum level seating distance is increased to 30 ft. This distance can be further increased again by raising the sound source. However, there are many very obvious limits to this procedure, especially the effect of a raised speaker on the neck-tilted front row audience! Beyond this distance d, raised seating is required to maintain the 4 in. minimum clearance between consecutive seating rows. A constant floor slope of approximately 5 to 15° is usually required for the main floor audience, while this slope has to be increased through the use of stepped rows in the balconies to achieve gradients of 15 to 25°, but never greater than 30° for safety reasons.

Figure 5.29 shows the geometry of a sloped main floor audience with the symbols having the same meaning as above. By simple geometry it is easy to show that in order to obtain a sight clearance of h ft at D ft from the source the slope must not be less than θ as given by

$$\tan\theta = \frac{hD - s(H_s - H_e)}{ds} \tag{5.44}$$

If the number of level rows (over a distance d, in Figure 5.29) has been chosen according to Equation 5.42 as $d = s(H_s - H_e)/h$, then Equation 5.44 becomes

$$\tan\theta = \frac{hD - s(H_s - H_e)}{s^2 (H_s - H_e)/h}. \tag{5.45}$$

By choosing an eye level height above the gradient level of $H_e = 44/12$ ft we obtain

$$\tan\theta = \frac{hp}{s^2(3H_s - 11)}, \tag{5.46}$$

where

p = horizontal distance, ft, between front and rear rows of sloped section of audience, with $p = D - d$.

As an example, consider a lecture theater where $H_s = 5$ ft, $D = 50$ ft, $s = 3$ ft, and $h = 1/3$ ft, then from Equation 5.43 it can be seen that the level rows shoud extend for $d = 12$ ft away from the speaker and so $p = 50 - 12 = 38$ ft. From Equation 5.46 the required slope of the raked audience is obtained as $\theta = \tan^{-1}(0.34) = 19°$. Consequently, the height of the last row above ground level will be 38 tan 19° = 13.4 ft. If we had taken $H_s = 7$ ft then $d = 30$ ft and $p = 20$ ft so that $\theta = 4°$ and the rear row height would have been reduced to 1.5 ft. In rooms where there are no level rows or where their number does not conform to Equation 5.43, then Equation 5.44 should be used directly.

Balcony design can be considered in a similar manner to raked floors. As shown in Figure 5.30, the geometry of the problem is the same as for a raked floor and Equation 5.44 can be used by putting the effective height of the source $H_s' = -(H_B - H_s)$ to give the required minimum gradient β of the balcony as

$$\tan\beta = \frac{hD + s(H_B - H_s + H_e)}{ds}, \tag{5.47}$$

where

H_B = height of balcony above ground level, ft
H_s = height of source above ground level, ft
h = minimum sight line clearance, ft
H_e = height of average eye level above balcony level, ft
s = row spacing, ft
d = horizontal distance from speaker to balcony front row, ft
D = horizontal distance from speaker to balcony rear row, ft,
P = depth of balcony, ft.

Note $P = D - d$.

As an example, consider a theater of length $D = 100$ ft with a balcony of height and depth = 30 ft (i.e., $H_B = p = 30$). If we take $H_s = 7$ ft, $d = 70$ ft, $s = 3$ ft and $h = 1/3$ ft then $\beta = \tan^{-1}(0.54) = 28°$. On the main floor we would have level seating for 30 ft from the speaker and then rake the floor at $\theta = 14.5°$ so that the rear seat of the main floor would be 18 ft above ground level and therefore 12 ft under the balcony. The height of the rear balcony seats would be 30 + 30 tan 28° = 46 ft above ground level. It can be seen that these critical dimensions largely govern the volume of the room. For a given seating capacity a room with a balcony usually occupies a lower volume than one that does not.

Since larger rooms generally contain more absorption than smaller ones, and since the reverberant sound level depends upon the source power level and room absorption, it is desirable to keep the volume of the room as low as possible. Although, as we shall see later, the signal to noise ratio (i.e., direct to reverberant sound level ratio) is independent of the source power output, one can clearly achieve greater loudness in a small

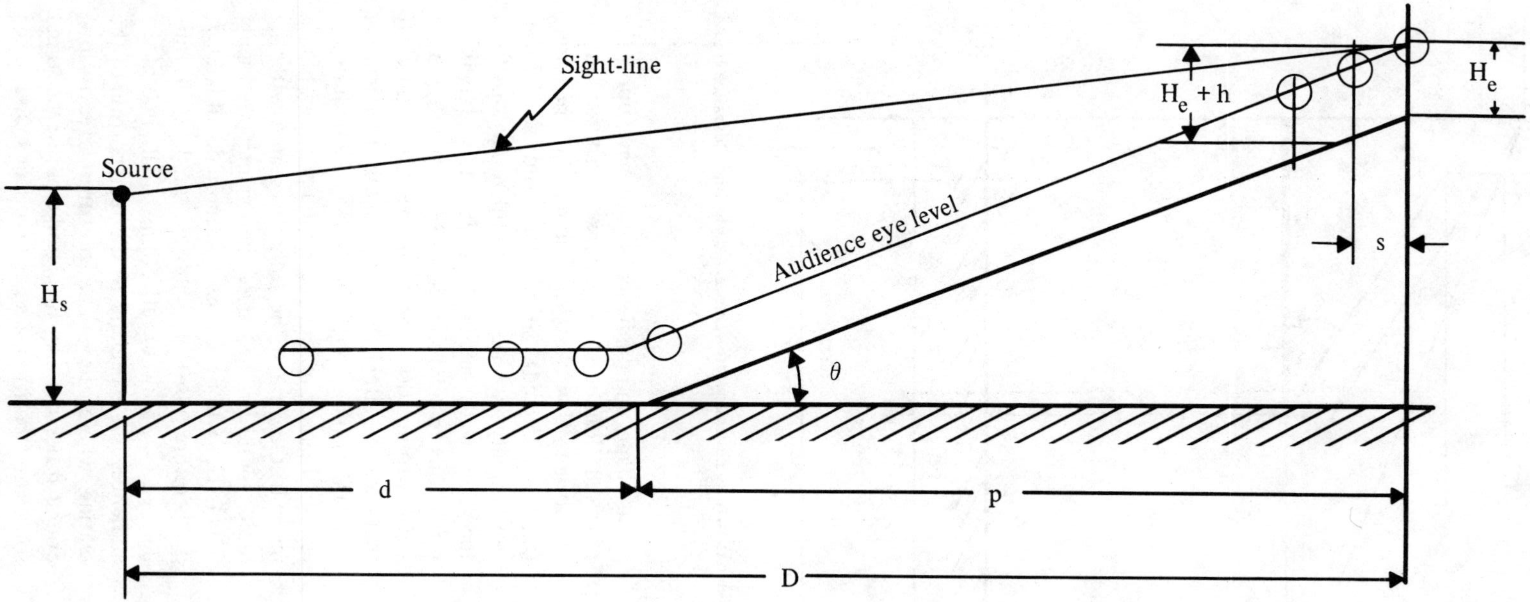

FIGURE 5.29 Audience members can be seated further away from the sound source if the seating is raked at an angle θ to the horizontal. However, a clearance height of h can only be fulfilled for a given angle θ up to a distance D from the source. The rake need only begin after a distance d – the maximum distance for sight-lines on a horizontal plane (as shown in Figure 5.28).

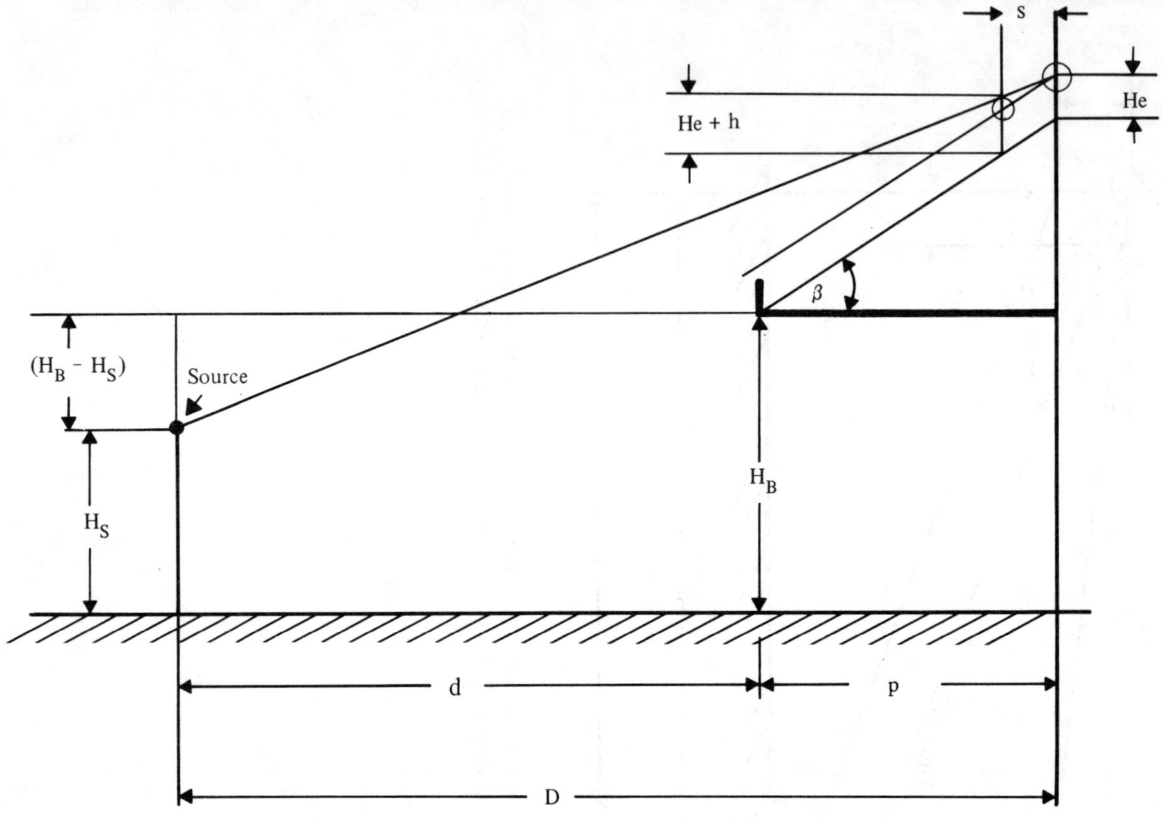

FIGURE 5.30 Sight-line requirements dictate the slope of a balcony B necessary to achieve a sight-line clearance h at a distance D from the source.

volume room for any given source. Smaller rooms also require less absorption to obtain the desired reverberation time and, in fact, the audience itself may provide all the required absorption with no need for additional expensive acoustical treatments. Small rooms also reduce the difference between the times of arrival of the direct and reflected sound. This leads to better intelligibility of speech and an improved assessment of intimacy for music reception. For rooms used for unamplified speech it is advisable to keep the volume per person in the range of 100 to 150 ft^3, and preferably nearer to 100 ft^3 per person when possible, while in large concert halls, volumes of 250 to 350 ft^3 per person are required. In order to achieve sufficient loudness from an unamplified human voice, the room volume is restricted to approximately 200,000 ft^3 and hence an audience size of close to 1,500 people. Concert halls may have larger volumes because of the greater acoustic power available from the orchestra; volumes usually fall in the range of 500,000 to 900,000 ft^3 with audience sizes of 2,000 to 3,000.

So far we have only considered the effect the room shape and audience layout have upon the observed loudness as far as direct sound is concerned. However, the loudness of the direct (i.e., unreflected) sound can, and should be, substantially increased by useful reflections from the side walls and ceiling. In order for these reflections to reinforce the direct sound they must arrive within 100 msec after the direct sound. If they arrive later than 100 msec the hearing mechanism judges the reflections to be undesirable echoes. In fact, reflections should arrive within 35 msec if they are to be fully integrated with the direct sound by the ear of the listener (see Figure 5.31). This will be discussed in more detail later. It is therefore most desirable to send short time delayed (<35 msec) reflections to all areas of the audience, but especially to the rear portion of the seating where the sound pressure level is low. Sound reflectors may be placed close to the sound source and positioned so as to reflect sound to the rear of the theater. The ceiling shape can be designed to achieve the same goal, although curved ceilings should be used with some caution since they tend to focus the sound on separate areas of

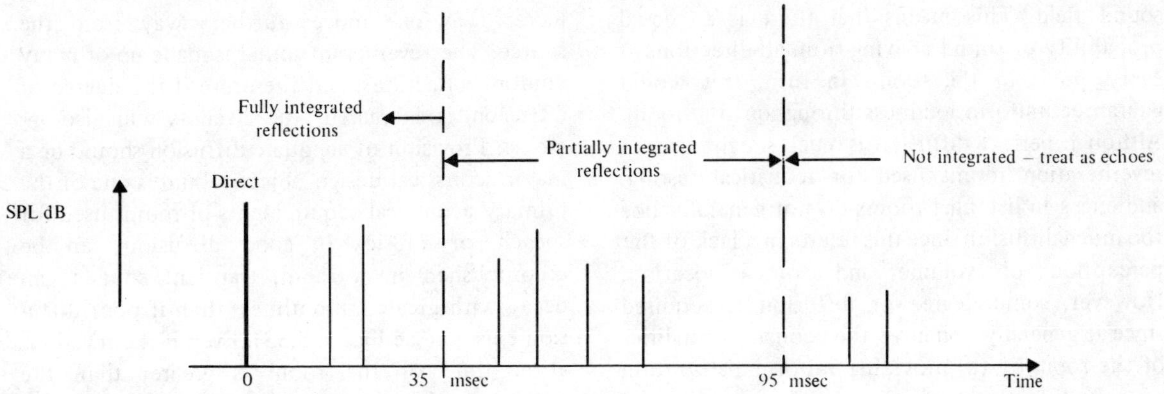

FIGURE 5.31 **Reflections arriving within 35 msec of the direct sound are fully integrated by the hearing mechanism. Those arriving after 35 msec are decreasingly integrated until 95 msec, after which no integration takes place and they are regarded as echoes.**

the audience. A variety of ceiling shapes are found in practice but their purpose is always to send short time delayed reflected sound to low sound pressure level areas of the audience to improve the overall loudness and speech intelligibility. It is also found that it is these short time delayed reflections which play a major role in the subjective assessment of the acoustical quality of rooms used for music.

In addition to using the ceiling shape as a useful sound reflector, the side walls may also be used to send reflections to the rear of the theater. If the side walls are splayed outwards (Figure 5.32) from the stage, more sound can be reflected to the back rows than if parallel side walls are employed. In fact, a side wall splay of only 2 to 3° will often give a substantial improvement. In addition, nonparallel side walls and floor-ceilings also eliminate flutter echoes.

The fact that only short time delayed reflections are useful (and long time delayed reflections cause adverse echoes) again places restrictions on the size and shape of listening rooms. For example, the room should not be too wide or the ceiling must not be too high for echoes will result. If side wall echoes are a problem and the room floor plan cannot be changed, then the long time delayed reflections must be eliminated by making the portion of the side walls responsible highly sound absorptive. If the ceiling height of a large room (i.e., concert hall or lecture theater) is sufficiently large to cause long time delayed reflections, but cannot be lowered without a serious decrease in the required reverberation time, then large panel reflectors may be hung from the ceiling at such heights and angles as to give useful short delayed reflections. These will be described in more detail later.

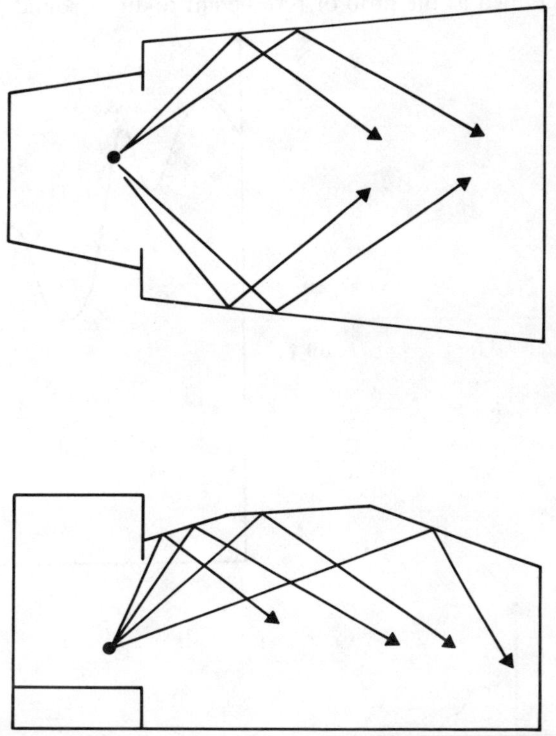

FIGURE 5.32 **Splayed walls and inclined ceiling planes can result in sending reflected sound to the center and rear sections of an auditorium where increased loudness is required.**

5.7.2 Provision of Diffusion

It may be remembered from Chapter 4 that the reverberation time equations for an enclosure are based on the assumption of a perfectly diffuse sound field. This means that there is an equal probability of sound arriving from all directions at every point in the room. In turn, this would guarantee uniform loudness throughout the room. Although perfect diffusion is much sought after in reverberation rooms used for acoustical testing, audiences in listening rooms do not generally like too much diffusion since this results in a lack of the perception of volume and source location. However, some degree of diffusion is required since it generally enhances the acoustical qualities of the room by (a) providing uniform distribution of sound levels at all seats, (b) improving the uniformity of the growth and decay of transient sound, and (c) improving the "liveness" of the room. Since the liveness at any point in a room is defined as the ratio of reverberant to direct sound pressure levels, it follows that the liveness will increase as one moves further away from the source. The reverberant sound is made up of many random reflections, and therefore if the degree of diffusion is increased, the liveness will also increase. Provision of adequate diffusion should be a major acoustical design objective and is one of the primary acoustical requirements of rooms used for speech or music. If good diffusion can be accomplished in a room, transient sounds can decay with greater smoothness than if poor diffusion existed (see Figure 5.33). Even if the reverberation time of the room is greater than the optimum value usually chosen for that particular room and function, increased listener acceptability can be expected if good diffusion is present.[49]

Diffusion in listening rooms is generally increased by:

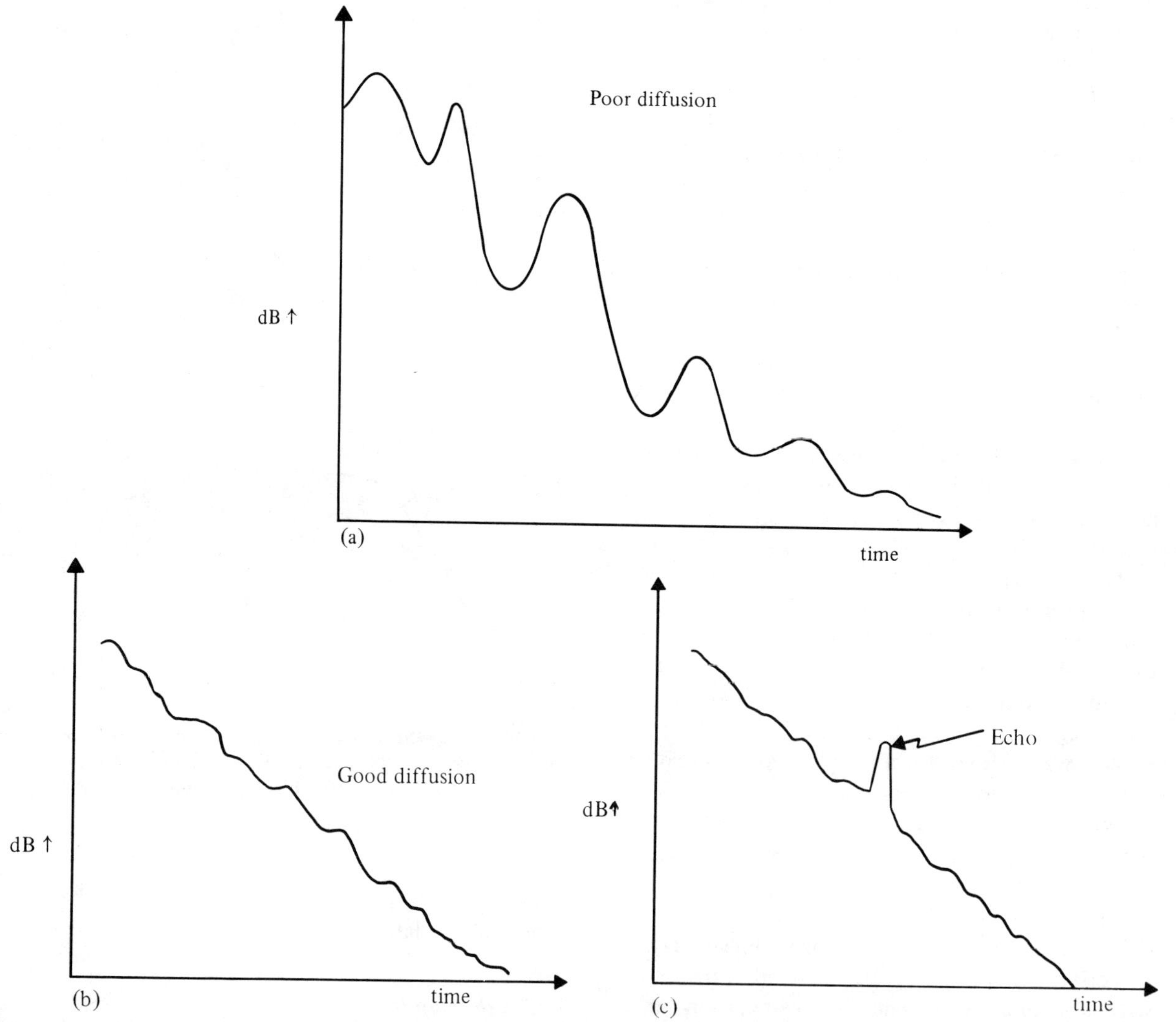

FIGURE 5.33 Sound level decays typical of rooms having (a) poor diffusion, (b) good diffusion, and (c) good diffusion but with an echo defect in the room.

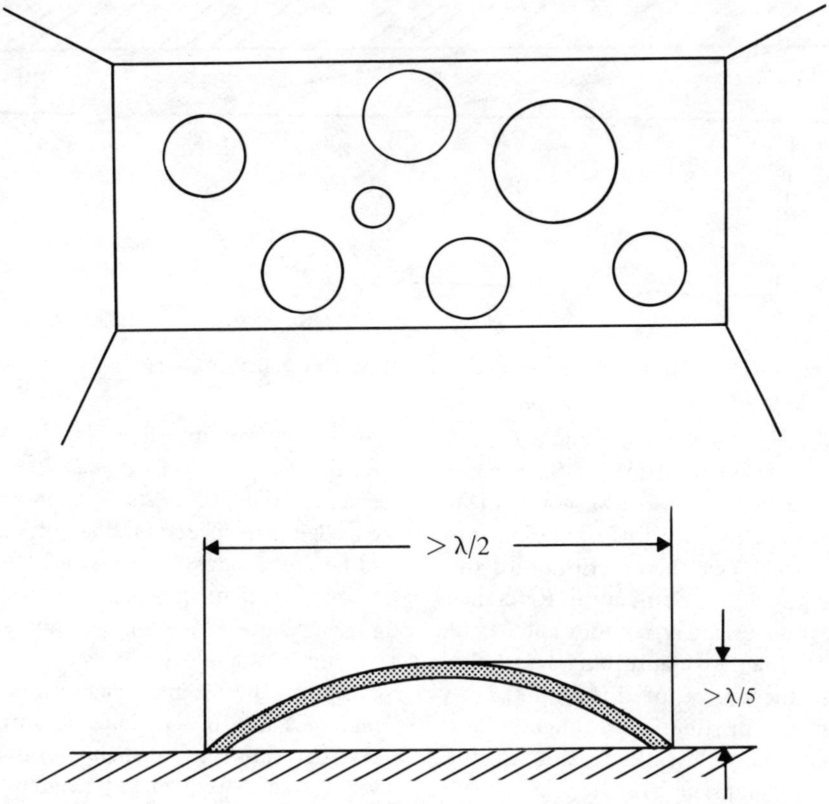

FIGURE 5.34 (a) Polycylindrical diffuser elements attached to a wall. (b) The chord length should exceed λ/2 and the projected height should exceed λ/5 in order to be effective.

(a) irregularities in wall surfaces – by introducing changes in shape and orientation sound is scattered (or diffracted) and reflected over a wide area of the audience.

(b) objects within the room – by scattering incident sound.

(c) irregular application of absorbing material on the room boundaries.

These techniques are clearly similar in nature to those described in Chapter 4 for use in reverberant testing rooms. However, many more restrictions are involved in the case of auditorium design since diffusing shapes and objects have to fit in with the architectural concept and other visual and acoustical requirements.

Irregular wall surfaces can be achieved in many ways. In all cases, however, the size of the irregularity must be greater than or at least comparable to a wavelength in order to achieve the desired scattering or diffraction effect. Consequently diffusion generally improves with increasing frequency (since wavelength is inversely proportional to frequency). Some examples of surface irregularities include:

(a) placing polycylindrical diffusers[50] made from plywood or plasterboard on the side walls. They should be made in a variety of sizes, since each contributes resonant absorption to the room. To obtain uniform absorption, different sizes must therefore be employed (see Figure 5.34).

(b) using wedge-shaped irregularities on the side walls (Figure 5.35). Again the size of these must be at least comparable to a wavelength and their projected height should be at least 1 to 2 ft in order to be effective at low frequencies.

(c) using ceiling irregularities such as beams to achieve diffusion by specular reflection and by diffraction (dependent upon the size of the irregularities).

(d) using side wall splay, as little as 5/8 in./linear ft, will considerably aid the diffusion.

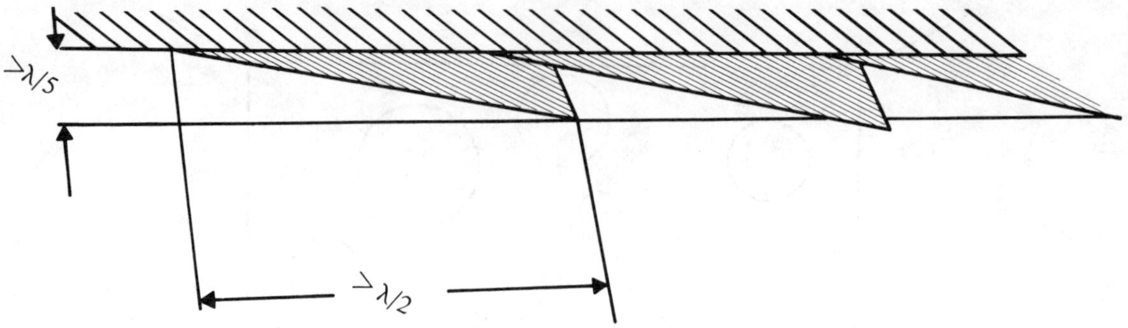

FIGURE 5.35 Wedge shape side wall or ceiling diffusers.

In most auditoria the audience and rows of seating provide additional diffusion. Suspended ceiling reflector panels can also give good diffusion. Some concert halls have diffusing elements attached to the ceiling (see the description of the "Beethovenhalle" at Bonn, Germany in Reference 51). The application of many random patches of absorption, rather than large absorbing areas, also greatly increases the degree of diffusion to give more uniform sound distribution throughout the room and substantially less fluctuation in the growth and decay of transient sounds.

Diffusion on and close to the stage area of an auditorium is also important since it ensures that the players can receive sound adequatly from each other.

5.7.3 Optimum Reverberation

The optimum reverberation time characteristics of a room can be simply defined as those which result in the highest intelligibility for rooms used for speech and produce maximum acoustical conditions for the rendition and appreciation of music. The term "optimum reverberation" refers not only to the reverberation time vs. frequency characteristics of a room, but also includes the nature of the growth and decay patterns observed throughout the room. The term also implies that the ratio of reverberant and direct sound will be optimum so as to produce the most satisfactory listening conditions.

Reverberation time has become recognized as the most important single parameter used to describe the acoustical performance of auditoria. However, achieving the optimum reverberation time for a room may not necessarily lead to good speech intelligibility or music appreciation. It is essential to adhere strictly to the other design rules for shape, volume, and time of arrival of early reflections. In fact, it has been found[52] that reverberation time itself is no measure of the excellence of speech intelligibility in a room, but is rather a secondary effect which happens to depend upon the same physical conditions and hearing mechanisms which, in combination, result in greater intelligibility of speech.

Since the sound absorption properties of materials are strongly dependent on frequency, the reverberation time of such a room is therefore also very much frequency-dependent. Most human listeners prefer to have a near uniform reverberation time vs. frequency characteristic above 500 Hz, with gradually increased values at lower frequencies as shown in Figure 5.36. A somewhat flatter response is acceptable for speech, but music rooms deficient in increased low frequency reverberation time are considered to lack "warmth." For design purposes, the 500-Hz reverberation time is taken as the optimum value (use Figure 5.36 to evaluate all other frequency values). A variety of authors have recommended optimum reverberation times for many functionally different listening rooms of increasing volume. Figure 5.37 shows a summary of many of those recommendations. Although an average music reverberation time function is shown, it must be remembered that this will vary depending upon the type of music being played. For example, organ and certain kinds of church music require very long reverberation times, while certain piano pieces require substantially lower values. In some auditoria a variety of activities and performances might be contemplated. Such rooms have become known as multipurpose halls, and therefore generally require multipurpose reverberation characteristics, which are not always easy to

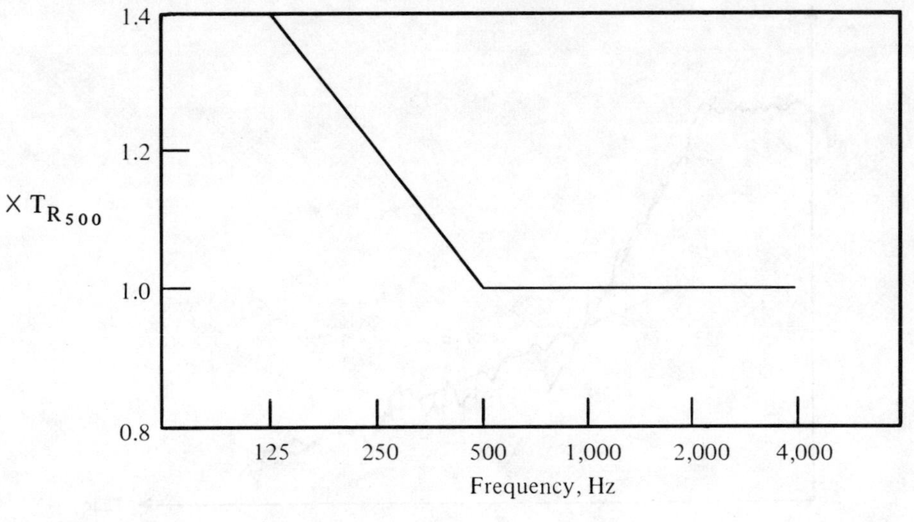

FIGURE 5.36 Chart for computation of the reverberation time at frequencies below 500 Hz. Simply multiply the optimum reverberation time $T_{R_{500}}$ by the ordinate value to obtain the required value at some other frequency.

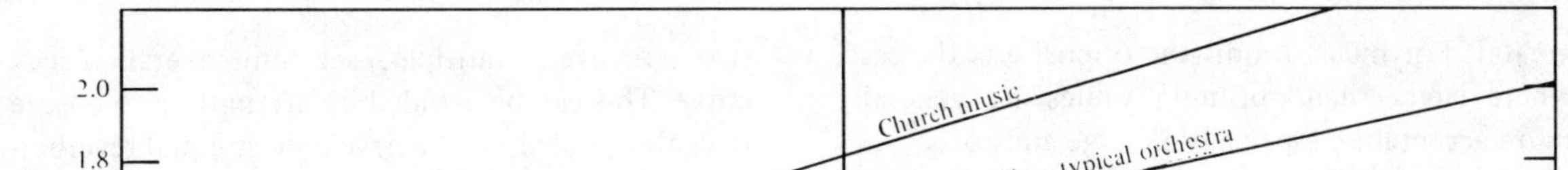

FIGURE 5.37 Optimum reverberation times at 500 Hz for different types of rooms as a function of their volume.

achieve. It is possible, however, to design the room to be fairly live and incorporate a number of surfaces which can be revolved or extended to reveal substantially higher sound-absorbing materials when lower reverberation is required.

Usually audiences will tolerate up to 10% or, in some cases, even 25% deviation[53] from the optimum reverberation time, unless they happen to be all trained musicians or have had wide and long experience listening to music in auditoria. In rooms designed for speech it is better to have a reverberation time less than optimum rather than

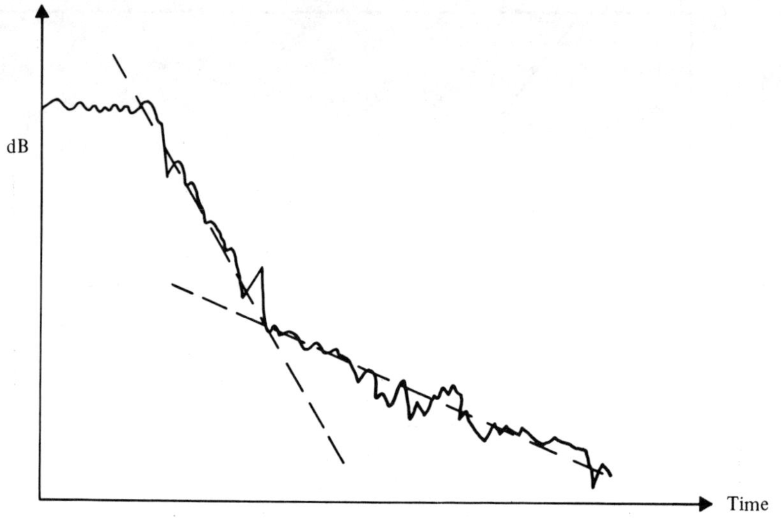

FIGURE 5.38 A typical two-slope reverberation decay as seen in a room containing a nonuniform distribution of absorbing material or in a room coupled to another relatively large space (i.e., a stage well) having a different reverberation time.

greater. For music rooms the opposite is the case, where larger than optimum values are generally more acceptable, especially in large auditoria.

Although a perfectly diffuse sound field in a single enclosure would be expected to decay exponentially so that the sound pressure level would decay linearly with time, such perfect decay rates are rarely observed. Two main factors contribute to the deviation from the ideal linear decay rate. Firstly, nonuniform distribution of absorption in a room leads to a sound field which is not diffuse and where certain room modes are subjected to large damping while others may have only very small damping presented to them. This results in the heavily damped modes decaying rapidly until the sound pressure level decreases to a point where more slowly decaying modes dominate. The overall decay rate of the sound field is then substantially reduced. This gives rise to the characteristic two-slope reverberation decay shown in Figure 5.38. The other common reason for such a decay is due to the acoustic power flow between two coupled spaces within the complete auditorium, such as the main section and the stage tower. If each of the sections have greatly different reverberation times, then the sound pressure level will clearly decay much faster in one space than in the other. When this level becomes less than in the dead space with the longer decay or reverberation time, acoustic power will flow into the space having the faster decay rate and again result in a two, or even multiple, segment overall decay curve. This can be avoided by attempting to ensure that all coupled spaces have almost equal reverberation time characteristics. Deeply recessed galleries also tend to become separate acoustical spaces coupled to the main body of the auditorium, and consequentially are often found to suffer from this phenomenon. For this and other reasons balcony heights and depths should be approximately equal.

By definition, reverberation time is the time taken for the sound pressure level to drop 60 dB (T_{60}) from its initial steady-state value (see Sections 1.6.2 and 4.2). However, we rarely have an opportunity to hear the sound decay a full 60 dB, either because of background noise or because of the arrival of a consecutive transient sound. The ear's assessment of reverberation time is therefore only rarely able to be made of the full 60 dB decay and, in fact, is based on the decay over a much shorter period of time. It has been found[54] that very good correlation exists between subjective values of reverberation time and those calculated from the initial 160 msec of the decay ($T_{160\,msec}$). In some circumstances the subjective value is only 50% of measured T_{60} values. If the decay is very slow so that the initial 160 msec decrease cannot be measured, then good correlation can still be obtained by measuring the time taken for the sound pressure level to decrease 15 dB from the steady-state level. The value of reverberation time calculated from this time (i.e.,

$\times$ 4) is known as T_{15} and is almost always in much closer agreement with the subjective value of reverberation time than T_{60}.

Optimum reverberation therefore, in fact, correlates better with T_{15} or T_{160} msec rather than to T_{60}. A rapid initial decay tends to produce good definition and clarity in an auditorium, as has been observed for rooms in which the rise time is also short. The rise and decay characteristics of a room are complementary and both are especially important parameters.[55] (See Sections 4.2.1 and 4.2.2.)

5.7.4 Acoustical Defects

Even if an auditorium is designed to have adequate loudness, diffusion, and reverberation, several acoustical defects may exist in the room which may completely destroy the perhaps otherwise excellent acoustics. As discussed previously in Section 5.7.1, short time delayed reflections ($<$ 95 msec) are sought to increase loudness and the sense of volume, while long delayed reflections are to be avoided. These long delayed reflections can result from the side walls and ceilings in large theaters but can be avoided by either splaying or breaking up the side walls or by applying absorption to them at the "guilty" portions. Long delayed ceiling reflections can be reduced by hanging large reflecting panels at some distance below the real ceiling. Many modern auditoria (e.g., theaters, movie halls) have a large radius curved rear wall or balcony rail and rows of seats which tend to reflect and focus the source sound back to the front of the room giving rise to echoes. If the focus of the concave surface is outside the room, then an echo or sound concentration (hot spot) problem may not occur at all. However, in many such theaters the radius of curvature of either the rear wall or balcony front is such that sound is focused to the front and center rows of the audience thereby creating a major acoustical defect. If the focus is at the speaker himself, then since he receives focused sound he is led to the impression of increased loudness and may be tempted to lower his voice! In order to be able to reflect sound, the curved reflector must have dimensions at least comparable to a wavelength. Therefore, if there were a portion of exposed rear wall above the audience's heads of a height of, for example, 6 ft, then sound down to approximately 180 Hz would be reflected.

In most cases the solution to this problem is simply to make the rear concave wall either highly sound absorbing or highly diffusive. The former of these two alternatives is usually chosen. Figure 5.39 shows the construction of the highly sound absorbing rear wall at Aula Magna (great hall) at the University of Caracas, Venezuela.[56] In some cases, however, the rear wall is left hard and reflective but built at such a slanting vertical angle so as to direct sound only to the rear section of the audience (see Figure 5.40) where they experience only a small delay time between the direct sound and the sound reflected from the rear wall. This is especially useful for the section of audiences under a deep balcony, where both the balcony soffit and the rear wall should be designed to reflect sound to the rear listeners. Furthermore, since the human ear has poor vertical localization (see Section 2.2.1), it is not able to detect the fact that the useful short delayed sound is coming from behind and integrates it with the direct sound coming from the stage.

In general, it is wise to avoid all large concave surfaces including the rear wall and ceiling domes. In the case of ceiling domes, focusing problems can be partially avoided if the radius of the dome is either greater than at least twice the ceiling height, or less than half the ceiling height. The most serious focusing problems occur with domes, arches, and barreled ceilings when the radius of curvature coincides with the ceiling height.[57] In this case, the audience at the focus will receive greatly increased loudness (hot spot) while other portions may receive decreased loudness levels (dead spot).

Plane reflecting parallel surfaces can also be an acoustical problem since transient sound produced between them gives a rapid series of noticeable echoes as the sound bounces back and forth. This is known as a *flutter echo* and can be created in most rooms, if the floor and ceiling are reflecting and parallel, by clapping one's hands and listening to the flutter of successive short time delay echoes produced. This phenomenon can be eliminated by avoiding parallel reflecting surface pairs such as floor – ceiling and side walls, and by avoiding the location of any sound source between them. This also helps to eliminate low frequency room resonances (where the separation of opposite parallel walls equals a multiple of half wavelengths) which, in turn, reduces the tendency for a narrow band of frequencies to become excessively loud. This effect is sometimes known as "coloration."[58]

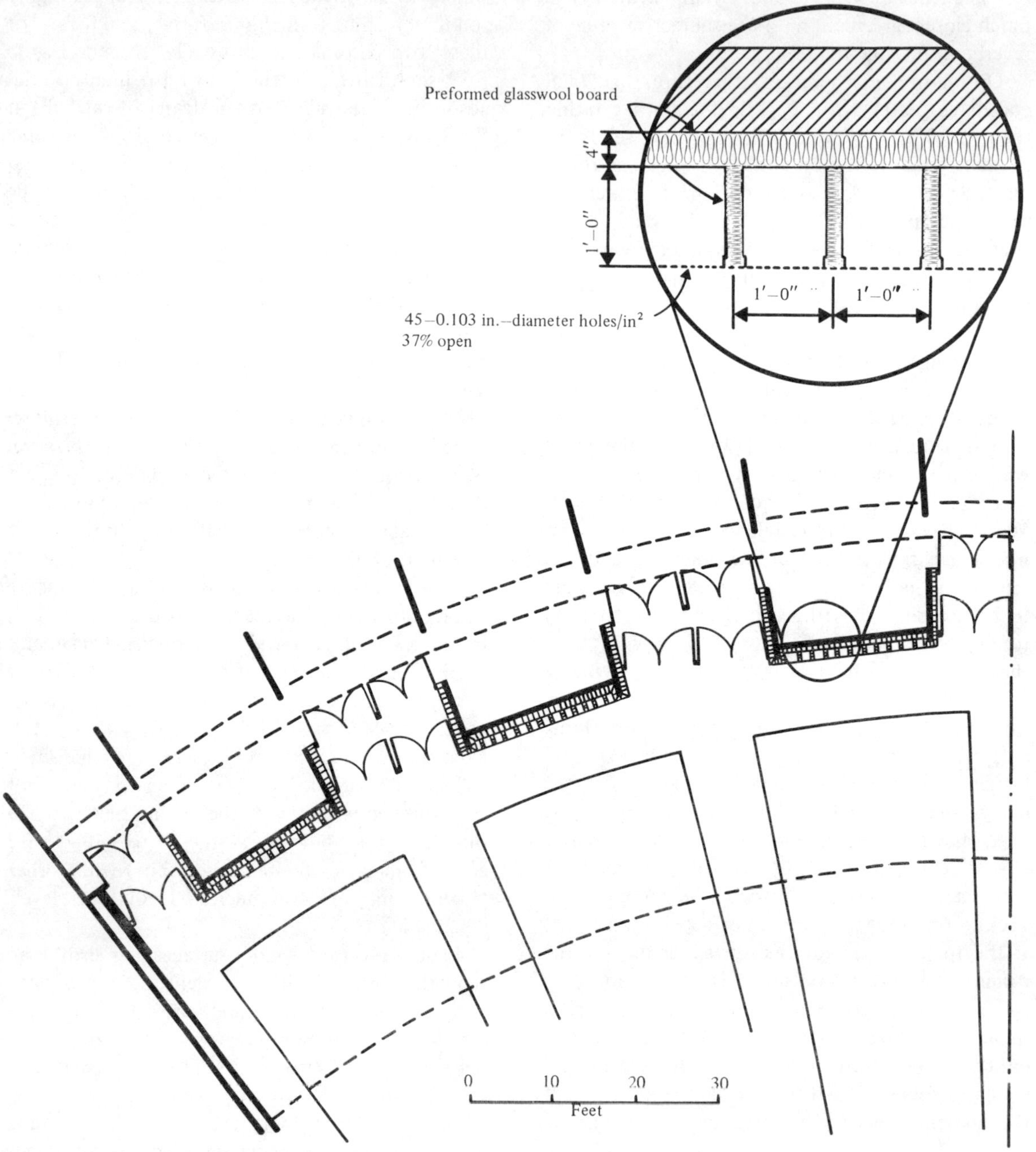

FIGURE 5.39 To prevent echo, a highly absorbing surface was created at the rear of the Aula Magna, Caracas, an auditorium with a long circularly curving rear wall whose focal point is on the stage. Sound-absorbing material is also inserted in the perforated metal faces of the doors. (From Beranek, L. L., *Music, Acoustics and Architecture,* John Wiley & Sons, New York, 1962, 454. With permission.)

5.8 DESIGN OF ROOMS FOR SPEECH

The ultimate acoustical requirement of rooms used for speech production and reception is that every member of the audience should be able to understand every word spoken; in other words, this is a problem of providing maximum speech intelligibility. At the same time, we also require that the sound should be as natural as possible to retain the character of the speech. This second requirement is especially important in drama theaters. Since, as we shall see, the physical

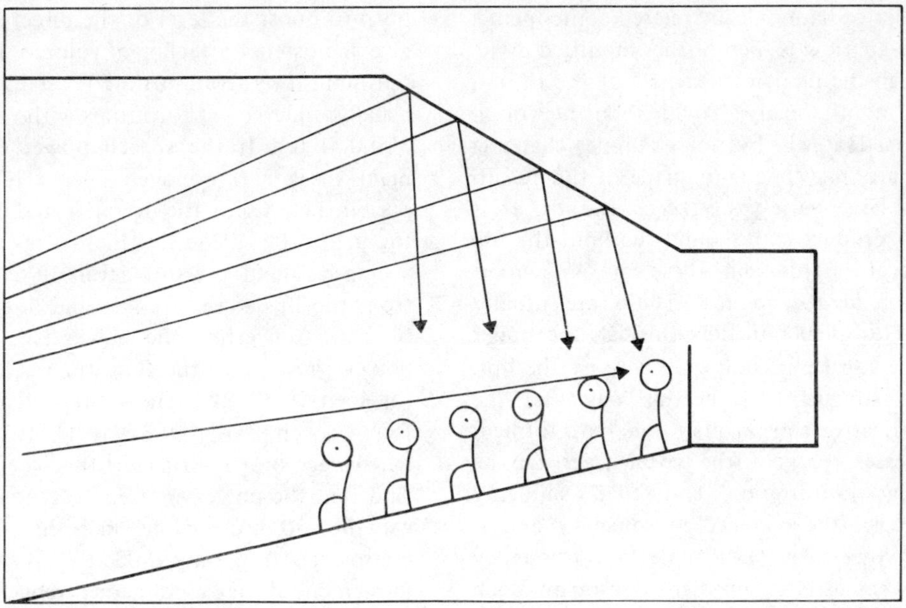

FIGURE 5.40 A sloping rear ceiling can increase the loudness of the useful reflections received by the rear rows of the audience.

properties of speech differ considerably from those of music, we find that the acoustical requirements of listening rooms used for either speech or music often differ considerably. In order to be able to design rooms for speech reception it is clearly necessary to be able to understand the physical characteristics and composition of the sounds most important to people – speech.

5.8.1 Physical Characteristics of Speech

Since speech and hearing must be compatible, it is not surprising to find that the speech frequency range corresponds to the most sensitive region of the ear's response (Section 2.3.1) and generally extends from 100 to 10,000 Hz. The general mechanism for speech generation involves the contraction of the chest muscles to force air out of the lungs and up through the vocal tract. This flow of air is modulated by various components of the vocal mechanism (Figure 5.41) to produce sounds which make up part of our speech pattern. The modulation effect first takes place at the larynx, across which are stretched the vocal chords. These are composed of two bands of membranes separated by a slit which can open and close to modulate the flow of air. The modulation frequency depends upon the tension in the muscles attached to the vocal chords, and on the size of the slit in the membranes (about 1 in. for males and 0.6 in. for females). The sound emitted by the vocal cords has a buzz-type sound corresponding to a sawtooth waveform containing a

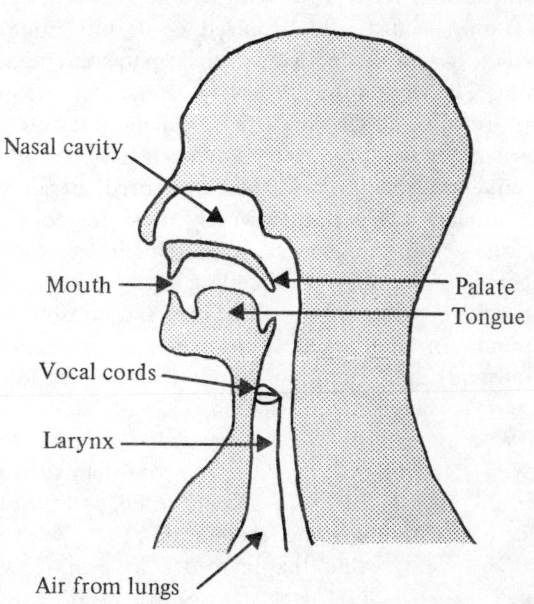

FIGURE 5.41 Sectional view of the head showing the important elements of the voice mechanism. (From Kinsler, L. E. and Frey, A. R., *Fundamentals of Acoustics,* John Wiley & Sons, New York, 1962, 381. With permission.)

large number of harmonically related components. This sound/air flow is then further modified by its flow through the numerous cavities of the throat, nose, and mouth, many of which can be voluntarily changed at will by, for example, changing the position of the tongue or shape of the lips, to produce a large variety of *voiced* sounds. It is possible to produce some sounds without the use of the vocal chords and these are known as *unvoiced* or *breath* sounds. These are usually caused by turbulent air flow through the upper parts of the vocal tract and especially by the lips, teeth, and tongue. It is in this way that the unvoiced fricative consonants *f* and *s* are formed. In some cases, part of the vocal tract can be blocked by constriction and then suddenly released to give the unvoiced consonants *p* and *g*. Generally, vowels have a fairly definite frequency spectrum whereas many unvoiced consonants such as *s* and *f* tend to exhibit very broad-band characteristics. Furthermore, when several vowels and/or consonants are joined together their individual spectra appear to change somewhat. The time duration of individual speech sounds also tends to vary widely over a range of 20 to 300 msec.

In the general context of speech, vowels and consonants become woven together to produce not only linguistically organized words, but sounds which have a distinctive personal characteristic as well. The vowels usually have greater energy than consonants and give the voice its character. This is probably due to the fact that vowels have definite frequency spectra with superimposed periodic short duration peaks. However, it is the consonants which give speech its intelligibility. It is therefore essential in the design of rooms for speech to preserve both the vowel and consonant sounds for all listeners. Consonants are generally transient, short duration sounds of relatively low energy. Therefore, for speech, it is necessary to have a room with a short reverberation time to avoid blurring of consecutive consonants; we would expect therefore speech intelligibility to decrease with increasing reverberation time. At the same time we find that in order to produce a speech signal level well above the reverberant sound level (i.e., high signal-to-noise ratio), we require increased sound absorption in the room. This necessitates a lower reverberation time. Although this may lead us to think that an *anechoic room* would be most suitable for speech intelligibility, some sound reflections are required both to boost the level of the direct sound and to give the listener a feeling of volume. Therefore, an optimum reverberation time is established. This is usually under 1 sec for rooms with volumes under 300,000 ft^3. If the speech power emitted by a male speaker is averaged over a relatively long period (i.e., 5 sec), the overall sound power level is found to be 75 dB. This corresponds to an averaged sound pressure level of 65 dB at 1 m from the lips of the speaker and directly in front of him. Converting the power level to acoustic power shows that the long time averaged power for men is 30 μW. The average female voice is found to emit approximately 18 μW. However, if we average over a very short time (i.e., 1/8 sec) we find that the power emitted in some vowel sounds can be 50 μW, while in other soft spoken consonants it is only 0.03 μW.[59] Generally, the human voice has a dynamic range of approximately 30 dB throughout its frequency range.[60] At maximum vocal effort (loud shouting) the sound power from the male voice may reach 3,000 μW.

Figure 5.42 shows the long time averaged value of the normal male voice sound spectrum levels measured 1 m from the speaker's lips as a function of frequency. This represents the sound pressure level at 1 m in 1-Hz frequency bands. Table 5.2 gives the long-term rms sound pressure levels at 1 m from the average male mouth for normal vocal effort as given by the American National Standards Institute[61] for both one third octave and octave bands.

Although approximately 80% of the energy in speech lies below 600 Hz[62] (including most vowels), it is in the higher frequencies that most consonants have their strongest energy. These low energy transient consonants contribute to the intelligibility perceived. For example, it has been found[63] that if speech is passed through a high pass filter having a cutoff frequency of 1,000 Hz, then 90% of the spoken words can be understood. However, if the same speech is passed through a low pass filter, then a cutoff frequency of 3,000 Hz is required to produce the same percentage word intelligibility. Figure 5.43 shows the relative contribution to speech intelligibility as a function of 1/3-octave frequency bands. Speech sounds below 200 Hz and above 6,000 Hz do not significantly contribute to intelligibility but they do add to the natural qualities of the voice.

Since speech is emitted from the mouth, it is not surprising to find that the acoustic radiation

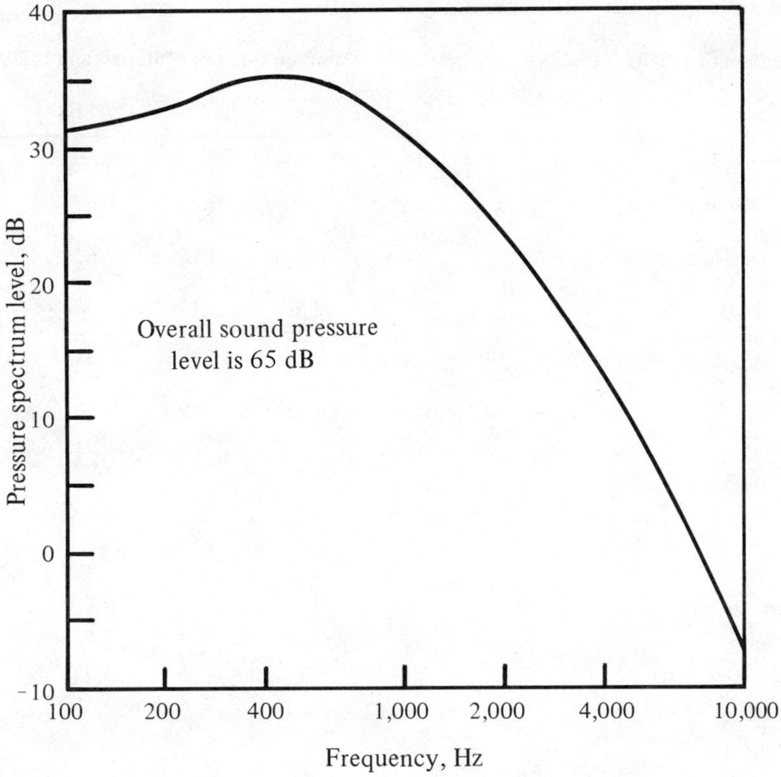

FIGURE 5.42 Spectrum level curve for average conversational speech measured at 1 meter from the mouth of the talker. (From Kinsler, L. E. and Frey, A. R., *Fundamentals of Acoustics,* John Wiley & Sons, New York, 1962, 382. With permission.)

TABLE 5.1

Absorption Coefficients (Sabine) for a Variety of Common Building Materials and Finishes

	Frequency, Hz					
Material	125	250	500	1,000	2,000	4,000
Acoustic tiles						
5/8-in. perforated aluminum-faced mineral fiber tiles #7 mounting (16-in. air space)	0.32	0.32	0.51	0.81	0.77	0.55
Mat-faced textured fiberglass, 5/8-in. #7 mounting	0.63	0.90	0.68	0.90	0.96	0.91
Pin perforated fiberglass, 5/8-in. #7 mounting	0.85	0.86	0.64	0.84	0.90	0.89
Sonocor fiberglass®, 1-in. #7 mounting	0.60	0.61	0.92	0.83	0.71	0.46
Sonoglas® – nubby 3/4-in. #7 mounting	0.89	0.78	0.83	0.97	0.99	0.91
Perforated 1/2-in. wood fiber tiles, #7 mounting	0.14	0.30	0.51	0.67	0.68	0.56
Audience and seats						
100% occupied audience – upholstered seats	0.52	0.68	0.85	0.97	0.93	0.85

TABLE 5.1 (continued)

Absorption Coefficients (Sabine) for a Variety of Common Building Materials and Finishes

Material	Frequency, Hz 125	250	500	1,000	2,000	4,000
Unoccupied – average well-upholstered seating areas	0.44	0.60	0.77	0.89	0.82	0.70
Unoccupied leather-covered upholstered seating areas	0.40	0.50	0.58	0.61	0.58	0.50
Wooden pews – 100% occupied	0.57	0.61	0.75	0.86	0.91	0.86
Wooden chairs – 100% occupied	0.60	0.74	0.88	0.96	0.93	0.85
Wooden chairs – 75% occupied	0.46	0.56	0.65	0.75	0.72	0.65
Brick wall – painted	0.01	0.01	0.02	0.02	0.02	0.02
Brick wall – unpainted	0.02	0.02	0.03	0.04	0.05	0.07
Carpets						
Ozonite	0.05	0.13	0.20	0.42	0.48	0.48
Heavy, on concrete	0.02	0.06	0.14	0.37	0.60	0.65
Heavy on 40 oz hairfelt or foam rubber	0.08	0.24	0.57	0.69	0.71	0.73
Heavy, with impermeable latex backing on 40 oz hairfelt or foam rubber	0.08	0.27	0.39	0.34	0.48	0.63
Concrete						
Poured/unpainted	0.36	0.44	0.31	0.29	0.39	0.25
Poured/painted	0.10	0.05	0.06	0.07	0.09	0.08
Blocks/unpainted	0.36	0.44	0.31	0.29	0.39	0.25
Blocks/painted	0.10	0.05	0.06	0.07	0.09	0.08
Cork						
Floor tiles (3/4-in. thick) glued down	0.08	0.02	0.08	0.19	0.21	0.22
Wall panels (1-in. thick)	0.25	0.55	0.70	0.75	0.75	0.75
Cotton						
Fabric, 14 oz/yd² draped to half area	0.07	0.31	0.49	0.81	0.66	0.54
Fabric, draped to 7/8 area	0.03	0.12	0.15	0.27	0.37	0.42
Curtains						
Light velour, 10 oz/yd², hung straight in contact with wall	0.03	0.04	0.11	0.17	0.24	0.35
Medium velour, 14 oz/yd² draped to half area	0.07	0.31	0.49	0.75	0.70	0.60
Heavy velour, 18 oz/yd² draped to half area	0.14	0.35	0.55	0.72	0.70	0.65
13 oz/yd² fiberglass full	0.29	0.38	0.57	0.53	0.72	0.89
Fiberglass insulation blankets and boards on #6 mounting						
0.5 lb/ft³ PF 334® 1/2 in.	0.09	0.40	0.32	0.43	0.64	0.70
3/4 in.	0.10	0.46	0.44	0.61	0.73	0.75
1 in.	0.10	0.51	0.54	0.73	0.79	0.79
1 1/2 in.	0.11	0.58	0.69	0.90	0.87	0.83

TABLE 5.1 (continued)

Absorption Coefficients (Sabine) for a Variety of Common Building Materials and Finishes

Material	Frequency, Hz 125	250	500	1,000	2,000	4,000
1 lb/ft³ PF 336® 1/2 in.	0.09	0.40	0.36	0.55	0.71	0.75
1 in.	0.10	0.51	0.58	0.86	0.85	0.83
1 1/2 in.	0.11	0.58	0.73	0.96	0.92	0.87
1.5 lb/ft³ PF 338® 1/2 in	0.09	0.40	0.38	0.63	0.76	0.78
1 in.	0.10	0.51	0.61	0.90	0.89	0.85
AF board AF 110® 1 in.	0.16	0.47	0.61	0.83	0.81	0.76
AF 160® 1 in.	0.19	0.51	0.63	0.88	0.83	0.78
AF 530® 1 in.	0.15	0.56	0.65	0.88	0.88	0.78
AF 545® 1 in.	0.17	0.62	0.75	0.91	0.91	0.84
AF 530® 1 in. behind burlap	0.15	0.30	0.65	0.85	0.75	0.40
Floors						
Concrete or terrazzo	0.01	0.01	0.01	0.02	0.02	0.02
Linoleum, asphalt, rubber, or cork tile on concrete	0.02	0.03	0.03	0.03	0.03	0.02
Parquet flooring with subfloor on sleepers	0.05	0.03	0.06	0.09	0.10	0.20
Parquet flooring in asphalt on concrete	0.04	0.04	0.07	0.06	0.06	0.07
Varnished wood joist floor	0.15	0.11	0.10	0.07	0.06	0.07
Wood platform with large space beneath	0.40	0.30	0.20	0.17	0.15	0.10
Glass						
Large panes of heavy plate	0.18	0.06	0.04	0.03	0.02	0.02
Windows glazed with up to 32-oz glass.	0.35	0.25	0.18	0.12	0.07	0.04
Gypsum wall board						
1/2 in. gypsum wall board on 2 × 4 studs at 16 in. on center	0.29	0.10	0.05	0.04	0.07	0.09
5/8 in. gypsum wall board on 2 × 4 studs at 16 in. on center	0.20	0.08	0.05	0.05	0.05	0.05
Glazed tile/marble	0.01	0.01	0.01	0.01	0.02	0.02
Panels						
3/8-in. plywood paneling	0.28	0.22	0.17	0.09	0.10	0.08
3/4-in. pine sheathing	0.10	0.11	0.10	0.08	0.08	0.05
1/4-in. plywood, 3-in. air space with absorptive blanket	0.60	0.30	0.10	0.09	0.09	0.09
1/4-in. plywood, no absorption in air space	0.13	0.15	0.10	0.09	0.09	0.09
Thin wood (0.2- to 0.4-in.) paneling with air space	0.42	0.21	0.06	0.05	0.04	0.04
Wood paneling 3/8 in. to 1/2 in. thick mounted over 2-in. to 4-in. air space	0.30	0.25	0.20	0.17	0.15	0.10
Cedar wall	0.20	0.15	0.15	0.10	0.10	0.10

Table 5.1 (continued)

Absorption Coefficients (Sabine) for a Variety of Common Building Materials and Finishes

Material	Frequency, Hz 125	250	500	1,000	2,000	4,000
Plaster						
Plaster, gypsum, or lime, rough finish on lath	0.14	0.10	0.06	0.05	0.04	0.03
Plaster, gypsum, or lime, smooth finish	0.14	0.10	0.06	0.04	0.04	0.03
Acoustic plaster, 1/2-in. trowel application	0.31	0.32	0.52	0.81	0.88	0.84
Acoustic plaster, 1 in. thick	0.25	0.45	0.78	0.92	0.89	0.87
1-in. damped plaster walls or ceiling	0.14	0.10	0.06	0.05	0.04	0.04
Spray-on materials						
Spray-acoustic, on metal lath, 35/8-in. air space	0.47	0.88	0.87	0.95	0.95	0.92
Spray-acoustic, on gypsum wall board	0.15	0.47	0.88	0.92	0.87	0.88
Limpet® asbestos, 3/4 in.	0.08	0.19	0.70	0.89	0.95	0.85
Limpet® asbestos, 1 in.	0.30	0.42	0.74	0.96	0.95	0.96
Limpet® asbestos, 3/4 in. on metal lath	0.41	0.88	0.90	0.88	0.91	0.81
K-13® 1 in. on metal lath	0.47	0.90	1.10	1.03	1.05	1.03
1 in., solid backing	0.08	0.29	0.75	0.98	0.93	0.76
5/8 in., solid backing	0.05	0.16	0.44	0.79	0.90	0.91
Stage Openings	0.30	0.40	0.50	0.60	0.60	0.50
Water (surface of pool)	0.01	0.01	0.01	0.015	0.02	0.03

from this small aperature set in a larger object (the head) is subject to fairly strong directivity effects. These directivity effects become more marked at high frequencies. Figures 5.44 and 5.45 show the radiation patterns for the male voice in the horizontal and vertical planes, respectively. Directivity effects can become important for audience members seated at the ends of the front rows of an auditorium, since they will receive considerably less of the direct sound at high frequencies. This can considerably reduce the intelligibility of speech.

5.8.2 Speech Intelligibility—Articulation Index

So far we have seen that speech has a dynamic range of approximately 30 dB in each 1/3-octave band from 200 to 6,000 Hz, and that the long-term rms overall sound pressure level at 1 m from the speaker's lips is 65 dB. The instantaneous level varies considerably with time. It has been found that 1% of the rms values of speech measured in 1/8-sec intervals exceeds the long-term rms value by 12 dB or more.[62] It is these peak levels, known as *long-term rms plus 12 dB,* which contribute to the degree of speech intelligibility realized. In the presence of a masking noise (i.e., background and/or reverberant) each frequency band contributes to the overall intelligibility depending upon (a) the difference (in dB) between the long-term rms plus 12 dB value and the total masking noise level, and (b) the relative importance of that frequency band (as shown in Figure 5.43). The difference between the signal peak and the masking rms sound pressure levels is known as the *peak signal-to-noise ratio* S–N given by:

$$S - N = L_{p_s}^{pk} - L_{p_m}^{rms}, \tag{5.48}$$

where

$L_{p_s}^{pk}$ = long-term rms + 12 dB, signal level

$L_{p_m}^{rms}$ = long-term rms, masking noise level.

The values for $L_{p_s}^{pk}$ can be obtained in Table 5.2

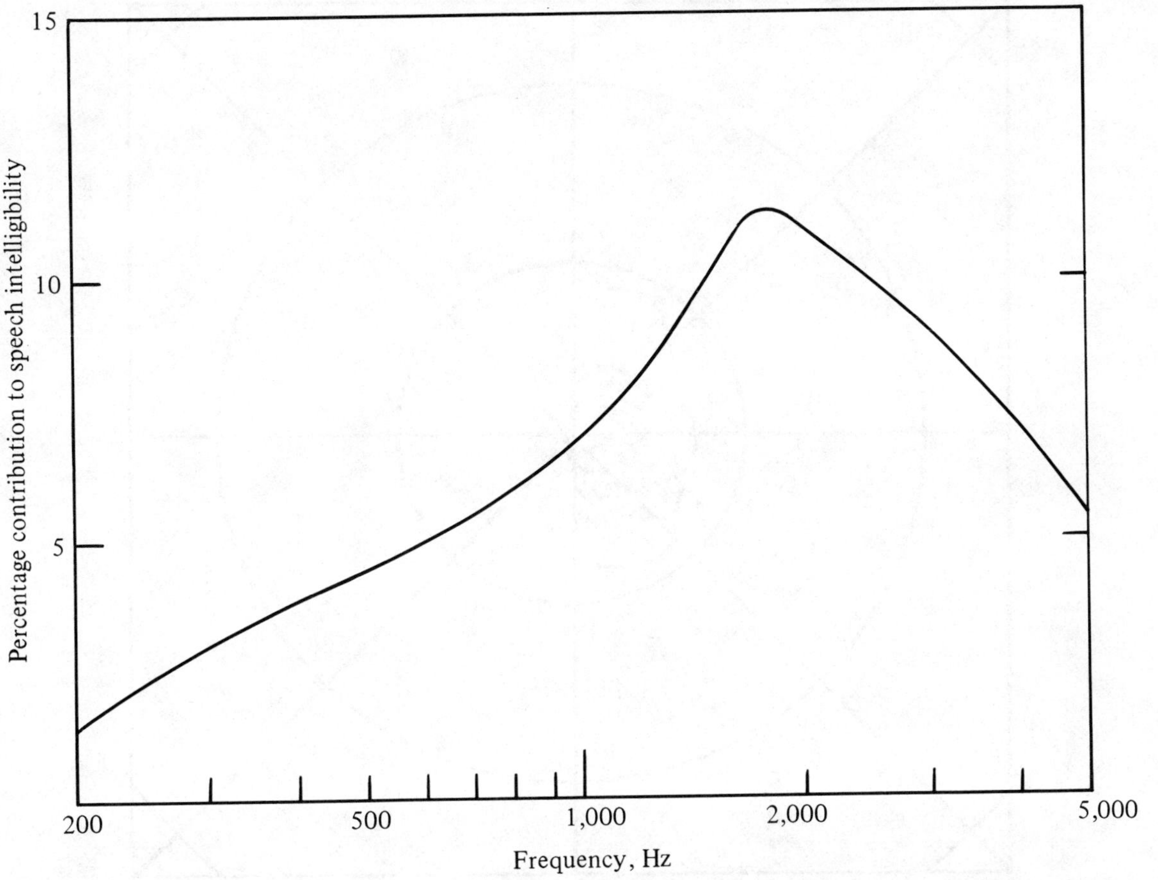

FIGURE 5.43 Percentage contribution to speech intelligibility vs. frequency.

for both 1/3-octave and octave bands. If the peak signal-to-noise ratios are evaluated for each contributing frequency band, they may be combined to give a single figure rating of intelligibility by applying a weighting factor to each band. Using the values shown in Table 5.3 (taken from ANSI S3.5–1969) for the appropriate weighting factors, a single rating known as the *Articulation Index AI* is obtained. This is defined as:

$$AI = \sum_{i=1}^{i=n} (S-N)_i w_i, \tag{5.49}$$

where

$(S-N)_i$ = peak signal-to-noise ratio for i^{th} frequency band, with i=1 for 200 Hz 1/3-octave band or 250-Hz octave band

w_i = weighting factor for i^{th} frequency band.

This calculation is performed over the range 200 to 5,000 Hz in the case of 1/3-octave bands and 250 to 4,000 Hz for octave bands. It is important to note that only frequency bands contributing positive (S–N) values are taken into account in the above summation, since AI cannot be less than 0. Any (S–N) values greater than 30 dB are set at (S–N) = 30 dB. The Articulation Index ranges from AI = 0 to 1.0 corresponding to 0% and 100% intelligibility, respectively.

Values of AI may be converted to estimated percentage speech intelligibility ratings[64,65] by the use of Figure 5.46. It immediately becomes obvious that the intelligibility score (i.e., percentage correct) will depend very much upon the nature of the information being conveyed by the speaker. For example, it is much easier to understand whole sentences than a list of everyday single words, which, in turn, is much easier than a list of nonsense syllables. The intelligibility score also depends upon the familiarity of the audience

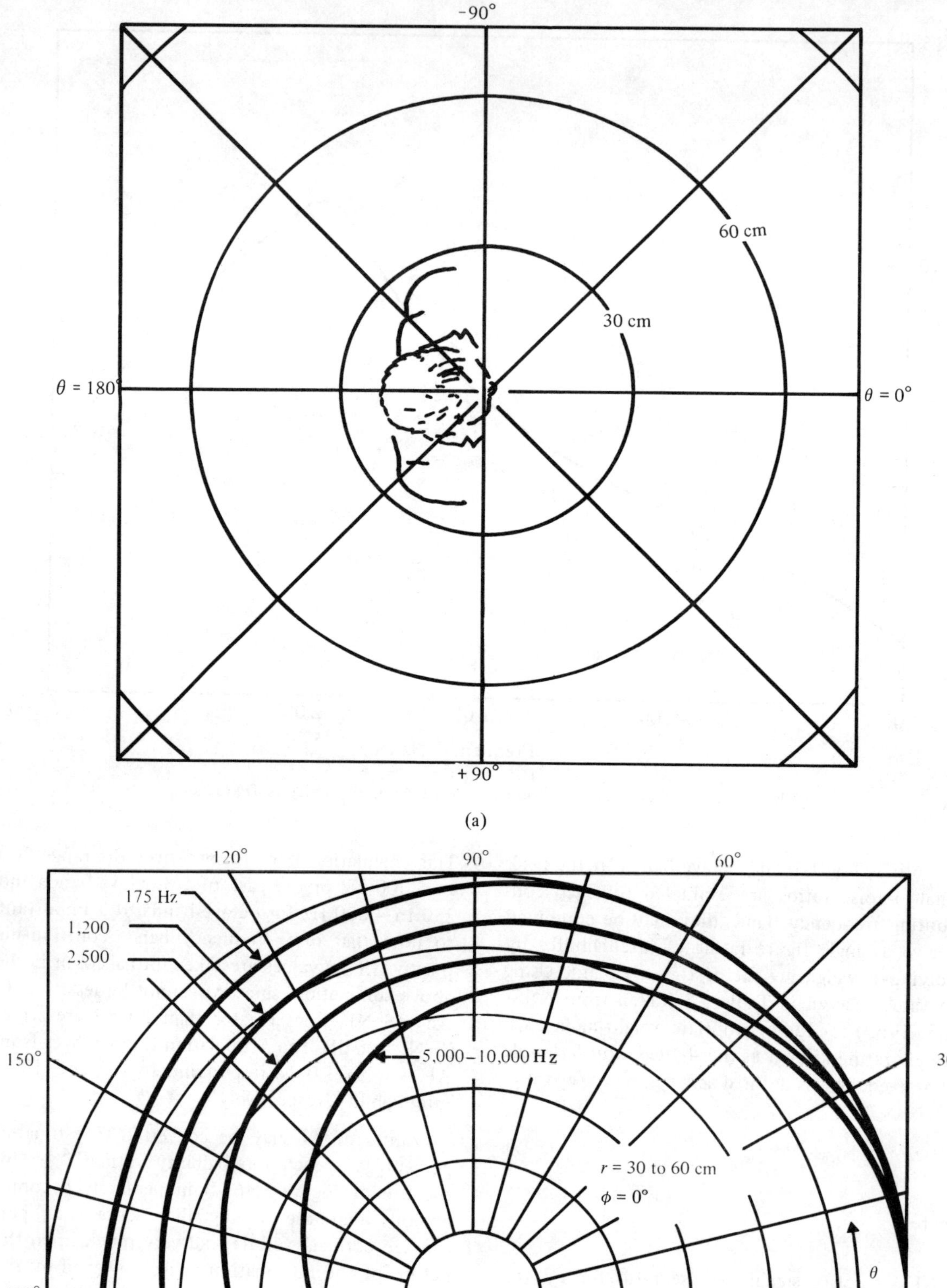

FIGURE 5.44 Directivity patterns for the human voice in a horizontal plane. (After Dunn, H. K. and Farnsworth, D. W., *J. Acoust. Soc. Am.*, 10, 184, 1939; from Beranek, L. L., *Acoustics*, McGraw-Hill, New York, 1954. With permission.)

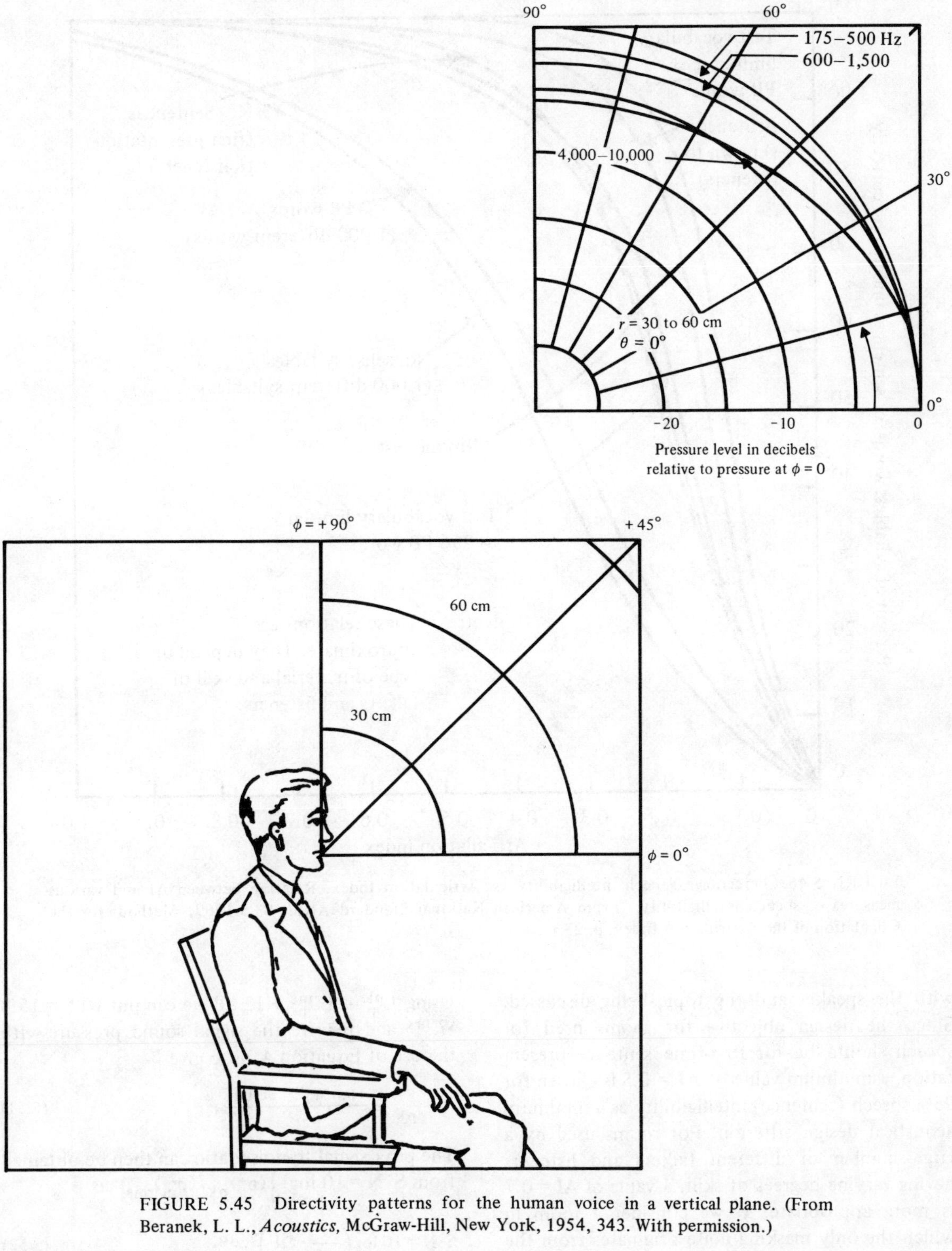

FIGURE 5.45 Directivity patterns for the human voice in a vertical plane. (From Beranek, L. L., *Acoustics,* McGraw-Hill, New York, 1954, 343. With permission.)

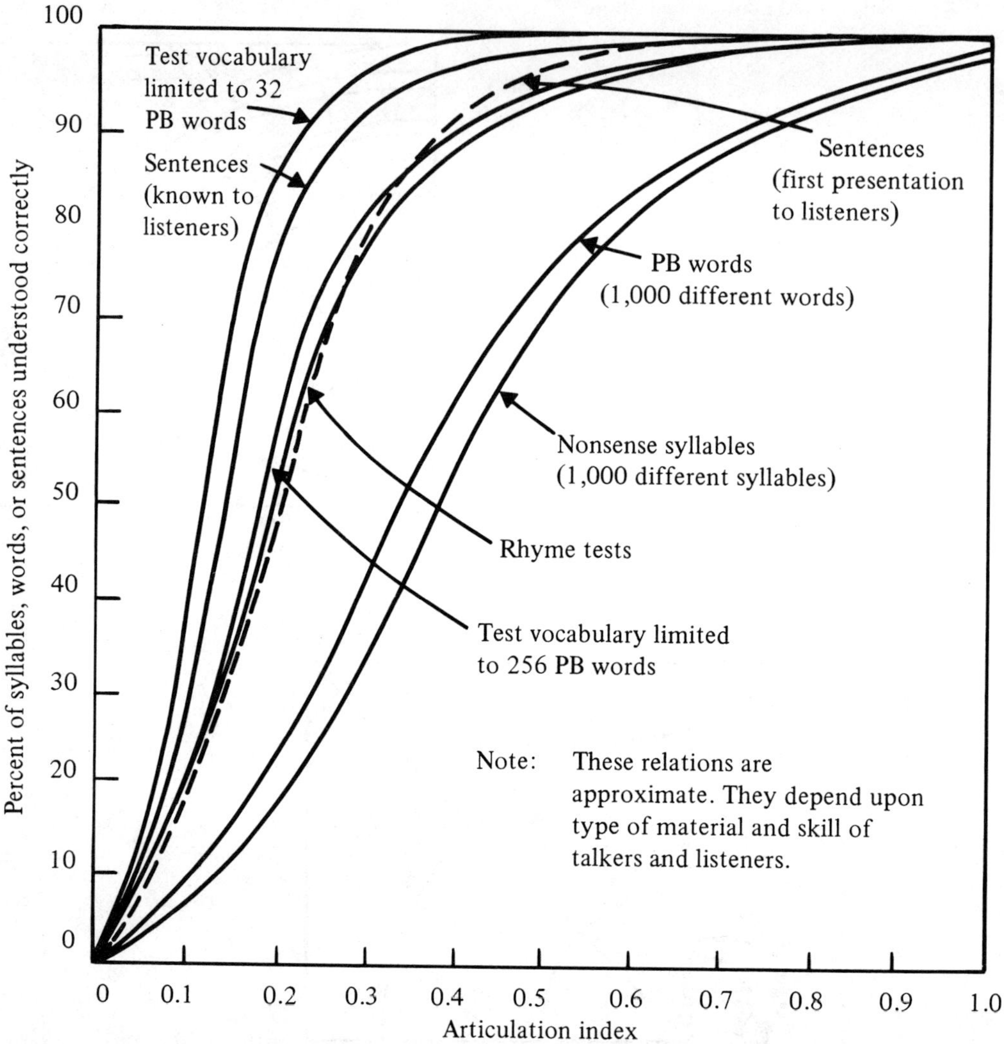

FIGURE 5.46 Percentage speech intelligibility vs. Articulation Index. Relation between AI and various measures of speech intelligibility. (From American National Standard ANSI S3.5-1969, Methods for the Calculation of the Articulation Index, p. 23.)

with the speaker and the topic being discussed. Since the design objective for rooms used for speech should be for first-time sentence presentation, a minimum value of AI = 0.5 is chosen for 95% speech (sentence) intelligibility as a minimum acoustical design criterion. For rooms used by a large number of different talkers and listeners having varying degrees of skill, a value of AI = 0.7 is more appropriate. If we consider a room in which the only masking noise originates from the reverberant sound field (i.e., no background noise), then the rms masking sound pressure can be written using Equation (4.31) as:

$$\langle p_m^2 \rangle_{rms} = 4\,\rho c W^{rms}/R. \tag{5.50}$$

Using $L_p^{pk} = L_p^{rms} + 12$ dB we can put $W^{pk} = 15.8\, W^{rms}$ and rewrite the signal sound pressure with the use of Equation 4.28 to give

$$\langle p_s^2 \rangle_{pk} = \frac{15.8\, W^{rms}\, \rho c\, DF}{4\pi r^2}, \tag{5.51}$$

The peak signal-to-noise ratio can then be obtained from $S\text{–}N = 10 \log \left\{ \langle p_s^2 \rangle_{pk} / \langle p_m^2 \rangle_{rms} \right.$ as

$$S\text{–}N = 10 \log \left(\frac{R}{\pi r^2} \cdot DF \right), \text{ dB}. \tag{5.52}$$

This shows that when the masking comes wholly from the reverberant field, the peak signal-to-noise ratio and hence AI will be independent of the source power output. In this case, Equations 5.49

TABLE 5.2

Male voice speech sound pressure levels +12 dB at 1 m from lips for both 1/3- and 1-octave bands according to Reference 61. These levels represent the speech peaks that contribute to intelligibility. The voice peak sound power levels, L_w^{pk}, can be evaluated by adding 10.8 to the above values as shown for octave bands.

Frequency	L_p^{pk} (1/3)	L_p^{pk} (1)	L_w^{pk}
200	67.0		
250	68.0	72.5	83.3
315	69.0		
400	70.0		
500	68.5	74.0	84.8
630	66.5		
800	65.0		
1,000	64.0	68.0	78.8
1,250	62.0		
1,600	60.5		
2,000	59.5	62.0	72.8
2,500	58.0		
3,150	56.0		
4,000	53.0	57.0	67.8
5,000	51.0		

TABLE 5.3

Weighting factors to be applied to speech signal-to-noise ratios to obtain articulation index*

Frequency	1/3-Octave	Octave
200	0.0004	
250	0.0010	0.0024
315	0.0010	
400	0.0014	
500	0.0014	0.0048
630	0.0020	
800	0.0020	
1,000	0.0024	0.0074
1,250	0.0030	
1,600	0.0037	
2,000	0.0038	0.0109
2,500	0.0034	
3,150	0.0034	
4,000	0.0024	0.0078
5,000	0.0020	

*See Reference 61

and 5.52 can be used to determine speech intelligibility.

Since reverberation in a room will cause speech components of phonemes to overlap, there will be a decrease of speech intelligibility with increasing reverberation time. Beranek' suggested a connection be made to the reverberant speech spectrum of approximately $-10^{(4\ -T_R)/3}$, dB to take account of these reverberant overlap effects. However the current ANSI 53.5 1969 standard ignor the reverberant speech level in the noise calculation and only takes general background noise into account. It then suggests that after the AI value has been calculated, a correction for the reverberation time of the room can be made using Figure 5.47. For rooms with reverberation times T_R between 1 and 4 sec, the following approximate correction may be made to the AI value obtained from Equation 5.52 to give the effective AI′ as

$$AI' = AI + 0.033(1 - 4T_R) \tag{5.53}$$

Table 5.4 shows an example of an octave band Articulation Index calculation for an audience member 70 ft from the speaker who is talking in a normal voice in a lecture theater.

In determining the power output of the noice we should consider that +6 dB is obtained for a "raised voice," +12 dB for "shouting," and +18 dB for maximum vocal effort.

In cases where the listener can clearly see the face and lips of the talker, higher values of AI than expected often result.[66] The effect of such "visual cues" upon the AI observed can be predicted using Figure 5.48.

So far, we have considered the direct sound as being the only useful information-carrying signal. However, there are many sound reflections which are able to reach a distant listener within a relatively short time period after the direct sound. As long as this time delay is small, the ear does not classify these reflections as echoes, and instead, the energy of these reflections contributes to the apparent signal level and intelligibility of the received sound. The way in which the ear integrates such short delay reflections to give an impression of a continuous signal is analogous to the way the eye integrates a flickering light to give apparent visual continuity. The fact that a listener in an auditorium does not notice short time delay reflections leads one to believe that the reflections

TABLE 5.4

Example of Speech Intelligibility Calculation for a Person Seated 100 Ft Directly in Front of a Male Speaker in an Auditorium, Assuming Hard Reflecting Surfaces for the Useful Reflections

Octave bands (Hz)	250	500	1,000	2,000	4,000
Given:					
Speech weighting factors (Table 5.1)	0.0024	0.0048	0.0074	0.0109	0.0078
Voice peak power levels, dB	83.3	84.8	78.8	72.8	67.8
(A) Signal:					
L_p direct at 70 ft	45.9	47.5	41.5	35.5	30.5
L_p reflected at 80 ft	44.8	46.3	40.3	34.3	29.3
L_p reflected at 90 ft	43.8	45.3	39.3	33.3	28.3
L_p reflected at 90 ft	43.8	45.3	39.3	33.3	28.3
Total Signal	50.7	52.2	46.2	40.2	35.2
(B) Noise:					
General background noise given by NC 30	41.5	36.0	32.0	29.5	28.0
(C) Signal-to-Noise Ratio (S-N)	9.2	16.2	14.2	10.7	7.2
(D) Weighted (S-N)	0.022	0.078	0.105	0.117	0.056

(E) Sum (D) = Articulation Index = 0.38 (or 90% sentence intelligibility – Figure 5.46)

(F) Correction to (E) for T_R = 1.0 sec, reverberation time is obtained from Figure 5.47 = –0.10 and so

Corrected Articulation Index = 0.28 (or 75% sentence intelligibility

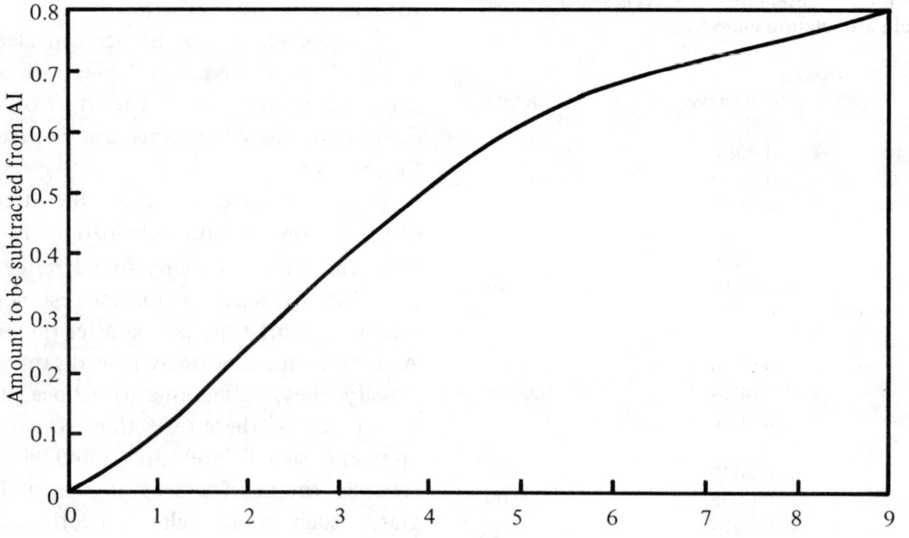

FIGURE 5.47 Articulation Index corrections vs. reverberation time. Note that the ordinate shows the amount to be subtracted from AI. AI cannot be less than 0.0. Corrected AI's less than 0.0 are to be called 0.0. (From American National Standard ANSI S3.5-1969, Methods for the Calculation of the Articulation Index, p. 20.)

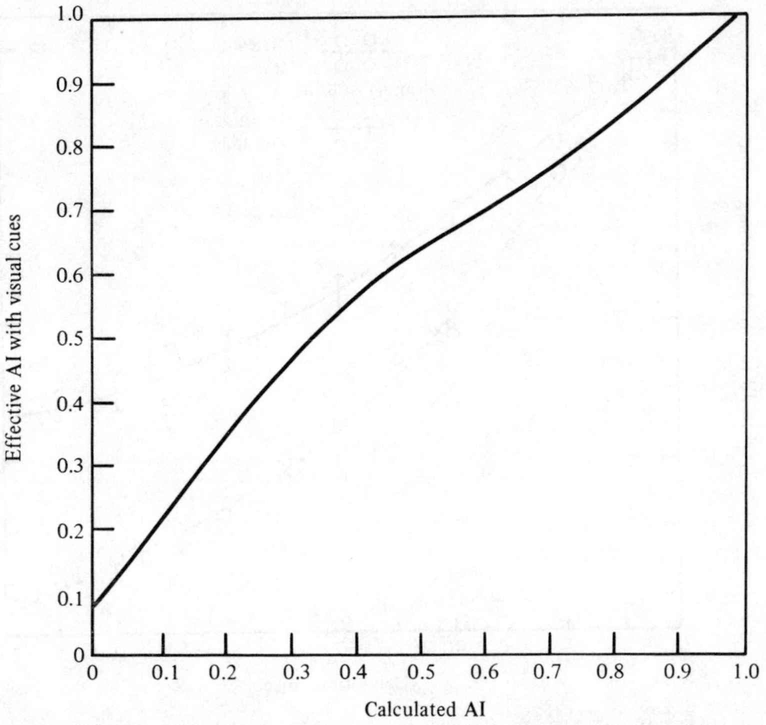

FIGURE 5.48 Relation between calculated AI and effective AI for a communication system wherein the listener can see the lips and face of the talker. (From American National Standard ANSI S3.5-1969, Methods for the calculation of the Articulation Index, p. 22.)

are masked by the primary (direct) sound. Indeed, Haas[67] found that such a masking effect does take place between a primary and a single reflection and is dependent upon the time delay. Later work by Lochner and Burger[52] has shown that this masking effect is also dependent upon the level of the primary sound. Figure 5.49 shows the minimum relative level of a single reflection required to be just perceived as a separate sound (from the primary sound), as a function of (a) the relative time delay and (b) the level of the primary sound. It can be seen that when the primary sound is 50 dB above the ear's threshold level, then the reflection becomes just noticeable (as a separate source) after 30 msec. If the primary sound is only 25 dB above threshold, then the time delay required to produce the feeling of an echo is increased to 50 msec. This finding would indicate that only reflections having time delays of less than 30 msec should be considered useful, and that longer delay reflections should be avoided. However, reflections having time delays greater than 30 msec can often usefully contribute to overall intelligibility. In fact, Lochner and Burger[52] reported that the arrival of the primary sound starts a complex integration process by the hearing mechanism during which the primary and reflected sounds are fully integrated to increase the loudness of the direct sound during the first 30 msec and then partially integrated up to 95 msec. This integration process also depends upon the relative intensity of the primary and subsequent reflections as shown in Figure 5.50. In effect, the ear's integration process produces a masking effect (triggered by the arrival of the primary sound) so that reflections arriving soon after the direct sound are not as well perceived as those arriving much later. It is clear from the above findings that the effect of reflections must be included in the calculation of Articulation Index. For example, we could take the ear integration mechanism into account by writing the total energy of the effective speech signal E_e as follows:

$$E_e = K \int_0^{95} \beta p^2 \, dt, \tag{5.54}$$

where for a pulse

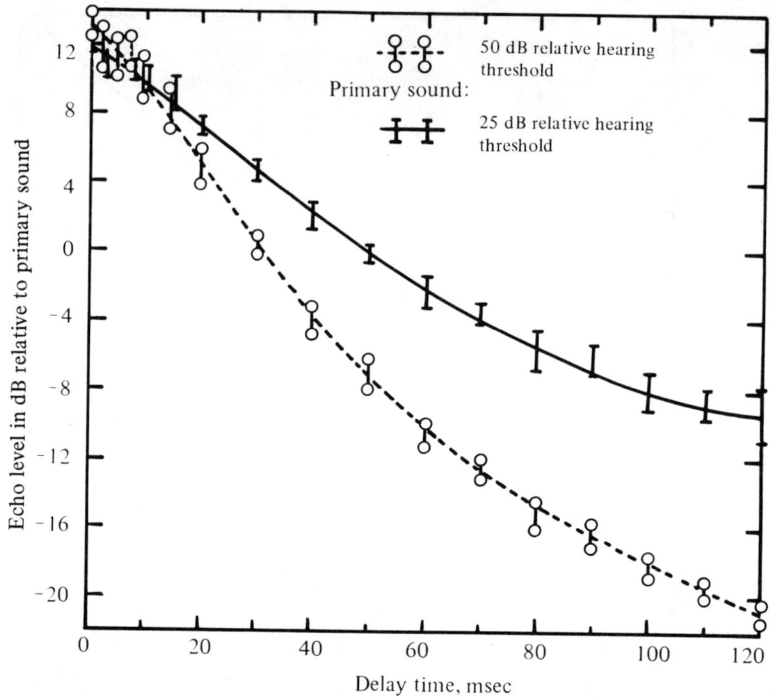

FIGURE 5.49 Curves showing the just perceptible level of a speech echo in dB relative to the primary sound, at different delay times. (From Lochner, J. P. A. and Burger, J. F., The influence of reflections on auditorium acoustics, *J. Sound Vib.*, 1(4), 426, 1964. With permission.)

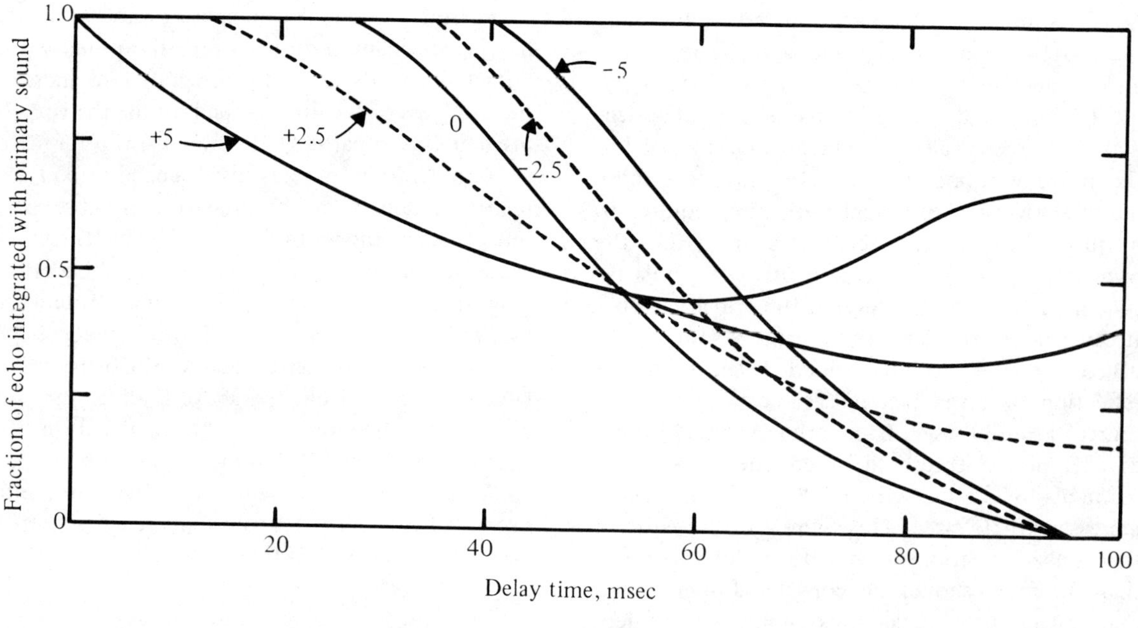

FIGURE 5.50 Fraction of echo integrated with primary sound for +5 dB, 0 dB, and –5 dB echoes. Dotted lines are interpolations for +2.5 dB and –2.5 dB echoes. (From Lochner, J. P. A. and Burger, J. F., The influence of reflections on auditorium acoustics, *J. Sound Vib.*, 1(4), 426, 1964. With permission.)

p = instantaneous sound pressure at time t msec
β = fraction of reflected sound integrated with primary sound obtained from Figure 5.50.

If the masking noise is only due to reflections arriving after 95 msec (i.e., no ambient noise), then the signal-to-noise (reverberant) ratio can be writen as

$$(S\text{–}N) = 10 \log \left[\frac{\int_0^{95} \beta p^2 \, dt}{\int_{95}^{\infty} p^2 \, dt} \right]. \tag{5.55}$$

If background noise is present, then it should be added to the reverberant sound pressure on an energy basis. For more practical purposes it is useful to consider equally all those reflections arriving within 50 msec of the primary sound (on an energy basis) to give the total signal. The peak sound pressure level of each reflection is given by

$$L_p^{pk} = L_W^{pk} - 20 \log d - 0.5 - 10 \log DF + 10 \log (1 - \alpha) \tag{5.56}$$

where

L_W^{pk} = long-term rms + 12 dB, speech power level re 10^{-12} W
α = net absorption coefficient for the reflecting surface(s)
DF = directivity factor for given reflected path from the speaker's voice in the horizontal and/or vertical planes (Figures 5.44 and 5.45)
d = total path length (ft).

In addition to contributing to the overall intelligibility, subsequent reflections lead to an auditory impression of the size of the room. This assessment of "presence" is related to the time interval separating secondary reflections arriving at the two ears of the listener,[52] while the impression of direction is obtained from the relative amplitudes and times of arrival of the primary (direct) sound at each ear. For further reading on speech intelligibility calculations the reader should consult References 52 and 67 to 73. The problem of designing rooms for speech privacy requires the Articulation Index to be very small– usually below AI = 0.05. Speech privacy will be separately dealt with later in Volume 2.

5.8.3 Physical Requirements

In all architectural spaces having requirements for speech reception the primary acoustical objective must be to achieve maximum speech intelligibility. From the preceding sections it is clear that this objective can only be fulfilled by (a) careful choice of optimum size and shape, (b) provision for useful short delayed (preferably less than 35 msec) reflections, (c) good direct sound, (d) adequate diffusion, (e) freedom from acoustical defects, and (f) optimum reverberation time for both intelligibility and enhancement of the "naturalness" of the speech. All of these requirements are equally important and the lack of provision of any one of them can lead to poor acoustical quality and intelligibility.

In general, the same initial design procedure may be applied to rooms used for music as those used for speech. For those rooms in which the only source of acoustical power is a single unamplified voice, the total power is obviously much lower than that of a full orchestra and consequently, the room must be considerably smaller to achieve the same loudness levels (for good speech intelligibility). Audience sizes in rooms used for speech are therefore severely restricted in their number by the voice power of the speaker to usually not more than 1,500 for drama theaters and preferably not more than 500 for lecture theaters.

Figure 5.51 shows an acoustical design flow chart for rooms used for speech. Having decided upon the function of the room, the required floor area can be established. In the case of lecture halls, movie theaters, and other audience/spectator orientated rooms where the people listening to the speaker are gathered together in one main body rather than spread out, as in law courts, for example, the floor area can be established by assuming 5 to 7 ft^2/person. The proposed floor plan can then be used to design the required seating rake by following the procedure described in Section 5.7.1. Having done this, the direct sound pressure level for the most remote listener can be determined to ensure adequate loudness of the direct sound for each of the octave bands from 125 to 4,000 Hz. If this loudness level is not found to be sufficient, then either the floor plan has to be changed, a balcony has to be introduced, or the audience size decreased. The only remaining alternative would be to introduce an amplification sound system. Each of these new alternatives can

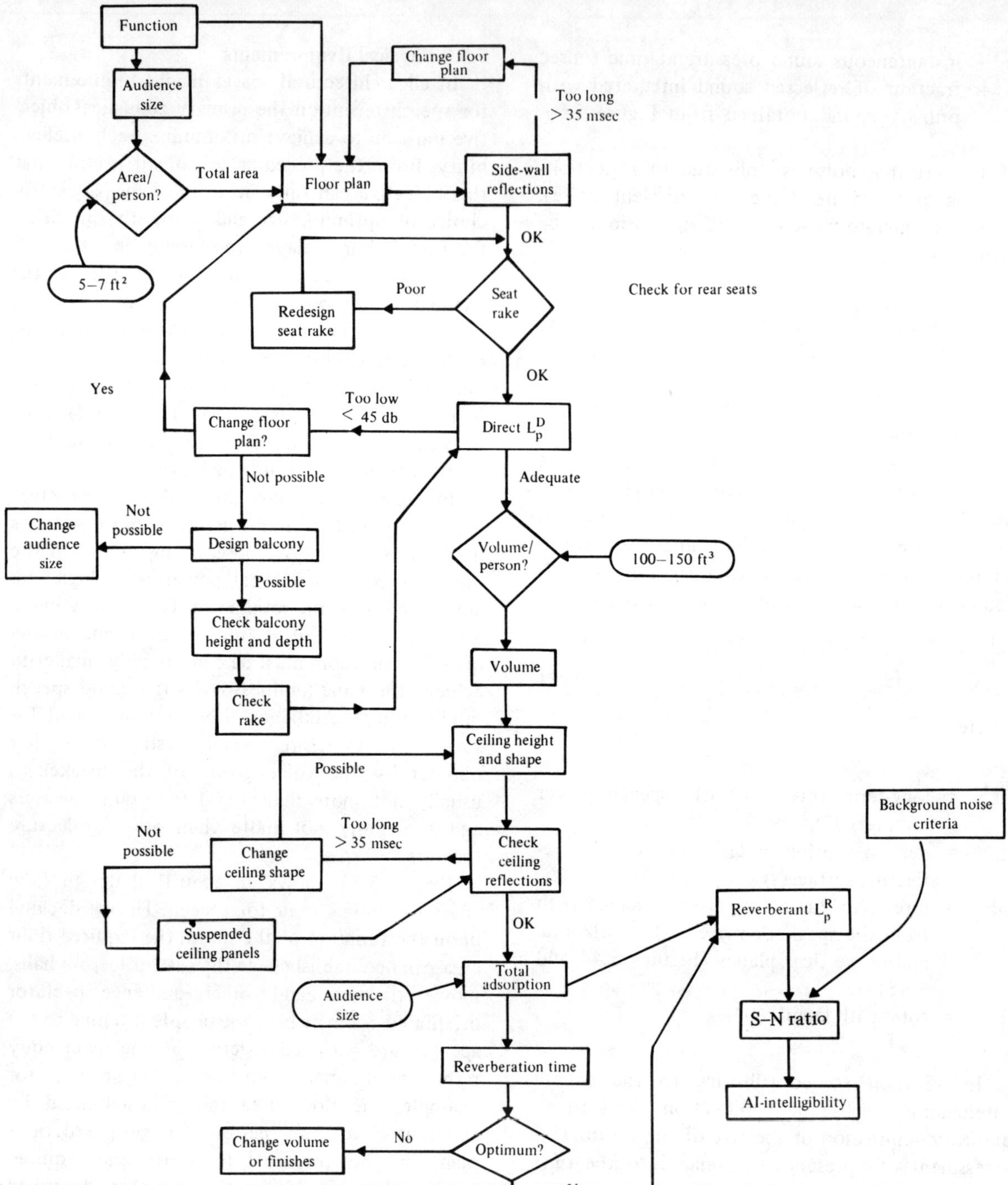

FIGURE 5.51 Design flow chart for rooms used for speech.

then be retested for adequate direct sound loudness (usually not less than 40 dB and preferably > 45 dB). Once this has been accomplished, the volume of the room can be decided by choosing a volume per person of 100 to 150 ft³, and preferably closer to 100 ft³/person. The volume then sets the approximate ceiling height, although this should be shaped to provide good reflection to the rear portion of the room. The maximum time delay of ceiling reflections (probably worse for the central audience) can be checked to ensure that they are preferably less than 35 msec and on no account

TABLE 5.5

Audience Absorption (in Sabins) Per Person for an Almost Fully Occupied Auditorium***

	63	125	250	500	1,000	2,000	4,000	Hz
*	1.5	2.0	3.0	3.5	3.8	4.0	4.2	sabins
**	1.6	2.0	4.3	5.0	5.0	5.5	4.0	sabins

*See Reference 26
**See Reference 14
***It is interesting to note that W. Sabine originally calculated 4.75 sabins at 512 Hz.

greater than 95 msec. If the ceiling reflection time delays are excessive, then suspended ceiling panels may have to be incorporated to reduce the time delay and improve the sound distribution without decreasing the effective volume of the room. However, arrays of such panels must be used with great care since their size must be sufficiently large as compared to the longest wavelength of interest in order to behave as infinite reflecting surfaces. Below a critical frequency, such reflector arrays are found to be essentially totally transparent[74,76] and therefore provide no low frequency reflections. For the size of panel reflectors often found in modern auditoria, (i.e., 6 X 12 ft) this frequency is found to be between 200 and 300 Hz.[74] This effect has resulted in a loss of low frequency response in some large auditoria and concert halls.[77] It is also important to ensure that the individual reflector panels are constructed to be heavy enough to avoid the creation of low frequency resonances. The next stage, and perhaps the most important in the whole process, is to determine the total absorption in the room when occupied by an audience. Since a large percentage of the total absorption in most auditoria is due to the audience itself, it is very important to be able to estimate accurately this absorption and to take the effect of empty seats into account. Published data have shown that the effective absorption per person in an audience varies considerably from 3.1 to 6.5 sabins at 500 Hz, depending upon the surface density and rake and type of the seating. Table 5.5 shows a comparison of some published values for audience absorption on a per person basis as a function of frequency. This method may be used for rooms used for speech with quite good success, but as we shall see later, a better approach, especially in large theaters and concert halls, is to evaluate the effective audience absorption based on the total area occupied. Auditoria used for speech are much less dependent upon their reverberation time for successful acoustical qualities than those used for music and consequently the "per person" approach to audience absorption will generally yield accurate values of the Articulation Index. However, for very large lecture theaters it is advisable to use the absorption per unit area method outlined in the next section. To overcome the problem of partial audiences it is necessary to use heavy upholstered seats so that they can contribute equal absorption whether occupied or empty. This is not always possible in many lecture halls, for example, where hard wood seats are used. This can severely change the reverberation characteristics of the room.

Once the total absorption has been established the reverberation time can be evaluated and compared to the optimum for the chosen volume using Figure 5.52. The estimated value should be within approximately 10% of the optimum value, and if it is not, then remedial sound absorbing/reflecting treatment will be required. Once this has been achieved, the signal-to-noise ratio can be evaluated and the Articulation Index checked for satisfactory conditions, as described in the previous section. In the process of this calculation it should be remembered that a speaker will generally tend to talk louder in a large room (see Figure 5.53) and this can be taken into account by increasing the voice sound power levels obtained from the levels given in Table 5.2 by +6 dB to correspond to a "raised voice" in rooms having a volume in excess of 100,000 ft^3. Since ambient noise has to be added to the reverberant noise level in the calculation of the signal-to-noise ratio, it is very important to keep this extra noise quite low. In practice, the ambient noise should not exceed PNC (Preferred Noise Criterion) 35 and preferably

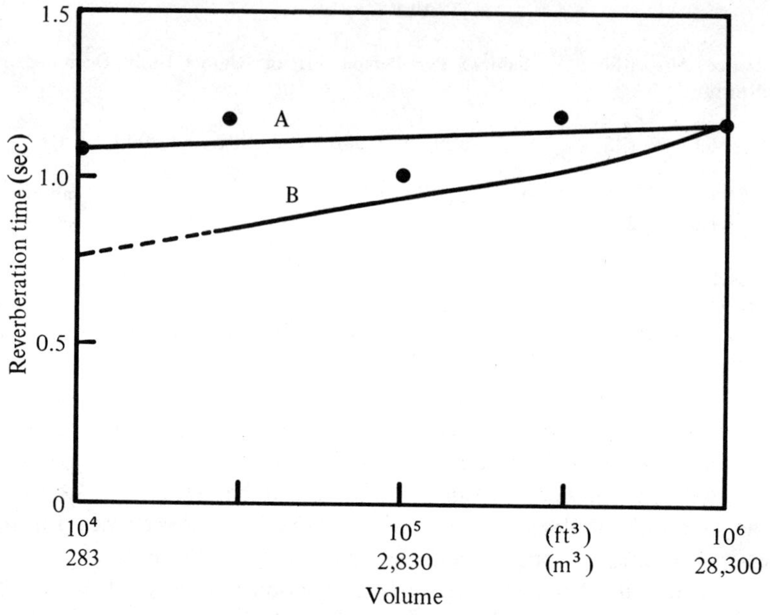

FIGURE 5.52 Curves of optimum reverberation time as a function of room volume: (A) as obtained from discrete reflections; (B) after Knudsen.[57] (From Lochner, J. P. A. and Burger, J. F., The influence of reflections on auditorium acoustics, *J. Sound Vib.*, 1(4), 426, 1964. With permission.)

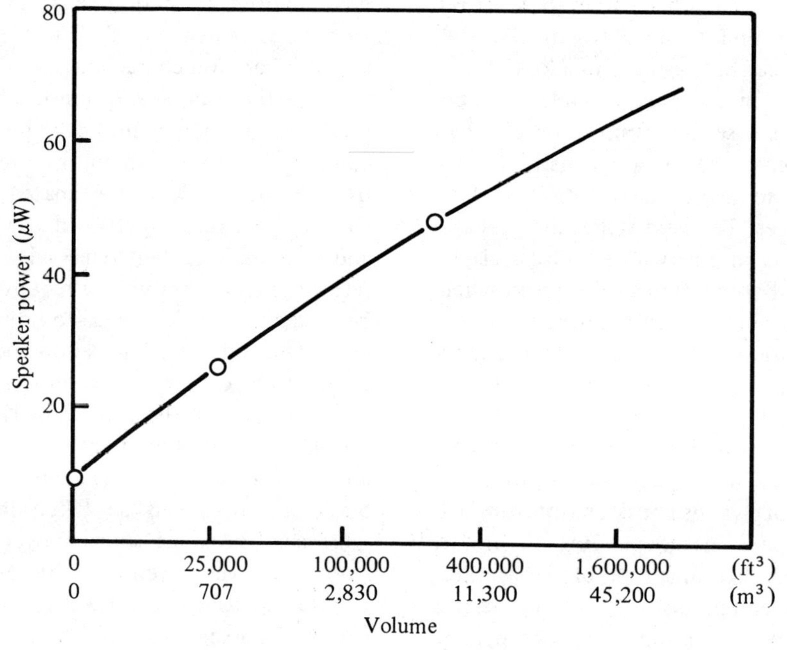

FIGURE 5.53 Curve showing the probable speaker power in auditoria of different sizes. (From Lochner, J. P. A. and Burger, J. F., The influence of reflections on auditorium acoustics, *J. Sound Vib.*, 1(4), 426, 1964. With permission.)

be close to PNC 30. For this reason a careful check has to be made of the acoustical performance of any air handling system that might be used in the auditorium. Furthermore, it is usually good practice to use a carpet floor covering where possible in the audience area since this will substantially reduce footstep noise. In public spaces, such as law courts, the use of a resilient floor covering material is highly recommended. Small classrooms of volumes between 5,000 and 10,000 ft^3 are usually not a major acoustical problem[78] provided the reverberation time can be kept below 1 sec[79] and the ceiling can act as a useful reflector. If sound-absorbing material is required, it should be kept to the upper parts of the side walls and possibly above the speaker and on the rear wall. Insulation from excessive exterior noise is also very important and acoustical double glazing is often required. Sound transmission from adjacent classrooms is often a potential problem but can be avoided if the dividing partition has a sound transmission class rating in excess of STC 50. Additional sound absorption in a classroom will lower the general noise level leading to better working conditions for the students and improved speech articulation.

5.9 DESIGN OF ROOMS FOR MUSIC

5.9.1 The Development of Music and Concert Halls

Throughout history, music has played an important role in the entertainment and enjoyment of mankind. Early in the development of music, man became aware of the enhancement effect of reverberation within enclosed spaces and made use of this, especially in religious buildings which in many cases had reverberation times greater than 10 sec. Such long reverberation times were ideal for the recitation of liturgy and chanting. One of the finest examples of this type of building would be Westminister Abbey in London which dates back to the 10th Century and has a reverberation time in excess of 6 sec. It should be pointed out, however, that these long reverberation times did not actually result in many problems of communication, due to the particularly poor speech intelligibility, since the original religious activity was usually restricted to a small number of worshippers and clergy (and a small choir) positioned around the high altar. The lay public only took part in major religious events when speech communication was probably not as important as the occasion itself. Today, of course, the great cathedrals have much more stringent requirements on communication due to the changes in the nature of the service and size of congregation. In such cases, the long reverberation times do become a problem as far as speech intelligibility is concerned and very often electronic acoustic reinforcement systems have to be employed.[80]

As the nature of music changed throughout the centuries, so the architecture and acoustic environments used for the creation and performance of the music also changed. In many cases, composers have specifically written music with a particular building and its acoustics in mind, while in countless cases the composer has no doubt unconsciously molded his music to suit the acoustics of the auditoria available to him. The creation, performance, and experience of music cannot be separated from the acoustics of the space in which it is played and musicians will adjust their performance (either consciously or unconsciously) to suit best the acoustics of the auditoria in which they find themselves. Concert halls that are claimed to have excellent acoustics for one type of performance may equally be rated as only reasonable or fair for another type. It is therefore not surprising to find that it is very difficult to design a large space for music performance and appreciation. The present state of knowledge of architectural acoustics enables the acoustical consultant to make recommendations that will almost certainly produce good acoustics, but it is almost impossible to ensure excellent acoustics. To make matters more difficult, it is often impossible to get even qualified musicians to agree on either what is meant by excellent acoustics or on the acoustical qualities of a given concert hall. Since modern concert halls have to be used with performances of many types of music for large audiences, this has resulted in the rules for excellent acoustics becoming even more elusive. Perhaps the only sure measure of success for a modern concert hall is time itself.

In the previous section, we saw that rooms can be designed for excellent speech intelligibility. Their performance can easily be verified by reading a set of phonetically balanced words as sentences and observing the scores of the listener. However, in rooms used for music, the criteria are considerably more complex and almost totally subjective, making them very difficult to define and often impossible to measure. These criteria may be totally different from the ones developed for speech intelligibility and certainly a room which is good for speech may be totally inadequate for music. The acoustical design of such spaces is therefore both an art as well as a science. It is only during the last 20 years that the acoustical design and performance has been more fully understood. This increase in knowledge has been due to many researchers, but it would be inappropriate not to mention Dr. L. L. Beranek's famous survey of some 60 concert halls from all over the world which was later published as *Music, Acoustics and Architecture* (see Reference 51). This work authoritatively defined the fundamental parameters and at last made clear their order of importance. He listed 18 acoustical factors which were found to influence the assessment of musical acoustics in a concert hall. Of these, reverberation time (as a function of frequency) stood out as the most important factor – as we saw in the previous section for rooms used for speech. However, reverberation by itself may be either important or unimportant since it depends so much on the style of music being played and upon several other acoustical parameters. As the development of music has progressed through the centuries, so the requirements for reverberation have also changed. During the Baroque period (1600–1750) composers such as Bach and Handel wrote music which placed its emphasis on being able to listen to independent melodic lines rather than chords and therefore required high musical articulation. This restricted auditoria to relatively small values of reverberation time (less than 1 to 5 sec).[51] The latter half of the 18th Century saw the Classical period of Hadyn, Mozart, and Beethoven and the first appearance of public concert halls which could accommodate 300 to 400 people. During this period the emphasis was on the form and structure of the music. By present subjective impressions, listeners prefer a reverberation time of 1.5 to 1.7 sec for this type of music.[51] As the Romantic period of the 19th Century developed, less emphasis was placed on the structure of the music and more on the emotional expression it could present to the listener. The works of Brahms, Wagner, Tchaikovsky, and Berlioz, for example, required less definition and clarity than music from the Classical period and more fullness of tone. This type of music therefore required longer reverberation times often as high as 1.3 sec, and many composers of this period (perhaps more than in any other) wrote with a particular hall in mind. The latter part of the 19th Century also witnessed the building of modern concert halls and attendance was often more social than musical. The old Boston Music Hall was built in 1863 and was able to accommodate an audience of 2,400 with a reverberation time of 1.8 sec. A little later the Grosser Musikvereinssal in Vienna was built in 1870 to house 1,680 people for a 2.0-sec reverberation time. This is still rated as one of the very best concert halls in the world today.

The 20th Century has seen a variety of styles and types of concert music. This has very often been written for large orchestras, and mainly for economic reasons has been played to large audiences of frequently close to 3,000 people. Because of the variety of types of music, taken from all the previous periods described above, which are now played in our modern 20th Century concert halls, the need is for great flexibility in the requirements for reverberation time. It is therefore exceedingly difficult to design a concert hall for all periods of music without having some means to vary the acoustics. Although such techniques as rotating absorbers (hard on one side and soft on the other) and whole ceilings which can be lowered and raised are possible, it is very difficult to achieve satisfactory results in large concert halls. A recent example of a variable ceiling is found in The Edwin Thomas Performing Arts Hall at the University of Akron, Ohio where a 44-ton ceiling composed of 3,600 interlocking steel panels, counterbalanced by 27 chrome plated steel cylinders totaling 47 tons in the lobby, can be raised and lowered with small motors and computerized circuitry. In addition to this volume control provided by the ceiling, 12,000 ft^2 of acoustical absorption can be added in stages further to vary the reverberation time.

5.9.2 Musical Qualities of a Room

In order to understand better the complex subjective assessment of music appreciation in listening rooms it is necessary first of all to review some of the more important musical qualities which are affected by the room acoustics. These are generally regarded as (1) Fullness of tone, (2) Definition, (3) Intimacy, and (4) Loudness, and have the following meaning and acoustical implications:

1. **Fullness of tone** – the expression used to

describe the blending effect that reverberation has upon successive notes and chords when heard in a room. This is an extra "continuity" effect added to the musical character by the room; it would be totally absent outdoors in an open space. Furthermore, a tone played in a reverberant room will sound louder, due to multiple short delayed reflections and the reverberant sound field, then it would if played in a totally dead space. As the ratio of the reverberant to direct sound pressure levels increases, the fullness of tone also becomes greater (as the blending process increases). As the fullness of tone becomes great, the blurring of tones results in a loss of clarity (or definition) and these two musical qualities are therefore contradictory. It is important to note that both the reverberation time and the ratio of reverberant to direct sound are responsible for the degree of fullness of tone obtained. Hence, it is not unusual to find concert halls in which the subjective reaction of the audience shows a large variation in their assessment of fullness of tone, yet measurements show only little variation in reverberation time.

2. **Definition (sometines *clarity*)** – refers to the degree of intelligibility obtained from consecutive musical sounds. In other words, how well individual notes or instruments can be heard distinctly from adjacent ones. Beranek refers to two kinds of definition,[51] namely *horizontal definition,* which relates to successive tones, and *vertical definition,* which relates to tones played simultaneously. Apart from the skill of both the composer and player, horizontal definition is affected by the reverberation time of the room and the ratio of direct to reverberant sound. Since this ratio has the inverse requirement for fullness of tone, an increase in horizontal definition results in a corresponding decrease in fullness of tone. Vertical definition depends upon the same acoustical factors as for horizontal definition, upon the balance of the sound from the different instruments both transmitted to the audience and among the orchestra, and upon the relative response of the hall as a function of frequency. Once again we can see that specific types of music demand specific acoustics of the room in which it is played. Some music requires high definition (i.e., Mozart) and low fullness of tone, while other pieces (i.e., Bach) may require high fullness of tone and low definition. The skill of the performer and the speed at which he plays the music also affect the degree of definition obtained.

3. **Intimacy (or presence)** – related to the subjective impression of the size of the room in which the music is being played. Beranek found[51] that intimacy was the most important single factor in the judged acceptability of a concert hall – reverberation time affecting several qualities. Primarily, intimacy is dependent upon the time delay between the direct sound and the first reflection. Hence narrow halls, such as the Grosse Musikvereinssaal in Vienna and the Boston Symphony Hall which have very small initial time delays of 12 and 15 msec, respectively, at the center of the main floor, exhibit extremely high intimacy. Another term often used to describe intimacy is *Spatial Impression* SI. It is now fairly well established that it is the early *lateral* reflections which determine the degree of SI, and that this is a function only of the width (and not height) of the hall. In multiple reflection situations, the degree of SI or intimacy is related to the ratio of lateral to nonlateral early energy by:[82]

$$SI = k \left[\sum_{t=5\text{msec}}^{80\text{msec}} P \cos\theta \right] \div \left[\sum_{t=0}^{t=80\text{msec}} P\,(1 - \cos\theta) \right], \quad (5.57)$$

where

k = constant
P = "reflection level" – including direct sound
θ = angle of incidence of reflection relative to an axis through the listener's ear.

Since many pieces of concert music were written with a particular hall in mind, or at least for a particular size room, there is an inherent requirement for intimacy in the music, and if it is played in a hall having a completely different degree of spatial impression, then it will sound very noticeably wrong and somehow unnatural.

4. **Timbre** – a term which refers to the tonal qualities that distinguish one instrument (or voice) from another. It depends upon the relative strengths and combination of harmonic overtones present in the sound produced. When a number of individual voices and instruments are put together, the effect of their combined timbres is known as *tone color.*[51] Beranek reported that the tone color at a particular seat in a concert hall depends upon how the sounds from the different sections of the orchestra reach the listener and that it is a complex quantity dependent upon many acoustical factors.

Each of the above musical factors is dependent

upon several physical acoustical parameters, often in a complex relationship. Figure 5.54 shows the interrelationships between the musical qualities heard in a hall and the corresponding acoustical qualities.

5.9.3 Acoustical Factors and Their Relative Importance

In an effort to define the vocabulary used to describe the subjective assessment of the musical qualities of an auditorium, Beranek developed an 18-word vocabulary of musical-acoustic quality.[51] This is shown in Figure 5.55. Most of these terms have already been defined in the previous section and in Section 5.8. For a detailed description the reader should consult Reference 51.

By studying over 60 concert and opera halls throughout the world, Beranek was able to develop a system for rating the acoustical qualities of a hall in terms of these 18 acoustical factors. A point system was developed whereby a maximum of 100 points could be obtained for an excellent hall. The scale was further divided into subjective categories as shown in Figure 5.56. The relative importance of the 18 acoustical factors was investigated and a rating scale set up for each of them. Of these, *intimacy, liveness, warmth, loudness of direct sound,* and *loudness of reverberant sound* were related to measureable physical acoustic parameters. Although cubic volume and reverberation time do not influence the subjective judgement of the quality by themselves, they do

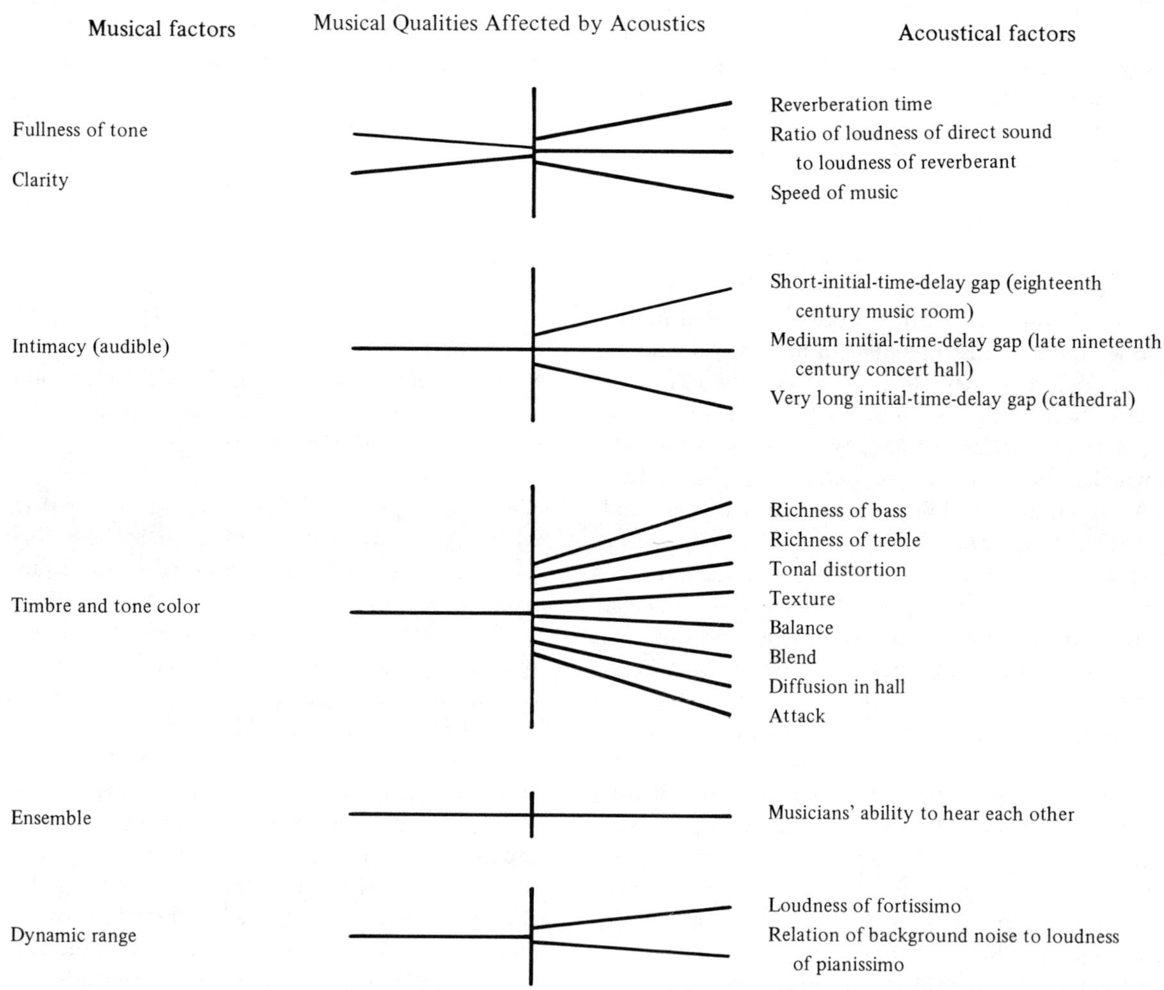

FIGURE 5.54 Chart showing the interrelations between the audible factors of music and the acoustical factors of the halls in which the music is performed. (From Beranek, L. L., *Music, Acoustics and Architecture,* John Wiley & Sons, New York, 1962, 43. With permission.)

Quality		Antithesis	
Noun form	*Adjectival form*	*Noun form*	*Adjectival form*
1 intimacy, presence	intimate	lack of intimacy lack of presence	non-intimate
2 liveness, fullness of tone	live	dryness deadness	dry dead
reverberation	reverberant	lack of reverberation	unreverberant
resonance	resonant	dryness	dry
3 warmth	warm	lack of bass	brittle
4 loudness of the direct sound	loud direct sound	faintness . . . weakness . . .	faint . . . weak . . .
5 loudness of the reverberant sound	loud . . .	faintness . . . weakness . . .	faint . . . weak . . .
6 definition, clarity	clear	poor definition	muddy
7 brilliance	brilliant	dullness	dull
8 diffusion	diffuse	poor diffusion	non-diffuse
9 balance	balanced	imbalance	unbalanced
10 blend	blended	poor blend	unblended
11 ensemble	–	poor ensemble	–
12 response, attack	responsive	poor attack	unresponsive
13 texture	–	poor texture	–
14 no echo	echo-free anechoic	echo	with echo echoic
15 quiet	quiet	noise	noisy
16 dynamic range	–	narrow dynamic range	–
17 no distortion	undistorted	distortion	distorted
18 uniformity	uniform	non-uniformity	non-uniform

FIGURE 5.55 Vocabulary of subjective attributes of musical-acoustic quality. (From Beranek, L. L., *Music, Acoustics and Archietcture,* John Wiley & Sons, New York, 1962, 64. With permission.)

play a contributing role in many of the above parameters. The rating scales developed by Beranek for orchestral concerts are shown in Figure 5.57. Of these, it can be seen that *intimacy* was found to be the most important single acoustical factor, contributing a maximum of 40 points to the 100-point overall scale. It was found that halls having initial reflection delay times of less than 20 msec were able to accumulate 40 points. Following intimacy came *liveness* (dependent upon midfrequency reverberation time) which could contribute a maximum of 15 points to the overall scale. As may be expected, the requirements for liveness depend upon the type of music being performed. For typical orchestral music, a value for $(T_{R500} + T_{R1000})/2 = 1.9$ sec gives 15 points while a value of 1.5 sec gives only 7 points. While liveness depends upon the midfrequency reverberation time, the lack or excess of low frequency reverberation is known to be equally important. The bass response is know as the *warmth* of the hall and is defined as the ratio of low to midfrequency reverberation times: $(T_{R125} + T_{R250})/(T_{R500} + T_{R1000})$. Beranek found that a bass ratio of 1.20 to 1.25 gave the best conditions for warmth and he gave this a value of 15 points.

The loudness of the direct sound was also found to be an important parameter in the overall subjective assessment of the acoustics of a concert hall. It is necessary that the direct sound should be neither too strong (causing listener discomfort) nor too weak (and hence lost in the audience and other background noises). It was found[51] that the best liked seats in the most highly rated concert halls were all in the middle of the main floor and at an average distance of 60 ft from the orchestra. Beranek therefore assigned 10 points to the value 60 ft, and decreased the number of points to 0 at 160 ft (too weak) and to 7 points at 30 ft (too

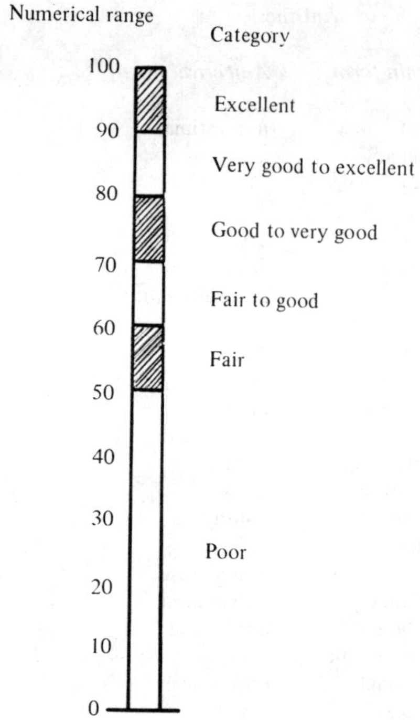

FIGURE 5.56 Range of numerical values arbitrarily assigned to the five categories into which the concert halls of this study are divided, plus the category "poor" according to Beranek's rating method. (From Beranek, L. L., *Music, Acoustics and Architecture,* John Wiley & Sons, New York, 1962, 412. With permission.)

strong). The rating scale shown in Figure 5.57 is to be applied only to seats on the main floor. For balconies, the same scale is used but corrections for useful short delayed (< 35 msec reflections from the side walls (+4 points) and from highly reflective ceilings (+4 points) must be applied, provided the overall rating for this parameter does not exceed 10 points.

The loudness of the reverberant sound is also important in the overall impression of acoustical quality. The loudness level of a reverberant tone is found[82] to be directly proportional to the ratio of the reverberation time and cubic volume, T_R/V. Beranek then multiplied this by 1,000,000 (since most concert halls have volumes between 300,000 and 1,500,000 ft^3) and obtained a rating scale of 1 to 6 for the reverberant sound pressure level. Examination of the concert halls in Beranek's study showed that a value for $(T_R/V) \times 10^6$ of between 2.5 and 3.5 gave maximum acceptability for reverberant sound loudness, and so was assigned the maximum 6 points on the rating scale.

Further ratings are given by Beranek to take into account diffusion, balance and blend, ensemble, and acoustical faults. As we have seen previously, diffusion depends upon reverberation time and scattering effects. Concert halls with ceiling coffers, deep beams, and other major surface irregularities (especially 19th Century halls) all appear to have adequate diffusion, giving the listener the feeling of being completely surrounded in sound. Concert halls that do have sufficient surface irregularities, and therefore presumably adequate diffusion, are given 4 rating points. Less diffusion results in a lower number of points as shown in Figure 5.57.

If the sound from the various sections of an orchestra is well mixed and blended together and is projected to the audience by the stage and front portion of the hall, then the room is said to possess good *balance and blend*. Beranek gives 6 rating points to concert halls used for symphony orchestras if the degree of balance and blend is good. This is obtainable by the proper architectural acoustical design of the stage, front hall, and orchestra pit. As well as being able to balance the sound from different sections of the orchestra to the audience, it is important that the performers be able to hear each other. By doing so, they are better able to play together and consequently are said to exhibit good *ensemble*. A maximum of 4 points is assigned to this parameter, as shown in Figure 5.57

In addition to the above comments on the physical parameters that influence the acoustics of a concert hall, we must remember to avoid all of the acoustical faults described in Section 5.8.3.

Since Beranek's 18-word vocabulary was developed in 1962, several other lists of descriptive acoustical terms have been suggested. In 1970, Watters and Beranek gave a revised list of rating scales to be used to describe auditorium acoustics.[82] For a detailed description of this rating method the reader is directed to Reference 82.

5.9.4 The Effect of the Audience

The audience in an auditorium represents the major contribution to the total sound absorption present. For example, it would not be unusual to see an empty hall exhibit perhaps five times the reverberation time observed when the hall is fully occupied. As discussed briefly in Section 5.8, the absorption provided by the audience can be

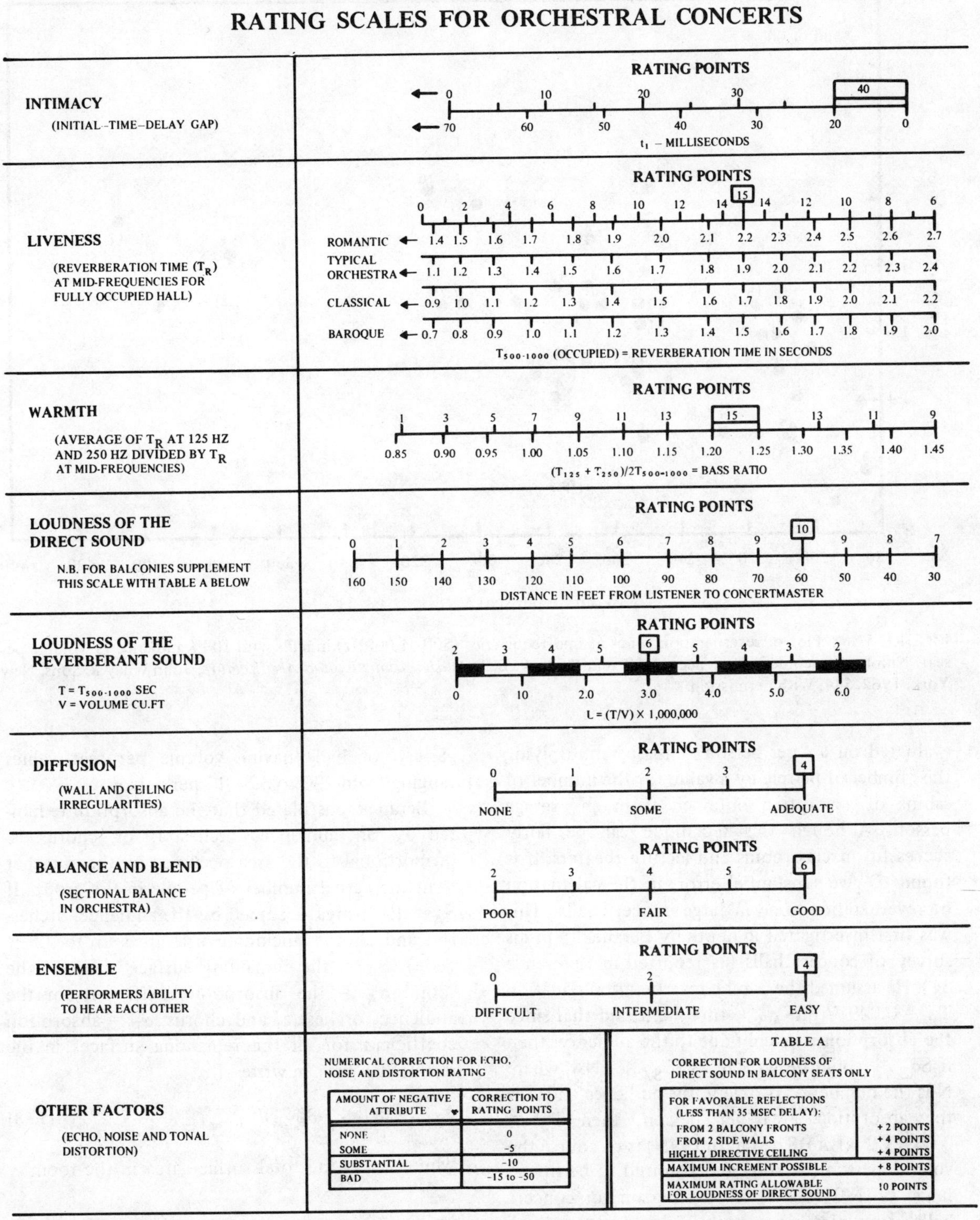

FIGURE 5.57 Beranek's rating scales for orchestral concerts. (From Beranek, L. L., *Music, Acoustics and Architecture*, John Wiley & Sons, New York, 1962, 472. With permission.)

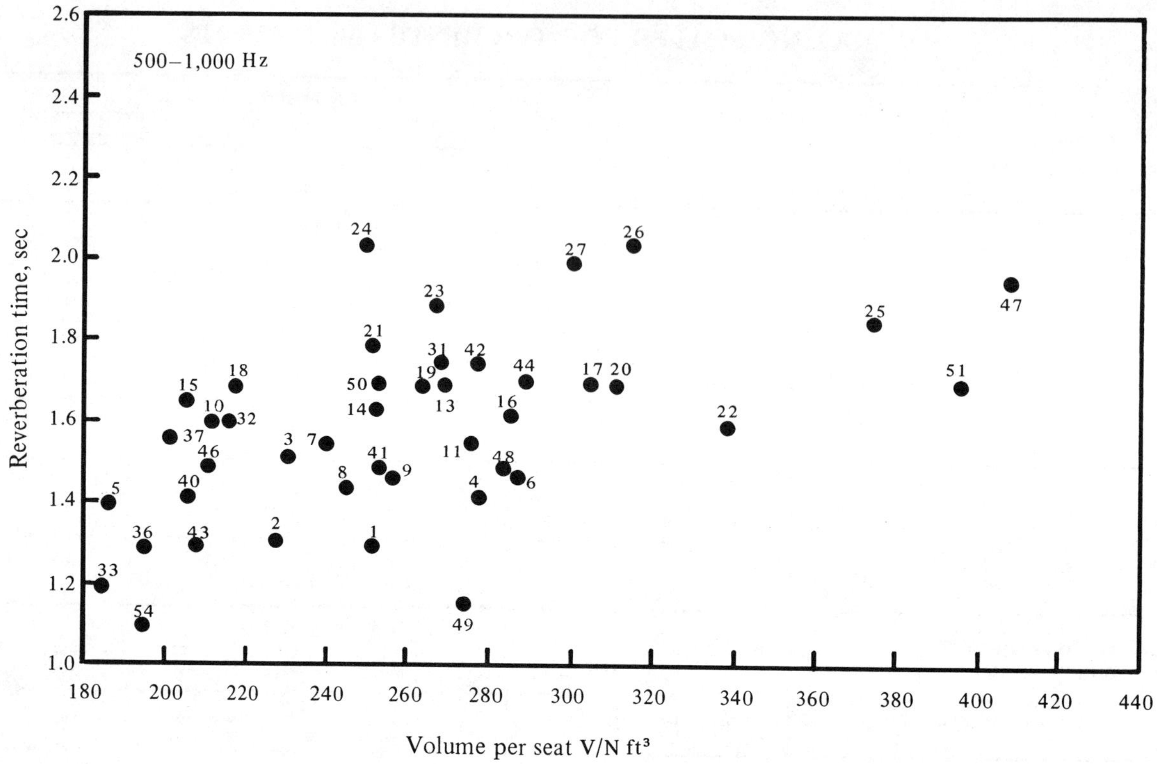

FIGURE 5.58 Plot of reverberation times at midfrequencies (500–1,000Hz) in halls with 100% audience vs. volume per seat. Number of people in audience = N. (From Beranek, L. L. *Music, Acoustics and Architecture,* John Wiley & Sons, New York, 1962, 544. With permission.).

evaluated on a "per person" basis by multiplying the number of people by a value for the number of sabins of absorption obtained from an average person. Although this technique can be fairly successful in classrooms and lecture theaters, it is found to give substantial errors in the calculation of reverberation time in large concert halls. This was first investigated in depth by Beranek[83] in his survey of concert halls (as reported in Reference 51). He assumed the Sabine reverberation equation $T_R = 0.049V/(S\bar{\alpha}_{SAB} + 4mV)$ and said that since the absorption is mainly due to the audience, then if $S\bar{\alpha}_{SAB}$ could be written as $S\bar{\alpha}_{SAB} = NA$, where N is the number of people in the audience and A is the absorption area per person, then $T_R \cong 0.049(V/N)(1/A)$. The factor (V/N) is of course the volume person (or seat) and is found to be in the range of 180 to 400 ft³ per person for concert halls.[51] By plotting a graph of reverberation time T_R against the volume per seat (V/N), Beranek found that the correlation was really very poor, indicating the error in the absorption per person assumption (see Figure 5.58). For example, a reverberation time of 1.7 sec was observed for a variety of halls having volume per seat values ranging from 220 to 320 ft³ per person.

Beranek postulated that the absorption exhibited by an audience, orchestra, or chorus is proportional to the area occupied by them, rather than the actual number of people in the group. If S_A = floor area occupied by the audience, orchestra, and chorus (including aisle areas up to 3.5 ft wide), S_R = the remaining surface areas of the room, α_A = the absorption coefficient for the audience, orchestra, and chorus, α_R = absorption coefficient for all the remaining surfaces in the room, then we can write

$$\alpha_{SAB} = (S_A\alpha_A + S_R\alpha_R)/S_T \qquad (5.58)$$

where S_T is the total surface area in the room:

$$S_T = S_A + S_R \qquad (5.59)$$

By rewriting the Sabine reverberation formula (Equation 4.24) with $A = S\bar{\alpha}_{SAB}$ as

$$\bar{\alpha}_{SAB} = 0.049\,(V/S_T)(1/T_R) - 4m\,(V/S_T), \qquad (5.60)$$

Description*†	Absorption coefficients, $\overline{\alpha}_{Sab}$ frequency, Hz shown 67	125	250	500	1,000	2,000	4,000	6,000
(1) Occupied audience, orchestra, and chorus areas, S_A	0.34	0.52	0.68	0.85	0.97	0.93	0.85	0.80
(2) Unoccupied "average" cloth-covered, well-upholstered seating areas (seats with perforated bottoms)	0.28	0.44	0.60	0.77	0.89	0.82	0.70	0.64
(3) Unoccupied leather-covered upholstered seating areas	–	0.40	0.50	0.58	0.61	0.58	0.50	0.46
(4) Thin (0.2- to 0.4-in.) wood paneling areas with air space behind	–	0.42	0.21	0.06	0.05	0.04	0.04	0.04
(5) 1.0-in. damped plaster or thick (more than 1.5-in. thick) well-fitted wood walls (these absorption coefficients have averaged in them the absorption of usual doors, ventilating grilles, etc. in the room) S_A	0.19	0.14	0.10	0.06	0.05	0.04	0.04	0.04

*The occupied areas include aisle widths up to, but not exceeding, 3.5 ft. Orchestra areas include men, instruments, and music racks.

†The values in Rows 1 to 3 include the absorption of a reflective floor beneath the seats. Carpet under the seats will not affect the values in Row 1. The increase in absorption for Row 2 will be significant.

FIGURE 5.59 Absorption coefficients after Beranek for (1) occupied audience, orchestra, and chorus areas; (2) unoccupied "average" well-upholstered seating areas; (3) unoccupied leather-covered thinly upholstered seating areas; (4) thin wood paneling areas with air space behind; and (5) 1.0-in. damped plaster walls and ceiling or 1.5-in. wood walls. The numbers above are for use in the Sabine equation:

$$T_R = \frac{0.049V}{S\overline{\alpha}_{SAB} + 4mV}$$

(From Beranek, L. L., *Music, Acoustics and Architecture,* John Wiley & Sons, New York, 1962, 547. With permission.)

we can equate Equations 5.58 and 5.60 to obtain

$$1/T_R = 20.4\ (S_A/V)(\alpha_A - \alpha_R) + 20.4\ (S_T/V)\ \alpha_R + 81.6\ m. \qquad (5.61)$$

A graph of $(1/T_R)$ against (S_A/V) should therefore be a straight line of slope = 20.4 $(\alpha_T - \alpha_R)$ and intercept $[20.4(S_T/V)\alpha_R + 81.6m]$. Beranek plotted a series of such graphs for octave band frequencies and found remarkable correlation, thereby proving his original hypothesis of a "per area" basis for audience absorption. Figure 5.59 shows the values so obtained for audience absorption coefficients for several types of seating. The occupied audience values are for almost 100% capacity. One would expect to observe some decrease in these values for very much lower capacity audiences. Using Beranek's value of α_A = 0.92 for the 500 and 1,000- Hz frequency bands, and the fact that the average area occupied per seat in concert halls is found to vary from 5.0 ft^2 to 7.8 ft^2, then it can be seen that this would yield an effective absorption per person of between 4.6 and 7.2 sabins.

Beranek[83] also investigated the relation between the reverberation times of occupied and unoccupied halls, since it is often very difficult to measure the reverberation time of an occupied concert hall. Figure 5.60 shows the relation found for the 500- to 1,000- Hz octave bands between the reverberation times for concert halls when occupied and unoccupied.

So far, we have discussed the effective absorption that an audience can contribute to the hall as a whole, i.e., its effect upon the reverberant sound field. However, it is found that the audience also affects the direct sound by absorbing the direct

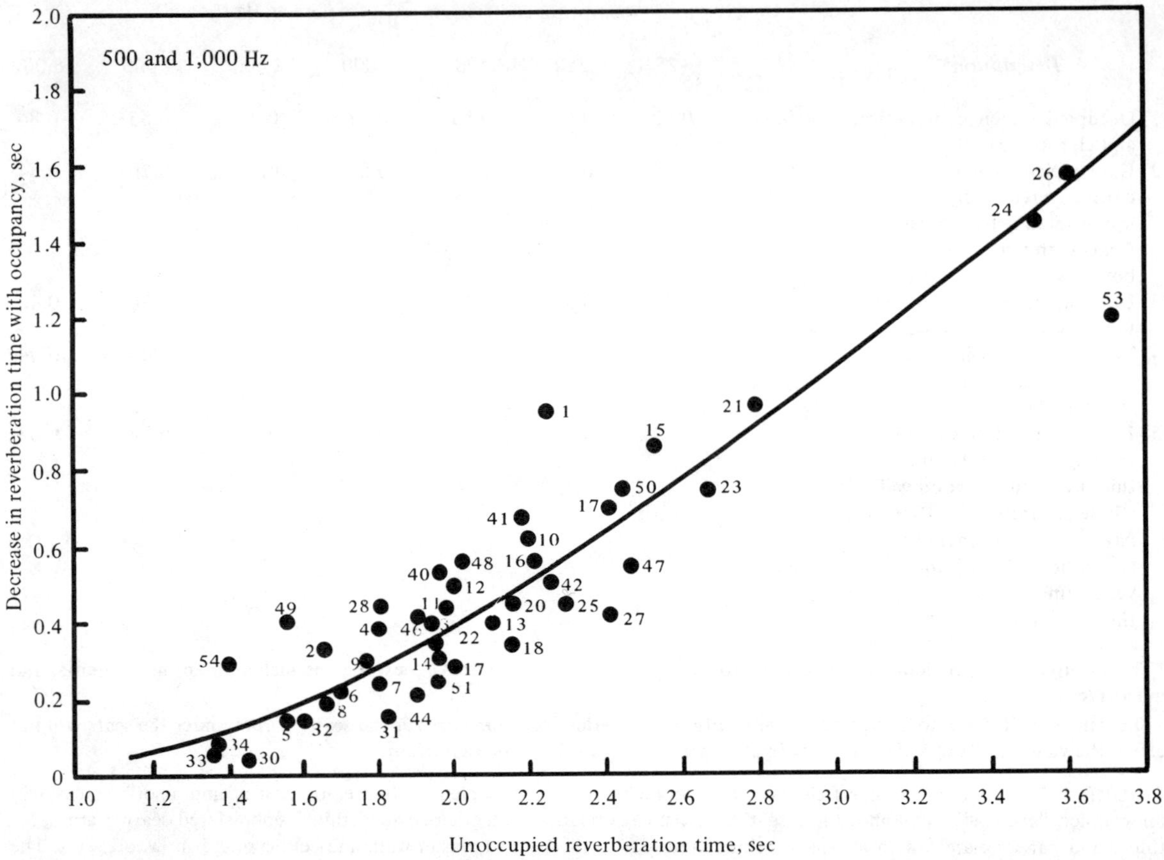

FIGURE 5.60 Plot for 500 and 1,000 Hz of decrease in reverberation time with full occupancy as a function of unoccupied reverberation time. All types of seats from folding chairs (No. 24) to exceptionally heavily upholstered seats (Nos. 3 and 16) and all shapes and volumes of halls are included. Halls No. 13, 19, 21, 23, 25, 26, and 27 are rectangular and very similar in shape and construction. (From Beranek, L. L., *Music, Acoustics and Architecture,* John Wiley & Sons, New York, 1962, 550. With permission.)

energy as it passes over the audience and rows of theater seats at grazing incidence. At low frequencies (400 Hz) this added attenuation can exceed the inverse square law loss by some 10 to 15 dB per doubling of distance. The excess attenuation at these frequencies is found to increase with the number of rows up to approximately 15 dB for 15 rows.[84,85] Previously, Beranek had found a characteristic dip in the reverberation times of unoccupied halls in the region of 250 Hz (see Figure 5.61), and had attributed this to the absorption properties of the exposed underside of the seats (unoccupied). However, the studies of Sessler[84] and Schultz[85] have shown that this is actually an interference effect between rows of seats and does not markedly depend upon the absorption properties of the seats. In fact, only the height of the seats seems to be important.[84] The spacing between rows appears to have only a minor influence on either the frequency or magnitude of this grazing incidence audience attenuation. The attenuation is caused by diffraction and by *vertical resonance*[84] in the gaps between rows of seats. The diffraction effect causes attenuation in a broad frequency range below 400 Hz, while the vertical resonance is responsible for selective attenuation, depending upon the height of the seats and the size of the gap beneath them. For an average seat of height 33 in., the maximum grazing incidence attenuation would be at approximately 200 Hz for a seat with a wide underpass and 100 Hz for a seat with a solid pedestal.[84]

Because of this additional attenuation of grazing incidence low frequency sounds, we might well expect the main floor of a concert hall to suffer from a deficiency of bass sound. However, since this is not generally found to be the case, we must conclude that the low frequency response of the

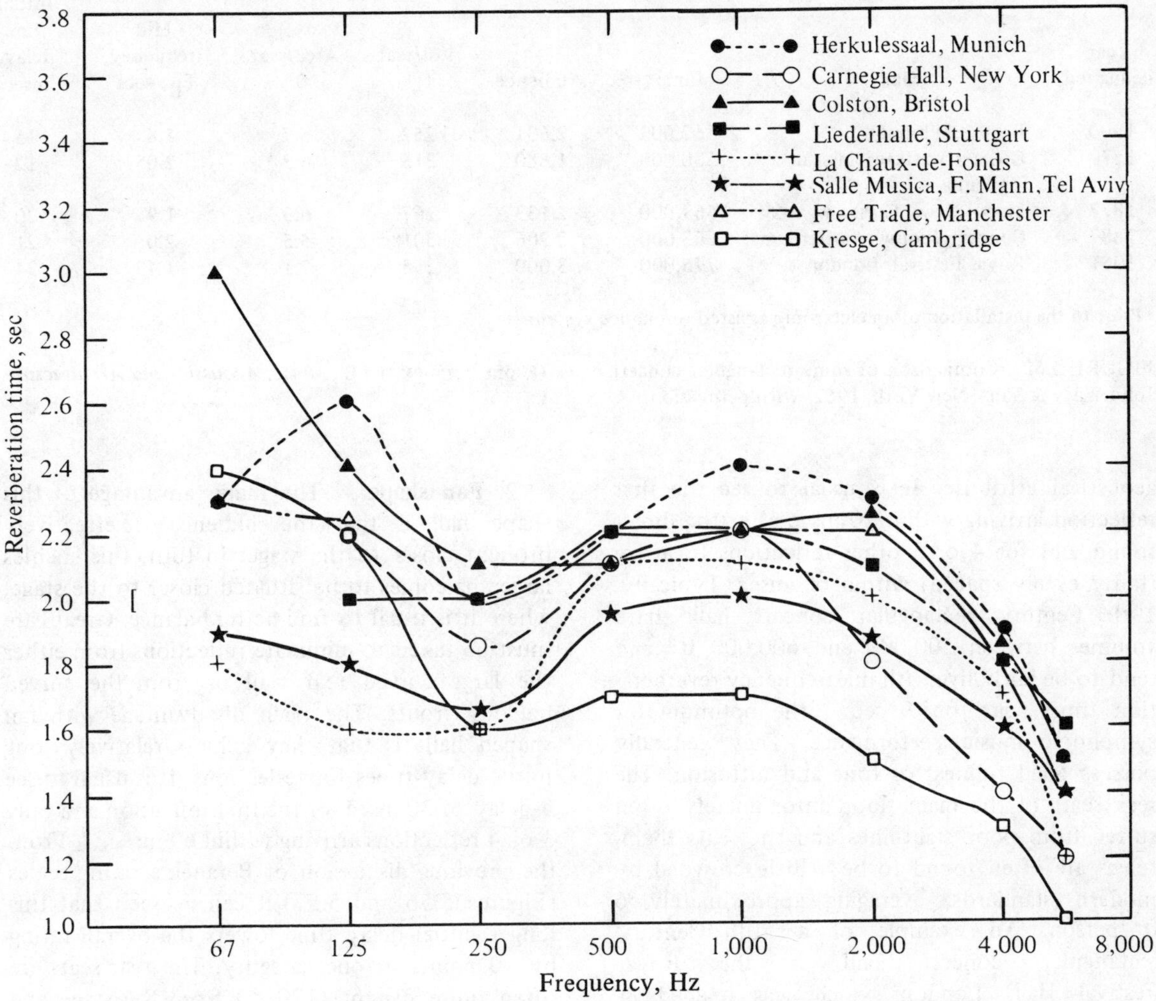

FIGURE 5.61 Reverberation time (unoccupied halls) vs. frequency for halls in which a peak in seating absorption occurred at about 250 Hz. (From Beranek, L. L., *Music, Acoustics and Architecture,* John Wiley & Sons, New York, 1962, 552. With permission.)

room to the direct sound is not too important and that it is more useful to be able to receive early reflections from the upper portions of the room and to have a strong low frequency reverberant sound field.

5.9.5 The Effect of Concert Hall Shape

As previously discussed, the shape and volume of an auditorium substantially governs its acoustics. While this was seen to be important in the design of rooms for speech intelligibility, it is even more crucial in the design of rooms for music. This is due to the complex nature and relationships of the music acoustical factors involved and also to the requirement of being able to accommodate large audiences of approximately 3,000 people. In addition to these complex acoustical requirements, the architect and acoustical consultant have to work within the framework of several other requirements including aesthetics, structural stability, lighting, air-handling systems, and economic feasibility both for construction and operation. It is therefore not surprising to find a variety of shapes of concert halls and opera houses in existence. These can, however, be divided into the following six groups:

1. **Rectangular** – This shape has been associated with traditional 19th Century European concert halls which in many cases tend to be relatively narrow and accommodate small audiences. Their small width enables the audience to receive many early lateral reflections and thereby give excellent presence or intimacy – the most important single

Year dedicated	Hall	Volume ft³	Audience	Vol/seat ft³	Area/seat ft²	Mid frequency, T_R, secs	Initial time delay, msec
1900	Boston Symphony	662,000	2,631	252	5.7	1.8	15
1870	Grosser Musikvereinssaal, Vienna	530,000	1,680	315	6.3	2.05	12
1877	St. Andrews, Glasgow	569,000	2,133	267	6.3	1.9	20
1887	Concertgebouw, Amsterdam	663,000	2,206	301	5.5	2.0	21
1951*	Royal Festival, London	775,000	3,000	258	7.1	1.47	34

*Prior to the installation of an electronic assisted resonance system.

FIGURE 5.62 Comparison of some rectangular concert halls. (From Beranek, L. L., *Music, Acoustics and Architecture*, John Wiley & Sons, New York, 1962. With permission.)

acoustical attribute. It is usual to see the first reflection arriving within 20 msec after the direct sound and for 4 or 5 other reflections to arrive (fairly evenly spaced) within 60 msec. Typically, 19th Century rectangular concert halls have volumes between 500,000 and 600,000 ft^3 and tend to be fairly live with midfrequency reverberation times near to 1.9 sec – the optimum for symphonic music performance. They generally possess good fullness of tone and diffusion. The rear seats of the main floor unfortunately often suffer from poor sight-lines and the seats themselves are often found to be a little crowded by modern standards, averaging approximately 6 ft^2/person. An example of a 20th Century rectangular concert hall is the Royal Festival Hall, London which was opened in 1951. This has an increased volume of 775,000 ft^3 and accommodates 3,000 people, having 7.1 ft^2 per person. The effect of this increased capacity, volume, and area per seat has led to generally inferior acoustics than experienced in, for example, the Boston Symphony Hall or the Grosser Musikvereinssaal, Vienna. Figure 5.62 shows a comparison of five rectangular concert halls. All of these are now rated as having excellent acoustics, although the Royal Festival Hall required the use of an artificial electronic reverberation system; before this was installed the hall was generally regarded as being too "dry". Figures 5.63 and 5.64 show the plan and elevation drawings for the Boston Symphony Hall[86] and the Royal Festival Hall,[87,88] respectively. The Orchestra Hall at Minneapolis which was opened in 1974 is rectangular in shape and has a highly diffusing ceiling. With a relatively small audience (2,573) it has been rated as excellent, having tremendous intimacy, balance, bass and clarity.

2. **Fan shape** – The major advantage of this shape hall is that the audience is effectively brought closer to the stage. In turn, this enables larger balconies to be situated closer to the stage, where it is usual to find better balance. Great care must be taken to eliminate reflections from either the large curved rear wall or from the curved balcony fronts. The main disadvantage with fan shaped halls is that they exhibit relatively long initial delay times for reflections. It is usual to see a delay of 30 msec for the first reflection and only 3 or 4 reflections arriving within 60 msec.[51] From the previous discussion of Beranek's rating scales (Figures 5.56 and 5.57) it can be seen that this longer initial delay time lowers the overall rating by 10 points or one category. The rear seats are often quite distant (130 ft) from the stage and consequentially suffer from a loss of loudness of direct sound. A comparison of some of the better fan shaped concert halls is shown in Figure 5.65 from which it can be seen that the average reverberation time tends to be approximately 1.4 sec and certainly lower than that for rectangular halls. They generally therefore tend to lack fullness of tone. Figure 5.66 shows the layout of the F.R. Mann Concert Hall, Tel Aviv.[89]

3. **Combination fan rectangular** – By combining the rectangular and fan shaped floor plans, several of the acoustical advantages of each shape can be maintained. This has been done in several highly rated concert halls such as the Philharmonic Hall, New York (see Figure 5.67), the Beethovenhalle, Bonn[90] and Konserthus, Gothenburg.[91] In general, this shape hall is able to provide better intimacy (time delay = 25–27 msec) and longer reverberation times (≅ 1.7 sec) than the fan shaped ones discussed above. By

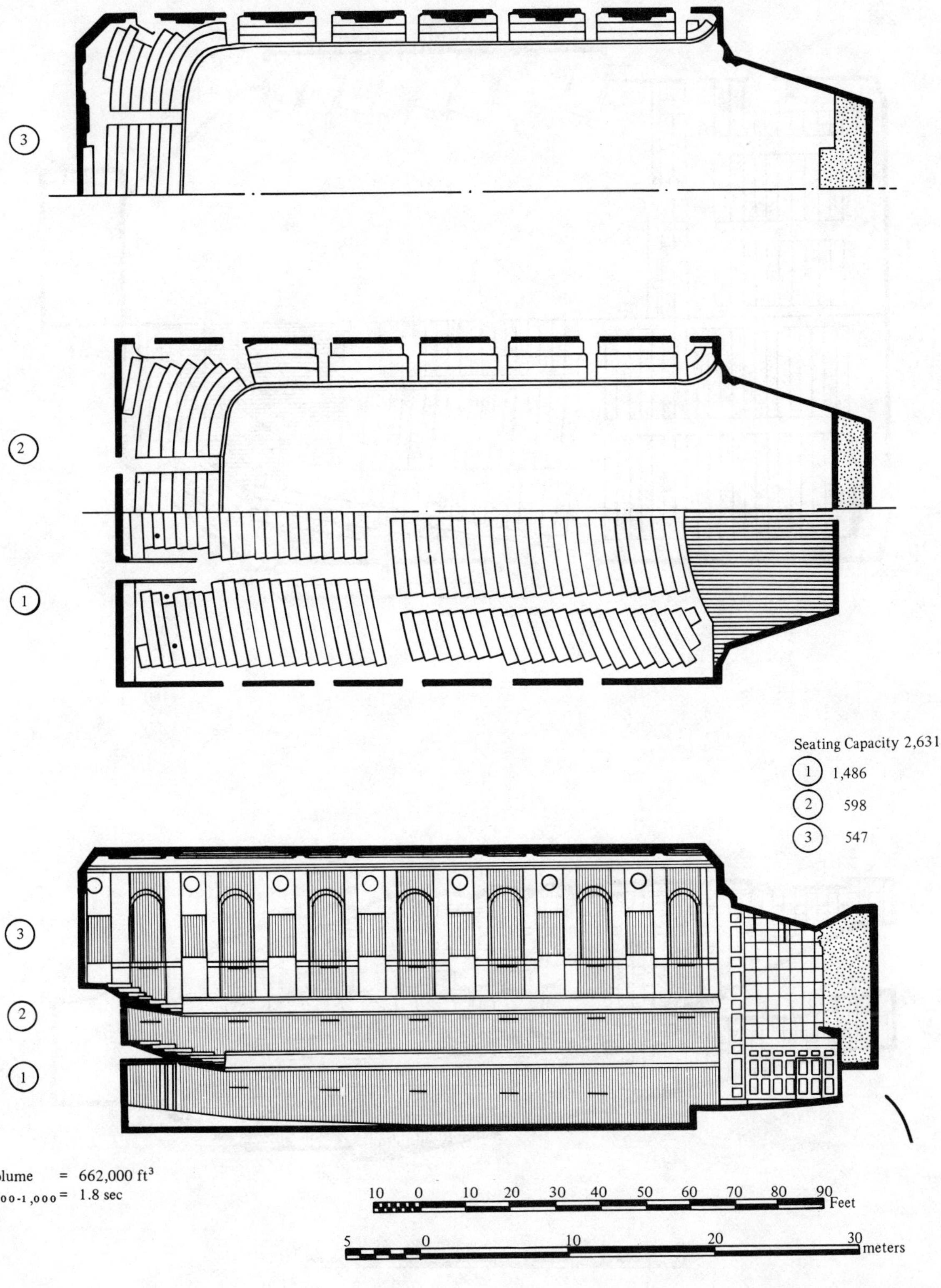

FIGURE 5.63 The Boston Symphony Hall built in 1900 shows a narrow rectangular shape which generates excellent intimacy. (From Beranek, L. L., *Music, Acoustics and Architecture,* John Wiley & Sons, New York, 1962, 95. With permission.)

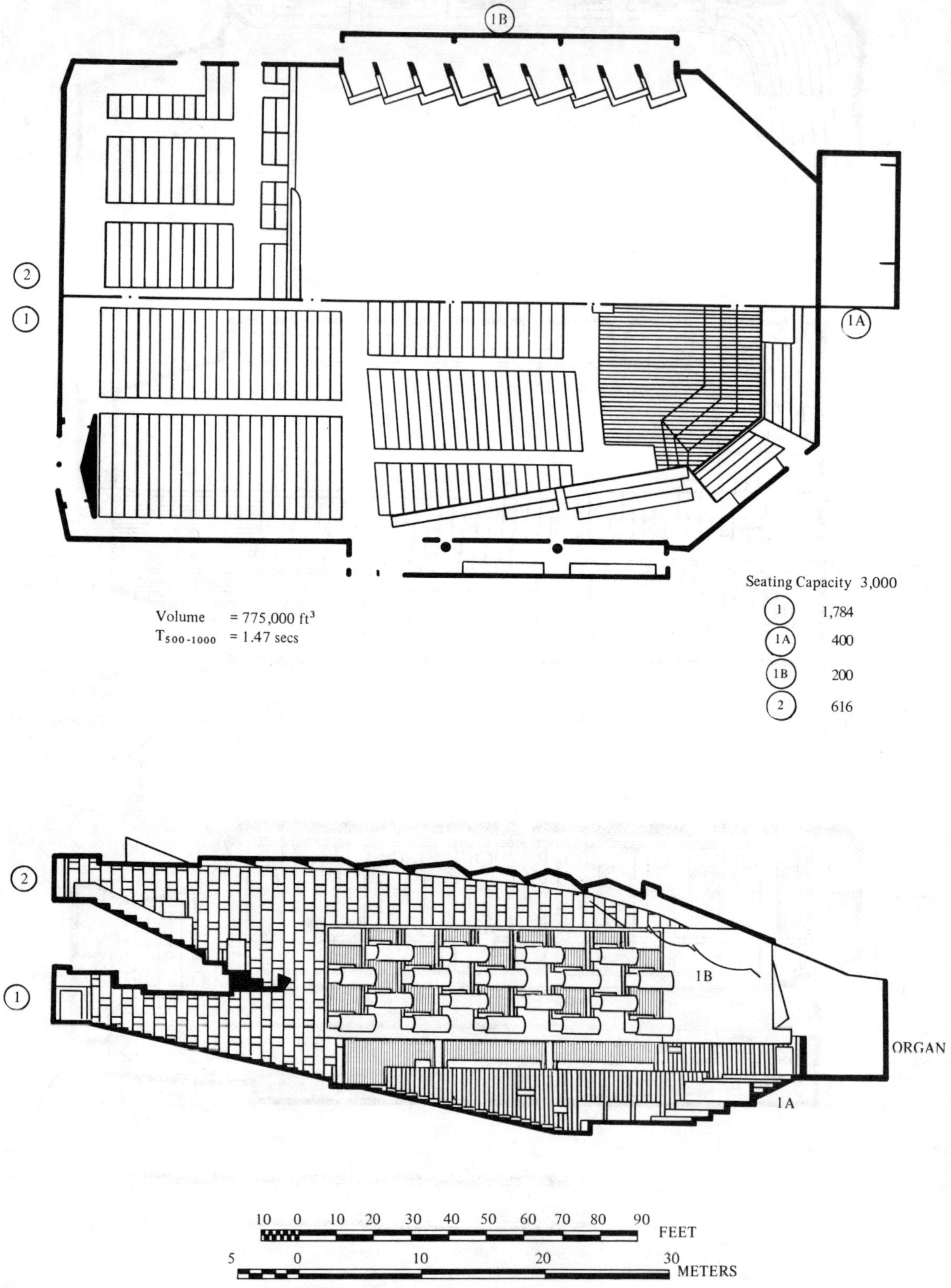

FIGURE 5.64 The Royal Festival Hall, London, England (1951). (From Beranek, L. L., *Music, Acoustics and Architecture,* John Wiley & Sons, New York, 1962, 327. With permission.)

Year dedicated	Hall	Volume ft³	Audience	Vol/seat ft³	Area/seat ft²	Mid frequency, R_T, secs	Initial time delay insec
1957	F. R. Mann, Tel Aviv	750,000	2,715	276	6.7	1.55	30
1938	Aula Magna, Caracas	880,000	2,660	331	7.6	1.35	30
1954	Edmonton/Calgary, Jubilee Auditoriums	759,000	2,731	278	7.7	1.42	31
1959	Liederhalle, Grosser Saal, Stuttgart	565,000	2,000	283	7.0	1.62	29
1951	Tanglewood Music Shed, Massachusetts	1,500,000	6,000	250	5.1	1.24	19

FIGURE 5.65 Comparison of some fan-shaped concert halls. (From Beranek, L. L., *Music, Acoustics and Architecture*, John Wiley & Sons, New York, 1962. With permission.)

careful shaping of the sidewalls (as well as the ceiling) very good diffusion can be obtained.

4. **Horseshoe shaped** – This has become the traditional shape for opera halls. The audience is seated close to the stage on the main floor and in several upper rows of boxes around the inside perimeter of the horseshoe. The reverberation times are relatively short, as they should be for opera, at an average of 1.3 sec and because of the room size and shape the initial delay times are fairly small averaging, 15 to 20 msec. This small initial delay time is due to the short distance between opposite balcony faces. However, because most opera halls have high ceilings, we find that only 2 or 3 reflections actually arrive within 60 msec. Diffusion is usually not a problem in opera halls since the audience tend to be more concerned with the direct sound. Excellent examples of horseshoe shaped opera halls can be found in La Scala, Milan, the Royal Opera House, London, the Metropolitan Opera House, New York, the Academy of Music, Philadelphia, and Teatro Colon, Buenos Aires. Figure 5.68 shows the layout of the famous La Scala, Milan built in 1778 and still considered to be the finest opera house in existence today. The theater has only 2,289 seats (at only 5.5 ft^2/seat), and with its narrow main floor gives a tremendous feeling of intimacy.

5. **Circular** – This shape of auditorium should be avoided at all costs since the circular shape normally requires a domed roof and the overall result is a completely acoustically deficient space, plagued by echoes, hot/dead spots and long delayed reflections. Despite all this, the Royal Albert Hall[92] in London built in 1882 was constructed in an elliptical shape, approximately 200 ft wide and 230 ft long.[51] It holds up to 6,000 people within its 3,000,000 ft^3 volume for a reverberation time of 2.5 sec and as expected, suffers badly from echoes and loss of direct sound. It is often referred to as the worst concert hall in the world, yet for some musical events such as military or religious music the space is ideal.

6. **Irregular** – Several concert hall designers have tried to avoid the fixed geometric shapes as outlined above and instead have designed an irregular shaped plan in the search for increased flexibility, intimacy, definition, and diffusion. The acoustics of such spaces are naturally very difficult to predict and several irregular shaped auditoria have consequently turned out to be very poor in their acoustical qualities. However, the Philharmonic Hall, Berlin[93] is an example of an irregular shaped concert hall which was able to achieve very good, if not excellent, acoustics. The Liederhalle, Grosser Saal, Stuttgart[94] is also irregular and its acoustics are rated as very good.

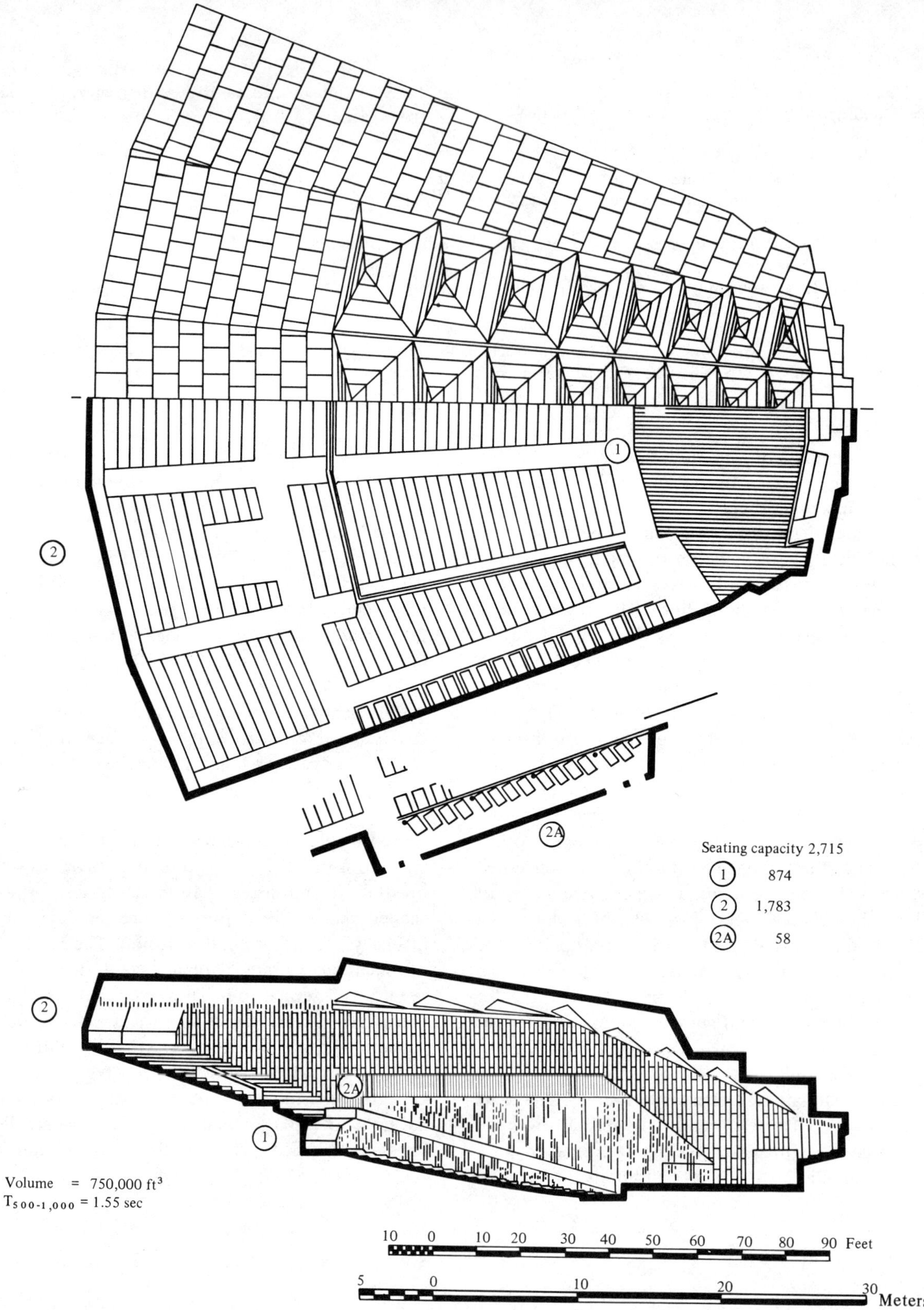

FIGURE 5.66 The F. R. Mann Auditorium, Tel Aviv, Israel (1957). (From Beranek, L. L., *Music, Acoustics and Architecture,* John Wiley & Sons, New York, 1962, 355. With permission.)

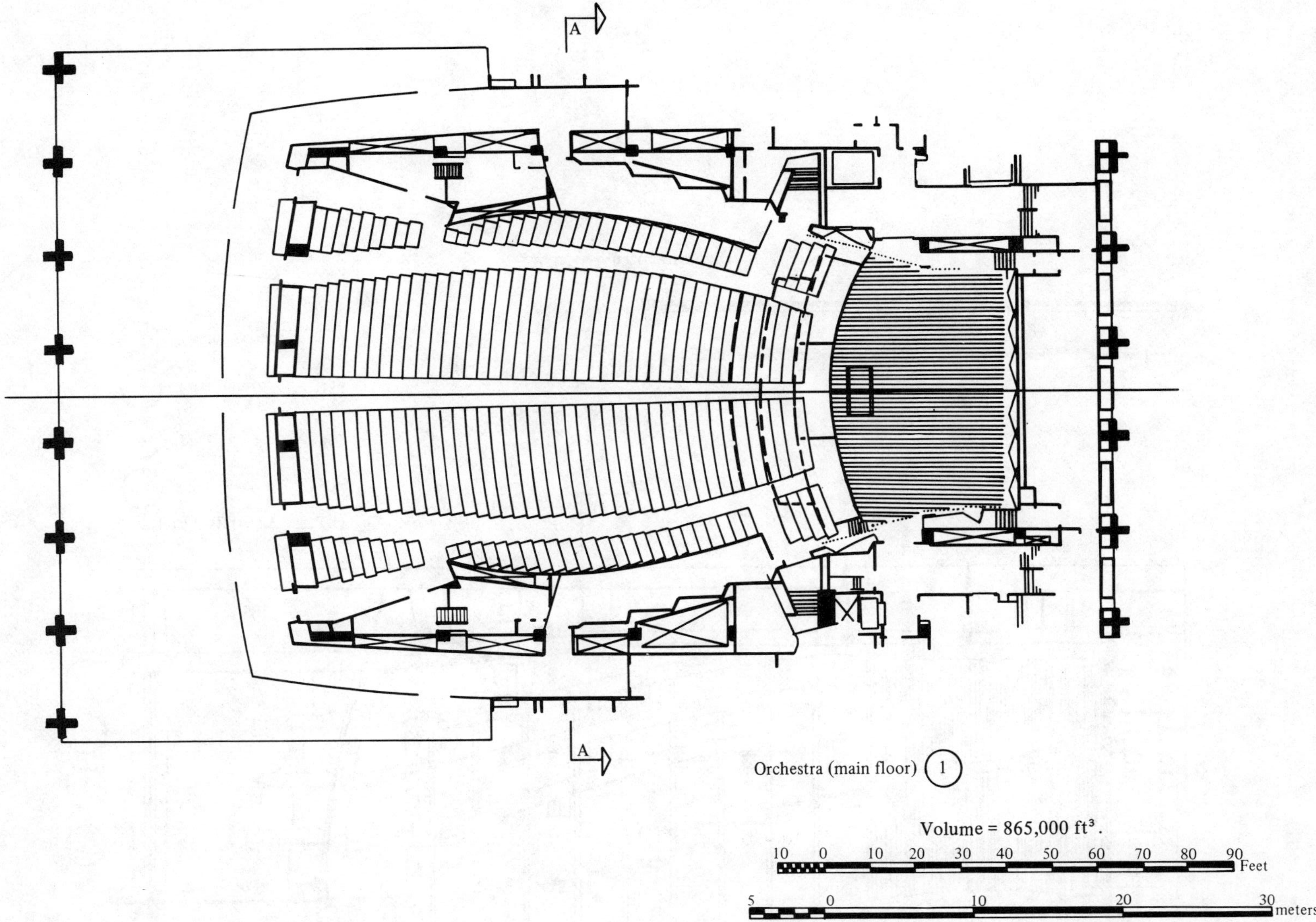

FIGURE 5.67 The Philharmonic (now Avery Fisher) Hall, New York (1962). (From Beranek, L. L., *Music, Acoustics and Architecture*, John Wiley & Sons, New York, 1962, 519, 521.

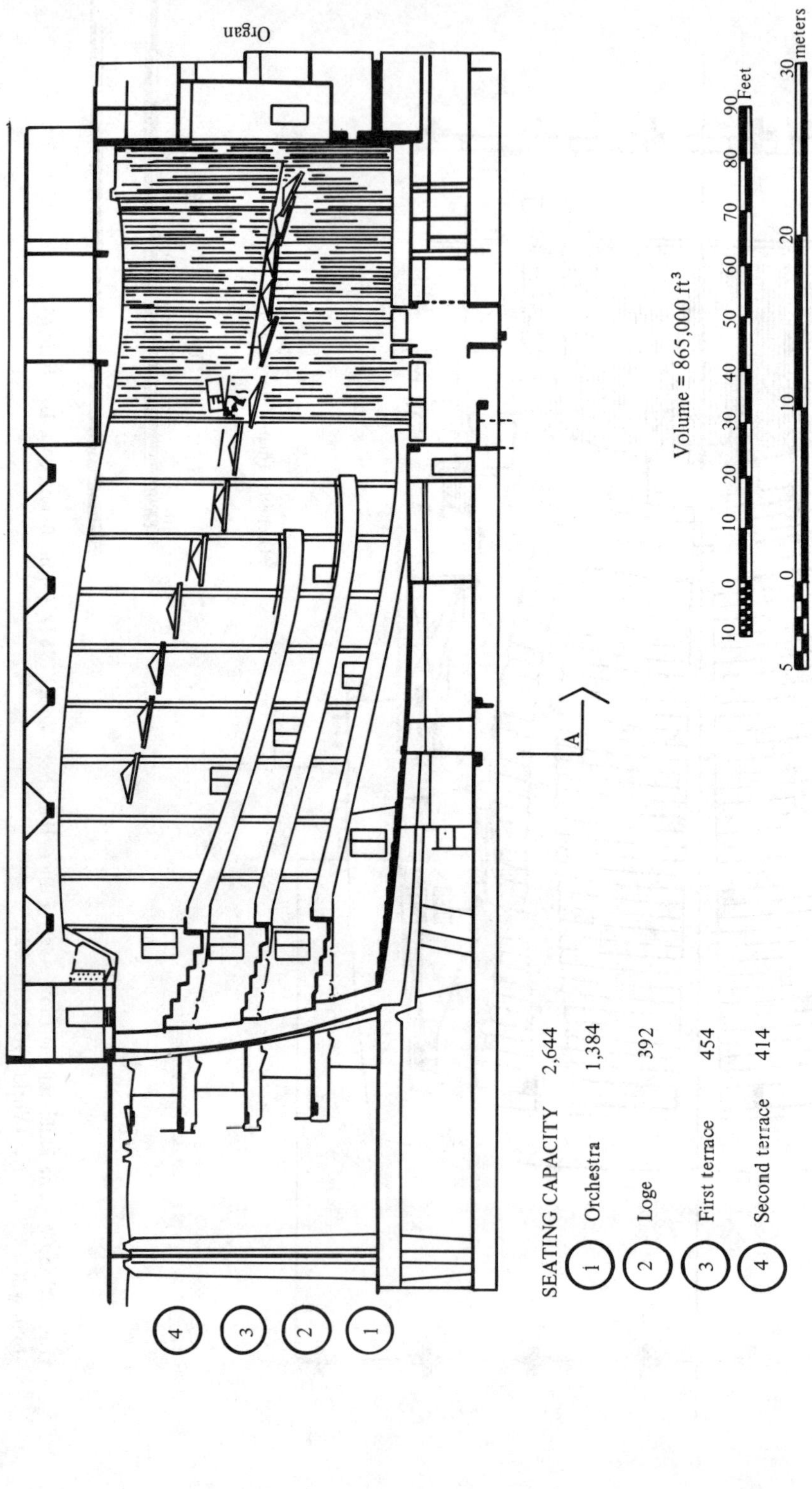

FIGURE 5.67 (continued)

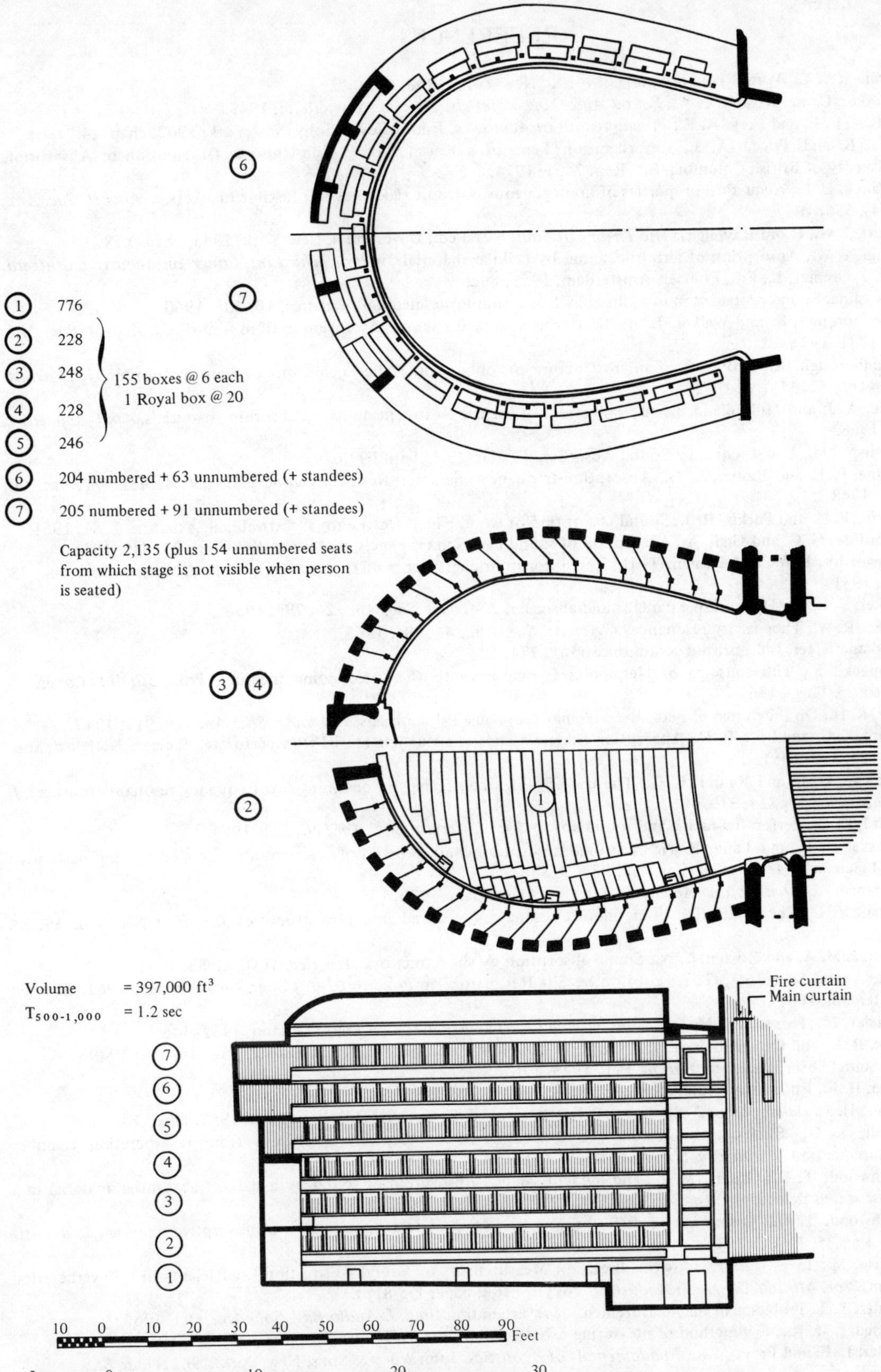

FIGURE 5.68 Teatro Alla Scala (La Scala), Milan, Italy (1778) – one of the most outstanding opera houses ever built. (From Beranek, L. L., *Music, Acoustics and Architecture*, John Wiley & Sons, New York, 1962, 361. With permission.)

REFERENCES

1. **Beranek, L. L.,** *Noise Reduction,* McGraw-Hill, New York, 1960, chap. 12.
2. **Zwikker, C. and Kosten, C. W.,** *Sound Absorbing Materials,* Elsevier, Amsterdam, 1949.
3. **Kinsler, L. E. and Frey, A. R.,** *Fundamentals of Acoustics,* John Wiley & Sons, New York, 1962, chap. 14.
4. **Riva, L. and Price, A. J.,** Reverberation Time of a Room Having Non-Uniform Distribution of Absorption, University of British Columbia, Int. Rep. 7401, 1974.
5. **Beranek, L. L.,** Acoustical properties of homogeneous isotropic rigid tiles and flexible blankets, *J. Acoust. Soc. Am.,* 19(4), 556, 1967.
6. **Strutt, J. W., (Lord Rayleigh),** *The Theory of Sound,* 2nd ed., Dover Press, New York, 1945, chap. XIX.
7. **Long, W. W.,** Absorption of airborne sound by cellular materials, in *Proc. Third Int. Congr. on Acoustics, Stutgard,* Vol. 2, Cremer, L., Ed., Elsevier, Amsterdam, 1959, 862.
8. **Kawasima,** Sound propagation in a fibre block as a composite medium, *Acustica,* 10, 208, 1960.
9. **Attenborough, K. and Walker, L. A.,** Scattering theory for sound absorption in fibrous media, *J. Acoust. Soc. Am.,* 49, 1331, 1971.
10. **Attenborough, K.,** Influence of microstructure on oblique incidence behaviour of fibrous absorbents, *J. Acoust. Soc. Am.,* 52, 119(A), 1972.
11. **Price, A. J. and Mulholland, K. A.,** The effect of surface treatment on sound absorbing materials, *Appl. Acoust.,* 1, 67, 1968.
12. **Rettinger, M.,** Low-Frequency Sound Absorber, dB Magazine, April 1970.
13. **Sabine, P. E. and Ramer, L. G.,** Absorption-frequency characteristics of plywood panels, *J. Acoust. Soc. Am.,* 20, 267, 1948.
14. **Parkin, P. H. and Purkis, H. J.,** Sound absorption of wood panels for the royal festival hall, *Acustica,* 1, 81, 1951.
15. **Sacerdote, G. G. and Gigli, A.,** Absorption of sound by resonant panels, *J. Acoust. Soc. Am.,* 23, 349, 1951.
16. **Kenworthy, R. W. and Buman, T. D.,** The absorption coefficients of fir plywood panels, *J. Acoust. Soc. Am.,* 23, 531, 1951.
17. **Becker, E. C. H.,** The multiple panel sound absorber, *J. Acoust. Soc. Am.,* 26, 798, 1954.
18. **Troke, R. W.,** Tube-cavity resonance, *J. Acoust. Soc. Am.,* 44, 684, 1968.
19. **Bruel and Kjaer, Ltd.,** private communication. 1974.
20. **Czamecki, S.,** The coupling of Helmholtz resonators with the surrounding medium, *Proc. 5th Int. Congr. on Acoustics,* Liege, 1965.
21. **Bolt, R. H.,** On the design of perforated facings for acoustical materials, *J. Acoust. Soc. Am.,* 19, 917, 1947.
22. **Ingard, K. U. and Bolt, R. H.,** Absorption characteristics of acoustic material with perforated facings, *J. Acoust. Soc. Am.,* 23, 533, 1951.
23. **Callaway, D. B. and Ramer, L. G.,** The use of perforated facings in designing low frequency resonant absorbers, *J. Acoust. Soc. Am.,* 24, 319, 1952.
24. **Ingard, K. U.,** Perforated facing and sound absorption, *J. Acoust. Soc. Am.,* 26, 151, 1954.
25. **Mikeska, E. E. and Lane, R. N.,** Measured absorption characteristics of resonant absorbers employing perforated panel facings, *J. Acoust. Soc. Am.,* 58, 987, 1958.
26. **Lawrence, A.,** *Architectural Acoustics,* Elsevier, Amsterdam, 1970, 173.
27. **Jordan, V. L.,** The application of Helmholtz resonators to sound absorbing structures, *J. Acoust. Soc. Am.,* 19, 92, 1947.
28. **Smith, J. M. A. and Kosten, C. W.,** Sound absorption by slit resonators, *Acustica,* 1(3), Ab 83, 1951.
29. **Gigli, A. and Sacerdote, G.,** Absorption by Slit Resonators, in *Proc. 3rd Int. Congr. on Acoustics,* 1961, Elsevier, Amsterdam, 1961.
30. **Ingerslev, F.,** *Acoustics in Modern Building Practice,* The Architectural Press, London, 1952, 146.
31. **Cook, R. K. and Chrazanowski, P.,** Absorption by sound absorbent spheres, *J. Acoust. Soc. Am.,* 21, 1949.
32. Unit sound absorbers, *Arch. Forum,* Feb. 1949, 126.
33. **Olson, H. F.,** Functional sound absorbers, *RCA Rev.,* 7, 503, Dec. 1946.
34. **Sabine, H. J.,** *Handbook of Noise Control,* Harris, C. M., Ed., McGraw-Hill, New York, 1957, chap. 18.
35. **Powell, J. G.,** Choosing a formula for calculating the absorption coefficient from reverberation chamber measurements, *J. Audio Eng. Soc. Am.,* Dec. 1970.
36. **Northwood, T. D., Grisara, M. T., and Medcof, M. A.,** Absorption of sound by a strip of absorptive material in a diffuse sound field, *J. Acoust. Soc. Am.,* 31, 595, 1959.
37. **Northwood, T. D.,** Absorption of diffuse sound by a strip or rectangular patch of absorptive material, *J. Acoust. Soc. Am.,* 35, 1173, 1963.
38. **Koyasu, M.,** Investigations into the Precision Measurement of Sound Absorption Coefficients in a Reverberation Room, *Proc. 6th Int. Congr. on Acoustics,* Tokyo, 1968, paper E-5-8, pE189.
39. **Schultz, T. J.,** Problems in the measurement of reverberation time, *J. Audio Eng. Soc. Am.,* Oct. 1963.
40. **Schroeder, M. R.,** New method of measuring reverberation time, *J. Acoust. Soc. Am.,* 37, 409, 1965.
41. **Kinsler, L. E. and Frey, A. R.,** *Fundamentals of Acoustics,* John Wiley & Sons, New York, 1962, chap. 6.

42. **Dubout, R. and Davem, W.,** Calculation of statistical absorption coefficient from acoustic impedance tube measurements, *Acustica,* 9, 53, 1959.
43. **Beranek, L. L.,** Precision measurement of acoustic impedance, *J. Acoust. Soc. Am.,* 12, 3, 1940.
44. **Scott, R. A.,** An apparatus for accurate measurement of acoustic impedance of sound absorbing materials, *Proc. Phys. Soc.,* 1946.
45. **Lippert, W. K. R.,** The practical representation of standing waves in an acoustic impedance tube, *Acustica,* 3, 153, 1953.
46. **Mawardi, O. K.,** Some notes on the measurement of acoustic impedance, *J. Acoust. Soc. Am.,* 28, 351, 1956.
47. **Shankland, R. S.,** Acoustics of Greek theaters, *Phys. Today,* 26(10), 30, October 1973.
48. **Papathanasopoulos, B.,** paper presented to the 5th Int. Congr. on Acoustics, 1965.
49. **Furrer, W.,** *Raum-und Banakustik–Lärmabwehr,* Birkhäuser, Basel, 1961, 258.
50. **Volkman, J. E.,** Polycylindrical diffusers in room acoustic design, *J. Acoust. Soc. Am.,* 13, 234, 1942.
51. **Beranek, L. L.,** *Music, Acoustics and Architecture,* John Wiley & Sons, New York, 1962.
52. **Lochner, J. P. A. and Burger, J. F.,** The influence of reflections on auditorium acoustics, *J. Sound Vib.,* 1(4), 426, 1964.
53. **Nickson, A. F. B. and Muncey, R. W.,** Criteria for room acoustics, *J. Sound Vib.,* 1, 292, 1964.
54. **Atol, B. S., Schroeder, M. R., and Sessler, G. M.,** Subjective reverberation time and its relation to sound decay, *Proc. 5th Int. Congr. on Acoustics,* Liège, 1965.
55. **Schroeder, M. R.,** Complementarity of sound buildup and decay, *J. Acoust. Soc. Am.,* 40, 29, 1966.
56. **Beranek, L. L.,** *Music, Acoustics and Architecture,* John Wiley & Sons, New York, 454.
57. **Knudsen, V. O. and Harris, C. M.,** *Acoustical Designing in Architecture,* John Wiley & Sons, London, 1950, 186.
58. **Doelle, L. L.,** Acoustics in Architectural Design, Bibliography No. 29, National Research Council, Canada, Division of Building Research, January 1965.
59. **Kinsler, L. E. and Frey, A. R.,** *Fundamentals of Acoustics,* 2nd Ed., John Wiley & Sons, New York, 1962, 383.
60. **Fletcher, H.,** *Speech and Hearing in Communication,* Van Nostrand, New York, 1953, 68.
61. American National Standard ANSI S3.5–1969, Methods for the Calculation of the Articulation Index, Table 8, 16.
62. **Dunn, H. K. and White, S. D.,** Statistical measurements on conversational speech, *J. Acoust. Soc. Am.,* 11, 278, 1940.
63. **Denes, P. B. and Pinson, E. N.,** *The Speech Chain, Bell Telephone Laboratories,* Williams & Wilkins, Baltimore, 1963, 139.
64. **French, N. R. and Steinberg, J. C.,** Factors governing the intelligibility of speech sounds, *J. Acoust. Soc. Am.,* 19, 90, 1949.
65. **Kryter, K. D.,** Some comparisons between rhyme and PB-word intelligibility tests, *J. Acoust. Soc. Am.,* 42, 820, 1967.
66. **Sumby, W. H. and Pollack, I.,** Visual contribution to speech intelligibility in noise, *J. Acoust. Soc. Am.,* 26, 212, 1954.
67. **Haas, H.,** Über den Einfluss eines Einfachechos auf die Hörsamkeit von Sprache, *Acustica,* 1, 49, 1951.
68. **Janssen, J. H.,** A method for the calculation of speech intelligibility under conditions of reverberation and noise, *Acustica,* 7, 305, 1957.
69. **Kryter, K. D.,** Methods for the calculation of the Articulation Index, *J. Acoust. Soc. Am.,* 34, 11, 1962.
70. **Lochner, J. P. A. and Burger, J. F.,** Optimum reverberation time for speech rooms based on hearing characteristics, *Acustica,* 10, 394, 1960.
71. **Lochner, J. P. A. and Meffert, P.,** Analysis of reflection patterns in auditoria, *Acustica,* 8, 1, 1958.
72. **Muncey, R. W., Nickson, A. F. G., and Dubout, P.,** The acceptability of speech and music with a single artificial echo, *Acustica,* 3, 168, 1953.
73. **Lochner, J. P. A. and Burger, J. F.,** The subjective masking of short time delayed echoes by their primary sound and their contributions to the intelligibility of speech, *Acustica,* 8, 1, 1958.
74. **Leonard, R. W., Delsasso, L. P., and Knudsen, V. O.,** Diffraction of sound by an array of rectangular reflective panels, *J. Acoust. Soc. Am.,* 36, 12, 2328, 1964.
75. **Meyer, E. W. and Kuttruff, H.,** Reflexionseigenschaften durchbrochener Decken (Modelluntersuchungen an der Reflecktoranordnung der neuen Philharmonic Hall in New York), *Acustica,* 13, 183, 1963.
76. **Watters, B. G., Beranek, L. L., Johnson, F. R., and Dyer, I.,** Reflectivity of panel arrays in concert halls, *Sound,* 2, 26, 1963.
77. **Beranek, L. L., Johnson, F. R., Schultz, T. J., and Watters, B. G.,** Acoustics of Philharmonic Hall, New York, during its first season, *J. Acoust. Soc. Am.,* 36, 1247, 1964.
78. **Knudsen, V. O. and Harris, C. M.,** *Acoustical Designing in Architecture,* John Wiley & Sons, London, 1950, 333.
79. **Kodaras, M. J.,** Reverberation times in typical elementary school classrooms, *Noise Contr.,* 6, 17, July 1960.
80. **Parkin, P. H. and Allen, W. A.,** Cathedral acoustic reinforcement design, *Proc. 8th Int. Congr. Acoustics, London,* 1974, Goldcrest Press, Trowbridge, England, 1974, 603.
81. **Barron, M.,** Early Lateral Reflection and Cross-Section Ratio in Concert Halls, Proc. 8th Int. Cong. Acoustics, London, 1974.
82. **Watters, B. G., Beranek, L. L., and Schultz, T. J.,** Rating Scales for Auditorium Acoustics, Presented to the 79th meeting of the Acoustical Society of America, Atlantic City, April 1970.

83. **Beranek, L. L.,** Audience and seat absorption in large halls, *J. Acoust. Soc. Am.,* 32, 6, 661, 1960.
84. **Sessler, G. M. and West, J. E.,** Sound transmission over theatre seats, *J. Acoust. Soc. Am.,* 36, 9, 1725, 1964.
85. **Schultz, T. J. and Watters, B. G.,** Propagation of sound across audience seating, *J. Acoust. Soc. Am.,* 36, 5, 885, 1964.
86. **Johnson, H. E.,** *Symphony Hall, Boston,* Little, Brown & Co., Boston, 1950.
87. **Parkin, P. H., Allen, W. A., Purkis, H. J., and Scholes, W. E.,** The acoustics of the Royal Festival Hall, London, *Acustica,* 3, 1, 1953.
88. **Parkin, P. H. and Morgan, K.,** Assisted resonance in the Royal Festival Hall, London: 1965–1969, *J. Acoust. Soc. Am.,* 48, 5(1), 1025, 1970.
89. **Beranek, L. L.,** Acoustics of the F. R. Mann Concert Hall, *J. Acoust. Soc. Am.,* 31, 882, 1959.
90. **Meyer, E. and Kuttruff, H.,** Zur akustischen Gestaltung der reverbauten Beethovenhalle in Bonn, *Acustica,* 9, 6, 465, 1959.
91. **Berlin, S. and Frieber, R.,** Measurement of reverberation time in Göthenburgs Konserthus, *Acustica,* 11, 119, 1961.
92. **Clark, R. W.,** *The Royal Albert Hall,* Hamish Hamilton, London, 1958.
93. **Sharoun, H.,** The Philharmonie Concert Hall Berlin, *Arch. Des.,* 33, 284, 1963.
94. **Cremer, L., Keidel, L., and Muella, H.,** Stuttgart, Liederhalle, *Acustica,* 6, 466, 1956.

INDEX

A

B

C

D

E

F

G

H

I

J

K

L

M

N

O

P

Q

R

S

T

U

V

W

Z